Principles of Enzymology for the Food Sciences

FOOD SCIENCE AND TECHNOLOGY

A Series of Monographs, Textbooks, and Reference Books

1. Flavor Research: Principles and Techniques, *R. Teranishi, I. Hornstein, P. Issenberg, and E. L. Wick*
2. Principles of Enzymology for the Food Sciences, *John R. Whitaker*
3. Low-Temperature Preservation of Foods and Living Matter, *Owen R. Fennema, William D. Powrie, and Elmer H. Marth*
4. Principles of Food Science
   Part I: Food Chemistry, *edited by Owen R. Fennema*
   Part II: Physical Methods of Food Preservation, *Marcus Karel, Owen R. Fennema, and Daryl B. Lund*
5. Food Emulsions, *edited by Stig E. Friberg*
6. Nutritional and Safety Aspects of Food Processing, *edited by Steven R. Tannenbaum*
7. Flavor Research: Recent Advances, *edited by R. Teranishi, Robert A. Flath, and Hiroshi Sugisawa*
8. Computer-Aided Techniques in Food Technology, *edited by Israel Saguy*
9. Handbook of Tropical Foods, *edited by Harvey T. Chan*
10. Antimicrobials in Foods, *edited by Alfred Larry Branen and P. Michael Davidson*
11. Food Constituents and Food Residues: Their Chromatographic Determination, *edited by James F. Lawrence*
12. Aspartame: Physiology and Biochemistry, *edited by Lewis D. Stegink and L. J. Filer, Jr.*
13. Handbook of Vitamins: Nutritional, Biochemical, and Clinical Aspects, *edited by Lawrence J. Machlin*
14. Starch Conversion Technology, *edited by G. M. A. van Beynum and J. A. Roels*
15. Food Chemistry: Second Edition, Revised and Expanded, *edited by Owen R. Fennema*

16. Sensory Evaluation of Food: Statistical Methods and Procedures, *Michael O'Mahony*
17. Alternative Sweetners, *edited by Lyn O'Brien Nabors and Robert C. Gelardi*
18. Citrus Fruits and Their Products: Analysis and Technology, *S. V. Ting and Russell L. Rouseff*
19. Engineering Properties of Foods, *edited by M. A. Rao and S. S. H. Rizvi*
20. Umami: A Basic Taste, *edited by Yojiro Kawamura and Morley R. Kare*
21. Food Biotechnology, *edited by Dietrich Knorr*
22. Food Texture: Instrumental and Sensory Measurement, *edited by Howard R. Moskowitz*
23. Seafoods and Fish Oils in Human Health and Disease, *John E. Kinsella*
24. Postharvest Physiology of Vegetables, *edited by J. Weichmann*
25. Handbook of Dietary Fiber: An Applied Approach, *Mark L. Dreher*
26. Food Toxicology, Parts A and B, *Jose M. Concon*
27. Modern Carbohydrate Chemistry, *Roger W. Binkley*
28. Trace Minerals in Foods, *edited by Kenneth T. Smith*
29. Protein Quality and the Effects of Processing, *edited by R. Dixon Phillips and John W. Finley*
30. Adulteration of Fruit Juice Beverages, *edited by Steven Nagy, John A. Attaway, and Martha E. Rhodes*
31. Foodborne Bacterial Pathogens, *edited by Michael P. Doyle*
32. Legumes: Chemistry, Technology, and Human Nutrition, *edited by Ruth H. Matthews*
33. Industrialization of Indigenous Fermented Foods, *edited by Keith H. Steinkraus*
34. International Food Regulation Handbook: Policy · Science · Law, *edited by Roger D. Middlekauff and Philippe Shubik*
35. Food Additives, *edited by A. Larry Branen, P. Michael Davidson, and Seppo Salminen*
36. Safety of Irradiated Foods, *J. F. Diehl*
37. Omega-3 Fatty Acids in Health and Disease, *edited by Robert S. Lees and Marcus Karel*
38. Food Emulsions: Second Edition, Revised and Expanded, *edited by Kåre Larsson and Stig E. Friberg*
39. Seafood: Effects of Technology on Nutrition, *George M. Pigott and Barbee W. Tucker*
40. Handbook of Vitamins: Second Edition, Revised and Expanded, *edited by Lawrence J. Machlin*
41. Handbook of Cereal Science and Technology, *Klaus J. Lorenz and Karel Kulp*
42. Food Processing Operations and Scale-Up, *Kenneth J. Valentas, Leon Levine, and J. Peter Clark*
43. Fish Quality Control by Computer Vision, *edited by L. F. Pau and R. Olafsson*
44. Volatile Compounds in Foods and Beverages, *edited by Henk Maarse*

45. Instrumental Methods for Quality Assurance in Foods, *edited by Daniel Y. C. Fung and Richard F. Matthews*
46. Listeria, Listeriosis, and Food Safety, *Elliot T. Ryser and Elmer H. Marth*
47. Acesulfame-K, *edited by D. G. Mayer and F. H. Kemper*
48. Alternative Sweeteners: Second Edition, Revised and Expanded, *edited by Lyn O'Brien Nabors and Robert C. Gelardi*
49. Food Extrusion Science and Technology, *edited by Jozef L. Kokini, Chi-Tang Ho, and Mukund V. Karwe*
50. Surimi Technology, *edited by Tyre C. Lanier and Chong M. Lee*
51. Handbook of Food Engineering, *edited by Dennis R. Heldman and Daryl B. Lund*
52. Food Analysis by HPLC, *edited by Leo M. L. Nollet*
53. Fatty Acids in Foods and Their Health Implications, *edited by Ching Kuang Chow*
54. *Clostridium botulinum*: Ecology and Control in Foods, *edited by Andreas H. W. Hauschild and Karen L. Dodds*
55. Cereals in Breadmaking: A Molecular Colloidal Approach, *Anne-Charlotte Eliasson and Kåre Larsson*
56. Low-Calorie Foods Handbook, *edited by Aaron M. Altschul*
57. Antimicrobials in Foods: Second Edition, Revised and Expanded, *edited by P. Michael Davidson and Alfred Larry Branen*
58. Lactic Acid Bacteria, *edited by Seppo Salminen and Atte von Wright*
59. Rice Science and Technology, *edited by Wayne E. Marshall and James I. Wadsworth*
60. Food Biosensor Analysis, *edited by Gabriele Wagner and George G. Guilbault*
61. Principles of Enzymology for the Food Sciences: Second Edition, *John R. Whitaker*

Additional Volumes in Preparation

Carbohydrate Polyesters as Fat Substitutes, *edited by Casimir C. Akoh and Barry G. Swanson*

Principles of Enzymology for the Food Sciences

Second Edition

John R. Whitaker

Department of Food Science and Technology
Davis College of Agricultural and Environmental Sciences
University of California
Davis, California

Marcel Dekker, Inc. New York • Basel • Hong Kong

Library of Congress Cataloging-in-Publication Data

Whitaker, John R.
Principles of enzymology for the food sciences / John R. Whitaker. -- 2nd ed.
p. cm. -- (Food science and technology; v. 61)
Includes bibliographical references and index.
ISBN 0-8247-9148-7 (acid-free paper)
1. Enzymes. I. Title. II. Series: Food science and technology (Marcel Dekker, Inc.); v. 61
QP601.W44 1994
612'.0151--dc20 93-37226
CIP

The publisher offers discounts on this book when ordered in bulk quantities. For more information, write to Special Sales/Professional Marketing at the address below.

This book is printed on acid-free paper. how disappointing

MARCEL DEKKER, INC.
270 Madison Avenue, New York, New York 10016

Current printing (last digit):
10 9 8 7 6 5 4 3 2 1

PRINTED IN THE UNITED STATES OF AMERICA

Preface to the Second Edition

There have been remarkable advances in biology, chemistry and enzymology since *Principles of Enzymology for the Food Sciences* was initially published in 1972. Through the applications of restriction enzymes, ligases, and polymerases, it is now possible to transfer a gene from any donor genome into almost any host plasmid genome and to have the new gene express a protein or enzyme in substantial quantities. This new field, called genetic engineering (more specifically recombinant DNA technology), makes it possible to produce transgenic plants, animals and microorganisms with new genotypic and phenotypic characteristics that improve or change the color, texture, flavor, aroma, nutrition, and stability of our raw food materials. Not only can genes be inserted from donor organisms, defective genes can be repaired or replaced and DNA or RNA expression turned off by antisense DNAs or RNAs. These methods are being used to produce better enzymes for commercial applications, including production of new chemicals, conversion of waste materials to glucose, ethanol, or proteins and to help clean up the environment. The foundation for these remarkable advances was laid in the 1950s and 1960s, when the restriction enzymes, ligases, and polymerases were discovered, purified and their properties determined.

No less significant for enzymology have been the advances in chemical analytical techniques, driven in part by the explosive advances in genetic engineering. Genetic engineering can be used to change systematically one amino acid residue at a time in enzymes. These highly specific chemical mutation methods have greatly influenced investigations of structure–function relationships in enzymes, but also the design of enzymes that are more stable, enzymes that can be used for chemical syntheses in organic solvents, enzymes that are active at more favorable pHs than originally, tailored enzymes from antibodies (abzymes), enzymes that can be purified more readily because of attachment of specific residues, and so on. Markedly improved spectrophotometers, automated protein and DNA sequencers, nuclear magnetic resonance (NMR) spectrometers, Fourier transform-infrared spectrometers, fast protein liquid chromotography (FPLC), and high performance liquid chromatography (HPLC) are readily available; all of these are computer interfaced. These advances provide the instrumentation needed to purify and characterize enzymes using 1 mg, or less, of pure protein. Availability of enzymes of high purity and known characteristics have markedly improved the use of enzymes as catalysts for specialized chemical syntheses and as highly specific analytical reagents.

These remarkable advances in genetic engineering and chemical analyses have not changed the basic principles of enzymology. It is how we use these principles, based on new opportunities and the new tools available, that has changed. The serious student and the user of enzymology continue to need a strong background in chemistry, protein structure, protein purification methods, kinetics, and thermodynamics. Both need a mechanistic appreciation of how and why enzymes are so incredibly efficient and specific, and how enzyme and substrate concentrations, pH, temperature, inhibitors, and activators influence enzyme activities. It is the purpose of this book to assist students of enzymology—whether undergraduates, graduate students, postdoctorates, or senior scientists—to acquire this knowledge. The book is designed to provide not only theory, but also an understanding of the experimental approach to the study and use of enzymes. At the University of California, Davis, students learn about enzymology by taking a three-hour-per-week lecture course and a four-hour-per-week laboratory course.

I acknowledge the contributions of all my undergraduate and graduate students and colleagues to my evolving understanding of enzymology, which led to writing this book. May future students, past graduates, and colleagues continue to advance our understanding of enzymology.

John R. Whitaker

Preface to the First Edition

> Truly, the most fundamental work in enzymes is of interest to food technologists and belongs in a department of food technology just as much as in a department of biochemistry. . . . By obtaining a thorough knowledge pertaining to the fundamentals of these reactions, we can do more in the way of preserving and producing better foods. [See Ref. 38, Chapter 1.]
>
> Emil M. Mrak

Now, as never before, emphasis is being placed on finding solutions to the problems of feeding the teeming multitudes. Involved in a solution of these problems are needs for higher yielding varieties and species, better storage and processing methods, better distribution methods and the means of adequately meeting the nutritional requirements of a variant population. Enzymes are of central importance in the complicated sequence of events leading from the growth, maturation, storage and processing of raw materials to their consumption by man as foods in order to supply his nutritional requirements. Application of a fundamental knowledge of the nature of these catalysts, which make all life possible, will lead to improvement at all levels of the food chain.

The author had several reasons for undertaking the arduous and challenging task of writing this book. These reasons include the following: first and foremost was the desire to encourage everyone working in the food sciences to become better acquainted with the fundamental principles of enzymology; secondly was the desire to encourage the teaching of enzymology in the food sciences by providing a textbook; thirdly, the need to provide a textbook for the author in teaching his own students; and last there is a personal satisfaction in having stated in a permanent record those fundamental aspects of enzymology which the author holds so dear.

This book is based on material presented in a one quarter, four unit course entitled "An Introduction to Enzymology" given by the author in Food Science and Technology since 1958. The course consists of three lectures and one laboratory session per week. The laboratory experiments (not a part of this book) are carefully integrated with the lecture part of the course so as to emphasize the protein nature of enzymes, the major parameters which affect enzyme activity and how enzyme activity can be measured. Major emphasis in the course is placed on the unifying features of enzymology with illustrative examples drawn from among the enzymes of importance to food science. The student is continuously challenged to use the principles of organic and physical chemistry in understanding the behavior of enzyme systems. The student benefits from and responds enthusiastically to this approach since he can, often for the first time, see a practical importance of these principles.

Each instructor will want to develop his own format of how best to utilize the material of this book in his teaching. In one arrangement, emphasis may be placed on an understanding of the parameters which affect enzyme activity (Chapters 4–13) while Chapters 1, 2, 3, and 14 to 25 are used as illustrative material. In another arrangement, emphasis might be placed on the materials in Chapters 1, 2, 3, and 14 to 25 with Chapters 4–13 used as source material to understand the influence of experimental parameters on the enzymes described within Chapters 14 to 25. Whatever the format of choice it is recommended that knowledge of the general nature of the active site of an enzyme and the factors which account for catalysis as described in Chapter 4 be included.

This book is not a comprehensive review nor is it intended as a source book for the broad field of enzymology. Rather it reflects the author's own views as to the fundamental areas and common features of enzymology that everyone in the food sciences should endeavor to master. There has been a tendency to stress those aspects of enzymology with which the author is most familiar. It is hoped he may be forgiven for having ignored other aspects of enzymology of equal importance but in which his knowledge is less adequate.

It is impossible to properly acknowledge the contributions to this book made by my students and colleagues. I should like to thank particularly Drs. Wallace Cleland, Owen Fennema, Roger McFeeters, Don Williams, and Roger Romani for their critical reading of the manuscript at various stages in its preparation and for their helpful suggestions. However, the author alone is responsible for the opinions and the errors in the book. Thanks are due to the editorial staff of Dekker, particularly Mrs. Rita Mullins, for their contributions, to Mrs. John Whitaker who struggled with the often illegible notes in preparation of the original manuscript, to Mrs. Clara Robison whose typing skill, diligence, undying spirit and pride in a job well done resulted in the attractive presentation, and to Charles Soderquist for his artistic work. I should like to thank also the several publishers (listed in text) who gave permission for reproduction of several of the tables and figures.

John R. Whitaker

Contents

1
Introduction

I. BREADTH OF ENZYMOLOGY

The area of enzymology is of special interest to both the biological and physical sciences. Enzymes are of universal occurrence in biological materials, and life itself depends on a complex network of chemical reactions brought about by specific enzymes. Any alteration in the normal enzyme pattern of an organism may have far-reaching consequences. Enzymes, as catalysts, are of great interest to the physical chemist, and investigation of the mechanisms of action of enzymes is a very important area of enzymology.

The area of enzymology has continued to grow rapidly for more than 60 years because of its importance to a large number of the sciences, especially biochemistry, physical chemistry, microbiology, genetics, botany, zoology, food science, nutrition, pharmacology, toxicology, pathology, physiology, medicine, and chemical engineering. Enzymology has important practical applications to activities as diverse as brewing and industrial fermentations, pest control and chemical warfare, dry cleaning, sizing and detergents, analytical determinations, and recombinant DNA technology.

Many research workers in universities, research institutes, and private industry around the world are devoting their attention to both the basic and applied aspects of enzymology. Symposia and general paper presentations on enzymes are substantial parts of all national and international professional meetings in biology and chemistry. Several journals, monographs, and textbooks devoted in part or completely to the subject exist and the literature in the field of enzymology is immense. While no one can know more than a fraction of the total definitive and descriptive information on enzymes, all must build their work on certain common, basic facts and principles concerning enzymes.

II. BRIEF HISTORY OF ENZYMOLOGY

A. Discovery of Enzymes

The major growth in enzymology has occurred relatively recently. Beginnings of enzymology can be traced back to the early nineteenth century, but the great developments have come during the last 60 years. Although the phenomena of fermentation and digestion were known before, the first clear recognition of enzyme involvement was by Payen and Persoz in 1833 [1], when they found that an alcohol precipitate of malt extract contained a thermolabile substance that converted starch into sugar. This substance, now called amylase, was named diastase by Payen and Persoz because of its ability to separate soluble dextrins from the insoluble envelopes of starch grains.

Enzymes such as catalase, pepsin, polyphenol oxidase, peroxidase, and invertase were known by the middle of the nineteenth century. In 1855, Schoenbein described an enzyme (peroxidase) in plants which, in the presence of hydrogen peroxide, caused a solution of gum guaiac to turn from brown to blue [2]. In 1856, Schoenbein described the presence of another enzyme (polyphenol oxidase) in mushrooms which, in the presence of molecular oxygen, brought about the aerobic oxidation of certain compounds [3]. In 1860, Berthelot [4] discovered an enzyme in yeast which was subsequently named invertase because of its ability to change the direction of optical rotation of a sucrose solution (by hydrolysis to glucose and fructose). Several other enzymes were discovered during the latter half of the nineteenth century.

B. Nature of Enzymes

Earlier researchers had observed a parallelism between the action of enzymes and that produced by yeast during fermentation. The name *ferment* was consequently used for enzymes. During this time, the opposing views of Justus Liebig, who held that fermentations and similar processes were due to the action of chemical substances, and those of Louis Pasteur [5], who maintained that fermentation was inseparable from living cells, caused considerable controversy. The names *unorganized ferments* and *organized ferments* were used to denote what would now be called extracted enzymes and microorganisms, respectively. It is easy to see how this division could come about since there were enzymes from malt (amylase) and from the stomach (pepsin) which were active in the absence of living organisms. On the other hand, fermentation was thought to take place only in the presence of living organisms but not in their absence.

C. Origin of Name Enzyme

In 1878, Kühne [6] proposed use of the word *enzyme* (Greek, "in yeast") to avoid the use of the names *unorganized* and *organized ferments*. Kühne's reason for suggesting the use of the term *enzyme* is best explained in his own words [translated in Ref. 7, p. 204, by courtesy of John Wiley & Sons, Inc.]:

> The latter designations (formed and unformed ferments) have not gained general acceptance, since on the one hand it was objected that chemical bodies, like pepsin, etc., could not be called ferments since the name was already given to yeast cells and other organisms; while on the other hand, it was said that yeast cells could not be called ferments, because then all organisms, including man, would have to be so designated. Without stopping to enquire further why the name has excited so much opposition, I have taken the opportunity to suggest a new one, and I give the name enzymes to some of the better known substances, called by many unformed ferments. This name is not intended to imply any particular hypothesis, it merely states that *en zyme* (Greek, in yeast) something occurs which exerts this or that activity, which is supposed to belong to the class fermentative. The name is not, however, intended to be limited to invertin of yeast, but it is intended to imply that more complex organisms, from which the enzymes pepsin, trypsin etc. can be obtained, are not so fundamentally different from the unicellular organisms as some people would have us believe.

The Liebig–Pasteur controversy came to an end when Büchner, in 1897, succeeded in obtaining the fermentation system from yeast in a cell-free extract [8].

D. Enzyme Specificity

The great German chemist Emil Fischer developed the concept of enzyme specificity and of the close steric relationship between an enzyme and its substrate. On the basis of careful studies with synthetic substrates, Fischer enunciated his famous *lock-and-key* analogy of enzyme–substrate interaction in 1894 [9]. This relationship is shown schematically in Fig. 1. Fischer's concept of the necessary close stereospecific fit between enzyme and substrate has influenced thinking on the nature of the enzyme·substrate complex to the present day. A necessary consequence of the close fit between enzyme and substrate is that each enzyme acts on a limited number of compounds. Proper studies of enzyme specificity require highly purified enzymes and substrates of known structure and purity. Based on the specificity concepts of Fischer, Bergmann and his students synthesized many peptides during the 1930s and 1940s for study of the action of proteolytic enzymes. They showed that a proteolytic enzyme hydrolyzes a given peptide bond only if the specificity of the enzyme is satisfied by the amino acid residues in proximity to the susceptible bond of the substrate. Furthermore, they

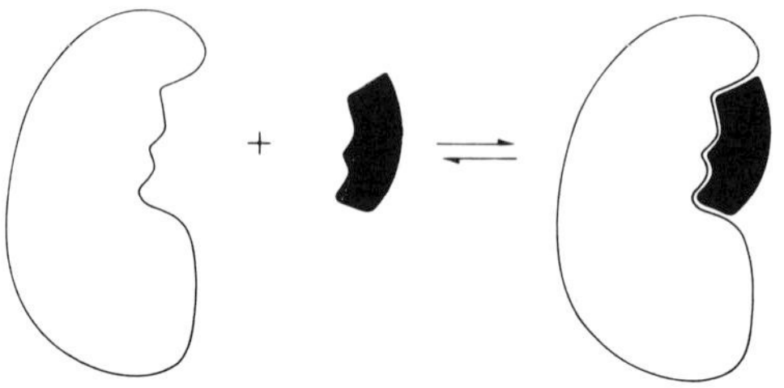

Figure 1 Schematic representation of the lock-and-key analogy of enzyme–substrate interaction.

were able to divide the proteolytic enzymes into *endopeptidases* and *exopeptidases*, depending on the enzymes' preference for hydrolyzing peptide bonds in the interior or at the terminal end of a peptide.

Another important contribution to the concept of enzyme specificity was Koshland's *induced fit* concept of enzyme–substrate combination in 1959 [10]. While Koshland's induced fit theory retained the concept of a stereospecific complex formation between enzyme and substrate, it rejected the idea that the binding locus of the active site is a rigid structure that maintains a configuration complementary to the substrate even in the absence of substrate. Rather, the presence of substrate near the active site could cause some changes in the active site so as to bring about a closer fit between substrate and enzyme and to bring the groups involved in the conversion of substrate to product into proper juxtaposition. The extent of flexibility of the active site probably varies markedly among different enzymes.

E. Quantitation of Rates and Reactions

The first part of the twentieth century saw the development of quantitative methods for describing the action of enzymes. In 1902, Henri [11] and Brown [12], independently, suggested that the saturation-type curve obtained when increasing amounts of substrate are added to a fixed amount of enzyme (Fig. 2) is the result of an obligate intermediate, the enzyme·substrate complex. In 1913, Michaelis and Menten [13] derived their now famous mathematical expression which describes quantitatively the saturation-like behavior:

$$v = \frac{V_{max}(A)}{K_m + (A)} \tag{1}$$

where v is the observed velocity, V_{max} the maximum velocity when the enzyme is fully saturated with substrate, (A) the substrate concentration, and K_m the

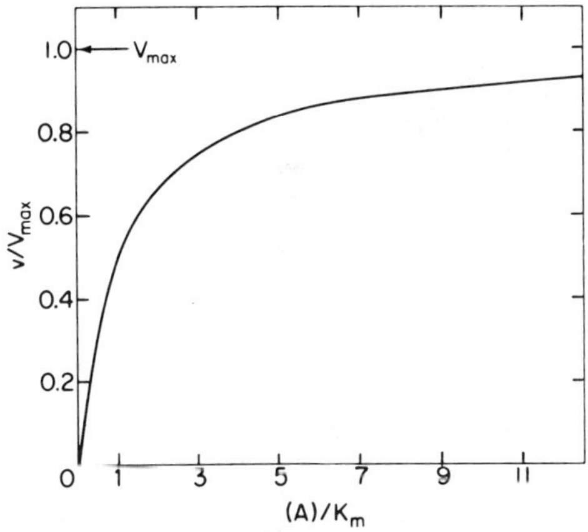

Figure 2 Relationship between substrate concentration, (A), and observed velocity, v, for an enzyme-catalyzed reaction. When (A) $= K_m$, $v = 0.5V_{max}$. Notice that even when (A) $= 12K_m$, v is not yet equal to V_{max}.

substrate concentration at which $v = 0.5V_{max}$. In 1909, Sörenson developed his classic paper on the effect of pH on enzyme activity.

Progress in understanding the kinetic functioning of enzyme systems paralleled understanding of the physicochemical properties. Quantitative treatment of enzyme kinetics has been put on a much more rigorous basis through the work of Laidler [14], Alberty [15], Cleland [16], Dalziel [17], Wong and Hanes [18], and others. Treatment of the kinetics of multisubstrate and multistep enzyme reactions was ill defined and almost nonexistent before 1960.

Kinetics of many of the regulatory enzymes do not follow the Michaelis–Menten kinetics shown in Fig. 2. The concept of Monod et al. [19] that ligands bound to an enzyme at sites other than the active site can influence the catalytic process has been of tremendous importance in understanding the kinetics of regulatory enzymes.

F. Purification of Enzymes

The rigorous purification of enzymes began after 1920. Some of the early purifications were carried out by Willstätter and his colleagues between 1922 and 1928. Although complete purity probably was not attained in any case, peroxidase was purified to the point where it showed appreciable activity at concentrations of protein below the detectibility limits of existing protein assays. This led Willstätter to conclude that enzymes were not proteins and that in those cases where protein was found it served only as a *trager* (carrier) for the true *ferment*.

One of the most significant steps in the history of enzymology came in 1926 when James B. Sumner, at Cornell University, succeeded in the crystallization of an enzyme, urease [20], and an unequivocal demonstration that it is a protein. An appreciation of the true significance of Sumner's accomplishment was slow in coming. Workers in the nineteenth century had generally considered enzymes to be proteins or associated with protein and it was only around the period of Willstätter that this concept was discarded. Willstätter's influence dominated the field of biochemistry during the 1920s and he was not alone in his belief that enzymes were not proteins. In 1914 the British biochemist Bayliss [21] had expressed a similar view, and as late as 1930 the outstanding biochemist J. B. S. Haldane expressed the view that with the single exception of Sumner's urease, next to nothing was known about the chemical nature of enzymes [22]. It was not until 1929 that there was an acceptance of the fact that urease is a protein. Sumner was later awarded a Nobel prize for his contributions to enzymology.

The 1930s saw a flurry of activity in the crystallization of enzymes, particularly by the team of Northrop, Kunitz, Herriott, and Anson at the Rockefeller Institute in New York City [23]. They crystallized such enzymes as pepsin, trypsin, chymotrypsin, and carboxypeptidase A. At the present time more than 100 enzymes have been crystallized, while more than 600 distinctive enzymes have been purified to some extent. Figure 3 shows crystals of two enzymes.

Much effort has gone into the development and refinement of methods for purification of enzymes. The work of Peterson and Sober in developing the ion-exchange celluloses [25] and of Ornstein and Davis in developing polyacrylamide

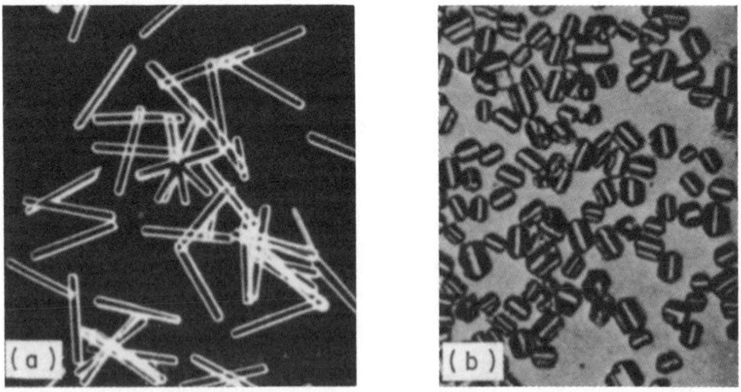

Figure 3 Photomicrographs of crystallized enzymes: (a) bovine pancreatic trypsin; (b) human salivary α-amylase. [Part (a) from Ref. 23, p. 994, by courtesy of Rockefeller University Press; part (b) from Ref. 24, plate facing p. 158, by courtesy of Archives of Sciences, Geneva.]

gels as stabilizing media for electrophoretic separations of enzymes [26] are of paramount importance. Applications of these methods have made it clear that many enzymes exist in multiple molecular forms (isozymes) and that crystallinity is not an adequate criterion for the purity of an enzyme.

Of equal importance were developments of gel filtration by Flodin [27], affinity chromatography by Cuatrecasas et al. [28], isoelectric focusing [29], chromatofocusing chromatography [30], and hydrophobic interaction chromatography [31]. Methods for purifying enzymes benefitted from methods of protein purification in general.

G. Enzyme Structure

The first breakthrough in determination of the primary sequences of proteins came in 1955 when Sanger and co-workers reported the complete amino acid sequence of insulin, a hormone of 6000 molecular weight [32]. The first determination of the primary sequence of an enzyme, that of ribonuclease of molecular weight 13,683, was in 1960 [33]. Subsequently, the amino acid sequences of many enzymes have been determined.

X-ray crystallography was developed during the early 1930s and was soon used to determine the three-dimensional structures of small peptides. However, determination of the three-dimensional structures of the much larger enzyme molecules proved too difficult a task until the advent of the computer. Not until 1967 was the first complete three-dimensional structure of an enzyme (ribonuclease) determined by x-ray crystallography [34]. This tool has led to an understanding of the secondary, tertiary, and in some cases quaternary structures of enzymes. Complete three-dimensional maps for several enzymes, including ribonuclease, lysozyme, chymotrypsin, trypsin, papain and carboxypeptidase A, have been published.

In 1969, the complete chemical synthesis of an enzyme, again ribonuclease, was reported by two groups [35]. Enzymes can now be synthesized by use of recombinant DNA techniques quite readily, as long as the gene can be isolated. These events not only inaugurated an entirely new approach to determining the relationship between structure and function in enzymes, but gave reality to the aspirations of many scientists of having available synthetic enzymes with which to perform innumerable catalytic functions.

Tremendous developments have been made in two other areas, that of modification of amino acid side chains of enzymes by chemical and *in vitro* mutagenesis and the quantitative interpretation of such modification on catalytic properties and structure [36], and that of elucidation of the detailed mechanisms [37] by which enzymes are able to perform their functions so efficiently. Despite all these many remarkable advances, it cannot be claimed that the complete mechanism of action is known for any enzyme.

III. GENERAL CHARACTERISTICS OF ENZYMES

An enzyme is *a protein with catalytic properties due to its power of specific activation*. Let us examine the meaning of each part of this concise, and precise, definition of an enzyme.

1. *Enzymes are proteins*. This assertion began to be generally accepted only after 1930; today it is taken as a fact. This does not imply that all biological catalysts are protein, only that all enzymes are proteins. All of the more than 100 crystallized enzymes are proteins. An important exception are the small catalytic RNAs, termed *ribozymes*. They are quasicatalysts, since they catalyze only a single turnover. *Abzymes* are catalytic antibodies. Scientists start with antibodies (proteins) produced in vivo with specific binding specificity and chemically modify the binding site to contain selective catalytic groups. Thus abzymes are one type of tailored enzymes. Another type is a mutant enzyme in which one or more amino acids have been changed to other amino acids. Wild-type enzymes are those occurring naturally in nature.

What are some of the experimental data supporting the concept that enzymes are proteins? They are susceptible to the same conditions that denature proteins, such as heat, strong acids, strong bases, heavy metals, and detergents. On prolonged treatment with acids and bases, enzymes give amino acids. Some give exclusively amino acids. Other enzymes contain, in addition, nonprotein components such as carbohydrate, phosphate, lipid, metal ions, or small organic moieties.

Enzymes are hydrolyzed by proteolytic enzymes which produce no other action than the hydrolysis of peptide bonds. Enzymes give all the typical protein tests, such as the biuret, Millon's, Hopkins–Cole, and Sakaguchi reactions, which detect peptide bonds, tyrosine, tryptophan, and arginine residues, respectively. X-ray crystallographic studies have confirmed that enzymes are composed of amino acids joined to each other through peptide bonds, just as in proteins.

A few definitions are in order at this point. The complete active enzyme system, including both the protein and any essential nonprotein moiety, is called the *holoenzyme*. The protein part of an enzyme is called the *apoenzyme*, and the essential nonprotein component is called a *cofactor*. The latter may be a low-molecular-weight organic compound or a metal ion. Both parts of the enzyme are essential for activity. For example, the enzyme peroxidase is composed of a protein portion (apoenzyme) of 39,600 molecular weight and a cofactor, ferriprotoporphyrin III, with a molecular weight of 652. The apoenzyme alone has no enzymatic activity, while ferriprotoporphyrin III has about 0.001% of the activity of the original enzyme. The two together give all the original enzymatic activity. Many enzymes, fully active, contain only a protein component.

A *substrate* is a compound which, after combination with an enzyme, is converted to a product catalytically. As shown schematically in Fig. 4, an

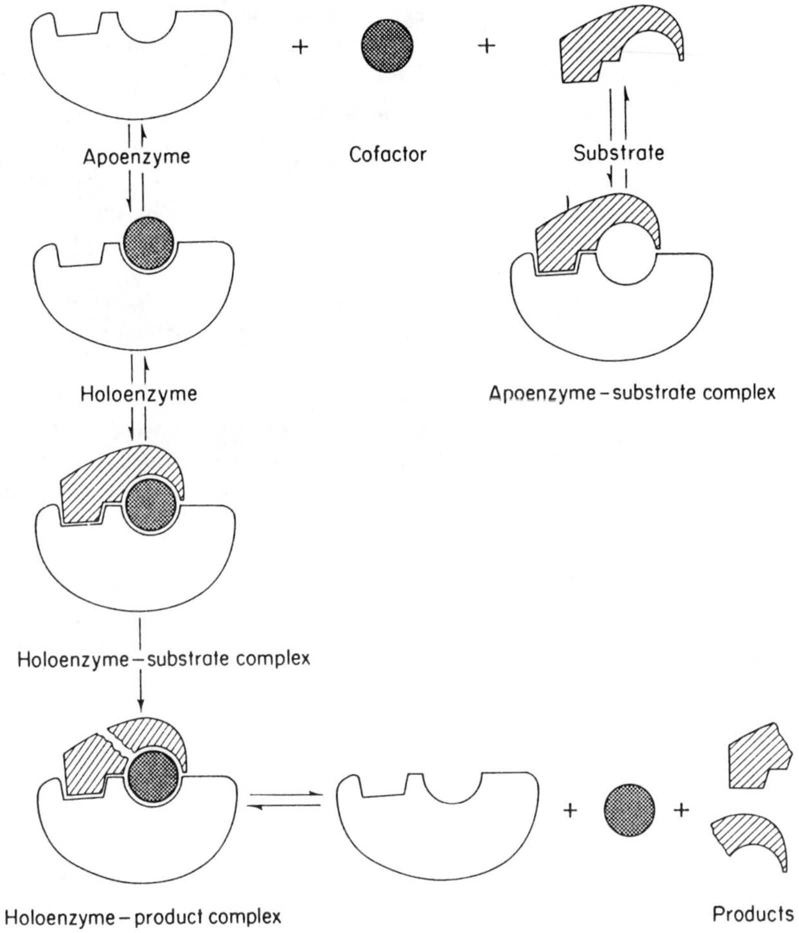

Figure 4 Apoenzyme–cofactor–holoenzyme–substrate relationships.

apoenzyme may combine with either a cofactor or a substrate to give a complex. When cofactor is bound, one has the holoenzyme. Binding of substrate only to apoenzyme gives an apoenzyme·substrate complex that is inactive. In some cases the substrate cannot combine with enzyme in the absence of cofactor. The holoenzyme is capable of combining with the substrate, which is then converted to product.

2. *Enzymes are catalytic*. A catalyst is a substance that affects the rate of a chemical reaction without itself appearing as a product of the reaction. Under ideal circumstances the catalyst is not used up during the reaction. However, only if an

enzyme is completely stable will its effective concentration not decrease during reaction. Even inorganic catalysts such as palladium and platinum become "poisoned" and thus are in essence "used up." However, enzymes catalyze the conversion of enormous amounts of substrate to product before losing their catalytic activity as a result of denaturation and/or side reactions.

Consider the reaction

$$A + B \underset{k_{-1}}{\overset{k_1}{\rightleftharpoons}} C + D \qquad (2)$$

The function of an enzyme is to speed up attainment of equilibrium as reflected by increases in the rate constants k_1 and k_{-1}. However, an enzyme does not change the ratio $k_1/k_{-1} = K_{eq}$. This point is discussed more thoroughly in Chapter 11. As a result of its catalytic properties an enzyme increases the rate of attainment of equilibrium at a temperature much lower than that possible without enzyme.

Enzymes are effective in small amounts. In in vitro experiments enzyme concentrations are usually of the order of 10^{-8} to 10^{-6} *M*; however, localized enzyme concentrations in living systems may be much higher. The efficiency of an enzyme is measured by the *turnover number*. The turnover number is defined as the number of moles of substrate converted to product by 1 mole of active enzyme site in 1 second (or 1 minute) under conditions where the enzyme is saturated with substrate. Turnover numbers vary widely among enzymes and are as low as 10^1 to 10^2 (sec^{-1}) for many of the hydrolytic enzymes and as high as 10^6 to 10^7 (sec^{-1}) for some of the oxidative enzymes. The turnover number is a function of pH, temperature, and other experimental parameters of the assay system.

A term used in our definition of an enzyme is *specific activation*. In subsequent chapters, we shall see many examples which indicate that an enzyme combines with substrate in a stereospecific fashion to cause changes in the electronic configuration around certain susceptible bonds. These bonds then are more easily changed. Physicochemical measurements indicate that the enzyme has lowered the activation energy (Fig. 5) and by so doing has increased the rate of reaction.

3. *Enzymes are specific*. A proteolytic enzyme will not hydrolyze the glycosyl or ester bonds in a carbohydrate or a lipid, nor will it hydrolyze all peptide bonds in a protein. Enzyme specificity is often dictated by the nature of the groups attached to the susceptible bonds. For example, both α-chymotrypsin and trypsin are proteolytic enzymes capable of hydrolyzing certain peptide bonds in proteins. α-Chymotrypsin will hydrolyze peptide bonds in which the carbonyl group of that bond is supplied by L-tyrosine, L-phenylalanine, or L-tryptophan. Trypsin will hydrolyze peptide bonds in which the carbonyl group of the peptide bond is supplied by L-arginine or L-lysine. Peptide bonds containing D-amino acids are not hydrolyzed. As another example of the specificity of enzymes, maltase hydrolyzes maltose to glucose, while cellobiase hydrolyzes cellobiose to glucose, but not vice versa. Differences between the two substrates appear to be slight in that maltose

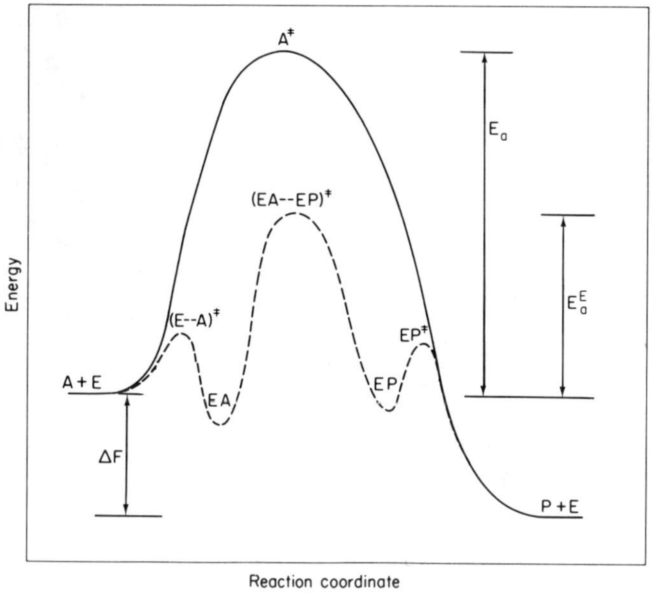

Figure 5 Schematic representation of the energy changes involved in conversion of A to P in the presence and absence of enzyme, E. E_a, and E_a^E are the observed activation energies for conversion of A to P in the absence and presence of enzyme, respectively.

is an α-glucoside whereas cellobiose is a β-glucoside (Fig. 6). Specificity of an enzyme is determined by the strict requirements to be met in order for the substrate to be bound at the active site of the enzyme and/or by the even more strict requirements that must be met after binding in order for the substrate to be converted to product. Many examples of this are given in subsequent chapters.

IV. IMPORTANCE OF ENZYMOLOGY TO THE FOOD, NUTRITIONAL, AND HEALTH SCIENCES

Use of enzymes to accomplish specific desirable changes in foods has been practiced for centuries. The techniques have been handed down from generation to generation without any knowledge, until recently, of the nature of the catalysts and reactions involved. Use of malted barley for starch conversions in brewing, addition of saliva to starchy products in preparing fermented liquors, and wrapping of meat in the bruised leaves of the papaya tree to tenderize it are only a few examples of the ancient use of enzymes. Enzymes involved in digestion, in fermentation, and in hydrolytic reactions were first studied in connection with their association with foods. For many years the food scientist has studied the

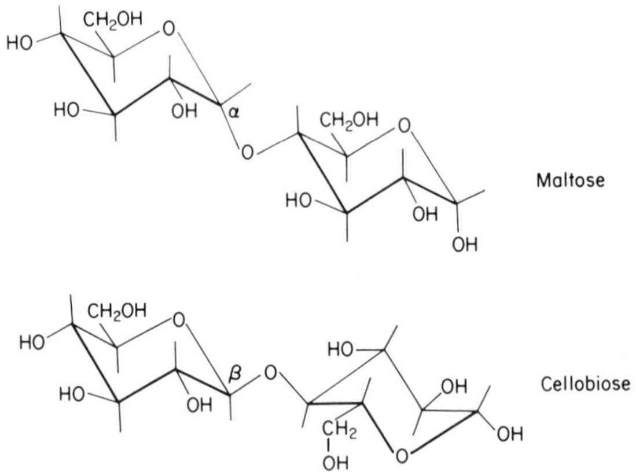

Figure 6 Structures of maltose and cellobiose.

actions of enzymes, particularly those that cause detrimental changes in foods. The actions of the pectic enzymes and polyphenol oxidase have received particular attention. Unfortunately, until recently the food scientist has been less concerned with the potential usefulness of enzymes in processing. More recently, chemical and food engineers have given considerable attention to the application of immobilized enzyme systems to the processing of foods and the production of food and chemical ingredients.

A. Need for Understanding Fundamental Nature of Enzymes

During the nineteenth and early twentieth centuries, scientists who described the action of enzymes in greatest detail were those concerned with foods; the organic chemists were generally content to study the behavior of other natural products. Why, given this head start, did the food scientists not contribute more of lasting importance to the field of enzymology? Is it really important that food scientists have an appreciation of enzymology? If so, how should they approach the study of enzymes?

In 1960, Emil Mrak, in his introductory address to a symposium on food enzymes held at Oregon State University, sounded the call for more fundamental work on enzymes by the food scientist [38]:

> Truly, the most fundamental work on enzymes is of interest to food technologists and belongs in a department of food technology just as much as in a department of biochemistry. Both groups should be working on enzymes; however, the point of view may be different. The biochemist seeks knowledge

> of the enzyme whereas the food scientist working in the fundamental area of enzymology will always have in the back of his/her mind the use or control of these systems, yet the contribution of the two groups might be similar. By obtaining a thorough knowledge pertaining to the fundamentals of these reactions, we can do more in the way of preserving and producing better foods.
>
> If we are to progress in our treatment of foods, in our production of better foods and the preservation of more foods, and if we are to create wealth by preservation, then we must get down to an understanding of the more fundamental aspects of enzymes. It is only the departments of food technology that look at these fundamental aspects that will make real progress in the future. (From Ref. 38, pp. 4 and 5, by courtesy of Avi Publishing Co.)

Too few food scientists have heard this call in the intervening years. Hopefully, this will not continue to be the case in the future.

Who are the food scientists? Are they the engineers or technologists involved with the design of equipment for better processing of foods or with the theory of heat transfer and mass flow? Are they the bacteriologists concerned with the role of microorganisms in the production, processing, storage, and safety of foods? Are they the chemists concerned with the chemical changes that go on in foods during storage, processing and distribution? Are they the biochemists involved in the study of biochemical events that occur in plants, animals, and microorganisms during their growth, maturation, storage, and processing? Could they be the psychologists or behavioral scientists concerned with factors that influence people in their preference and selection of foods? Could they be the geneticists who through careful manipulation of the genes produce a superior new food material? Could they be the plant physiologists or the animal scientists involved in the production of the raw food materials? Could they be the toxicologists concerned with the chemicals needed for control of insects, vermin, and diseases so as to assure a more plentiful and safe food supply? Could they be the nutritionists concerned with the proper balance of essential vitamins, minerals, amino acids, fatty acids, and calories? Could they be the clinicians concerned with the relationship between food intake and disease?

Given a broad definition of food scientists, many specific examples could be used to illustrate the importance of enzymes. We shall restrict ourselves to a brief discussion of the role of enzymes during storage and processing of raw food materials, in digestion and assimilation of foods, in health and disease, and in analytical procedures, as examples of importance of enzymes to the food scientists.

B. Enzymes in Storage and Processing of Foods

Many important enzyme reactions are involved from the very beginning of the growth process of plants, animals, and microorganisms. The kinds and quantities of these enzymes change continually during development and maturation. Enzyme activities are different in different organs, tissues, and cells. These enzymes are thought of as being generally beneficial since they lead to the growth and development of the raw food material, although imbalance in their activities may lead to abnormal states and hence to unnatural or undesirable foods.

When biological materials are gathered for food the enzymatic processes continue. Even though an animal is dead the enzyme systems continue to function until there is an exhaustion of substrates or the pH has changed until it is unfavorable for enzyme actions. An unfavorable pH may result in decreased enzyme activity either as a result of effect on rate or in denaturation of the enzyme. As an example, the enzymes of the Krebs cycle and electron transport system will become inactive soon after death because of the unavailability of the terminal substrate, O_2. The glycolytic enzymes, which function anaerobically, continue to act on glycogen and glucose to produce lactic acid until activity is terminated by exhaustion of substrate or unfavorable pH. The hydrolytic enzymes—such as the proteolytic enzymes, esterases, phosphatases, and glycosidases—continue to act on the cellular components long after the animal is dead. In fact, activities of these enzymes increase because of gradual destruction of the cellular organization and denaturation of the proteins (acted on by proteolytic enzymes). Since there is a major cessation of anabolic reactions after death and a continuation and even acceleration of catabolic reactions, the net result is a deterioration of the tissue. This is not necessarily bad since it may contribute to desirable food-quality attributes, such as flavor and texture.

The specific details of changes in enzymatic activity in most ripening fruits are different than in animal tissues. At the time of ripening there is increased activity and actual quantity of certain enzymes. The respiratory rate increases markedly, there is a conversion of starches to sugars, a degradation of chlorophyll, and a rapid increase in size of the cells. These activities are all considered to be beneficial in fruit, whereas degradation of chlorophyll in a green vegetable would be undesirable. Following ripening of fruit, activity of the hydrolytic enzymes continue at an ever-accelerating rate, so that without controlled atmospheric conditions, the fruit becomes overripe and mushy. With green vegetables, changes in turgor of the cells after harvest markedly influence the rate of deterioration of the tissues, due to the action of hydrolytic and oxidative enzymes. In fresh green beans, peas, and corn, reactions catalyzed by lipoxygenase lead to loss in flavor and color, for example.

An understanding of factors that influence enzyme activity can be a great asset in controlling and manipulating the activity of enzymes postharvest. The variables most easily manipulated are temperature and substrate. Storage at lower temperatures slows down the rate of enzyme activities and delays the time at which the quality of the product becomes unacceptable. It does not follow, however, that for all raw materials the lower the temperature, the longer the product can be stored.

An example will illustrate this. When potatoes are stored at too high a temperature, the sucrose and glucose are converted to starch and the texture and flavor are unacceptable. When the temperature is too low, conversion of starch to sugars occurs and the potato becomes objectionally sweet. Storage of some foods at or just below 0°C actually leads to an observed increase in enzymatic activity. This is a result of increasing accessibility of substrates to enzyme either as a result of concentration of the aqueous medium of the tissue and/or through damage to the cellular components by the large ice crystals formed by the slow freezing process. Storage should be either above or several degrees below 0°C, never at 0 to −5°C unless there is a desire to increase enzymatic activity.

In fruits the marked increase in respiration characteristic of the onset of ripening is due to the action of oxidative enzymes. Activities of these enzymes can be decreased by lowering the temperature and by changing the atmosphere from one of 21% oxygen, 0.03% CO_2, and 79% nitrogen (normal composition of air) to one that has a much lower oxygen content. Carbon dioxide or nitrogen are often used for this purpose. Complete exclusion of oxygen is undesirable as there is a need for certain oxidative processes to continue at a minimum rate to maintain the integrity of cells. Storage at too low oxygen concentrations results in internal browning in potatoes and in some fruits. Some fruits and vegetables rapidly develop off-flavors under suboptimum oxygen concentrations.

The two major methods of controlling enzyme activity in foods are heat processing and freezing. The former, if properly done, leads to complete destruction of all enzymatic activity, including that due to microorganisms. Since heat is generally destructive of the quality attributes of foods, it is important to use just the amount of heat needed to accomplish destruction of all enzyme activity. Proper heat treatment in fruits and vegetables is determined by using loss of peroxidase activity as an index of adequate treatment. Complete inactivation of peroxidase assures that all other enzymes will have been destroyed. Peroxidase is found in all plant tissues, thus making it a generally useful indicator. Ideally, however, the enzyme primarily responsible for loss of quality should serve as the best indicator.

Freezing does not destroy enzyme activity. It only slows down the activity, thereby prolonging storage life. When food has not received a short blanching treatment before freezing there is a marked acceleration of enzyme activity immediately upon thawing the food, as seen in strawberries.

Table 1 gives a summary of some of the beneficial as well as detrimental activities of enzymes in foods. This is by no means an exhaustive compilation. What may be classified as an undesirable enzymatic activity in one raw material may be beneficial in another. An appropriate example is polyphenol oxidase. This enzyme causes marked losses in bananas, apples, and potatoes because of browning reactions. Yet polyphenol oxidase is necessary for the development of proper color in tea, coffee, unbleached raisins, and prunes.

Table 1 Some Uses and Suggested Uses of Enzymes in Foods and Food Processing

Enzyme	Food	Purpose or action
Amylases	Baked goods	Increase in maltose content for yeast fermentation.
	Brewing	Conversion of starch to maltose for fermentation. Removal of starch turbidities.
	Cereals	Conversion of starch to dextrins and maltose. Increase in water absorption.
	Chocolate/cocoa	Liquefication of starches for free flow.
	Confectionery	Recovery of sugars from candy scraps.
	Fruit juices	Removal of starches to increase sparkling properties.
	Jellies	Removal of starches to increase sparkling properties.
	Pectin	Aid in preparation of pectin from apple pomace.
	Syrups and sugars	Conversion of starches to low-molecular-weight dextrins (corn syrup).
	Vegetables	Hydrolysis of starch as in tenderization of peas.
Cellulase	Brewing	Hydrolysis of complex carbohydrate cell walls.
	Coffee	Hydrolysis of cellulose during drying of beans.
	Fruits	Removal of graininess of pears, peeling of apricots, tomatoes.
Dextran–	Sugar syrups	Thickening of syrup.
sucrase	Ice cream	Addition of dextan as thickening agent, for body.
Invertase	Artificial honey	Conversion of sucrose to glucose and fructose.
	Candy	Manufacture of chocolate-coated soft-cream candies
Lactase	Ice cream	Prevention of crystallizatin of lactose, which results in grainy, sandy texture.
	Feeds	Conversion of lactose to galactose and glucose.
	Milk	Stabilization of milk proteins in frozen milk by removal of lactose. Hydrolysis of lactose, permitting use by adults deficient in intestinal lactase and infants with congenital lactase deficiency.
Tannase	Brewing	Removal of polyphenolic compounds.
Pentosanase	Milling	Recovery of starch from wheat flour.
Naringinase	Citrus	Debittering citrus juice by hydrolysis of the glucoside, naringin.
Pectic enzymes		
Useful	Chocolate/cocoa	Hydrolytic activity during fermentation of cocoa.
	Coffee	Hydrolysis of gelatinous coating during fermentation of beans.
	Fruits	Softening.
	Fruit juices	Improving yield of press juices, preventing cloudiness, improving concentration processes.
	Olives	Extracting oil.
	Wines	Clarification.
Deteriorative	Citrus juice	Destruction and separation of pectic substances.

Table 1 Continued

Enzyme	Food	Purpose or action
Pectic enzymes		
	Fruits	Excessive softening action.
Proteases		
Useful	Baked goods	Softening action in doughs. Cut mixing time, increase extensibility of doughs. Improvement in grain, texture, loaf volume. Liberate β-amylase.
	Brewing	Body, flavor, and nutrient development during fermentation. Aid in filtration and clarification, chill-proofing.
	Cereals	Modification of proteins to increase drying rate, improve product handling characteristics. Manufacture of miso and tofu.
	Cheese	Casein coagulation. Characteristic flavors during aging.
	Chocolate/cocoa	Action on beans during fermentation.
	Eggs, egg products	Improve drying properties.
	Feeds	Use in treatment of waste products for conversion to feeds.
	Meats and fish	Tenderization. Recovery of protein from bones, trash fish. Liberation of oils.
	Milk	In preparation of soybean milk.
	Protein hydrolysates	Condiments such as soy sauce and tamari sauce. Specific diets. Bouillon, dehydrated soups, gravy powders, processed meats.
	Wines	Clarification.
Deteriorative	Eggs	Shelf life of fresh and dried whole eggs.
	Crab, lobster	Overtenderization if not inactivated rapidly.
	Flour	Influence on loaf volume and texture if too active.
Lipase		
Useful	Cheeses	Aging, ripening, and general flavor characteristics.
	Oils	Conversion of lipids to glycerol, fatty acids, and monoglycerides
	Milk	Production of milk with slightly cured flavor for use in milk chocolates.
Deteriorative	Cereals	Overbrowning of oat cakes. Brown discoloration of wheat bran.
	Milk and dairy products	Hydrolytic rancidity.
	Oils	Hydrolytic rancidity.
Phosphatases	Baby foods	Increase available phosphate
	Brewing	Hydrolysis of phosphate compounds.

Table 1 Continued

Enzyme	Food	Purpose or action
Phosphatases	Milk	Monitoring progress of pasteurization.
Nucleases	Flavor enhancers	Production of nucleotides and nucleosides.
Peroxidases		
Useful	Vegetables, fruits	Monitoring progress of blanching.
	Glucose determinations	In combination with glucose oxidase.
	Doughs	Cross-linking of proteins
Deteriorative	Fruits	Contribution to browning action.
Catalase	Milk	Destruction of H_2O_2 in cold pasteurization.
	Variety of products	To remove glucose and/or oxygen to prevent browning and/or oxidation. Used in conjunction with glucose oxidase
	Vegetables	Monitoring progress of blanching.
Glucose oxidase	Variety of products	Removal of oxygen and/or glucose from products such as beer, cheese, carbonated beverages, dried eggs, fruit juices, meat and fish, milk powder, wine to prevent oxidation and/or browning. Used in conjunction with catalase.
	Glucose determination	Specific determination of glucose. Used in conjunction with peroxidase.
Polyphenol oxidase		
Useful	Tea, coffee, prunes, unbleached raisins	Development of browning during ripening, fermentation, and/or aging process
Deteriorative	Fruits, vegetables	Browning, off-flavor development, loss of vitamins.
Lipoxygenase	Vegetables	Destruction of essential fatty acids and vitamin A, development of off-flavors.
Ascorbic acid oxidase	Vegetables, fruits	Destruction of vitamin C (ascorbic acid).
Thiaminase	Meats, fish	Destruction of thiamine.

C. Enzymes in Digestion and Assimilation of Foods

Details of the specific enzymes of importance in the digestion and assimilation of foods is beyond the scope of this introductory chapter. However, it is instructive to point out the general importance of enzymes to this aspect of the food, nutritional, and health sciences. The major steps in digestion and assimilation of foods are shown diagrammatically in Table 2.

Food, which consists of complex proteins, carbohydrates, lipids, and nucleic acids often with organization at the tissue, cellular, and subcellular level still intact, is masticated in the mouth and mixed with saliva. Saliva contains α-amylase, which initiates degradation of starch and glycogen. This action ceases when the masticated food enters the very acid environment of the stomach (pH between 1 and 2). The low pH contributes to denaturation and swelling of proteins and swelling of starch. In the stomach the proteolytic enzyme, pepsin, and a lipase begin digestion of proteins and triglycerides, respectively.

After a suitable time the contents of the stomach are passed into the small intestine, where the pH soon becomes 7.5 to 8 as a result of secretions from the pancreas, gallbladder, and intestinal mucosa. The pancreatic secretions contain several proteolytic enzymes—trypsin, chymotrypsin, and carboxypeptidases A and B—as well as ribonuclease, α-amylase, and lipase. Some of the enzymes are secreted as proenzymes. Secretions from the intestinal mucosa contain a number of exopeptidases, dipeptidases, and glycosidases. The combined action of these enzymes leads to conversion of proteins to amino acids, triglycerides to glycerol

Table 2 Enzymes Involved in Digestion and Assimilation of Foods

Process	Location	Enzymes
Ingestion	Mouth	Amylase
Digestion	Gastrointestinal tract	Proteolytic enzymes, glycosidases, esterases, lipases, phosphatases, nucleases
Absorption	Mucosa of gastrointestinal tract	Enzymes in active transport of amino acids, monosaccharides, fatty acids, glycerol, purines, pyrimidines, etc.
Transport	Plasma	Enzymes in biosynthesis of lipids at inner membrane of mucosa
Metabolism	Cells (liver, storage, specialized)	(a) Enzymes in catabolic reactions leading to production of energy and heat (b) Enzymes in biosynthesis of proteins, nucleic acids, lipids, and complex carbohydrates
Digestion	Lysosomes (intracellular)	Proteolytic enzymes, glycosidases, esterases, phosphatases, sulfatases, lipases, nucleases
Excretion	Kidney, large intestine	Enzymes of active transport in reabsorption

and fatty acids, starch and glycogen to glucose, and nucleic acids to pyrimidines and purines. The function of this process is to convert the complex food compounds to low-molecular-weight compounds that can be absorbed by the intestinal mucosa. Complex carbohydrates such as cellulose, lignin, and pectin are not hydrolyzed in the intestine and are excreted in the feces along with other undigested materials. They are valuable dietary fiber.

Transport of compounds through the cells of the intestinal mucosa and into the bloodstream generally is not a passive process but involves active transport. Active transport requires energy and the involvement of a host of specific transport mechanisms that probably include many enzymes.

Once in the bloodstream the amino acids, lipids (resynthesized from fatty acids and glycerol in the mucosa), monosaccharides, pyrimidines and purines, vitamins, minerals, and so on, are transported to the cells (often in the liver). At the cellular level these compounds may be oxidized to provide energy for mechanical work, maintenance of the temperature of the organism, and synthesis of complex chemical compounds. Some of these compounds may go into reserve storage pools of amino acids, lipids, and glycogen. Others may be used in biosynthetic processes to maintain the integrity of the cells and for growth. The proportion of the ingested food that enters each of these separate pathways will depend on the nutritional, health, and age status of the individual.

Cellular constituents are continually turned over at variable rates, and in the normal adult the rate of catabolism (degradation) is just balanced by the rate of anabolism (biosynthesis). Subcellular organelles, the lysosomes, perform the function of degradation of constituents at the cellular level. They contain a variety of hydrolytic enzymes capable of degrading all the major compounds.

While the body is rather efficient in its reuse of compounds, there is an excess of certain compounds and an accumulation of certain toxic compounds that it must excrete. The two most important toxic compounds (in excessive amounts) are carbon dioxide, which is removed through the lungs, and urea, which is excreted by the kidney in the urine.

Digestion and assimilation of foods are under the general control of hormones, which perform their function by acting on key enzymes involved in the process.

D. Enzymes in Health and Disease

Digestion and assimilation of food are complex processes. Therefore, abnormalities that influence the overall nutritional and health status of the individual can occur at many stages. A number of diseases are associated with an imbalance in one or more enzymes. Perhaps the first metabolic disease to be associated with lack of a specific enzyme was *alcaptonuria* [39], which results in excretion of large amounts of homogentisate (2,5-dihydroxyphenylacetate). In alcaptonuria the enzyme responsible for conversion of homogentisate to acetoacetate and succinate

[Eq. (3)] is missing. The disease is characterized by excretion of a dark urine and pigmentation of cartilage and connective tissues due to deposition of homogentisate. This deposition leads to arthritis in later life.

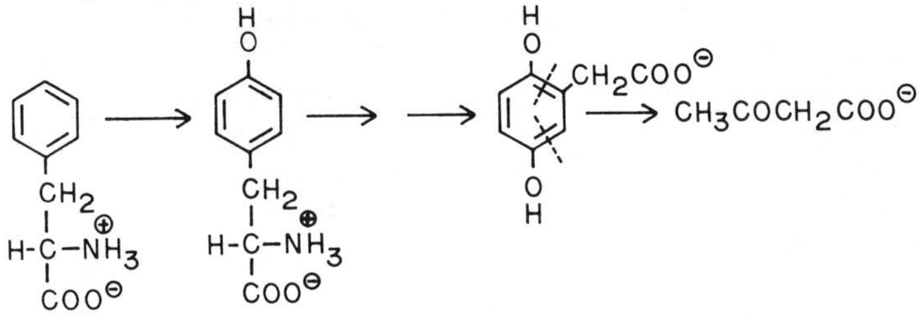

(3)

L-Phenylalanine L-Tyrosine Homogentisate Acetoacetate

Phenylketonuria is associated with certain types of mentally retarded patients who excrete large amounts of phenylpyruvate in the urine and who have elevated levels of phenylalanine in the blood. The high levels of phenylalanine result in defective myelinization of the brain and resultant mental retardation. The specific enzymatic defect is the absence in the liver of phenylalanine hydroxylase, which converts phenylalanine to tyrosine [the first step in Equation (3)]. Mental abnormalities can be prevented by use of a diet low in phenylalanine; thus this becomes as much a problem for the nutritionist as for the clinician.

Galactosemia is characterized by vomiting and diarrhea following milk ingestion. If not corrected promptly dehydration, jaundice, hypoglycemia, cataract formation, and mental retardation follow. The abnormality is detected by the presence of galactose in the urine and can be controlled by feeding diets free of galactose. In such cases, babies must be fed on soybean milk, not mother's or cow's milk.

Congenital lactase deficiency is characterized by an early intolerance to milk which produces severe diarrhea and wasting. The disease is due to the absence of lactase (β-galactosidase) and can be controlled by excluding milk from the diet or by pretreating the milk with β-galactosidase.

A large number of diseases are known to be associated with an absence or abnormality of a specific metabolic enzyme. Some of these are listed in Table 3. A comprehensive review is available on this subject [40]. These diseases result from genetic defects, and the only prevention and control is their early recognition by the clinician and their control by proper diet. Food scientists have been remiss in not recognizing their role in this important area. Recombinant DNA technology may play a major role in this area in the near future.

E. Enzymes in Analytical Procedures

Because of their specificity, enzymes are ideally suited for measurement of individual compounds in plant and animal material [41]. Generally, it is not

Table 3 Some Genetic Diseases with Established Enzymatic Defects

Disease	Defective enzyme
Alcaptonuria	Homogentisic acid oxidase
Phenylketonuria	Phenylalanine hydroxylase
Hyperammonemia	Ornithine transcarbamylase
Hemolytic anemia	Erythrocyte glucose 6-phosphate dehydrogenase
Acatalasia	Erythrocyte catalase
Congenital lysine intolerance	L-Lysine-NAD oxidoreductase
Gaucher's disease	Glucocerebrosidase
Refsum's disease	Phytanic acid oxidase
McArdle's syndrome	Muscle phosphorylase
Hypophosphatasia	Alkaline phosphatase
Congenital lactase deficiency	Lactase
Gout	Hypoxanthine-guanine phosphoribosyl transferase
Xanthinuria	Xanthine oxidase
Hereditary fructose intolerance	Fructose 1-phosphate aldolase

Source: Ref. 40.

necessary to purify a compound partially before it can be quantitatively determined by use of an enzyme.

Consider the determination of glucose. Many methods are available which determine glucose on the basis of its reduction of compounds such as alkaline Cu^{2+} solutions, 3,5-dinitrosalicylate, potassium ferricyanide, and ceric sulfate. These methods all suffer from the disadvantage that they do not distinguish between glucose and other reducing compounds, including other sugars. By combining the very specific action of glucose oxidase on glucose with the quantitative colorimetric determination of the hydrogen peroxide produced by use of peroxidase [Eqs. (4) and (5)], microgram quantities of glucose are determined accurately and specifically in just a few minutes.

$$\text{Glucose} + O_2 \xrightarrow[\text{oxidase}]{\text{glucose}} \delta\text{-D-gluconolactone} + H_2O_2 \qquad (4)$$

$$H_2O_2 + \text{chromogen} \xrightarrow{\text{peroxidase}} \text{colored compound} + H_2O \qquad (5)$$

The advantage of coupled enzyme assays, of which the glucose oxidase–peroxidase assay for glucose is one example, should not be overlooked. As shown in Eq. (4), the products of the glucose oxidase reaction are δ-D-gluconolactone and hydrogen peroxide, neither of which can be measured with precision at low concentrations. However, by coupling the reaction with peroxidase, as little as 5 to 20 μg/mL of glucose can readily be determined.

An even more convenient method of determining glucose concentrations with glucose oxidase is available [42]. Glucose oxidase is immobilized in polyacrylamide gel and a small column is packed with the material. The sample to be analyzed for glucose is passed through the column and the oxygen concentration of the solution just as it enters and as it leaves the column is measured with oxygen-sensitive electrodes. The decrease in oxygen concentration due to the action of glucose oxidase on glucose [Eq. (4)] is proportional to the concentration of glucose in the solution. The column packed with immobilized enzyme can be used repeatedly and analyses are rapid.

Enzyme electrodes capable of measuring many compounds, including glucose, in a solution have been described [43]. An example is the urea electrode. The electrode is prepared with immobilized urease and a glass sensitive to ammonium ions. As the enzyme acts on urea, ammonium ions are produced [Eq. (6)] which

$$\begin{array}{l} H_2N{-}CO{-}NH_2 + H_2O \xrightarrow{\text{urease}} CO_2 + 2NH_3 \\ \qquad\qquad\qquad\qquad\qquad -2H_2O \upharpoonleft\downharpoonright 2H_2O \\ \qquad\qquad\qquad\qquad\qquad 2NH_4OH \rightleftharpoons 2NH_4^{\oplus} + 2OH^{\ominus} \end{array} \tag{6}$$

are detected by the electrode. The convenience and precision of the method is equal to that of the glass electrode used for pH determination. With a little ingenuity the general concept has been adapted to many types of analytical procedures [43].

Medical clinicians have recognized the convenience of enzymes in analytical procedures for some time [44]. Again, food scientists have been remiss in not adapting this technology to their use. For many years they have used the destruction of peroxidase and alkaline phosphatase to determine the adequacy of heat treatment of fruits and vegetables and milk, respectively. However, they have not generally adapted the use of enzymes to other vital analytical procedures in food science. Biosensors are now being developed at ever-accelerating rates using microchips and amplifiers.

V. LITERATURE OF ENZYMOLOGY

Because of the importance of enzymology to so many fields, papers on the subject can be found in nearly all journals, including those devoted to engineering, physics, and chemistry as well as all the life sciences. The only way of keeping abreast of relevant work on an enzyme is by use of abstracting services such as *Chemical Abstracts* and *Biological Abstracts*. The depth of consideration of an enzyme as an entity varies from its use as a tool to accomplish some goal to detailed descriptions of its protein structure and mechanism of action.

Some of the more recent books, monographs, and treatises devoted to enzymology along with a list of journals are given below. The list is incomplete but it will serve as a guideline to the literature. Many specific references will be found at the end of each chapter of this book.

A. Journals

Archives of Biochemistry and Biophysics, *Biochemical and Biophysical Research Communications*, *Biochemical Journal*, *Biochemistry*, *Biochimica et Biophysica Acta*, *Bioenzymology*, *European Journal of Biochemistry*, *FEBS Letters*, *Journal of Biochemistry*, *Journal of Biological Chemistry*, *Journal of Food Biochemistry*.

B. Books, Monographs, and Treatises

Methods in Enzymology, S. P. Colowick and N. O. Kaplan, eds., Academic Press, New York, 1955–1992, more than 200 volumes. Several volumes are devoted to methods of purification of enzymes. Other volumes are devoted to methods of preparing substrates and techniques of studying enzymes both as proteins and as catalysts.

The Enzymes, 3rd Ed., P. D. Boyer, H. Lardy, and K. Myrbäck, eds., Academic Press, New York, 1970–1990, 19 volumes. Comprehensive treatise of the available information about enzymes and enzyme action at the molecular level.

Annual Review of Biochemistry, Annual Reviews, Palo Alto, Calif. While not devoted entirely to enzymes, this annual review is an excellent way of keeping abreast of work on enzymes outside one's own area of work. Some reviews on enzymes also are found in *Annual Review of Plant Physiology*, *Annual Review of Physiology*, and so on.

Advances in Enzymology, F. F. Nord, ed., (Wiley) Interscience, New York. Detailed treatment of a limited area of enzymology each year. The 65th volume was published in 1992.

Enzymes, M. Dixon and E. C. Webb, Academic Press, New York, 1980. One of the best books available on the general behavior of enzymes. (Previous editions published in 1958, 1964, and 1972).

Enzymes in Food Processing, G. Reed, Academic Press, New York, 1975. Devoted primarily to the technology of enzymes.

Handbook of Enzymatic Methods of Analysis, G. G. Guilbault, Marcel Dekker, New York, 1976.

Source Book of Enzymology, S. Schwimmer, AVI, Westport, Conn., 1981.

Methods of Enzymatic Analysis, 3rd Ed., H. U. Bergmeyer, ed., Academic Press, New York, 1983 on, 10 volumes.

Enzyme Mechanisms, M. I. Page and A. Williams, eds., Royal Society of Chemistry, London, 1987.

Enzyme Engineering, H. W. Blanch and A. M. Klibanov, eds., Vol. 542, Annals of the New York Academy of Sciences, New York, 1988.

Mechanism-Based Enzyme Inactivation: Chemistry and Enzymology, R. B. Silverman, Vols. 1 and 2, CRC Press, Boca Raton, Fla., 1988.

Proteins and Enzymes, J. E. Bell and E. T. Bell, Prentice Hall, Englewood Cliffs, N.J., 1988.

Biocatalysis in Agricultural Biotechnology, J. R. Whitaker and P. E. Sonnet, eds., ACS Symposium Series 389, American Chemical Society, Washington, D.C., 1989.

Enzyme Chemistry: Impact and Applications, 2nd Ed., C. J. Suckling, ed., Chapman & Hall, New York, 1990.

Enzyme Technology, M. F. Chaplin and C. Bucke, Cambridge University Press, Cambridge, 1990.

Microbial Enzymes and Biotechnology, 2nd Ed., W. M. Fogarty and C. T. Kelly, eds., Elsevier Applied Science, New York, 1990.

A Study of Enzymes, S. A. Kuby, Vols. 1 and 2, CRC Press, Boca Raton, Fla., 1991.

Food Enzymology, P. F. Fox, ed., Vols. 1 and 2, Elsevier Applied Science, New York, 1991.

Oxidative Enzymes in Foods, D. S. Robinson and N. A. M. Eskin, eds., Elsevier Applied Science, New York, 1991.

Biocatalysis at Extreme Temperatures: Enzyme Systems Near and Above 100°C, M. W. W. Adams and R. M. Kelly, eds., ACS Symposium Series 498, American Chemical Society, Washington, D.C, 1992.

Small catalytic RNAs, R. H. Symons, *Ann. Rev. Biochem.*, **61**, 641 (1992).

REVIEW QUESTIONS

1. Why does the energy-reaction coordinate diagram in Figure 5 show three peaks for the enzyme-catalyzed reaction and only one for the nonenzyme-catalyzed reaction?
2. Based on the discussion of enzymes as tools for analytical analyses, devise an experimental procedure that would permit you to determine the glucose and sucrose content of a plant tissue.
3. Speculate as to the reasons why much progress has been made in enzymology during the past 20 years.
4. What are some of the characteristics of an enzyme that would permit you to distinguish it from other types of molecules that occur in vivo?
5. Schematically, depict the major difference between Fischer's lock and key theory and Koshland's induced fit theory for enzyme–substrate interaction.
6. A relationship between substrate concentration and observed velocity of reaction is shown in Figure 2. Predict what concentration of substrate, relative to K_m, would be necessary for the observed velocity to equal the maximum velocity.
7. The chemical synthesis of ribonuclease in the laboratory required 18 months, while it takes a cell about 3 minutes to accomplish the same synthesis. Does this mean that the cell is more intelligent than the scientists?
8. Of the highlights of the history of enzymology described in this chapter, which do you consider to be *the* outstanding contribution? Explain.

9. In your opinion what should be the level of understanding of enzymology that a food scientist should strive to achieve?

10. Suggest some other practical uses of enzymes in food processing that are not listed in Table 1.

REFERENCES

1. A. Payen and J. F. Persoz, *Ann. Chim. (Phys.)*, **53**, 73 (1833).
2. C. F. Schoenbein, *Verhandl. Naturforsch. Ges. Basel*, **1**, 339 (1855).
3. C. F. Schoenbein, *Phil. Mag.*, **11**, 137 (1856).
4. M. Berthelot, *Compt. Rend. Acad. Sci.*, **50**, 980 (1860).
5. L. Pasteur, *Compt. Rend. Acad. Sci.*, **80**, 452 (1875).
6. W. Kühne, *Unters. Physiol. Inst. Univ. Heidelberg*, **1**, 291 (1878).
7. J. S. Fruton and S. Simmonds, *General Biochemistry*, Wiley, New York, 1953.
8. E. Büchner, *Ber.*, **30**, 117 (1897).
9. E. Fischer, *Ber.*, **27**, 2985 (1894).
10. D. E. Koshland, Jr., in *The Enzymes* (P. D. Boyer, H. Lardy, and K. Myrbäck, eds.), Vol. 1, Academic Press, New York, 1959, p. 305.
11. V. Henri, *Acad. Sci. Paris*, **135**, 916 (1902); *Lois générales de l'action des diastases*, Hermann, Paris, 1903.
12. A. Brown, *J. Chem. Soc.*, **81**, 373 (1902).
13. L. Michaelis and M. L. Menten, *Biochem. Z.*, **49**, 333 (1913).
14. K. J. Laidler, *The Chemical Kinetics of Enzyme Action*, Oxford, Fair Lawn, N.J., 1958.
15. R. A. Alberty, in *The Enzymes*, 2nd Ed. (P. D. Boyer, H. Lardy, and K. Myrbäck, eds.), Vol. 1, Academic Press, New York, 1959, p. 143.
16. W. W. Cleland, in *The Enzymes*, 3rd Ed. (P. D. Boyer, H. Lardy, and K. Myrbäck, eds.), Vol. 1, Academic Press, New York, 1970, p. 1.
17. K. Dalziel, *Acta Chem. Scand.*, **11**, 1706 (1957).
18. J. T. F. Wong and C. S. Hanes, *Can. J. Biochem. Physiol.*, **40**, 763 (1962).
19. J. Monod, J. Wyman, and J. P. Changeux, *J. Mol. Biol.*, **12**, 88 (1965).
20. J. B. Sumner, *J. Biol. Chem.*, **69**, 435 (1926).
21. W. M. Bayliss, *The Nature of Enzyme Action*, 3rd Ed., Longmans, Green, London, 1914.
22. J. B. S. Haldane, *Enzymes*, Longmans, Green, London, 1930.
23. M. Kunitz and J. H. Northrop, *J. Gen. Physiol.*, **19**, 991 (1936).
24. E. H. Fischer and E. A. Stein, *Arch. Sci. (Geneva)*, **7**, 131 (1954).
25. E. A. Peterson and H. A. Sober, *J. Am. Chem. Soc.*, **78**, 751 (1956).
26. L. Ornstein, *Ann. N.Y. Acad. Sci.*, **121**, 321 (1964); B. J. Davis, *Ann. N.Y. Acad. Sci.*, **121**, 404 (1964).
27. P. Flodin, Dextran gels and their applications in gel filtration, dissertation, University of Uppsala, Sweden (1962).
28. P. Cuatrecasas, M. Wilchek, and C. B. Anfinsen, *Proc. Natl. Acad. Sci. USA*, **61**, 636 (1968).

29. O. Vesterberg, *Methods Enzymol.*, **22**, 389 (1971).
30. L. Fägerstam, J. Lizana, U.-B. Axiö-Fredriksson and L. Wahlström, *J. Chromatog.*, **266**, 523 (1983).
31. I. Gillam, S. Millward, D. Blew, M. V. Tigerstrom, E. Wimmer, and G. Tener, *Biochemistry*, **6**, 3043 (1967).
32. A. P. Ryle, F. Sanger, L. F. Smith, and R. Kitai, *Biochem. J.*, **60**, 541 (1955).
33. D. H. Spackman, W. H. Stein, and S. Moore, *J. Biol. Chem.*, **235**, 648 (1960); C. H. W. Hirs, S. Moore, and W. H. Stein, *J. Biol. Chem.*, **235**, 633 (1960).
34. G. Kartha, J. Bello, and D. Harker, *Nature*, **213**, 862 (1967).
35. B. Gutte and R. B. Merrifield, *J. Am. Chem. Soc.*, **91**, 501 (1969); R. Hirschmann, R. F. Nutt, D. F. Veber, R. A. Vitali, S. L. Varga, T. A. Jacob, F. W. Holly, and R. G. Denkewalter, *J. Am. Chem. Soc.*, **91**, 507 (1969).
36. B. L. Vallee and J. F. Riordan, *Ann. Rev. Biochem.*, **38**, 733 (1969).
37. D. E. Koshland, Jr., and K. E. Neet, *Ann. Rev. Biochem.*, **37**, 359 (1968).
38. E. M. Mrak, in *Food Enzymes* (H. W. Schultz, ed.), AVI, Westport, Conn., 1960, p. 1.
39. A. E. Garrod, *Proc. Roy. Soc. Med. Chir.*, **2**, 130 (1899).
40. W. E. C. Wacker and T. L. Coombs, *Ann. Rev. Biochem.*, **38**, 539 (1969).
41. G. G. Guilbault, *Anal. Chem.*, **40**, 459R (1968).
42. S. J. Updike and G. P. Hicks, *Science*, **158**, 270 (1967).
43. G. G. Guilbault, R. K. Smith, and J. G. Montalvo, Jr., *Anal. Chem.*, **41**, 600 (1969).
44. H. U. Bergmeyer, ed., *Methods of Enzymatic Analysis*, 3rd Ed., Academic Press, New York, 1983–1990.

2

The Protein Nature of Enzymes

I. ROLE OF PROTEINS IN ORGANISMS

All enzymes are proteins (except for catalytic RNAs), but not all proteins are enzymes. Proteins occupy a central role in both the dynamic and structural aspects of living organisms. They perform many important functions in living organisms. Insulin and glucagon are hormones involved in regulation of metabolic pathways. Antibodies are involved in protection of organisms against various types of foreign materials (antigens), which in some cases are also proteins. Proteins of the human skin, fingernails, hair, and the protein coats of viruses perform protective functions. Some proteins are highly toxic, such as the *Clostridium* toxins, while other proteins are nutritious and are used as food. Certain proteins (actin and myosin) convert chemical energy into mechanical energy so that we may move; still others are elastic connectors (tendons) between the muscle and the bones so that muscular activity is converted to work. Hemoglobin and myoglobin are involved in transport of oxygen and carbon dioxide. Other proteins maintain the pH and osmotic pressure of the body fluids constant. Enzymes are biological catalysts permitting reactions to go at body temperature. This last group of proteins are the subject of this book.

The number of different proteins in nature are tremendous since the proteins from one species are different from those of another species. α-Amylases from human saliva, human pancreas, hog pancreas, bovine pancreas, and from bacteria are identical in their catalytic action in hydrolyzing the α-1,4-glucosidic linkages of starch, but each is different in terms of protein structure. These α-amylases can be distinguished from each other on the basis of amino acid composition, electrophoretic and chromatographic behavior, and reactions with specific antigens prepared against one or more of the α-amylases.

To appreciate fully what an enzyme is and how it functions, we must understand it both as a protein and as a catalyst. For proper understanding of the parameters that affect the catalytic site, we must understand fully the protein nature of the catalyst.

Proteins have several types and levels of molecular organization: *primary structure*, *secondary structure*, *tertiary structure*, *quaternary structure*, and *macromolecular structure*.

II. PRIMARY STRUCTURE OF PROTEINS

The building blocks of all proteins are the amino acids. While more than 175 amino acids have been isolated from various natural sources, only 20 of these are found regularly in proteins. Posttranslational modifications of amino acid side chains number about 135; however, not all 135 types of modification occur in each protein. In fact, only one or a few occur in any one protein. Some proteins, such as ribonuclease, chymotrypsin, and trypsin, are composed only of amino acids. Other proteins contain, in addition to amino acids, other organic and/or inorganic components and are called conjugated proteins. Some examples of conjugated proteins include the (a) chromoproteins, such as hemoglobin, myoglobin, catalase, and peroxidase, which contain iron–porphyrin cofactors; (b) nonheme metalloproteins, such as ferritin, carboxypeptidases A and B, alkaline phosphatase, and alcohol dehydrogenase; (c) lipoproteins, such as α_1- and β_1-lipoproteins of blood and rhodopsin, which is a carotenoid–protein complex present in the retinal rods of the eye and necessary for color vision; (d) glycoproteins, such as the blood group–specific substances responsible for the A, B, AB, and O factors, and the mucins, which are major components of certain body secretions, such as saliva, gastric juice, and intestinal secretions; (e) phosphoproteins, such as the caseins and pepsin; and (f) the large group of enzymes that require various cofactors (see Chapter 12).

A. Amino Acids and Their Properties

As the name implies, the amino acids contain a carboxyl and an amino group. With the exception of proline and hydroxyproline, all amino acids found in proteins are α-amino acids (i.e., the carboxyl and amino groups are on the α-carbon of the compounds). Proline and hydroxyproline are α-imino acids in that the nitrogen of the amino group is involved in a heterocyclic five-membered ring.

The general structure of all the α-amino acids may be written as shown in Fig. 1. The differences among the amino acids reside in the R group. This is

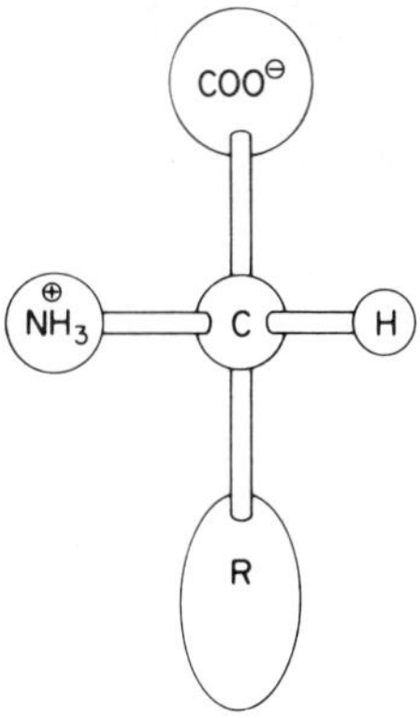

Figure 1 General structural formula of an L-amino acid. The size of the R group is variable depending on the amino acid.

apparent from examination of Table 1. In the case of proline and hydroxyproline the R group is bonded not only to the α-carbon but also to nitrogen of the imino group. The R groups of the 20 amino acids differ from each other in size, shape, charge, hydrogen- and hydrophobic-bonding capacity, and chemical reactivity.

All of the amino acids, with the exception of glycine, are optically active. This may be seen from Fig. 2, where the compound is represented as a tetrahedron with the tetrahedral α-carbon atom, with its four valence bonds, in the center and the four different groups at the corners.

For ease in rapid determination of the configuration about the asymmetric carbon atom, it is the usual convention (according to Fischer projection) to orient the carbon chain in the vertical direction with carbon atom 1 at the top. The horizontal bonds are considered to extend out of the plane of the paper, while those in the vertical direction extend behind the plane of the paper. For the D-isomer the functional group ($NH_3^{\oplus}$ in Fig. 2) is oriented to the right of the carbon chain, and for the L-isomer the functional group is oriented to the left of the carbon chain. However, the molecule may be oriented with carbon 1 at any one of the four corners of the tetrahedron. Once this has been done, the positions of the other three constituents are fixed. All of the amino acids in proteins are of the L-configuration.

Four amino acids have a second center of asymmetry in addition to that around the α-carbon: threonine, hydroxyproline, isoleucine, and hydroxylysine. The isomers that occur in vivo have the configurations shown in Table 1.

Exercise 1. Construct projection formulas for all possible isomers of threonine, hydroxyproline, isoleucine, and hydroxylysine.

In Figs. 1 and 2 the amino acids are written as dipolar ions. This represents the

Table 1 Amino Acids Commonly Found in Proteins (All Are L-configuration)

Group	Trival name	Abbreviation	p*K* values	p*I* value	Formula
Monoaminomono-carboxylic	Glycine	Gly	2.34; 9.78	6.06	$H—CH(NH_2)—COOH$
	Alanine	Ala	2.35; 9.69	6.02	$CH_3—CH(NH_2)—COOH$
	Valine	Val	2.32; 9.62	5.97	$CH_3CH(CH_3)—CH(NH_2)—COOH$
	Leucine	Leu	2.36; 9.60	5.98	$CH_3CH(CH_3)CH_2—CH(NH_2)—COOH$
	Isoleucine	Ile	2.36; 9.68	6.02	$CH_3CH_2CH(CH_3)—CH(NH_2)—COOH$
Hydroxymonoamino-monocarboxylic	Serine	Ser	2.21; 9.15	5.68	$CH_2(OH)—CH(NH_2)—COOH$
	Threonine	Thr	2.63; 10.43	6.53	$CH_3CH(OH)—CH(NH_2)—COOH$

Monoaminodicarboxylic (including the amides)					
	Aspartic acid	Asp	2.09 (α-Carboxyl) 3.86 (β-Carboxyl) 9.82	2.98	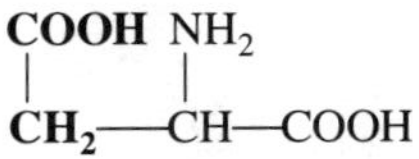
	Asparagine	Asn	2.02; 8.8	5.41	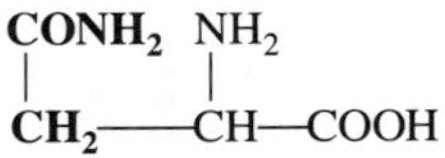
	Glutamic acid	Glu	2.19 (α-Carboxyl) 4.25 (γ-Carboxyl) 9.67	3.22	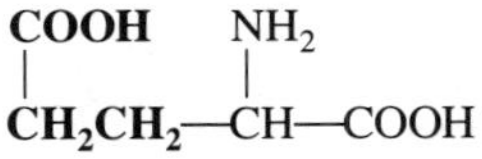
	Glutamine	Gln	2.17; 9.13	5.65	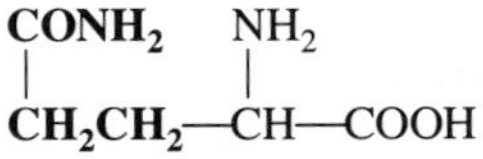
Diaminomonocarboxylic					
	Lysine	Lys	2.18 8.95 (α-Amino) 10.53 (ε-Amino)	9.74	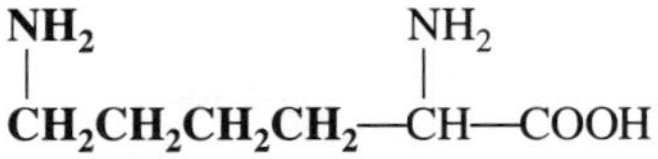
	Hydroxylysine	Hyl	2.13 8.62 (α-Amino) 9.67 (ε-Amino)	9.14	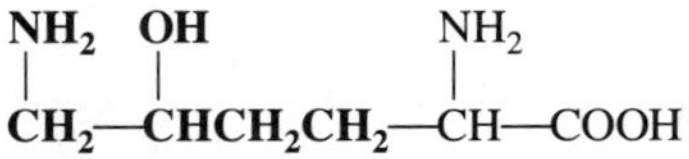
	Arginine	Arg	2.17 9.04 (α-Amino) 12.48 (Guanidino)	10.76	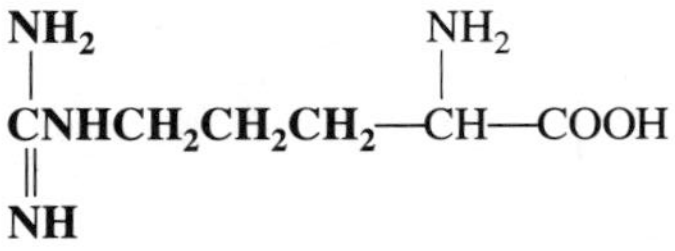

Table 1 Continued

Group	Trival name	Abbreviation	p*K* values	p*I* value	Formula
Sulfur-containing	Cysteine	Cys	1.71 8.33 (—SH) 10.78 (α-Amino)	5.02	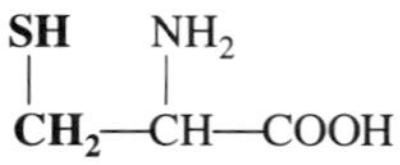
	Cystine	CySSCy	1.65; 2.26 (Carboxyls) 7.85; 9.85 (Aminos)	5.06	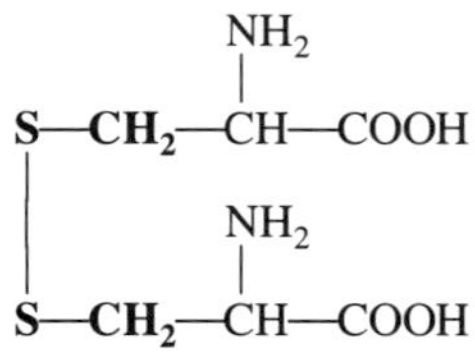
	Methionine	Met	2.28; 9.21	5.74	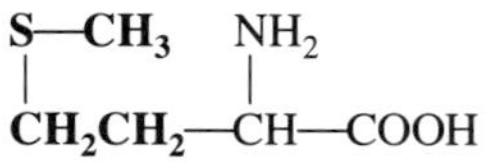
Cyclic (including aromatic)	Phenylalanine	Phe	1.83; 9.13	5.48	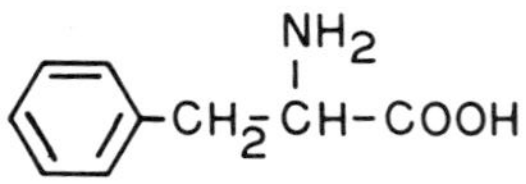
	Tyrosine	Tyr	2.20 9.11 (α-Amino) 10.07 (Phenolic)	5.66	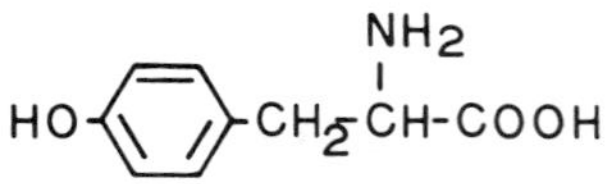

Tryptophan	Trp	2.38; 9.39	5.88	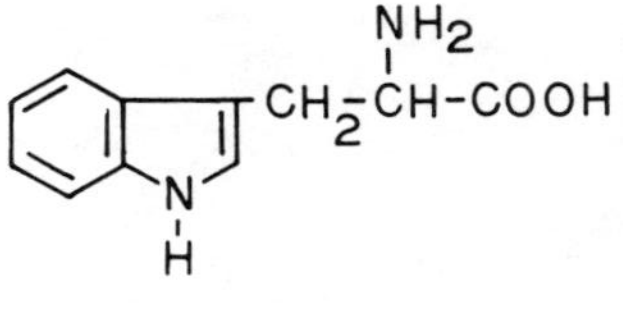
Histidine	His	1.82 6.0 (Imidazole) 9.17	7.58	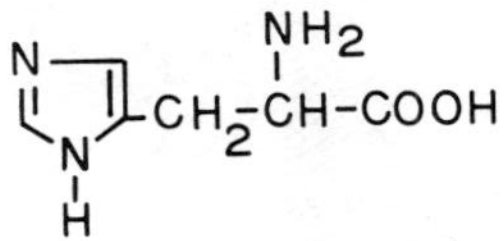
Proline	Pro	1.99; 10.60	6.30	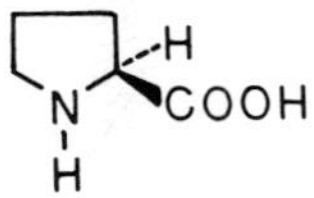
Hydroxyproline	Hyp	1.92; 9.73	5.82	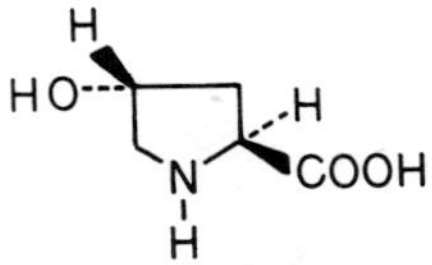

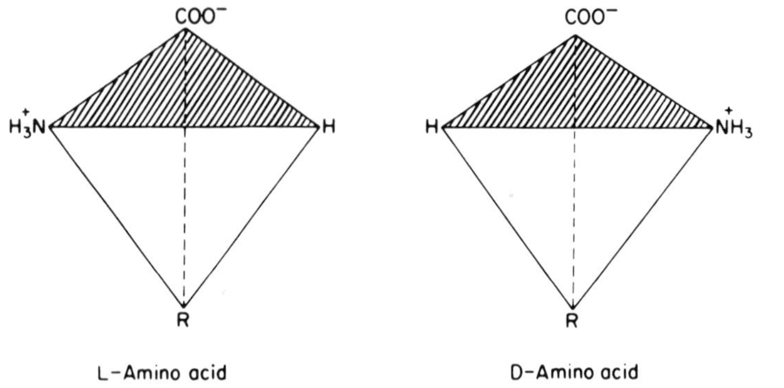

Figure 2 Fischer projection formulas for an L-amino acid and a D-amino acid.

form in aqueous solution at the isoelectric point of a particular amino acid. At other pH values other species of the amino acid will be present:

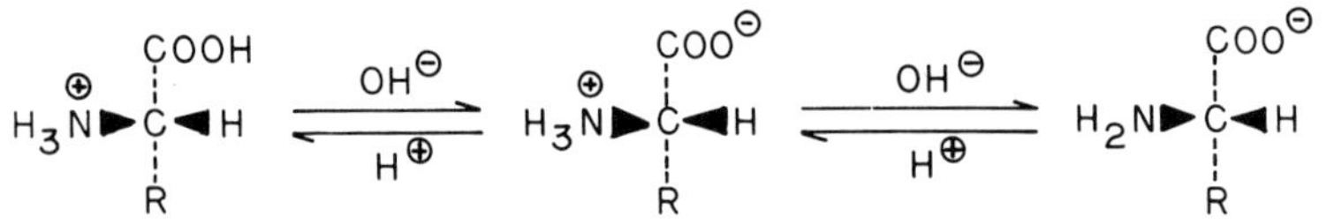

(1)

There are two prototropic groups on a neutral amino acid, the carboxyl and the ammonium group. As the pH is increased, the protons on the carboxyl groups of the molecules are titrated. When half of the protons have been titrated, the pH is equal to pK_1. When all the protons of the carboxyl groups have been titrated, there will be an equal number of carboxylate ions and ammonium groups so that the net charge is zero (isoelectric point). Continued addition of base will titrate the protons of the ammonium groups. When half of these have been titrated, the pH is equal to pK_2. When all the protons of the ammonium groups have been titrated, the amino acid will have a net negative charge of 1. The isoelectric point of a neutral amino acid is given by the average of the two pK values:

$$\frac{pK_1 + pK_2}{2} = \text{isoelectric point} = pI \qquad (2)$$

The titration curve for glycine is shown in Fig. 3.

Not all amino acids have neutral side-chain groups. For example, aspartic and glutamic acids each have a second acidic group, whereas lysine, arginine, and histidine have basic groups. The pK values and isoelectric points of the several amino acids are shown in Table 1.

Exercise 2. Formulate a rule for calculating the isoelectric points of the acidic and basic amino acids and show why the rule is justified.

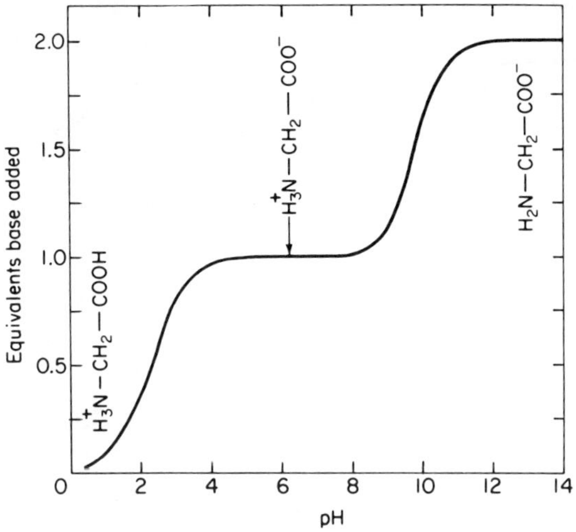

Figure 3 Titration curve for glycine based on pK_1 and pK_2 of 2.34 and 9.78, respectively.

B. Interaction of Amino Acids to Form Peptide Bonds

A protein is built from the 20 amino acids by reaction of the α-carboxyl group of one amino acid with the α-amino group of a second amino acid [Eq. (3)]. The

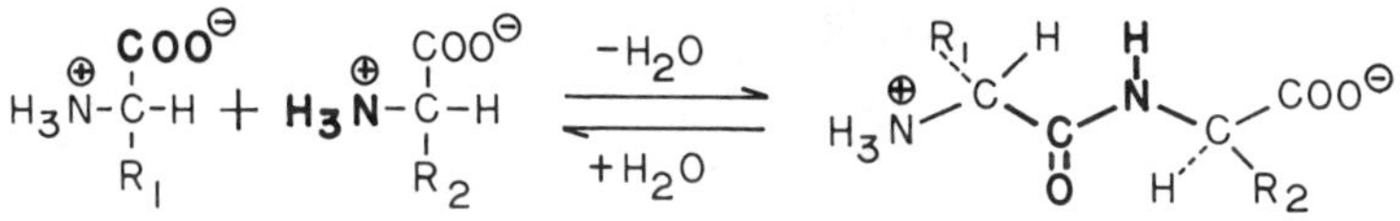

(3)

linkage, —CONH—, joining the two amino acids is called the peptide bond. This condensation reaction is repeated over and over again to build a polymer whose amino acid sequence and size is characteristic of a given protein. The peptide bond is neutral and planar. This planar nature of the peptide bond, coupled with the tetrahedral nature and asymmetry of the α-carbon atoms of the two amino acid residues involved in the peptide bond, places severe limitations on the shape of the polypeptide chain and is partly responsible for the secondary structure of proteins (see below). The polypeptide chain cannot be fully extended like a clothesline but has a corrugated appearance (Fig. 4). If the chain were fully extended, the length of two amino acid residues would be 8.64 Å rather than 7.23 Å as shown in Fig. 4. Another important feature of the peptide chain is that the R groups (and α-hydrogens) are alternately above and below the plane of the paper. The bond distances and angles are also shown.

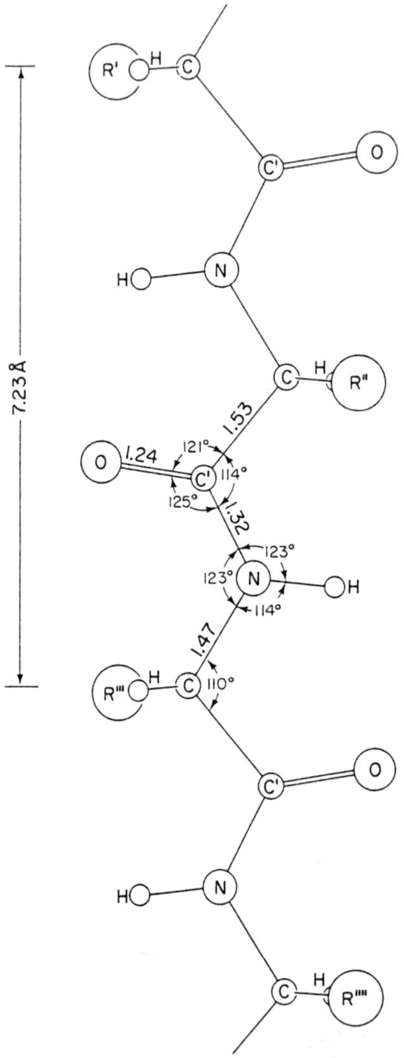

Figure 4 Primary structure of a polypeptide chain, including bond angles and distances. (From Ref. 1, p. 17, by courtesy of the Royal Society.)

C. Biosynthesis of Proteins

The amino acid sequence of each type of protein is different, while all molecules of a given protein synthesized by an organism are identical. This remarkable feat is accomplished in vivo as shown schematically in Fig. 5. Information for the faithful reproduction of each type of protein molecule is contained in the sequence of purine and pyrimidine bases in a DNA molecule shown as a double helix. For use, this information is transcribed into a complementary sequence of bases in the messenger RNA molecule. Each set of three bases in the messenger RNA constitutes a codon that specifies a specific amino acid. The messenger RNA is released from the master template (DNA) and is used as the secondary template on which the polypeptide chain is assembled. Another type of RNA, called transfer

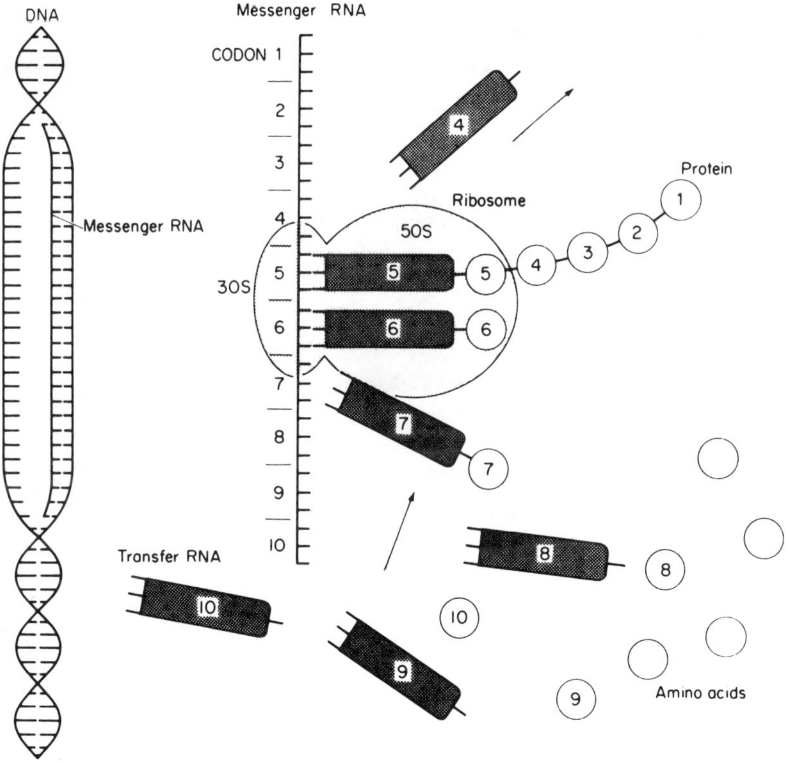

Figure 5 Schematic representation of the processes involved in the biosynthesis of proteins showing the role of DNA, messenger RNA, transfer RNA, ribosome, and amino acid. Not shown are the many enzymes involved in the process. (From Ref. 2, p. 30, by courtesy of Scientific American, Inc.)

RNA, combines with the amino acids. There is a specific transfer RNA, and a specific amino acid–activating enzyme, for each of the 20 amino acids. Each particular transfer RNA–amino acid compound recognizes a complementary site on the messenger RNA molecule through its complementary anticodons (sequence of three bases complementary to a sequence of three bases on the messenger RNA molecule). In the presence of ribosomes the transfer RNA–amino acid compound is attached to the messenger RNA template, and while so attached a series of enzymes in the ribosome couple one amino acid to another through the peptide bond. The message is read only in one direction, always starting and stopping at a specific point. As shown in Fig. 5, the first peptide bond is formed between amino acids 1 and 2, followed in sequence by the others. This process is called translation.

Although many amino acids are found in a cell and some may be D-amino acids, there are specific transfer RNAs and activating enzymes for only 20 of these. Thus the other amino acids do not get incorporated into a protein. A cell can carry out the complete assembly of a protein in 1 to 3 min. The complete chemical synthesis of ribonuclease (MW 13,683; 124 amino acids) was carried out simultaneously by two groups of workers, one at Rockefeller University [3] and the other from the Merck Sharp and Dohme Research Laboratories [3]. It took the group at Merck (eight authors on paper and 15 others on accompanying papers) 18 months to synthesize 5 mg of ribonuclease which was not enzymatically pure, active material. Recombinant DNA technology now permits laboratory synthesis of proteins as shown in Fig. 5 in a short time. Automated chemical synthesis of large peptides can now be done efficiently.

D. Distinguishing Feature of Primary Structures of Proteins

Every protein has in common the amino acid backbone. The distinguishing features among proteins, then, are the side chains of the amino acid residues and their sequence within the polypeptide chain. As indicated in Table 1, most of these side chains are neutral but differ one from the other in the size of the side chains, whether they are aliphatic or aromatic or contain hydroxyl groups. Other side chains of the amino acid residues contain basic or acidic groups. It is the nature and sequential arrangement of these side chains that determine what a protein will be.

How can the millions of different proteins found in nature be composed of just 20 amino acids and yet all be different? Theoretically, there are 20^{n-1} possible arrangements of amino acids in a protein that contains a total of n amino acids without having any two of the same sequence! For a protein the size of ribonuclease this would be 20^{123} possibilities!

Proteins vary in molecular weight from 5737 for insulin to well over several million for some other proteins (Table 2). At one time it was considered a hopeless task to attempt to elucidate the primary sequence of a protein. However, by 1955

Table 2 Molecular Weights of a Few Selected Proteins

Protein	Molecular weight	Protein	Molecular weight
Insulin	5,737	Alkaline phosphatase	80,000
Ribonuclease	13,683	Polyphenol oxidase	128,000
Lysozyme	14,100	Fumarase	194,000
Chymotrypsinogen	23,200	Catalase	232,000
β-Lactoglobulin	35,000	Aspartate transcarbamylase	310,000
Ovalbumin	45,000	Urease	483,000
Serum albumin	65,000	β-Galactosidase	520,000
Hemoglobin	64,500	Glutamate dehydrogenase	2,000,000

Sanger and his co-workers had accomplished this task for insulin [5]. Since that time the primary sequences for a large number of proteins have been determined. Amino acid sequences can be determined by the automated Edman degradation technique, but even more rapidly and cheaper by sequencing the nucleotides of the gene that encodes the protein.

E. Peptide Mapping of Proteins

It is not necessary to determine the complete amino acid sequences of two proteins to see how similar they are. This may be accomplished by peptide mapping ("fingerprinting") techniques. The proteins are subjected to partial hydrolysis with a protease; generally, trypsin is used because of its high specificity. The hydrolysate is spotted on cellulose paper and the peptides separated in one direction by high-voltage electrophoresis on the basis of charge followed by separation in a second direction perpendicular to the first by chromatography on the basis of differential solubility in a stationary phase, water, and a mobile phase, organic solvent. The peptide maps of the two proteins are compared to determine identical peptides and those that are different. The procedure, first applied by Ingram to normal and sickle-cell hemoglobins [6], is routine today in biochemical laboratories. Most frequently now, peptide maps are done by automated high-performance liquid chromatography or by automated capillary electrophoresis. Peptide maps of normal and sickle-cell hemoglobins are shown in Fig. 6. The two hemoglobins differ only in one peptide out of a total of 30. The difference between the two is due to replacement of a glutamyl residue with a valyl residue in each of the two β-chains. This change in only two amino acids of a total of 574 amino acids is of sufficient importance to cause the hemoglobins to have quite different O_2-carrying capabilities and to cause gross structural changes in the red blood cells. The complete primary sequence of bovine pancreatic ribonuclease is shown in Fig. 7. Note that numbering of amino acid

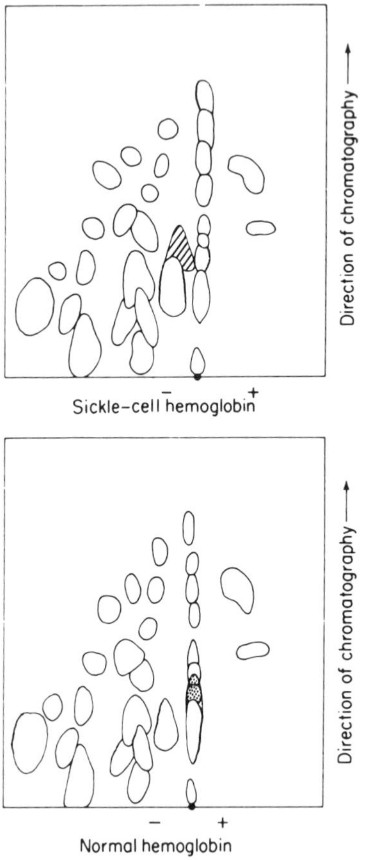

Figure 6 Peptide maps of tryptic digest of normal and sickle-cell hemoglobins. Electrophoresis was performed in the horizontal direction and chromatography in the vertical direction. The single peptide difference between the two hemoglobins is indicated by shading and by stippling. [From Ref. 6, p. 793, by courtesy of Macmillan (Journals) Ltd.]

residues is from the amino terminal end (Lys-1 for ribonuclease) to the carboxyl terminal end (Val-124 for ribonuclease). This is the same direction in which the protein is synthesized (Fig. 5).

III. SECONDARY STRUCTURE OF PROTEINS

Based on primary structure alone, a protein would be an exceedingly long, thin molecule that would flop around in solution in every direction as a random coil and

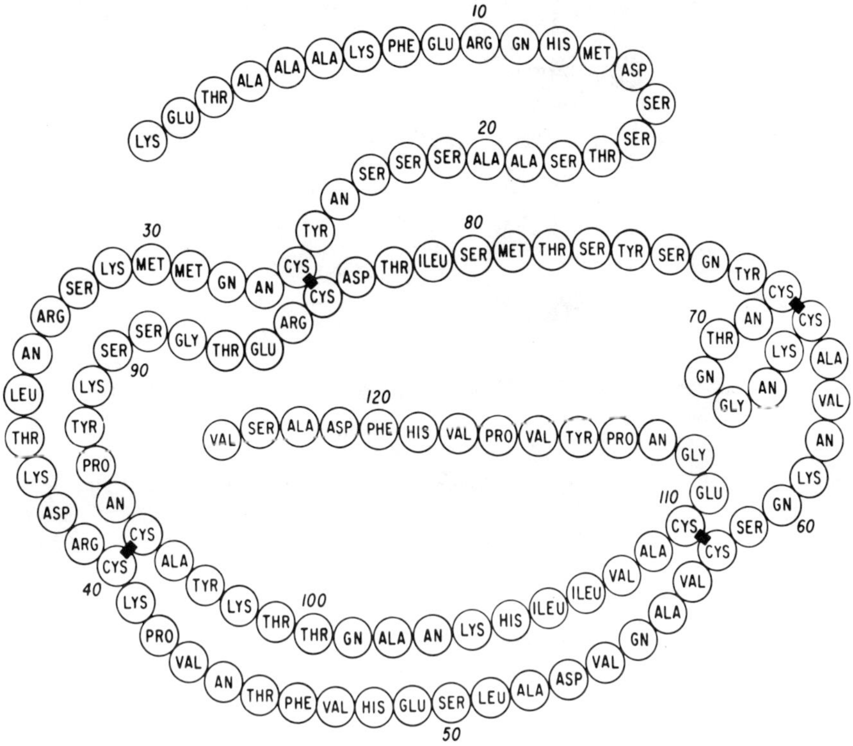

Figure 7 Primary structure of bovine pancreatic ribonuclease. AN, asparagine; GN, glutamine. A solid square, disulfide bond. (From Ref. 7, p. 358, by courtesy of Academic Press.)

become enmeshed with itself and other polymeric molecules in solution. For example, ribonuclease with 124 amino acids (MW = 13,683) would be 448 Å long and about 3.7 Å wide. A protein of 130,000 (MW) would be 4480 Å long and still be 3.7 Å wide. Figure 8 shows schematically the relative sizes and shapes of some protein molecules. Most enzymes are close to spherical molecules.

There must be additional structures imposed on the primary structure to cause a protein to be so compact (Fig. 8). These additional structures are reproducible and result from the unique primary amino acid sequence. This has been shown by unfolding proteins and permitting them to refold. When the four disulfide bonds in ribonuclease (Fig. 7) are reduced by β-mercaptoethanol in urea solution the polypeptide chain assumes a random coil conformation and loses all enzymatic activity. When the urea and reducing agent are removed and the disulfide bonds allowed to re-form slowly by oxidation in air, better than 90% of the enzymatic

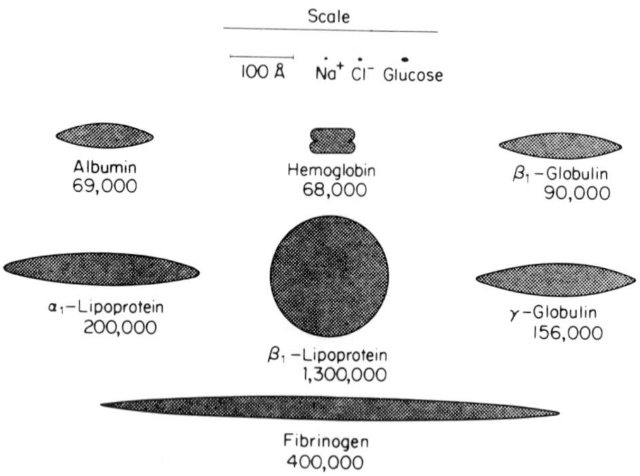

Figure 8 Relative dimensions of various proteins. The size of Na^+, Cl^-, and glucose are presented for comparison. (From Ref. 8, p. 79, by courtesy of the American National Red Cross.)

activity is regained and the properties of the molecule are in every way identical to those of the original molecule [9]. This is quite astonishing, as the disulfide bonds could have been re-formed at random to give 105 different forms of ribonuclease; only one form is active. Truly, the primary amino acid sequence must determine the higher structures. The higher structures of a protein must be considered to be as unique as the primary structure.

A. α-Helix

The secondary structure of proteins consists of helices, pleated sheets, β-turns, and random coils involving the *backbone* of the polypeptide chain. While there are other helical configurations, only the α-helix satisfies all restrictions placed on such a configuration by the planar nature of the peptide bond, the bond lengths of the —RCH—CO—NH— repeating unit, the bond angles around the C—C and C—N bonds, and the need to have the maximum number of hydrogen bonds involved for stabilization. Glutamate, methionine, alanine, and leucine residues favor α-helix formation. The α-helix is stabilized in solution by hydrogen bonds (=CO · · · HN—) involving the peptide nitrogen and the carbonyl oxygen of the third-nearest residue (Figs. 9 and 10). The strength of each hydrogen bond is low (1 to 5 kcal/mol), but there are a large number of such bonds. At least three successive hydrogen bonds must be broken before there can be any movement. The "hole" running down the center of the α-helix is too narrow to permit water to enter, so the α-helix is stabilized against disruption by hydrogen-bonding agents such as water.

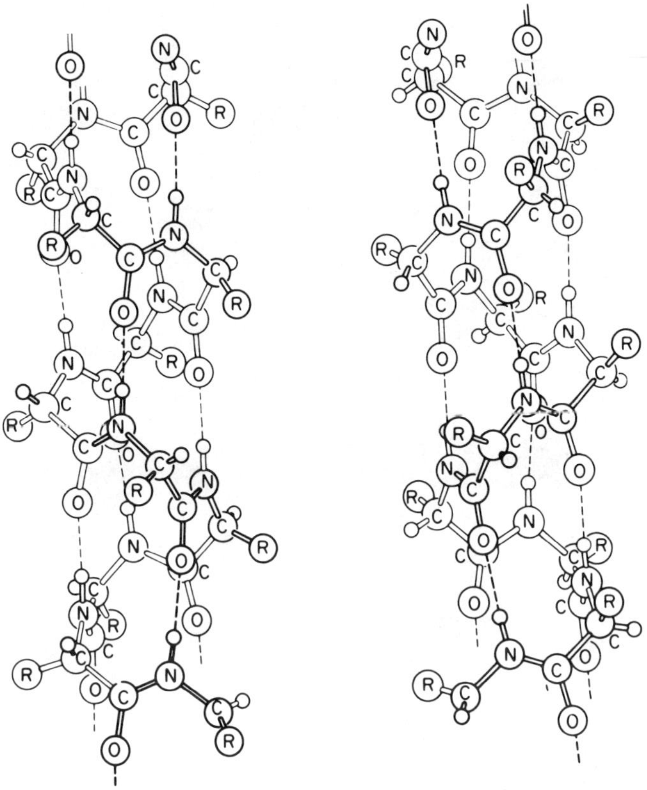

Figure 9 The α-helix: left-handed (left) and right-handed (right). All residues are L-amino acids. (From Ref. 10, p. 707, by courtesy of Academic Press.)

It is clear from x-ray crystallographic studies as well as rates of hydrogen–deuterium exchange and optical rotatory dispersion studies that enzymes contain α-helix contents ranging from near zero to 100%.

B. β-Pleated Sheet

The maximally extended polypeptide chain shown in Fig. 4 is referred to as the β-configuration. Such a maximally stretched form cannot exist in an α-helical structure, as the ═CO and —NH would not be in a correct orientation with respect to each other to permit hydrogen bonding. As noted in Fig. 4, all the potential hydrogen-bonding groups (═CO and —NH) lie in a plane and extend in a direction roughly perpendicular to the direction of the chain. As a consequence, another peptide chain, or a part of the first one folded back on itself, can be linked by lateral hydrogen bonds. Through lateral hydrogen bonding between

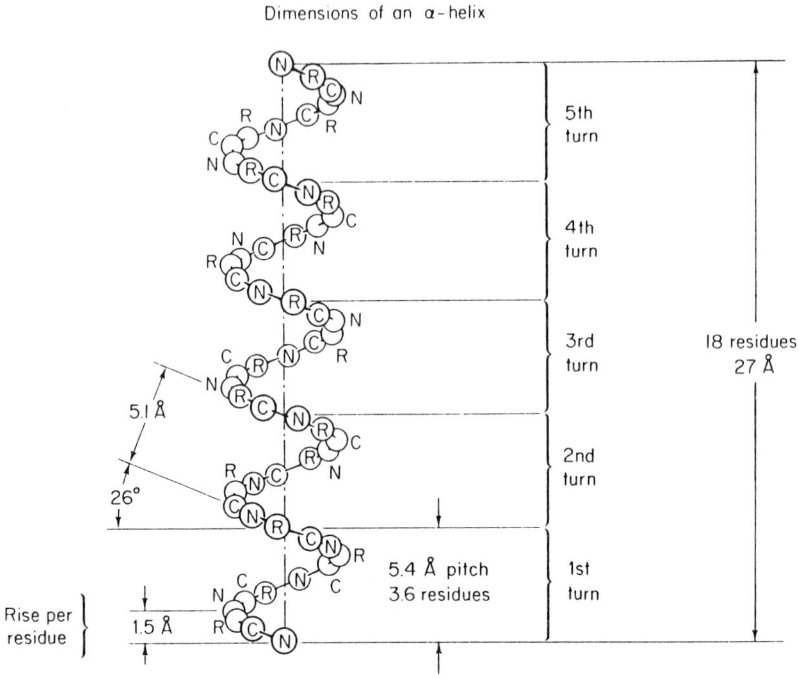

Figure 10 Another representation of the α-helix, showing the dimensions involved. (From Ref. 11, p. 101, by courtesy of C. B. Anfinsen.)

chains, infinite planar sheets can be formed. This is the *pleated-sheet* structure of Pauling and Corey. Two possible configurations are possible. In the *parallel* pleated sheet all the polypeptide chains run in the same direction, while in the *antiparallel* pleated sheet the chains run alternatively reversed in direction (Fig. 11). β-Sheet formation is enhanced by valine, isoleucine, and tyrosine residues. Enzymes contain variable amounts of β-pleated-sheet structure.

C. β-Bends and Random Coils

Because of the unique imino bonds involving proline (and hydroxyproline) residues, the polypeptide chain cannot form α-helices and β-pleated sheets in the lateral proximity of proline residues. This leads to random coil regions. Also, as a result of the imino bonds, a β-bend occurs wherever there is a proline residue, resulting in a change in direction of the primary structure.

The best evidence for α-helical, β-pleated-sheet, β-turn, and random coil structures in proteins comes from x-ray diffraction studies. Prior to this, data suggestive of α-helical regions in proteins came from indirect methods such as optical rotatory dispersion studies and from relative rates of hydrogen–deuterium

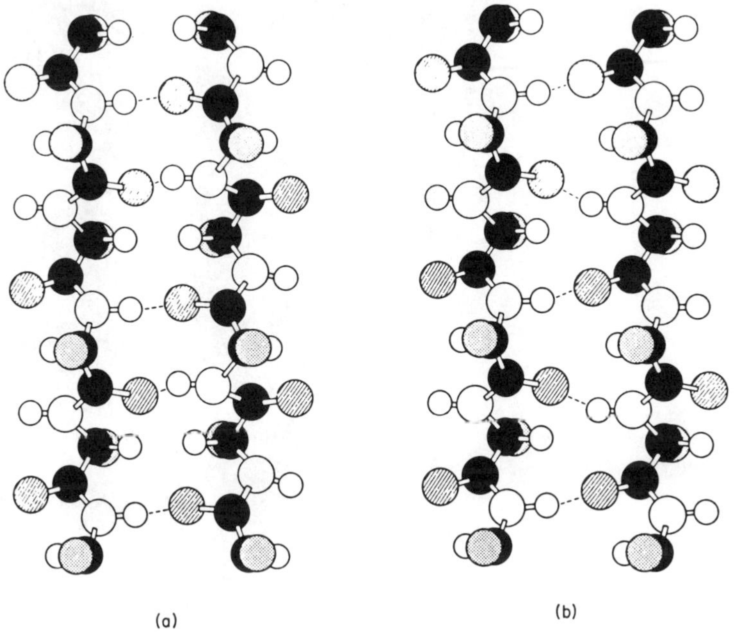

Figure 11 Schematic representation of parallel (b) and antiparallel (a) β-pleated sheets. Dimensions according to Ref. 1.

exchange. Optical rotatory dispersion and circular dichroism methods are very important in determining the percentage of each type of secondary structure of a protein.

X-ray crystallography analyses of more than 100 proteins reveal four common secondary structure patterns: the α/α, β/β/ α/β, and α + β structural types. Proteins, such as hemoglobin β-subunit, with α/α structure have mainly α-helices and few or no β-sheets. Proteins with β/β structure, such as immunoglobulin constant domain, have several strands of β-sheets but few or no α-helix strands. In α/β structure proteins, such as flavodoxin, α-helix and β-sheet strands alternate. Often the β-sheet strands form parallel sheets, surrounded by α-helix strands. Proteins with α + β structure, such as hen egg-white lysozyme, contain both α-helical and β-sheet strands but the two tend to segregate separately in different regions of the polypeptide chain.

IV. TERTIARY STRUCTURE OF PROTEINS

The picture we have developed so far of a protein is that it consists of a long polypeptide backbone interspersed with segments of α-helices, β-pleated sheets, β-turns, and random coils. This is shown schematically in Fig. 12. The secondary

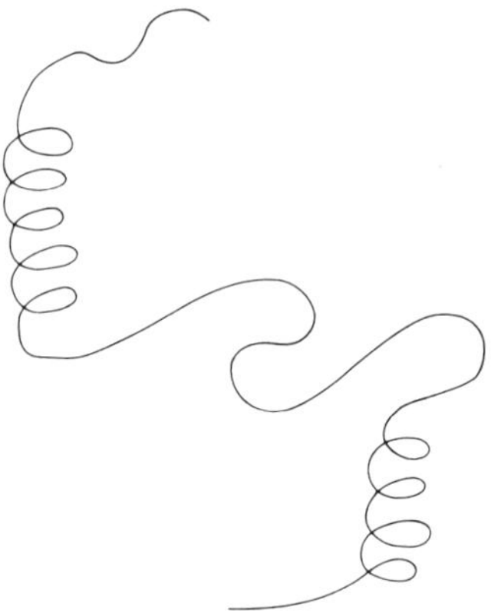

Figure 12 Schematic representation of a polypeptide chain consisting of segments of random coils and of α-helices.

structure alone is not sufficient to account for the compact, nearly spherical nature of most enzymes. X-ray diffraction patterns of proteins show that the polypeptide chain is folded back on itself in a unique and reproducible fashion. This is shown in Fig. 13 for myoglobin. Myoglobin has no β-pleated-sheet structure.

In myoglobin, 118 of the 153 amino acid residues are involved in eight helical segments of lengths ranging from seven to 24 amino acid residues each. The α-helical segments are separated by two sharp turns (β-turns) and five nonhelical regions (random coils) of one to eight amino acid residues. There is another nonhelical region involving the five amino acid residues at the carboxyl terminal end of the chain. In the tertiary structure, the protein is folded in a complex and irregular manner to form a compact, rather triangular, flattened prism with dimensions about 45 by 35 by 25 Å. This folding is such that the heme group and almost all of the polar amino acid residues are on the surface and almost all of the nonpolar residues are in the interior of the molecule. Thus the hydrophilic residues are exposed to the solvent, water, while the hydrophobic residues are removed from water as much as possible.

A. Bonds Involved in Tertiary Structure

A number of different types of bonds are involved in maintaining the tertiary structure of proteins. These include electrostatic bonds, hydrogen bonds, hydrophobic bonds, dipolar bonds, and disulfide bonds (Fig. 14). Not every protein has all of these types of bonds in its tertiary structure. The most stable bond is the disulfide bond (a covalent bond), as shown in Table 3. Many of the extracellular low-molecular-weight proteins contain disulfide bonds, and in general these

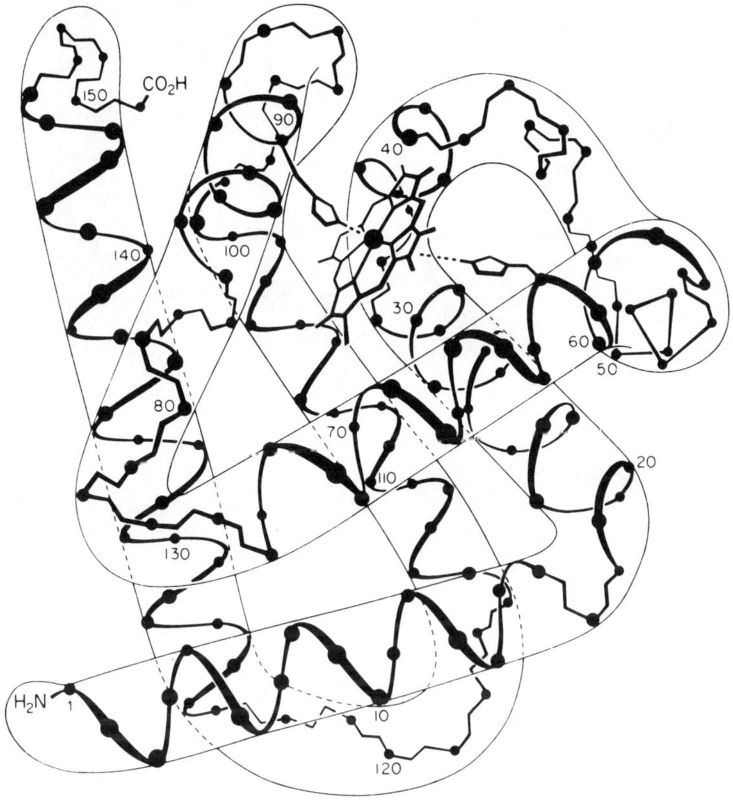

Figure 13 Schematic drawing of the primary, secondary, and tertiary structures of myoglobin as determined by x-ray crystallography. Only the α-carbons are shown. The heme is located in the upper center part of the molecule. (From Ref. 10, p. 634, by courtesy of Academic Press.)

proteins are quite resistant to denaturation. However, α-amylase and myoglobin, among other extracellular proteins, do not contain disulfide bonds and they are also reasonably stable. Maintenance of the tertiary structure of α-amylase is in part due to the one atom of Ca^{2+} per molecule; that of myoglobin is due in part to the heme group. In general, the intracellular proteins do not contain disulfide bonds. Disulfide bonds are disrupted by reducing agents that convert the disulfide bond to sulfhydryl groups.

Electrostatic bonds are the result of two opposite charges being in close juxtaposition. They may occur between the α-ammonium and α-carboxylate groups at the terminal ends of the polypeptide chain or more frequently by

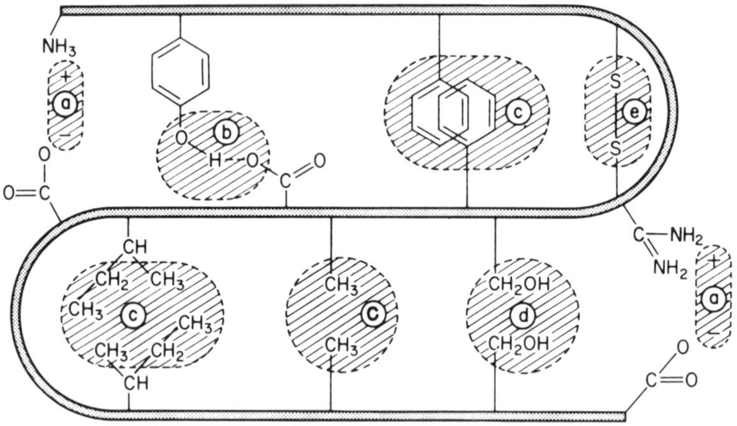

Figure 14 Bonds that stabilize protein structure. (a) Electrostatic interaction; (b) hydrogen bonding between tyrosine hydroxyl and carboxylate groups on side chains; (c) hydrophobic interaction of nonpolar side chains caused by mutual repulsion of solvent; (d) dipole–dipole interaction; (e) disulfide linkage, a covalent bond. Hydrogen bonding between carbonyl and amide groupings of peptide bonds is shown in Fig. 9. (From Ref. 12, p. 102, by courtesy of C. B. Anfinsen.)

interaction of the carboxylate groups of aspartyl and glutamyl residues and the ε-ammonium groups of lysyl residues or the guanidinium groups of arginyl residues or the imidazolium group of histidine residues. These rather strong electrostatic bonds are broken by high concentrations of salt, which suppress or disrupt the electrostatic interactions, and by adjusting the pH to high or low values. At a pH value 2 or more units below the p*K* value of the carboxyl group or 2 or more units above the p*K* value of the amino groups, these groups will no longer

Table 3 Energies and Bond Distances of Several Types of Bonds Found in Proteins

Type	Energy (kcal/mol)	Distance of interaction (Å)
Covalent bond	30–100	1–2
Ionic bond	10–20	2–3
Hydrogen bond	2–10	2–3
Van der Waals attractive forces (hydrophobic bonds)	1–3	3–5

Source: Ref. 13.

participate in electrostatic bonds. This is the reason that strong acids and strong bases denature proteins.

Exercise 3. Assuming 2.50 and 8.36 as average pK_1 and pK_2 values for the carboxyl and amino groups on a protein, plot the relative potential contribution of electrostatic bonding as a function of pH.

The nature of hydrophobic bonds may best be understood as an attempt of water to maintain its integrity. As a result, the hydrophobic groups are removed from contact with water and interact with themselves. This is what happens when an attempt is made to mix benzene and water. Despite the electronic repulsion due to overlap of the π bonds of the benzene ring in tyrosyl, phenylalanyl, and tryptophan residues, interaction between these types of groups is favored as the driving force for such interactions is the entropy factor, ΔS.

Hydrogen bonds are formed between atoms that have unshared pairs of electrons, primarily O, N, and H, which can share these electrons. Although it has an unshared pair of electrons, sulfur does not form hydrogen bonds to any large extent.

Dipole–dipole interactions such as those between the hydroxyl groups of two serine residues, and charge dipole interactions may occur in some cases, although evidence as to their importance in maintaining the structure of proteins is lacking. Even though there is not a formal positive charge on the molecule, there is an imbalance in charge distribution within some molecules. The dipoles then may interact with one another. Interaction between two different molecules may cause induced dipole moments.

B. Evidence for Tertiary Structure

The best evidence for the tertiary structures of proteins comes from x-ray diffraction studies. The data for the structure of myoglobin shown in Fig. 13 were obtained by this method. Additional suggestive evidence for the tertiary structures of proteins comes from studies on viscosity, chemical modification, and changes in reactivity of groups in the native versus the denatured enzymes. However, these methods cannot give the definitive data that are obtained by x-ray diffraction studies.

V. SOLVATION OF PROTEINS

Not all the properties of a protein can be accounted for on the basis of its primary, secondary and tertiary structures. As indicated above for myoglobin, all proteins whose tertiary structures are known are folded so that their polar groups are primarily on the outside of the molecules and exposed to the solvent molecules. There is binding of appreciable water molecules to the surface of the protein molecule. For many purposes, these water molecules are as much a part of the

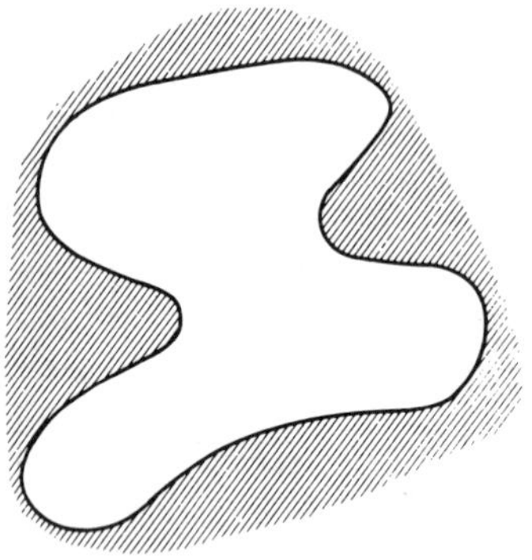

Figure 15 Schematic representation of the binding of water (lines) to the surface of a protein (white). No sharp outer boundary is shown for the water molecules which is in keeping with the uncertainty of where bound water stops and free water begins.

protein molecule as are the amino acid residues. It is difficult to measure directly the amount of water bound to a particular protein because there is not just a monolayer of water molecules. The first layer of water molecules will be bound more tightly than the second layer and the strength of binding will decrease with subsequent layers. Where does one decide that the protein no longer has an effect on the free water molecules, or vice versa? Indirect evidence, by nuclear magnetic resonance (NMR) for example, indicates that globular proteins immobilize 0.2 to 0.3 g of water per gram of protein. For a protein of molecular weight 18,000 this would represent 200 to 300 water molecules per molecule of protein. Random-coiled proteins, such as gelatin, can immobilize as much as 99 times their weight of water (at 4°C) due to entrapment of water inside a thatchwork of random-coiled protein molecules.

Water molecules are not dispersed uniformly over the surface of protein molecules but rather are concentrated in areas that are more hydrophilic (Fig. 15). Therefore, the shape of a protein molecule in solution is in part due to the unequal binding of water molecules to its surface.

VI. QUATERNARY STRUCTURE OF PROTEINS

A. Number of Polypeptide Chains

Many enzymes and other proteins are composed of a single polypeptide chain. This is the case for ribonuclease, lysozyme, trypsin, pepsin, and some α-amylases, for example. On the other hand, a large number of enzymes (and other proteins) are composed of more than one polypeptide chain (Table 4). A schematic representative of the quaternary structure of a protein is shown in Fig. 16.

Table 4 Subunit Constitution of Some Proteins

Proteins	Molecular weight	Subunits No.[a]	Subunits Weight
Insulin	5,737	1	2,338
		1	3,399
Thrombin	31,000	(3)	(10,000)
Alkaline phosphatase	80,000	2	40,000
Alcohol dehydrogenase			
Liver	80,000	4	20,000
Yeast	150,000	4	37,000
Procarboxypeptidase A			
Bovine	87,000	1	34,500
		1	25,000
		1	25,000
Pacific spiny dogfish	34,500	1	34,500
Polyphenol oxidase (mushroom)	128,000	4	(32,000)
Lactate dehydrogenase	150,000	4	35,000
		(or 8	18,000)
Catalase	232,000	4	57,500
Aspartate transcarbamylase (*E. coli*)	310,000	6	20,000
		6	35,000
Urease[b]	483,000	30	16,000

Source: Ref. 14. Specific references for each enzyme are included in the reference cited.

[a]Where it is certain the subunits are different, they are listed separately along with their weights. Values enclosed in parentheses are less certain.

[b]There is still uncertainty about the number and size of the subunits. Various reports range from 6 with molecular weight of 83,000 to 30 with molecular weight of 16,000.

B. Arrangement of Polypeptide Chains

In the case of insulin and chymotrypsin, the precursor of the active protein is synthesized as a single polypeptide chain, but in the activation process peptide fragments are removed from the chain, giving rise to two and three polypeptide chains, respectively (see Chapter 20). Integrity of the molecules is maintained by disulfide bonds (along with noncovalent bonds). In other proteins, the polypeptides are synthesized separately and then assembled (presumably as dictated by the primary, secondary, and tertiary structures of the protein) into the final molecule. In adult human hemoglobin there is a separate gene controlling the synthesis of the α- and β-chains. Two α- and two β-chains are then assembled in a unique

Figure 16 Schematic representation of the quaternary structure of proteins. The example shown is composed of four polypeptide chains of two different types. However, as is so often the case, both types of chains have the same tertiary structure.

arrangement to give the active tetrameric hemoglobin molecules. The monomer chains are completely inactive. All tetrameric molecules (hemoglobin) have exactly the same composition of two α- and two β-chains and the same spatial arrangement of the four polypeptide chains.

Lactate dehydrogenase is composed of four polypeptide chains. Two different types of polypeptide chains, designated H and M, are produced under the control of separate genes. In the assemblage of the tetramers H_4, H_3M, H_2M_2, HM_3, and M_4 are formed; the relative amounts of each isozyme formed are determined by the relative amounts of H and M produced by a given organ. The monomeric units are inactive until assembled into the tetramer. There seems to be no general biological preference for proteins in which all the subunits are identical and those which are composed of more than one type of subunit.

Aspartate transcarbamylase is an example of a type of arrangement of subunits different from that found in hemoglobin or lactate dehydrogenase. Aspartate transcarbamylase (MW 310,000) is composed of two different types of polypeptide chains. In the intact molecule there appear to be six polypeptide chains of each type. One type of polypeptide binds substrate and catalyzes the reaction of carbamyl phosphate and L-aspartate to form carbamylaspartate, while a second type of polypeptide is inactive as an enzyme but binds the allosteric inhibitor cytidine triphosphate (CTP). These two types of polypeptides are referred to as catalytic polypeptide (MW 3.5×10^4 per chain) and regulatory polypeptides (MW 2×10^4 for each chain). An arrangement of the polypeptide chains consistent with the data is shown in Fig. 17 [15]. The two types of subunits can be separated. The catalytic subunits are fully active alone but there is no allosteric behavior of the enzyme in the absence of the regulatory subunits.

C. Bonds Involved and Evidence for Quaternary Structure

Disulfide bonds are involved in maintaining quaternary structure in some post-translationally processed proteins, such as trypsin and chymotrypsin. Hydrophobic and electrostatic bonds, however, probably have predominant roles in maintaining the integrity of protein molecules comprised of two or more subunits. A good correlation is found between the combined percentage of certain hydrophobic side-chain residues and quaternary structure [16]. According to this theory, if the combined percentages of certain hydrophobic residues (leucine, isoleucine, proline, valine, and phenylalanine) are greater than 30%, the protein will have quaternary structure, whereas it will exist as a single polypeptide protein if the percentage is less than 30%.

Evidence for subunit structure of proteins is obtained by several methods. In the case of hemoglobin and aspartate transcarbamylase, for example, x-ray diffraction studies have shown the molecules to be composed of several polypeptide chains. Many of the data for subunit structure have been obtained by treating the protein with urea, guanidinium·HCl, high salt concentrations, *p*-chloromercuribenzoate (PCMB), or dithiothreitol or by changing the pH of the solution. These treatments are designed to cause the polypeptide chains to dissociate. Thus

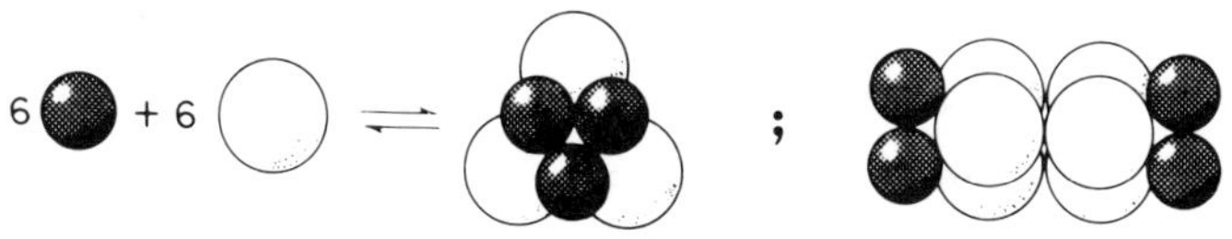

Figure 17 Diagrammatic representation of the quaternary structure of aspartate transcarbamylase. The molecule is composed of six polypeptide chains each of two types. The view of the molecule shown on the left is looking from one of the ends containing half of the regulatory chains, that shown on the right is at 90° removed from the first. The molecule has two axes of symmetry. The regulatory polypeptide chains are shaded, the catalytic polypeptide chains are white. An alternative arrangement of the regulatory subunits involves three clusters of two subunits each in the grooves formed by catalytic subunits.

the molecular weight decreases and the individual polypeptide chains may be separable by gel filtration or by gel electrophoresis, for example. Apparent molecular weights obtained by regular gel electrophoresis (PAGE) and by electrophoresis in presence of sodium dodecylsulfate (SDS-PAGE), following denaturation and reduction of disulfide bonds, are routinely used to determine the subunit structures of proteins [17]. The reagents listed above are not equally effective on all proteins and from the nature of the reagents required to cause dissociation one may learn something of the forces that contribute to association of the polypeptide chains. For example, urea and guanidinium·HCl are involved primarily in breakage of hydrogen bonds, pH adjustment and high salt concentrations in breakage of electrostatic bonds. PCMB probably acts by increasing the negative charges on the subunits through reaction with sulfhydryl groups and dithiothreitol and mercaptoethanol reduce disulfide bonds to sulfhydryl groups.

VII. MULTIMOLECULAR ORGANIZATION OF ENZYMES

Enzymes are usually either membrane bound or fully soluble. The enzymes involved in electron transport are attached to a structural protein in an ordered fashion. When this ordered complex of enzymes, structural protein, and lipid material is broken apart by detergents, the ability to transport electrons to O_2 becomes lost, although several fragments can be isolated that carry out transport part of the way. On the other hand, it is often assumed that the soluble enzymes do not have multimolecular organization. Recently, a number of soluble enzyme systems have been described which are composed of several different types of protein molecules [18,19]. Two of the systems are described to indicate the importance of such organization among soluble enzymes.

A. Pyruvate Dehydrogenase Complex

Animal tissues and many microorganisms carry out oxidative decarboxylation of α-keto acids, shown specifically for pyruvate as

$$CH_3COCO_2^- + CoASH + NAD^+ \rightarrow CH_3COSCoA + CO_2 + NADH \quad (4)$$

This reaction, which represents the main pathway of α-keto acid oxidation in animals, is thought to occur according to the sequence shown in Eqs. (5)–(9),

$$CH_3COCO_2^- + TPP\text{-}E_1 + H^+ \rightarrow [CH_3CHOH\text{-}TPP]\text{-}E_1 + CO_2 \quad (5)$$

$$[CH_3CHOH\text{-}TPP]\text{-}E_1 + LipS_2\text{-}E_2 \rightarrow [CH_3COS\text{-}LipSH]\text{-}E_2 + TPP\text{-}E_1 \quad (6)$$

$$[CH_3COS\text{-}LipSH]\text{-}E_2 + CoASH \rightarrow [Lip(SH)_2]\text{-}E_2 + CH_3COS\text{-}CoA \quad (7)$$

$$[Lip(SH)_2]\text{-}E_2 + E_3\text{-}FAD \rightarrow LipS_2\text{-}E_2 + E_3\text{-}FADH_2 \quad (8)$$

$$E_3\text{-}FADH_2 + NAD^+ \rightarrow E_3\text{-}FAD + NADH + H^+ \quad (9)$$

where the brackets indicate enzyme-bound intermediates. TPP, Lip, CoASH, FAD, NAD^+ and E_{1-3} are thiamine pyrophosphate, lipoic acid, coenzyme A, flavin adenine dinucleotide, oxidized form of nicotine adenine dinucleotide, and the three enzymes, respectively. The net result of reactions (5)–(9) is that shown in Eq. (4). At least three enzymes and five cofactors are involved in what appears to be a rather simple decarboxylation of pyruvate. The three enzymes are lipoyl reductase-transacetylase (E_2 with lipoic acid covalently bound to enzyme), pyruvate decarboxylase (E_1 with α-hydroxyethyl thiamine pyrophosphate attached), and dihydrolipoyl dehydrogenase (E_3 with FAD attached). CoASH is probably attached to lipoyl reductase-transacetylase along with lipoic acid.

The pyruvate dehydrogenase complex isolated from *E. coli*, with a particle weight of 4.44 million, has been dissociated into the three constituent enzymes. The complex contains 24 molecules of pyruvate decarboxylase (MW 90,000, with each molecule composed of two different polypeptide chains), 24 molecules of dihydrolipoyl dehydrogenase (MW 55,000), and 8 molecules of lipoyl reductase-transacetylase (MW 120,000, with each molecule composed of three polypeptide chains). The complex is large enough to be visible by electron microscopy. Based on these data as well as on micrographs of individual enzyme molecules, the organization of the molecules of the multimolecular complex may be as shown in Fig. 18 [20]. E_2 forms the framework on which the organization of the superstructure is built. The eight molecules of E_2 (shaded balls) form a cube. Four molecules of E_3 (black balls) are attached to each of the six faces of the cube. Three molecules of E_1 (white balls) are attached at each of the eight corners of the cube formed by E_2. E_1 and E_3 do not appear to bind to each other.

The multimolecular complex is dissociated in 4 *M* urea and the individual

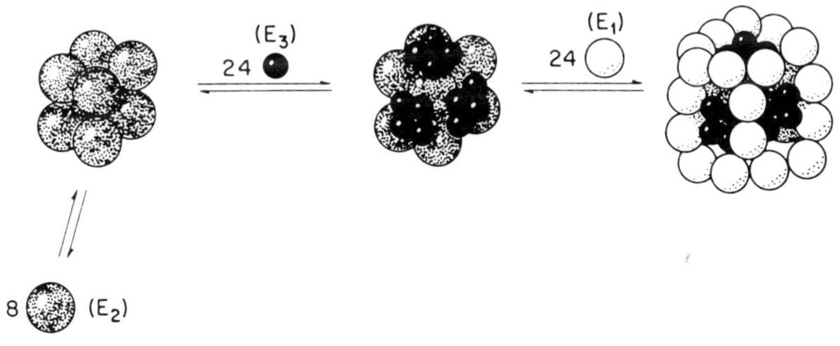

Figure 18 Diagrammatic representation of the macromolecular organization of the pyruvate dehydrogenase complex composed of 8 molecules of lipoyl reductase-transacetylase (E_2), 24 molecules of pyruvate decarboxylase (E_1), and 24 molecules of dihydrolipoyl dehydrogenase (E_3). Combined particle weight is 4.44 million.

enzymes can be separated by column chromatography. Individually, the enzymes cannot carry out the complete reaction shown in Eq. (4). On mixing, the enzymes reassociate into a fully active multimolecular complex that resembles the native complex by electron microscopy. On the basis of these data it is concluded that the enzymes are held together by noncovalent bonds, that the primary structure of each enzyme is sufficient to establish the secondary and tertiary structures of the protein, and that all three, acting together, are sufficient to establish the quaternary and multimolecular structures of the complex.

B. Acetyl-CoA Carboxylase

A second important biological reaction performed by a multimolecular complex is the reaction of CO_2 with acetyl CoA to give malonyl CoA [21]:

$$\begin{aligned} &MgATP^{2-} + HCO_3^- + CH_3COSCoA \rightleftharpoons \\ &\quad ^-O_2CCH_2COSCoA + MgADP^- + P_i^- \end{aligned} \tag{10}$$

The overall reaction is composed of two half-reactions catalyzed by two enzymes, E_A and E_B:

$$MgATP^{2-} + HCO_3^- + E_A \rightleftharpoons E_A{\sim}CO_2^- + MgADP^- + P_i^- \tag{11}$$

$$E_A{\sim}CO_2^- + CH_3COSCoA \overset{E_B}{\rightleftharpoons} {}^-O_2CCH_2COSCoA + E_A \tag{12}$$

The complex of E_A and E_B is known as acetyl-CoA carboxylase and involves three cofactors: $MgATP^{2-}$, CoA, and biotin. Acetyl-CoA carboxylase of *E. coli* is composed of at least three proteins that have been separated. At pH 9, E_A dissociates into two proteins which, when combined, give the reaction

$$MgATP^{2-} + HCO_3^- + \text{free biotin} \xrightarrow{E_C} \text{biotin-}CO_2^- + MgADP^- + P_i^- \tag{13}$$

One protein, E_C, has no activity alone but when free biotin is added it carries out the reaction. Therefore, E_C catalyzes the covalent attachment of CO_2 to the cofactor, biotin.

The second protein from dissociation of E_A has a molecular weight of 10,000 and contains one molecule of biotin. It has all the properties of an acyl carrier protein (ACP) and is called carboxyl carrier protein (CCP). It functions as the biotin-containing protein responsible for carrying the covalently bound CO_2.

C. Importance of Multimolecular Organization

At this moment one can only speculate on the advantages macromolecular organization of enzymes affords an organism. There is some evidence that aggregation of enzymes produces complexes that have overall activities unlike any of the enzymes separately. In the case of electron transport particle I (ETP I), the flavoprotein can be solubilized from the complex. However, the solubilized protein has quite different kinetic and reactive properties from those it has in ETP I. In a complex, the overall efficiency of the process can be altered since the substrate and intermediates appear to remain bound until the final product is released [See Eqs. (5) to (9)]. Such a process could allow increased efficiency and more control over the rate of reaction since the complex is probably subject to control by environmental factors, including hormones and allosteric effectors. It also could exclude water from the reaction, which is important in such reactions as electron transport and oxidative phosphorylation.

REVIEW QUESTIONS

1. Summarize several methods that may be used to determine the types of bonds involved in maintaining the tertiary structure of a protein. Rank the types of bonds involved in terms of the relative effect of heat on them.

2. Carefully draw a chemical structure for the polypeptide

Lys·Tyr·Val·Pro·Glu·Arg·His·Leu·Ala·Gly

Name this polypeptide.

3. From your outside reading make a list of several enzyme systems that exist in macromolecular structure.

4. Discuss why, in your opinion, one protein might have only 10% α-helix while another might have 50% α-helix.

5. Van Holde [16] has formulated a hypothesis that those proteins with less than 30% of certain hydrophobic amino acid residues (leucine, isoleucine, proline, valine, and phenylalanine) exist as single polypeptide molecules in solution, while those with more than 30% of these residues will have quaternary structure. Speculate as to what happens if a protein has exactly 30% of these residues.

6. Collagen, a structural protein of all animal tissues, has several unique properties. Describe several of these properties, including a unique feature of each type of structure.

7. Glycoproteins contain, in addition to amino acids, several residues of monosaccharides. How are these monosaccharides attached to the protein?

8. Several methods exist for estimating, more or less accurately, the amount of water bound to a protein. List several of these and discuss the principle of each method.

9. A large number of proteins are composed of subunits that dissociate when treated with various reagents. The same can be said for macromolecular organization. Formulate some general guidelines that would be helpful in deciding when a polypeptide is a molecule and when it is one polypeptide of several of a larger molecule.

10. Detergents are among the best reagents for dissociating quaternary and multimolecular structures of proteins. What can you conclude about the structure of those proteolytic enzymes that are used in enzyme–detergent preparations?

11. One of the major forces involved in quaternary structures is their hydrophobic character. Based on this factor alone one might expect to find all sorts of protein–protein interactions to form hybrid molecules with several unrelated activities. Experimentally, one does not find this. Suggest several reasons why such hybrids are not formed.

REFERENCES

General References

Braden, C., and J. Tooze, *Introduction to Protein Structure*, Garland, New York, 1991.

Brooks, C. L., III., M. Karplus, and B. M. Pettitt, *Proteins: A Theoretical Perspective of Dynamics, Structure, and Thermodynamics*, Wiley-Interscience, New York, 1988.

Chothia, C., and A. V. Finkelstein, Classification and origins of protein folding patterns, *Ann. Rev. Biochem.*, **59**, 1007 (1990).

Dickerson, R. E., and I. Geis, *The Structure and Action of Proteins*, Harper & Row, New York, 1969.

Ginsburg, A., and E. R. Stadtman, Multienzyme Systems, *Ann. Rev. Biochem.*, **39**, 429 (1970).

Klotz, I. M., N. R. Langerman, and D. W. Darnell, Quaternary structure of proteins, *Ann. Rev. Biochem.*, **39**, 25 (1970).

Oxender, D. L., and C. F. Fox, eds., *Protein Engineering*, Alan R. Liss, New York, 1987.

Robson, B., and J. Garnier, *Introduction to Proteins and Protein Engineering*, Elsevier, Amsterdam, 1988.

Schulz, G. E., and R. H. Schirmer, *Principles of Protein Structure*, Springer-Verlag, New York, 1979; 9th printing in 1991.

Stryer, L., *Biochemistry*, 2nd Ed., W.H. Freeman, San Francisco, 1981.

Cited References

1. R. B. Corey and L. Pauling, *Proc. Roy. Soc. (London)*, **B141**, 10 (1953).
2. M. Nomura, *Sci. Am.*, **221**(4), 28 (1969).
3. B. Gutte and R. B. Merrifield, *J. Am. Chem. Soc.*, **91**, 501 (1969).
4. R. Hirschmann, R. F. Nutt, D. F. Veber, R. A. Vitali, S. L. Varga, T. A. Jacob, F. W. Holly, and R. G. Denkewalter, *J. Am. Chem. Soc.*, **91**, 507 (1969).
5. A. P. Ryle, F. Sanger, L. F. Smith, and R. Kitai, *Biochem. J.*, **60**, 541 (1955).
6. V. M. Ingram, *Nature*, **178**, 792 (1956).
7. R. E. Canfield and C. B. Anfinsen, in *The Proteins* (H. Neurath, ed.), Vol. 1, Academic Press, New York, 1963, p. 311.
8. J. L. Oncley, in *Conference on the Preservation of the Cellular and Protein Components of Blood*, American National Red Cross, Washington, D.C., 1949.

9. J. Bello, D. Harker, and E. Dejarnette, *J. Biol. Chem.*, **236**, 1358 (1961).
10. R. E. Dickerson, in *The Proteins* (H. Neurath, ed.), Vol. 2, Academic Press, New York, 1964, p. 603.
11. L. Pauling and R. B. Corey, *Proc. Intern. Wool Textile Res. Conf.*, **B**, 249 (1955), as redrawn in C. B. Anfinsen, *The Molecular Basis of Evolution*, Wiley, New York, 1959, p. 101.
12. C. B. Anfinsen, *The Molecular Basis of Evolution*, Wiley, New York, 1959.
13. R. T. Jones, in *Symposium on Foods: Proteins and Their Interactions* (H. W. Schultz and A. F. Anglemeir, eds.), AVI, Westport, Conn., 1964, p. 33.
14. I. M. Klotz and D. W. Darnell, *Science*, **166**, 126 (1969).
15. D. C. Wiley and W. N. Lipscomb, *Nature*, **218**, 1119 (1968).
16. K. E. Van Holde, in *Molecular Architecture in Cell Physiology* (T. Hayashi and A. G. Szent-Györgyi, eds.), Prentice Hall, Englewood Cliffs, N.J., 1966, p. 81.
17. K. Weber and M. Osborn, in *The Proteins*, 3rd Ed. (H. Neurath, R. Hill, and C. Beede, eds.), Vol., 1, Academic Press, New York, 1975.
18. L. J. Reed and D. J. Cox, *Ann. Rev. Biochem.*, **35**, 57 (1966).
19. A. Ginsburg and E. R. Stadtman, *Ann. Rev. Biochem.*, **39**, 429 (1970).
20. L. J. Reed and R. M. Oliver, *Brookhaven Symp. Biol.*, **21**, 397 (1968).
21. P. K. Stumpf, *Ann. Rev. Biochem.*, **38**, 159 (1969).

3
Enzyme Purification

A living organism contains literally thousands of different molecules ranging from complex, high-molecular-weight compounds such as proteins, nucleic acids, carbohydrates, and lipids to small molecules of organic acids, bases, sugars, and salts. A large number of the proteins may be enzymatically active. The organism is precisely organized at the cellular level so that enzyme distribution is not uniform.

I. NEED FOR PURIFICATION

How much purification is necessary? An answer to this depends on the type of questions asked. If one wishes to measure the effect of temperature on inactivation of certain enzymes in fruits and vegetables so as to design proper blanching conditions, heat treatment should be performed on the tissue in the state normally used in processing. After heat treatment the sample is homogenized in an appropriate buffer, insoluble material is removed by centrifugation, and the supernatant liquid is assayed for remaining enzyme activity. Since enzymes are quite specific, use of various substrates permits a determination of different types of enzymes present in the mixture.

Such assays are crude because a number of enzymes may act on a common complex substrate. Assay of proteolytic activity using a protein substrate, for example, measures the combined effect of all proteolytic enzymes of the system. However, by changes in pH or by use of more specific substrates, one can distinguish among these activities. To illustrate the differential pH approach, both pepsin and trypsin hydrolyze proteins. However, their activities on a common protein can be measured separately even when they are together in a solution, because pepsin has a pH optimum of 2 while trypsin has a pH optimum of 8. Pepsin has no activity at pH 8 and trypsin has no activity at pH 2. An example of trypsin and chymotrypsin may be used to illustrate the utility of specific substrates. Both hydrolyze proteins and both have a pH optimum near 8. By use of specific substrates, such as *N*-acetyl-L-tyrosine ethyl ester for chymotrypsin and α-*N*-benzoyl-L-arginine ethyl ester for trypsin, one measures the activities separately. Even in those cases where only a gross enzyme activity assay is needed, it may be necessary to do a preliminary fractionation to remove inhibitors or other materials that interfere with the assay.

When there is a need to study the amino acid composition, molecular weight, tertiary structure, substrate specificity, or kinetic parameters of a specific enzyme, the enzyme must be completely separated from all other substances. In this case purification of the enzyme may represent substantial work.

II. SELECTION OF STARTING MATERIAL

A. Objectives

What material should be chosen for purification of an enzyme? The answer is based on the type of study to be done. To study a particular type of enzyme regardless of its source, a starting material with large amounts of the enzyme and in plentiful supply should be chosen. For example, compared to other plant tissues, horseradish and fig latex contain very high concentrations of peroxidase. About 20% of the total protein of the exudate from homogenized muscle is triose phosphate dehydrogenase. The relative amounts of several enzymes from a number of sources are shown in Table 1. On the other hand, if the study is to deal with a specific enzyme from a specific source, one has no choice but to start with that material regardless of enzyme concentration.

B. Enrichment by Cultural Practices

There may be growth conditions that increase the enzyme concentration. In the case of microorganisms and to some extent animals, the amount of enzyme may be increased by varying the composition of the growth medium. Many microbial enzymes are inducible. Growing bacteria on a culture rich in protein or in carbohydrate may induce the production of proteolytic enzymes and carbohydrases, respectively. The induction may be quite specific. Growing bacteria in the presence of maltose as the sole carbon source, for example, may induce high concentrations of maltase. Therefore, an enzyme that may be present in <0.1% of total protein in the absence of induction may have a concentration of 1 to 20% of that enzyme when induced.

With microorganisms advantage may be taken of mutants in which the control-

Table 1 Relative Amounts of Some Enzymes in Various Sources

Enzyme	Source	Relative amounts
Polygalacturonase[a]	Tomato	1.00
	Avocado	0.065
	Medlar	0.027
	Pear	0.016
	Pineapple	0.024
	Cranberry	0
	Carrot	0
	Grape	0
Lipoxygenase[b]	Soybeans	1.00
	Urd beans	0.60
	Mung beans	0.47
	Peas	0.35
	Wheat	0.02
	Peanuts	0.01
Peroxidase[c]	Green peas	1.00
	Pea pods	0.72
	String beans	0.62
	Spinach	0.32
	Lima beans	0.15

[a]Ref. 1.
[b]Ref. 2.
[c]Ref. 3.

ler gene is not functioning for the biosynthesis of a particular enzyme. In this case, the gene controlling production of a particular enzyme is not repressed in the usual fashion by high levels of products of the enzyme action; therefore, high concentrations of the enzyme accumulate.

C. Variation of Enzyme Concentration with Age

The level of enzymes is not the same throughout the life cycle of an organism. An enzyme may be essentially absent at an early stage, rise to a maximum level, and then decrease again. This is shown in Fig. 1 for production of polygalacturonic acid lyase by *Bacillus pumilus*. Cell population reaches a maximum at 16 hours, but there is no detectable enzyme production until 48 hours even though there has been a marked cell death.

Certain enzymes are present in green fruit but not in ripe fruit, and vice versa. One of the problems encountered in the assay of enzyme activity in green fruits is the high level of phenolic compounds present. When cells are broken during

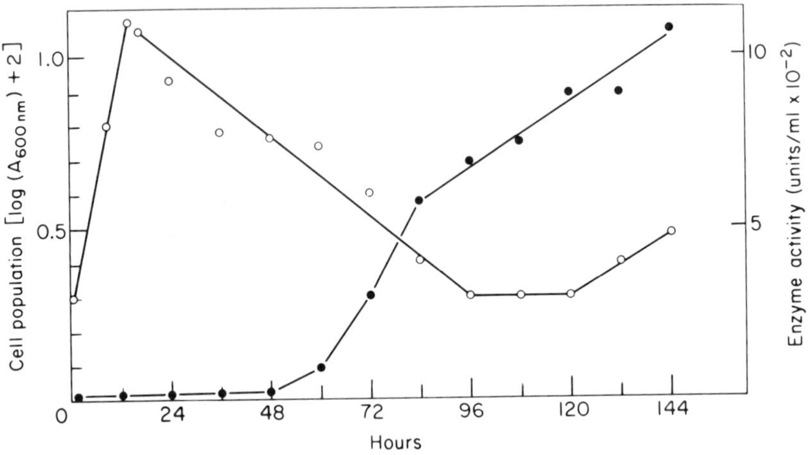

Figure 1 Changes in the polygalacturonic acid lyase activity of *Bacillus pumilus* as a function of age of culture. ○, absorbance at 600 nm of a 1:20 dilution as a measure of cell population; ●, activity of enzyme (units/mL). (From Ref. 4, p. 41, by courtesy of B. A. Dave.)

homogenization and other operations, the phenolic compounds combine with and inactivate enzymes. This problem will be dealt with later. At this point it is important to note that green fruit may contain high levels of certain enzymes but that based on the assay, the enzyme activity is low.

D. Localization of Enzymes

Enzymes are localized in organisms. Some enzymes in fruit are concentrated near the skin and/or near the pit with low concentrations of enzymes in the bulk of the tissue. It is well worth the effort to examine the concentration of enzyme in each of the parts of an organism and to use the richest source, with the realization that an enzyme from one organ is probably not quite the same as the enzyme with the same specificity from another organ. This selection process may lead to several-fold (10 to 100) enrichment over that obtained with the whole organism.

Within an organ there is not a uniform concentration of enzymes throughout. In a seed, a specific enzyme may be concentrated in the aleurone layer, in the endosperm, or in another part of the seed. In a leaf, the enzyme may be in the grana, chloroplasts, mitochondria, and so on. The liver contains different types of cells with different enzyme complements. The cells contain subcellular organelles: the lysosomes, the mitochondria, the nucleus, cell membrane, and so on (Fig. 2). Each of these organelles contains specific kinds of enzymes (Table 2). Differential centrifugation methods are used for separating the various organ-

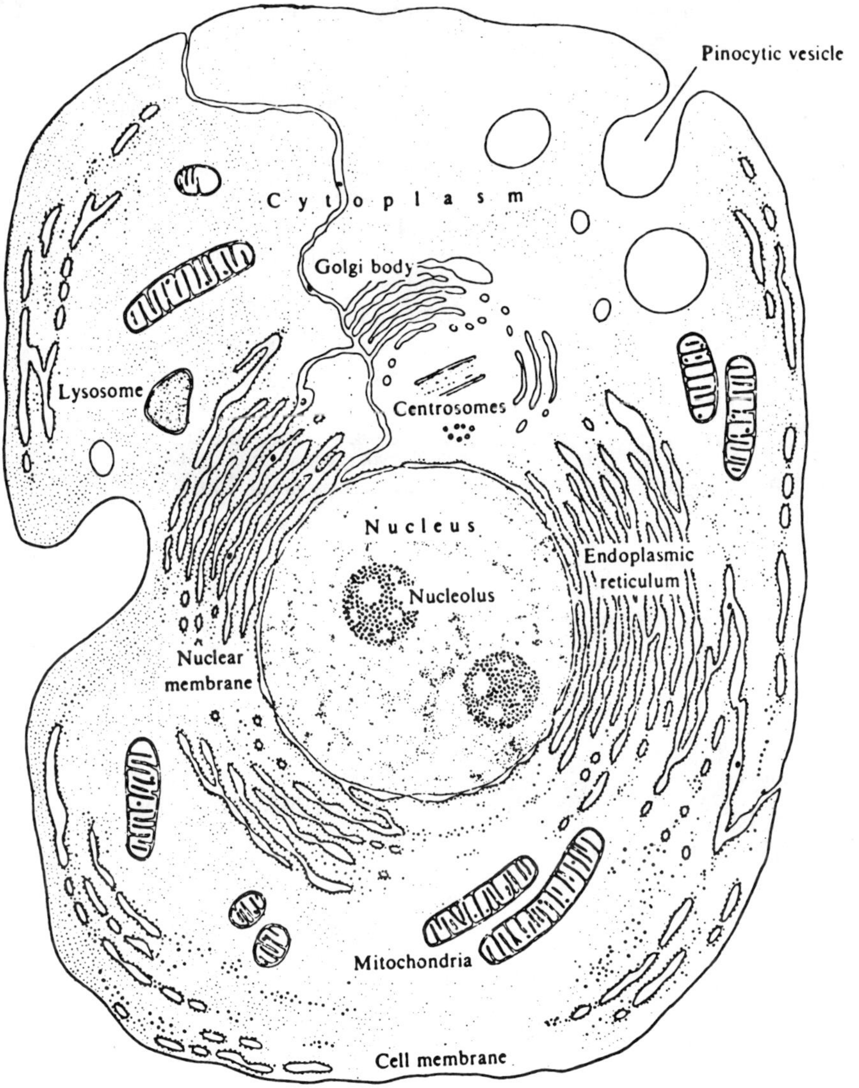

Figure 2 "Typical" cell. Diagram of a typical cell based on data from electron micrographs. (From Ref. 5, p. 55, by courtesy of Scientific American, Inc.)

Table 2 Subcellular Location of Several Enzymes in Animal Cells

Location	Enzymes
Nucleus	DNA-dependent RNA polymerase,[a] polyadenylate synthetase
Mitochondria	Succinate dehydrogenase, cytochrome oxidase, glutamate dehydrogenase, malate dehydrogenase, α-ketoglutarate dehydrogenase, α-glycerol phosphate dehydrogenase, pyruvate decarboxylase
Lysosomes	Cathepsins A, B, C, D and E, collagenase, acid ribonuclease, acid phosphatase, β-galactosidase, sialidase, lysozyme, triglyceride lipase
Peroxisomes (microbodies)	Catalase, urate oxidase, D-amino acid oxidase
Reticulum, golgi, etc.	Glucose 6-phosphatase, nucleoside diphosphatase, TPNH-linked lipid peroxidase, nucleoside phosphatases
Soluble	Lactate dehydrogenase, phosphofructokinase, glucose 6-phosphate dehydrogenase, transketolase, transaldolase
Pancreatic zymogen granules	Trypsinogen,[b] chymotrypsinogen,[b] lipase, amylase, ribonuclease

[a]Underlined enzymes most frequently used as indicator enzymes for type of cellular organelles present in a homogenate.
[b]Zymogens activated to trypsin and chymotrypsin, respectively.

elles of the cell. Fractionation at the subcellular level may lead to 10- to 20-fold enrichment of enzyme.

Localization of enzymes in a tissue may be readily seen by a technique called histochemistry. Different compounds—proteins, carbohydrates, lipids, and nucleic acids—are selectively stained with different dyes, making it easy to localize these compounds in properly prepared tissue slices. The same general technique, but more specific in application, may be applied to detection of enzymes. Figure 3 shows the localization of some enzymes in tissues. Histochemical techniques are available for many enzymes. So far, food scientists have made little use of this invaluable technique.

III. PRELIMINARY PURIFICATION

After a decision has been made as to source of enzyme to be used, the next step is to extract the enzyme from the tissue and to proceed with purification. For purification of enzymes, ice baths, cold rooms, and refrigerated centrifuges are a necessity. In general, most enzymes are more stable near 0°C and microbial growth is delayed. It is best to work up a homogenate as soon as possible. If work must be delayed, protection against microorganisms can usually be attained with a thin layer of toluene or with 0.1% sodium oxide.

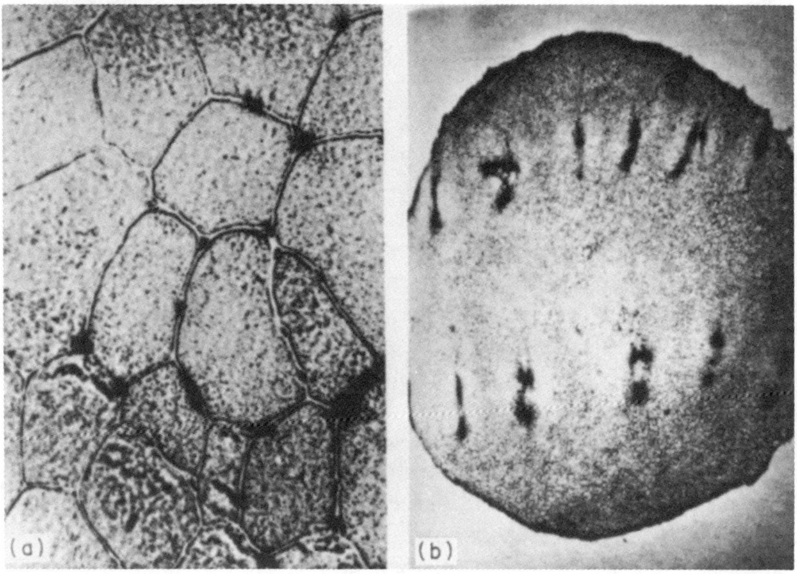

Figure 3 Histochemical localization of enzyme activity. (a) Alkaline phosphatase activity in fresh frozen section of porcine longissimus muscle detected with α-naphthyl phosphate and diazotized 4′-amino-2′,5′-diethoxybenzanilide. (From Ref. 6, p. 300, by courtesy of Institute of Food Technologists.) (b) Transverse section of fresh beet tissue after treatment with dihydroxyphenylalanine for detection of polyphenol oxidase. (From Ref. 7, p. 579, by courtesy of the Institute of Food Technologists.)

A. Breakage of Membranes and Cell Walls

If the enzyme is in the extracellular fluid, the cells are removed by centrifugation and the fluid is reserved for purification. If the enzyme is intracellular, the first step in purification is breakage of cells and membranes and extraction of enzyme from the tissue. Breakage of cell and organelle membranes of animal tissues is easily accomplished by homogenization in a Waring Blendor or a Potter–Elvejhem homogenizer with the extracting buffer (see Section III.B). Rupture of plant and microbial cell walls and organelles is not so readily accomplished. With plant tissues, preparation of acetone powders is sometimes useful. Freezing the tissue in liquid nitrogen and grinding, while still frozen, through a Wiley mill is an excellent way of rupturing plant cells.

Microbial cells may be disrupted by grinding with sand, by homogenization in a Waring Blendor containing glass beads, by sonic oscillation, or by forcing the cells through a small orifice under high pressure (French press). Yeast cells are often broken by autolysis in the presence of toluene. So many microbial enzymes are extracellular that it seems essential to caution that the extracellular broth should be examined for the desired enzyme before cell rupture is undertaken.

B. Extraction of Enzyme

Choice of extracting fluid depends on the particular enzyme, whether it is soluble or membrane bound, and the type of tissue. The usual volume of extracting fluid should be approximately two to three times that of the volume of tissue.

Extraction fluids for plant enzymes often contain sufficient buffer and salt to maintain the pH near 7.5 and at an ionic strength of 0.1 to 0.5. The buffer is needed to protect the enzymes from large quantities of acids released from the vacuoles on rupture of the cell. The high ionic strength is needed to desorb the enzyme from its attachment to the cell wall. In plants that contain large amounts of phenolic compounds, polyethylene glycol or polyvinylpyrrolidone may be incorporated into the extracting fluid to protect against inactivation of enzyme through reaction with the released phenolic compounds. Plant tissues that contain a large amount of pectic substances also present a problem because the extract may gel, especially if pectin methylesterase is present.

Many tissues contain relatively large amounts of proteolytic enzymes. The possibility of proteolysis, including that of the enzyme under purification, is very real. Proteolysis can be minimized by working at low temperatures, working as rapidly as possible, or by adding inhibitors of the proteolytic enzymes.

Extraction of enzymes that are bound to lipid or phospholipid membranes presents a special problem. Preparation of acetone powders may overcome this difficulty. Some tissues and fluids are not readily amendable to cold acetone treatment. This is particularly true of secretions such as milk. Morton introduced the use of isobutanol, usually at 33%, for solubilization of membrane-bound enzymes [8]. This technique has been widely adopted. Sometimes detergents must be added to solubilize membrane-bound enzymes.

Some enzymes are parts of systems and the activity of these enzymes depends on system integrity. Therefore, extraction procedures that disrupt this integrity destroy enzyme activity. In the last few years successful isolations of some of these complexes by special techniques, including the use of detergents such as deoxycholate, sodium dodecylsulfate, and other chaotropic reagents, have been made.

After the enzyme has been extracted from the tissue, the cellular debris is removed by high-speed centrifugation (10,000 to 20,000 *g* for 20 to 30 min), usually in a refrigerated centrifuge.

C. Removal of Nucleic Acids

When an extract is made from a microbial or plant source or from the nucleus, ribosomes, or microsomes of animals, it is likely to contain large amounts of nucleic acids (DNA and RNA). These polyanionic macromolecules interfere with protein separations and should be removed at this stage. Adjustment to pH 5.0, or addition of basic compounds such as protamine sulfate or streptomyocin to the solution, will precipitate the nucleic acids. Deoxyribonuclease and ribonuclease can be used to hydrolyze the nucleic acids to small nucleotides that will be removed by subsequent precipitation and dialysis steps, but this is not generally recommended because of possible contamination of the preparation with other enzymes, particularly proteolytic enzymes. Use of pure preparations of nucleases would alleviate this difficulty.

D. Differential Heat Treatment

Heat treatment may be used effectively at this point, where there is still a large volume of fluid. Effectiveness of this step will depend on the desired enzyme being relatively heat stable while a reasonable amount of the unwanted proteins is denatured and precipitated from solution. At least three variables can be manipulated to advantage: pH, temperature, and time of heat treatment. To determine the conditions to use, small aliquots of the extract are adjusted to different pH values, placed into plastic centrifuge tubes, and in turn into a water bath at selected temperatures for various periods of time. As the desired treatment is completed the tube containing the aliquot is removed into an ice bath. After centrifugation the supernatant liquid is analyzed both for protein remaining and for activity. Some results are presented in Table 3. The correct treatment conditions to use for the

Table 3 Data Illustrating Use of Heat as Step in Purification of an Enzyme[a]

	Supernatant after heating				
Temperature (for 10 min) (°C)	Protein (mg/mL)	Activity (units/mL)	Activity recovery (%)	Specific activity (units/mg)	Purification (fold)
Original	40.0	875	100	21.9	1.00
35	39.9	876	100	22.0	1.00
40	36.6	862	98.5	23.6	1.08
45	29.6	850	97.1	28.7	1.31
50	16.2	800	91.4	49.4	2.26
55	11.2	606	69.2	54.1	2.47
60	9.3	176	20.1	18.9	0.86
Time at 50°C (min)					
3	29.9	850	97.1	28.4	1.30
5	22.3	830	94.8	37.2	1.70
10	16.2	800	91.4	49.4	2.26
15	13.5	780	89.1	57.8	2.64
20	13.1	760	86.8	58.0	2.65
25	13.0	735	84.0	56.5	2.58

[a]These are hypothetical data designed to illustrate a point.

remainder of the extract are based on the percentage of activity recovery and the fold purification. A large amount of activity loss at this stage cannot be tolerated unless there is a correspondingly large increase in fold purification. From the data in Table 3 the best conditions to use are treatment at 50°C for 15 min.

IV. PURIFICATION

A. Differences Among Proteins

After the preceding preliminary steps the enzyme preparation is still relatively impure in most cases and may contain anywhere from <0.1 to 50% of the desired enzyme. At a given pH proteins may differ in (a) solubility, (b) charge density, (c) charge distribution, (d) hydration, (e) size, (f) shape, (g) density (specific volume), (h) specific reactive groups, (i) hydrophobicity, and (j) relative stability (Fig. 4). Each molecule in Fig. 4 is different from every other molecule. All molecules of the same protein will have identical properties. We have already discussed how differences in pH and temperature stability of proteins may be used in purification.

B. Stereospecific Adsorption Complexes (Affinity Chromatography)

The most specific method of distinguishing among and separating a group of proteins is on the basis of their specificity. Each enzyme will form an adsorptive complex with a compound which stereospecifically fits the active site of that enzyme. Therefore, when a crude solution of an enzyme is passed through a column prepared of an insoluble matrix to which a group for which the enzyme has

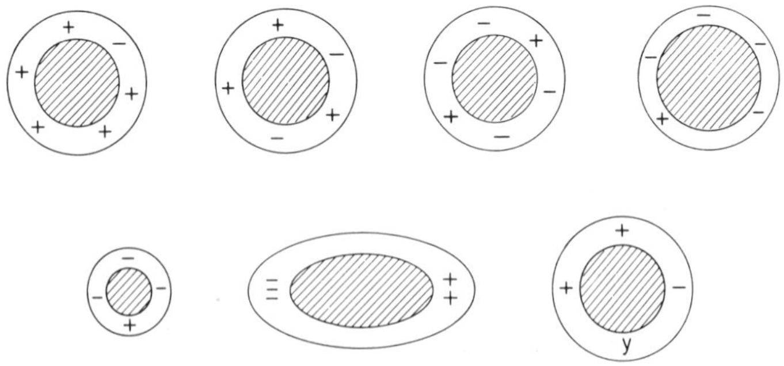

Figure 4 . Schematic representation of some of the differences found among proteins. The shaded area represents protein, the white area water of hydration. Shown are differences in size, shape, hydration, charge density, charge distribution, and specific reactive group (Y).

specificity is attached, the enzyme will be retarded because of its interaction with the attached substrate-like compound (i.e., competitive inhibitor), while the other proteins will pass directly through the column. The desired enzyme can be eluted from the column using a soluble competitive inhibitor, or by change in ionic strength, pH, or dielectric constant of the eluting solvent. This procedure, now called affinity chromatography, was understood and utilized as early as 1953 [9]; now it has achieved the importance it deserves [9–12]. Affinity chromatography is used increasingly in enzyme purification where possible because it can lead to pure enzyme with a minimum of purification steps. Affinity chromatography does not generally separate isozymes, so other methods are needed.

C. Precipitation Techniques

Separation of proteins on the basis of their different solubilities is a classic method. Solubility of a protein is determined by three main factors: (a) charge density and charge distribution, (b) degree of hydration, and (c) presence of nonprotein components of the molecule, such as phosphate, carbohydrate, or lipid.

1. Control of Charge Density

The charge density of a protein is determined by the number of acidic and basic amino acid residues on the surface of the protein. These charged groups are contributed by the side chains of aspartyl, glutamyl, cysteinyl, tyrosyl, histidyl, lysyl, and arginyl residues as well as the amino and carboxyl groups at the ends of the polypeptide chain. Because of the hydrophilic nature of these groups they are generally oriented on the outside of a protein molecule exposed to the aqueous phase.

The charge density of a protein is affected by changing the pH. At low pH all the prototropic (ionizable) groups will be protonated and the protein will have a net positive charge (Fig. 5). As the pH is increased, more and more of the carboxyl

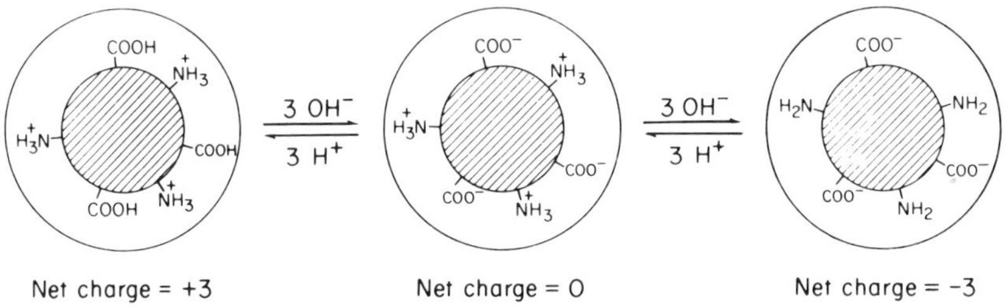

Figure 5 Schematic representation of effect of pH on the net charge of a protein. The three protein species differ only in ionization of prototropic groups, depicted as carboxyl and amino, and therefore in net charge.

group protons are neutralized until, at a pH characteristic for each protein, there are equal numbers of positive and negative charges. The net charge is then zero and this is called the isoelectric point. There are still many charges on the protein at the isoelectric point, but the net charge is zero. As the pH is increased further, the protein contains a net negative charge. This is shown schematically in Fig. 5.

Exercise 1. Consider a protein that contains 10 carboxyl groups with an average p*K* value of 3.50, two histidyl groups of average p*K* 6.0, and 14 ϵ-amino groups of average p*K* 10.0. Starting at pH 2.50, calculate the titration curve for this protein with NaOH and indicate the position of the isoelectric point.

Most proteins are least soluble at their isoelectric point (Fig. 6). Solubility increases as the pH is decreased below and increased above the isoelectric point. Proteins differ in the pH at which they have minimum solubility. There is also a difference among proteins in the degree of solubility at their minimum point of solubility. Proteins A, B, and C in Figure 6 have minimum solubilities at pH 4, 6, and 8, respectively. At their respective minimum solubility points protein B is less soluble than proteins A and C. At pH 6, proteins A and C are at least 15 times more soluble than protein B. Fractionation experiments involving precipitation at pH 6 should be quite successful in purifying protein B from proteins A and C.

Whether an individual protein precipitates at its point of minimum solubility

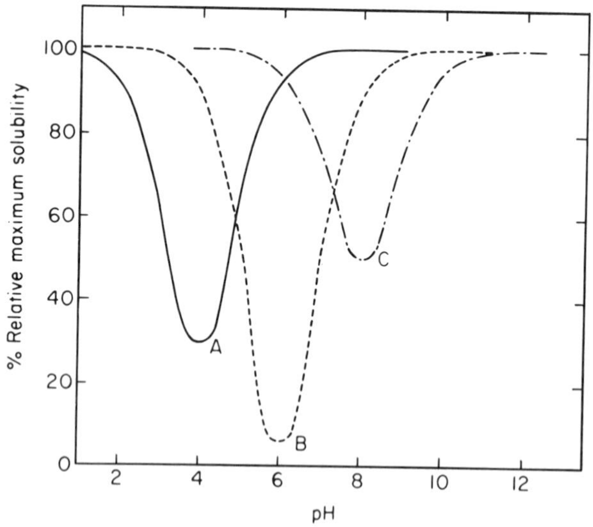

Figure 6 Schematic representation of effect of pH on the solubilities of proteins A, B, and C with isoelectric points of 4, 6, and 8, respectively.

depends on its concentration and solubility at that point. Many proteins are precipitated simply by adjusting the pH. For example, the caseins of milk have a minimum solubility between pH 4 and 5 and can be separated from the other proteins of milk by adjusting the pH of the solution.

It is the usual practice to purify a protein by precipitation at or near its point of minimum solubility. To determine the pH for minimum solubility, the pH of small aliquots of a crude enzyme preparation is adjusted to various values. If no precipitate is formed, sufficient saturated ammonium sulfate is added to each aliquot with continuous stirring to give 10% saturation at 0°C. Additional aliquots of saturated ammonium sulfate are added to each tube until there is a noticeable precipitate formed in at least half of the tubes. The tubes are centrifuged and the supernatant liquid is decanted and analyzed for remaining enzymatic activity and protein.

To form a protein precipitate, several molecules of that protein must come close enough together to form an aggregate whose combined solubility, because of mass size and masking of charged groups, is less than that of the individual molecules. This can be achieved by adjusting the pH of the solution to the isoelectric point of the protein (above) or by increasing the effective attractive forces between molecules. The latter can be accomplished by lowering the dielectric constant:

$$\text{attractive force, } F = \frac{Z^{+}Z^{-}}{Dr^{2}} \tag{1}$$

That is, the attractive force between two molecules is directly proportional to the product of the positive, Z^{+}, and negative, Z^{-}, charges on the molecules. The attractive force is at a maximum at the isoelectric point and inversely proportional to the dielectric constant, D, of the medium and the square of the distance, r, separating two molecules. All other factors being constant, lowering the dielectric constant of the solution increases this attractive force. Decreasing the dielectric constant from 80 (that of water) to 40 would double the attractive force. For this reason, acetone and alcohols precipitate proteins.

In precipitation of proteins by organic solvents it is essential to control the pH as well as the amount of inorganic ions presence. By increasing the effective charge on a protein, inorganic ions at low concentrations counteract the effect of lowering the dielectric constant.

To minimize denaturation of proteins by disruption of hydrophobic bonds, organic solvent precipitation should be done at low temperatures. The organic solvent should be precooled to −15°C and the enzyme-containing solution to 0°C. The solvent is added slowly with continuous stirring to the desired concentration. The precipitate should be removed from the solution soon after its formation and redissolved in buffer or dried at low temperature to remove solvent. Otherwise, considerable enzyme activity may be lost due to solvent denaturation.

2. Control of Hydration

The solubility of a protein is also determined by the extent of hydration of the molecule. The more charged groups on a protein molecule, the more highly hydrated it will be in general. The presence of other non-amino acid residues also affects the degree of hydration. Phosphate, sulfate, and carbohydrate groups increase, while lipids decrease, the degree of hydration.

The degree of hydration may be changed to some extent by varying the pH. However, the usual way of decreasing the degree of hydration is to add compounds that disrupt the water of hydration around a protein. The compounds most used for this purpose are ammonium sulfate, sodium chloride, sodium sulfate, and magnesium chloride. Of these, ammonium sulfate is by far the most often used because of its high solubility in water and the effect of the divalent SO_4^{2-} ion. While quite soluble, uni-univalent salts are relatively ineffective in precipitating proteins as compared with polyvalent salts (Fig. 7).

The effect of salts on solubility of proteins varies from one salt to another and is different with different proteins (Figs. 7 and 8). In the system shown in Fig. 8, fibrinogen is precipitated at an ammonium sulfate concentration where none of the

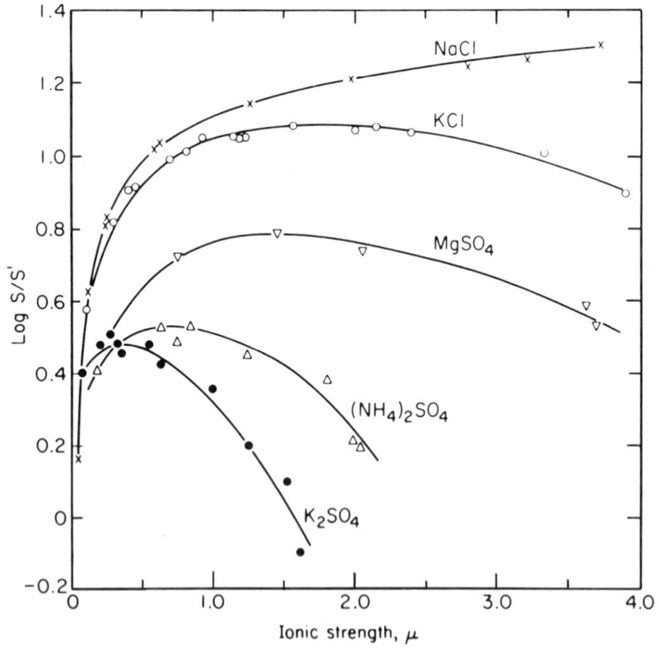

Figure 7 Solubility of carboxyhemoglobin at its isoelectric pH as a function of ionic strength and ion type. (From data of Ref. 13.)

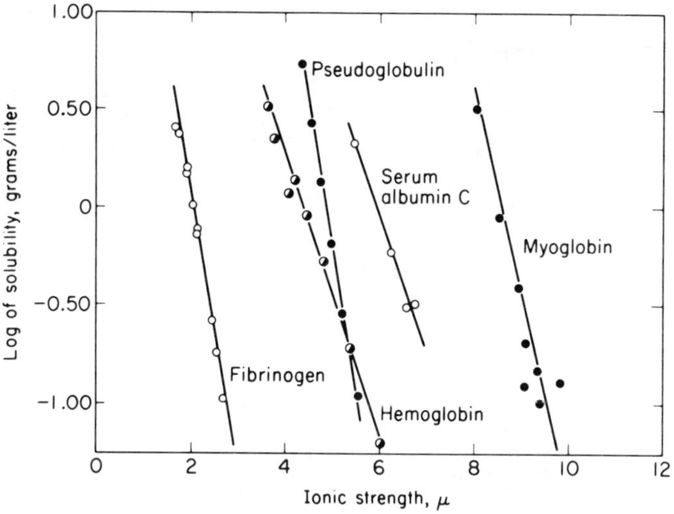

Figure 8 Solubility of several proteins in ammonium sulfate solutions. (From Ref. 14, p. 602, by courtesy of Reinhold Publishing Co.)

other proteins precipitate. Myoglobin is precipitated after all the other proteins. It would be relatively easy to obtain fibrinogen and myoglobin free of contamination. Other factors also play a role in determining protein solubility. With the hemoglobins (MW 64,000), for example, slight changes in ligands attached to the iron of the porphyrin nucleus have a marked effect on solubility. Oxygenation of human deoxyhemoglobin increases its solubility manyfold but has the opposite effect on horse deoxyhemoglobin. Changing the valence of iron from the ferrous to the ferric state also has a marked effect.

In general, the relationship between solubility of a protein and salt concentration is expressed by

$$\log S = \beta - K_s'\mu \tag{2}$$

where S is solubility of the protein, β the Y-intercept constant (and indicates solubility of the protein in absence of salt), K_s' the "salting out" constant, and μ the ionic strength. The ionic strength is given by the relationship $\mu = \Sigma C_i Z_i^2/2$, where C_i is the concentration of an ion and Z_i is its valence. A plot of data according to Eq. (2) is given in Fig. 9.

Exercise 2. Calculate the ionic strength of a 50% saturated ammonium sulfate solution at 0°C. A saturated solution contains 514.7 g/L.

At low ionic strength there is a salting-in effect, as illustrated in Fig. 10 for β-lactoglobulin. β-Lactoglobulin is much more soluble in 0.02 N sodium chloride

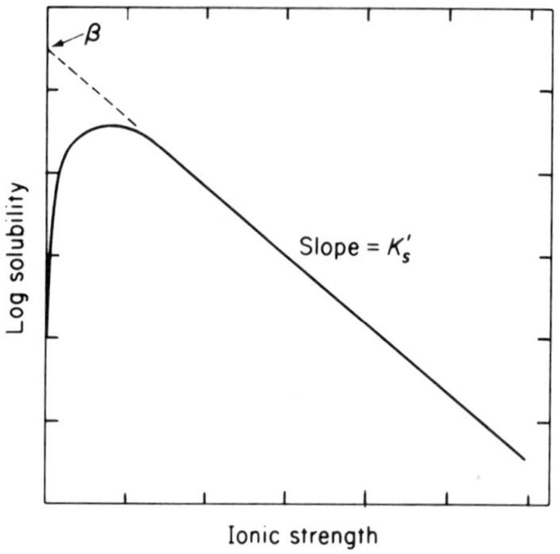

Figure 9 Relationship between solubility of a protein and salt concentration expressed as ionic strength.

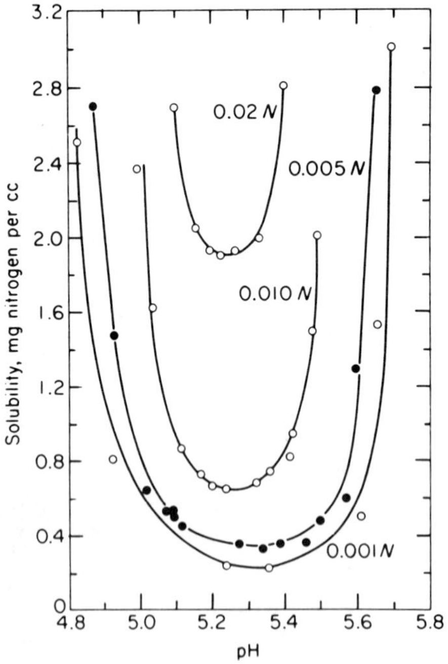

Figure 10 Solubility of β-lactoglobulin as a function of pH at different concentrations of sodium chloride. (From Ref. 15, p. 190, by courtesy of Carlsberg Laboratory.)

than in 0.001 *N*. As the ionic strength is increased, a point is reached where the protein begins to be less soluble. Above this point, solubility of the protein decreases as ionic strength is increased.

Experimental variables such as (a) temperature, (b) pH, (c) nature of protein, (d) nature of salt used for precipitation, and (e) protein concentration affect the values of β and K_s'. Temperature and pH affect only the β term. In general, as temperature is increased or pH is changed away from the pH of minimum solubility, β is increased. The entire line in Fig. 9 is raised, but the slope of the line, K_s', remains constant. The nature of the protein and protein concentration affect both β and K_s'. One takes advantage of this to separate one protein from another on the basis of their differential solubilities (Fig. 8). The nature of the salt used affects only K_s' since β is the solubility in the absence of salt.

In purification of proteins by precipitation it is necessary for investigators to establish the salt concentration to use under their conditions. Data reported in the literature should be used as a guideline only. The pH and temperature of the solution can be reproduced, but the exact concentration of the protein in a mixture is less easy to reproduce.

The most efficient way to determine the correct salt concentration needed is to place small aliquots of the crude protein solution into several tubes. Earlier, the pH should be adjusted to that value where the protein has minimum solubility, if known. The temperature is kept at 0°C by running the experiment in an ice bucket. The salt may be added as a solid with continual stirring until it has dissolved, or it may be added as a saturated solution (0°C). The salt concentration is varied in steps of 5 to 10% over the range from no precipitation to that of 80% saturation. After addition and dissolution of salt, the tubes are allowed to stand for 30 to 60 min before centrifugation. The precipitate is redissolved in an appropriate buffer solution. Both the supernatant liquid and the dissolved precipitate are analyzed for activity and protein concentration, usually after dialysis. The combined activity values for the two fractions should give 100% recovery since, in general, ammonium sulfate does not denature proteins. If recovery is much less or much greater than 100%, it is necessary to dialyze the solutions before analysis to remove the interfering ammonium sulfate. A pure protein is not obtained by a single precipitation procedure because all molecules of a given protein are not precipitated at a single salt concentration and there is overlap in precipitation with other proteins.

Proteins are precipitated by heavy metal ions (Pb^{2+}, Zn^{2+}, Hg^{2+}, Ca^{2+}) and by polyionic materials. All proteins are precipitated at high concentrations of heavy metal ions, but certain proteins are precipitated at quite low levels. For example, some of the plasma proteins are precipitated at 0.5 to 20 m*M* Zn^{2+} above pH 6. Casein is much less soluble in the presence of Ca^{2+}, insulin can readily be crystallized in the presence of Zn^{2+}, and enolase and papain may be crystallized as the Hg^{2+} derivatives. These metal ions probably function to precipitate proteins by neutralization of charges on the protein and/or by forming cross-linkages between protein molecules.

Nucleic acids and protamines precipitate proteins by electrostatic interaction with the protein, causing a net reduction in charged properties and causing large aggregates of polyion and proteins to form.

D. Adsorption

We shall limit our discussion under this topic to use of the adsorbent gels aluminum hydroxide and tricalcium phosphate (hydroxyapatite). Ion-exchange chromatography is also an adsorption process but will be treated under a separate heading. A number of procedures have been used to produce these gels, but the procedures of Willstätter and Kraut for aluminum hydroxide gel [16] and of Keilin and Hartree for tricalcium phosphate [17] are used most frequently. The adsorptive properties of the gels vary depending on the procedure used for preparation and the time of storage before use. There is sufficient variation from one batch of gel to another that some modification in the adsorption procedure may be needed in the use of different batches of gel. These gels are available from biochemical supply companies.

The gels are often used in batchwise purification procedures. Best results are obtained by systematically determining the effect of various experimental factors. The following procedure will serve as a starting point for the use of tricalcium phosphate gel. A suspension of the gel is centrifuged to remove excess liquid and a solution of the protein is added to the gel and stirred at intervals over a period of about 15 min. The protein concentration should be in the range of about 1%, the pH 5 to 5.5, the protein/gel ratio from 0.1 to 2, and the soluble phosphate and other ionic concentration low. After 15 min the suspension is centrifuged and the supernatant liquid decanted and analyzed for enzyme activity and protein content. Purification may be achieved by preferential adsorption of the contaminating proteins or of the desired enzyme. If the desired enzyme is adsorbed to the gel, the gel is washed with increasing concentrations of sodium phosphate buffer and each wash is collected by centrifugation. Phosphate ions displace the protein from the gel, and by use of a stepwise elution severalfold purification of a protein may be obtained.

Columns may also be prepared with tricalcium phosphate (hydroxyapatite) of sufficient particle size to give good flow rates. The protein is then eluted from the column by stepwise or gradient changes in the phosphate concentration of the eluting solution.

E. Ion-Exchange Chromatography

Separations based on differential behavior of proteins on ion-exchange materials have been very successful. The relative degree of discrimination among proteins on the basis of their charged properties is subject to experimental control and manipulation and the procedures can be performed under conditions where proteins are most stable.

1. Theory of Ion-Exchange Separations

The theory of ion-exchange chromatography is quite simple. In practice, one needs an insoluble material with hydrophilic properties that has ionizable groups of charge opposite to the net charge on the proteins. In addition, the proteins must be capable of being readily removed from the insoluble material in a selective fashion. This selectivity of elution may be achieved by varying the salt concentration of the solution or by changing the pH. Schematically, this is shown as

$$(R^{-}\cdot Na^{+})^{0} + \text{protein}^{+} \rightleftharpoons (R^{-}\cdot \text{protein}^{+})^{0} + Na^{+} \quad (3)$$

$$(R^{+}\cdot Cl^{-})^{0} + \text{protein}^{-} \rightleftharpoons (R^{+}\cdot \text{protein}^{-})^{0} + Cl^{-} \quad (4)$$

where R^- and R^+ indicate a cation- and an anion-exchange insoluble material, respectively. In the first case the positively charged protein interacts electrostatically with the cation-exchange material. Since Na^+ (and other cations including H^+) can also interact electrostatically with the cation-exchange material, it is in competition with the protein for binding. In order to have the proteins adsorbed the Na^+ concentration should be low. Proteins then are selectively eluted from the material by a systematic increase in salt concentration. A protein with a higher charge density requires a higher concentration of Na^+ to effect elution than would a protein with a lower charge density, everything else being equal. Charge density is molecular weight/charge. By this definition the lower the quotient, the higher the charge density.

Another method of eluting the protein is to change the pH. The ion-exchange material must have an opposite charge to the protein for adsorption to occur. After adsorption, the protein can be eluted by changing the pH of the eluting solution so that the charge on either the ion-exchange material or the protein is changed (positive to negative, or vice versa).

The same rationale holds for the use of an anion-exchange material. Which of these two types of materials to use will depend on the nature of the charge on the protein at the pH at which the separation is to be done. This pH is dictated, among other things, by the pH range in which the protein is stable. In actual practice other factors, such as hydrogen bonding, charge distribution, and shape of the protein molecule, may play a role in the behavior of a protein on ion-exchange columns.

2. Types of Ion-Exchange Materials

A number of ion-exchange materials are available for protein chromatography. While Dowex-50, a sulfonated polystyrene cation-exchange resin, has been of greatest utility in the separation of small cationic molecules, including amino acids, it has not been generally applicable to protein separations. This is because

(a) of limited capacity of the resin for proteins, (b) proteins are often denatured by the resin, and (c) recovery of protein is often quite low.

a. Amberlite IRC-50

In 1950, Paleus and Nielands showed that Amberlite IRC-50 could be used for the purification of cytochrome C [18]. Amberlite IRC-50 (also sold under the trade names Amberlite XE-64, Duolite CS 101, and Bio-Rad 70) is formed by copolymerization of methacrylic acid (2-methylpropenoic acid) with divinylbenzene in the presence of 1% benzoyl peroxide, which serves as a free-radical initiator. The resin is an insoluble matrix, cross-linked by divinylbenzene, containing free carboxylate groups which serve as cation-exchangers. Typically, Amberlite IRC-50 contains around 10 mEq of COOH groups per gram.

For chromatography of proteins, the resin is used in a very fine mesh size (100 to 400 mesh) in order to get sufficient exchange capacity. Proper equilibration of the resin before use is absolutely essential, as the success or failure of a procedure may be determined by a difference of as little as 0.1 pH unit.

In practically all cases in which Amberlite IRC-50 has been used effectively for protein separation, the pH has been in the region 6.0 to 6.5. It has been used effectively only in the separation of low-molecular-weight basic proteins such as cytochrome C, lysozyme, ribonuclease, chymotrypsin, and bromelain. Its capacity for proteins is quite low. Why use Amberlite IRC-50 in view of its limitations? It has the advantages that it is completely inert to enzymatic attack and that higher ionic strength solutions may be used with it. Higher ionic strength of the initial eluting solution may be necessary to keep a protein in solution or to prevent interaction among the protein components.

b. Cellulose Ion-Exchange Materials

The cellulose ion-exchange derivatives introduced in 1956 by Peterson and Sober [19] have revolutionized protein separation. The materials are readily prepared by chemical modification of cellulose, are innocuous to proteins, can be used under a wide variety of conditions, can be used over and over again following regeneration, and have a high capacity for protein. Under ideal conditions 1 g of cellulose ion exchanger (dry weight) is capable of binding approximately 1 g of protein. Under the conditions used for definitive separation of proteins the columns are not loaded this heavily, of course. But 100 to 500 mg of protein may be used on a column 2 cm in diameter.

Both cation- and anion-exchange cellulose resins are available (Table 4). The most frequently used materials are carboxymethyl (CM) cellulose and diethylaminoethyl (DEAE) cellulose, but others may be preferred for certain specific types of separations.

These cellulose ion-exchange materials can be bought from most biochemical supply companies. The content of ion-exchange groups and the particle size of the

Table 4 Cellulose Ion-Exchange Materials

Name	Structure	Approx. pK_a	Content ion-exchange groups (mEq/g)
Cation exchange			
Carboxymethyl (CM-) cellulose	$^{-}OOCCH_2OCH_2$cellulose	4.3	0.5–0.75
Sulfoethyl (SE-) cellulose	$^{-}O\overset{O}{\underset{O}{S}}OCH_2CH_2OCH_2$cellulose	<1	
Phosphoethyl (PE-) cellulose	$^{-}O\overset{O}{\underset{O}{P}}OCH_2CH_2OCH_2$cellulose	2 and 7	
Anion exchange			
Diethylaminoethyl (DEAE-) cellulose	$(CH_3CH_2)_2\overset{H}{\underset{+}{N}}CH_2CH_2O$cellulose	10	0.7–0.9
Triethylaminoethyl (TEAE-) cellulose	$(CH_3CH_2)_3\underset{+}{N}CH_2CH_2O$cellulose	12	

material will vary from batch to batch and from one supplier to another. One test of a material that will give good chromatography is that it be reasonably uniform in particle size. Fibrous ion-exchange materials can be improved by grinding through an 80-mesh sieve of a Wiley mill or by forcing manually through an 80-mesh sieve.

The ion-exchange groups listed in Table 4 have been covalently bound to Sephadex, which is a cross-linked dextran (see Section IV.F), Sepharose (neutral fraction of agar), and to polyacrylamide. All these ion-exchange materials are quite useful for purification of proteins. They have certain advantages over cellulose derivatives: that very few enzymes can degrade the support material, the particles are more uniform in size, and their capacity for proteins is higher.

3. Experimental Procedure for Ion-Exchange Chromatography

a. Determination of Conditions for Adsorption and Elution of Enzyme

The conditions to use in purification of a protein by chromatography should be established by a preliminary experiment that defines the best conditions for adsorption and elution of the protein. It is more efficient to do this by the following test tube experiments. Small amounts of the ion-exchange material are equilibrated with buffers as described in the next section at several different pH values and one or two different buffer concentrations, usually 0.005 and 0.01 *M*. Constant weighed amounts of the equilibrated material are added to test tubes and small aliquots of the crude protein solution, previously *dialyzed* against the buffers used for equilibration of the ion-exchange material, are added to each tube. The solution and ion-exchange material are mixed by stirring, and after standing for 15 min at 0°C the tubes are centrifuged. The supernatant liquid is decanted and analyzed for enzyme activity and protein concentration. This will indicate under what conditions the enzyme is preferentially adsorbed. The precipitate, consisting of ion-exchange material and adsorbed enzyme, from the tube that gives the best results is then suspended in buffer containing a low concentration of sodium chloride. After 5 to 15 min the tube is centrifuged and the supernatant liquid is analyzed for enzyme activity and protein concentration. The sodium chloride concentration is increased stepwise in subsequent washes until all the enzyme activity has been removed from the ion-exchange material.

b. Preparation of Column

The following procedures have been useful in the author's laboratory for the cellulose ion-exchange materials. If the material is not already granular it is ground through an 80- to 100-mesh sieve in a Wiley mill. The material is then suspended in a large volume of water in a graduate cylinder (~40 g of material to 2 L of water) and permitted to settle until 70 to 90% has reached the bottom of the cylinder. The supernatant liquid is siphoned off to remove the fine particles. This washing procedure is continued until the material settles uniformly and the supernatant liquid is clear. With DEAE-cellulose, the sizing is best performed by initial suspension in 0.25 *M* HCl. Water is used for subsequent suspensions. As much as 25 to 50% of the initial weight of material may be removed during this sizing operation, depending on the manufacturer.

The following description is based on 30 to 40 g dry weight of material. CM-cellulose is treated with 500 mL of 1×10^{-3} *M* cysteine·HCl to remove any monochloroacetic acid that may be present. (Monochloroacetic acid is used in carboxymethylation of the cellulose.) After filtration on a Büchner funnel, the resin is suspended in 500 mL of 0.50 *M* NaCl–0.50 *M* NaOH and filtered again. The filter cake is returned to the beaker, suspended in 500 mL of deionized water, and filtered again. This is repeated until the wash is free of chloride ions (usually, five washes are needed). The filter cake is suspended in 500 mL of 0.1 *M* buffer of the pH to be used, then filtered. It is then suspended in the buffer to be used at the start of the chromatography and filtered. This step is repeated five times.

To pour a column of the ion-exchange material, an extension tube is placed on top of the empty column to be filled, and with the outlet closed the extension tube and column are filled with a suspension of the material. The CM-cellulose is allowed to settle until a layer 3 to 5 cm thick has formed. The outlet is then opened to permit the liquid to flow out at a rate approximately that desired when the column is packed. The extension tube is kept filled so that the column is poured in one segment. It is more convenient to prepare the ion-exchange material and pour the column at room temperature. After the column is packed it is moved to a cold room and equilibrated with buffer overnight before use.

Preparation of DEAE-cellulose differs from that of CM-cellulose in that treatment with cysteine·HCl is omitted. The material is first washed with 0.25 *M* NaCl–0.25 *M* NaOH solution, with two water washings, with 0.25 *M* HCl, with water until chloride-free, and then is equilibrated against the starting buffer as described for CM-cellulose. With DEAE-cellulose, the rate of equilibration is slower than with CM-cellulose. It is necessary to adjust the pH of the suspension to the desired value when suspended in the first buffer wash and to check the pH at an intermediate wash and at the end to make sure the pH level is still that desired.

c. Choice of Elution Conditions

Should one use a stepwise, linear, curvilinear, or a constant buffer gradient in eluting an enzyme from a column? Of the stepwise and linear gradient, the latter is recommended because of the information it provides and the minimum amount of work involved. To achieve a linear gradient, two identical vessels are connected through an opening at the bottom of each or by a siphon arrangement between the two. Let us suppose that preliminary data (Section IV.E.3.a) indicated that adsorption of enzyme was optimum at pH 7 and 0.05 *M* NaCl and that it was eluted at 0.2 *M* NaCl. The mixing vessel is filled with 0.01 *M* phosphate buffer, pH 7.0, containing 0.05 *M* NaCl, and the other vessel is filled with an equal amount of 0.01 *M* phosphate buffer, pH 7.0, containing 0.20 *M* NaCl. The arrangement is shown in Fig. 11. When the containers are left open at the top, the levels in both fall at the same rate. This produces a linear gradient. The amounts of buffer used in each container will depend on the diameter of column used. For a 1-cm-diameter column, 150 to 300 mL in each container is adequate and 3-mL fractions are collected. For a 2-cm-diameter column the volume should be 1 L in each container and 10-mL fractions are collected.

Results of chromatography under these conditions are those shown in Fig. 12. Five protein peaks are obtained. The first protein peak comes through very rapidly in the void volume without being adsorbed to the CM-cellulose. The void volume, V_0, is defined as the volume of liquid required to transport a nonadsorbed compound from the top to the bottom of the column. The other four protein peaks are obtained as the NaCl concentration is increased. Only one of these peaks, C,

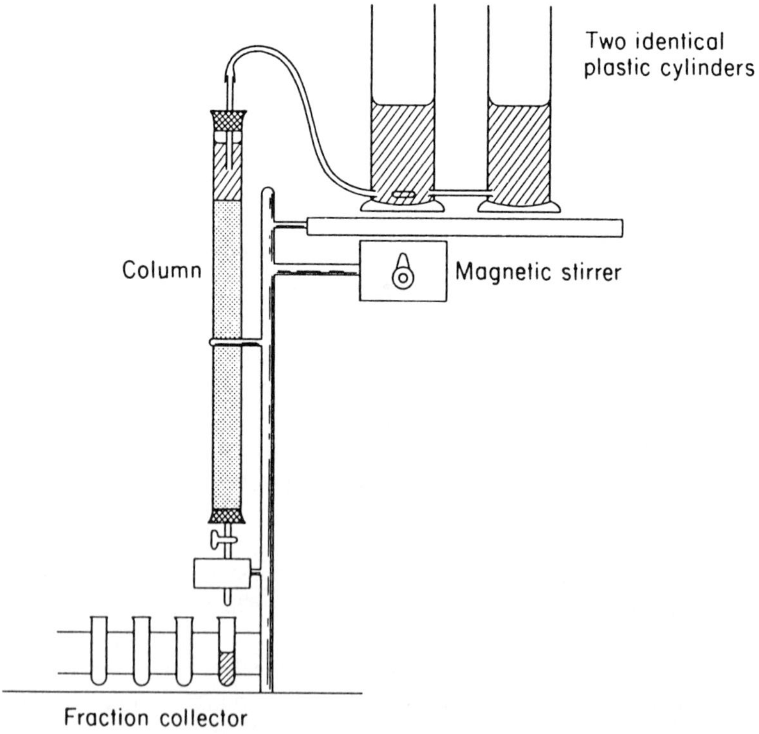

Figure 11 Arrangement for linear elution gradient and column chromatography. The mixing vessel is on the left.

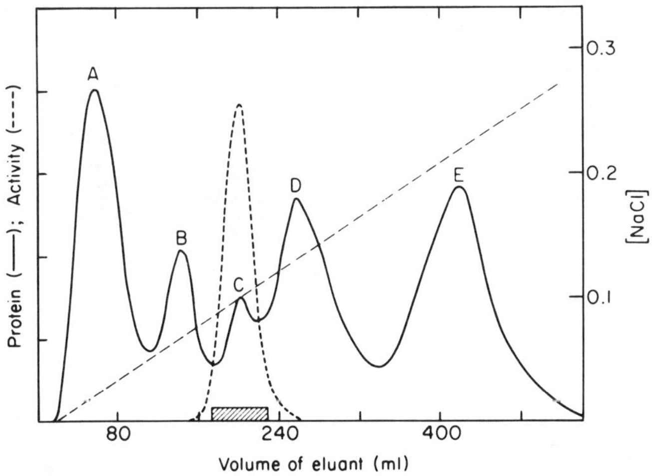

Figure 12 Separation of a hypothetical mixture of proteins by linear gradient elution chromatography. Only protein C has the enzyme activity being monitored. The diagonal dashed line indicates the change in NaCl concentration.

has any of the enzymatic activity being monitored. Pure enzyme is not obtained since there is overlap with protein peaks B and D. It will be necessary to rechromatograph the enzyme under a different set of conditions.

For rechromatography, fractions indicated by the horizontal bar in Fig. 12 are pooled, concentrated (see Section V), dialyzed against 0.01 *M* phosphate buffer containing 0.08 *M* NaCl, and placed on a column of CM-cellulose equilibrated against the same buffer. The 0.08 *M* NaCl was chosen on the following basis. As shown in Fig. 12, the enzyme is eluted at 0.10 *M* NaCl. Experience has indicated that when a starting buffer of approximately 20% lower concentration of NaCl is used the enzyme will be eluted within two to three void volumes of the column. It is under these conditions that CM-cellulose chromatography has its greatest resolving power. When the enzyme is eluted too near the void volume of the column it will be contaminated with other materials that come directly through the column without adsorption. The protein must not be eluted too late since the peak will be broad. Results of rechromatography of the enzyme under starting buffer conditions are shown in Fig. 13. Contamination due to protein peak B (Fig. 12) comes off the column in the void volume, while that due to protein peak D comes off the column much later than does the enzyme.

Other arrangements of the elution program may be used. Based on prior knowledge of the isoelectric points of the proteins in a preparation, a stepwise pH gradient may be used with advantage. A chromatographic separation of the egg white proteins based on a stepwise pH elution scheme is shown in Fig. 14.

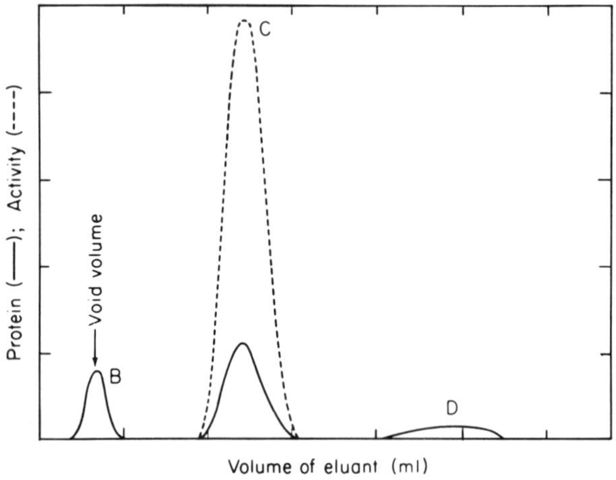

Figure 13 Rechromatography of protein C obtained from separation shown in Figure 12 by starting buffer elution. The elution volume for protein C was 3.6 times the void volume.

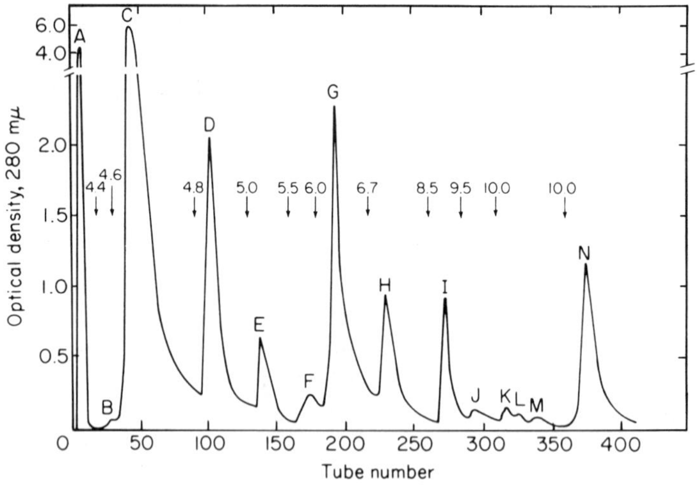

Figure 14 Fractionation of egg white proteins on CM-cellulose using a stepwise pH gradient. The vertical arrows show the pH and point of change of the eluting buffer. Fraction size was 15 mL. Identity of peaks: A, ovomucoid + flavoprotein; B, ovomucoid; C, D, and E, ovalbumin; F, "globulin"; G and H, conalbumin; I, J, and K, "globulin"; L, avidin; M, "globulin"; N, lysozyme. (From Ref. 20, p. 402, by courtesy of the American Society of Biological Chemists, Inc.)

Figure 15 shows the separation of the proteolytic enzymes of fig latex by a stepwise salt gradient on CM-cellulose. One of the inherent difficulties with stepwise gradients is knowing just where to change the gradient. If the gradient is changed too soon, the higher salt concentration will cause the trailing edge of the previous peak to come off more rapidly as an apparently new peak. This appears to be the case with the results shown in Fig. 14. Peaks C, D, and E all contain ovalbumin, and G and H contain conalbumin. Note the stair-step relation of these two sets of peaks.

We conclude our discussion of gradients by stating that one of the criteria to be used in selecting a gradient is that it change the least in the region where the desired protein is being eluted. The best resolution is obtained when the gradient is zero in this region. Consider as an example the behavior of the same mixture of proteins under two sets of gradients, as shown in Figs. 16 and 17. In the concave gradient (Fig. 16) the NaCl concentration is changing slowly in the region where the enzyme (peak 2) is eluted so that it is well separated from the other two protein peaks. With the convex gradient (Fig. 17) the NaCl concentration is changing too rapidly and the proteins are not well separated from each other.

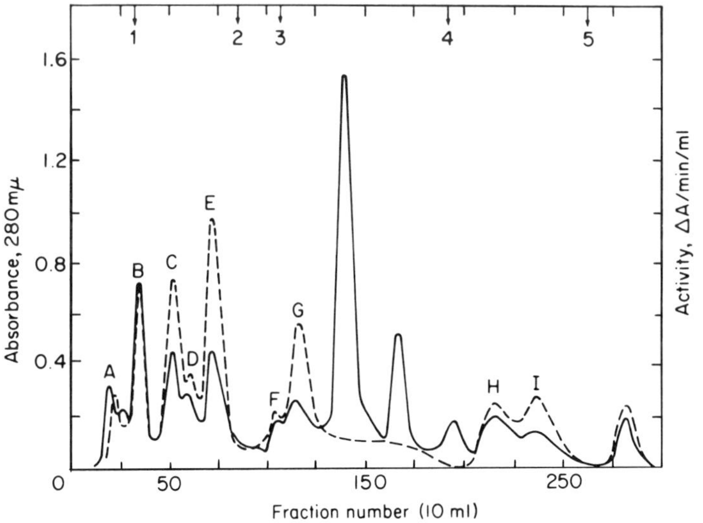

Figure 15 Chromatography of *Ficus glabrata* latex on CM-cellulose. Chromatography was performed in 0.01 *M* phosphate buffer, pH 7.0, with the following stepwise changes in sodium chloride concentration as indicated by the numbered arrows: 1, 0.04 *M*; 2, 0.10 *M*; 3, 0.12 *M*; 4, 0.20 *M*; and 5, 1.0 *M*. The dashed lines indicate proteolytic activity, the solid lines absorbance at 280 nm. (From Ref. 21, p. 2173, by courtesy of the American Society of Biological Chemists, Inc.)

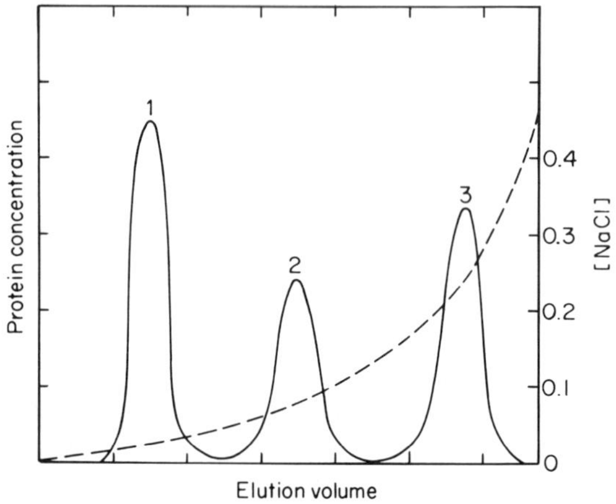

Figure 16 Chromatographic separation of a hypothetical mixture of three proteins by a concave sodium chloride gradient shown by the dashed line. Protein concentration is given by the solid line.

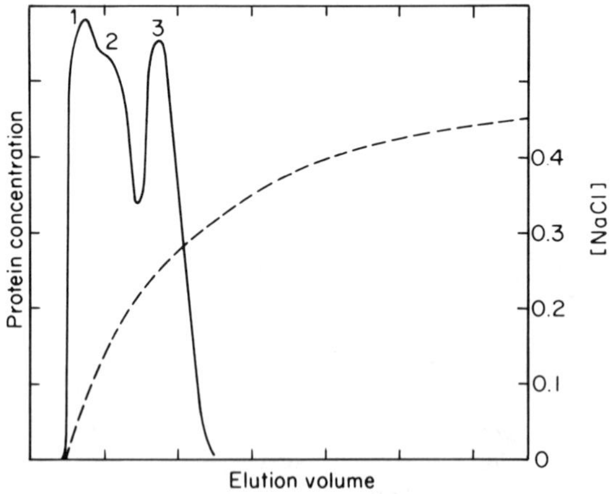

Figure 17 Chromatographic separation of the same mixture of three proteins shown in Fig. 16 by use of a convex gradient. All conditions were exactly the same as in Fig. 16 except for the different sodium chloride gradient.

What effect does the dimensions of a column have on separation? The diameter of a column determines the capacity. So the more protein to be added to a column, the larger should be the diameter. A 2-cm-diameter column will handle up to 1 g of protein quite well. The length of a column may or may not affect the resolution of the protein. In a stepwise gradient where one protein peak is obtained each time a change is made in the eluting buffer, the length of the column has little importance in determining the resolution. On the other hand, if more than one peak is obtained for each gradient change, resolution of the peaks within that gradient change will be affected by the length of the column. In starting buffer elution where true chromatography is achieved, the resolution of two peaks should be directly proportional to length of the column.

Chromatofocusing chromatography was developed more recently [22]. Separation is based on the isoelectric point (p*I*) of the proteins. The column packing material is either positively or negatively charged. For illustration, consider the packing material to be positively charged. The pH of the starting buffer is high enough that all the proteins are initially negatively charged; therefore, they bind electrostatically to the packing material at the top. A buffer of lower pH is then run slowly through the column. A pH gradient is formed at the interface between the two buffers. As the pH decreases, each protein is eluted from the packing material in order of its p*I*, with the one of higher p*I* elution first, and so on. The eluant is collected in fractions containing the separated proteins.

F. Gel Filtration (Permeation) Chromatography

1. Separation Based on Size

Proteins differ in molecular weight (Tables 2 and 6, Chapter 2). As we shall see later, proteins can be separated based on their differential sedimentation in a gravitational field related to their molecular weight. In this section we discuss separation of proteins by gel filtration techniques based on molecular size.

The first reported attempt to separate proteins on this basis was in 1950 on agar gel [23]. Two problems prevented agar from being a successful material for this purpose. First, agar is composed of two fractions, a neutral agarose and a fraction that contains ionic groups. Second, different batches of agar contain different amounts of these two fractions.

Around 1957, workers at the University of Uppsala cross-linked dextran molecules to decrease their water solubility [24] as a means of preparing a more suitable inert support for zone electrophoresis. It soon became apparent that columns packed with Sephadex (trade name of Pharmacia Fine Chemicals, Inc. for cross-linked dextrans) could be used to separate proteins on the basis of their size as well as to give good molecular weight determinations (Figs. 18 and 19). When the ratio of elution volume of a protein, taken as the volume of eluant at the peak fraction, to the elution volume of a molecule that is too large to penetrate

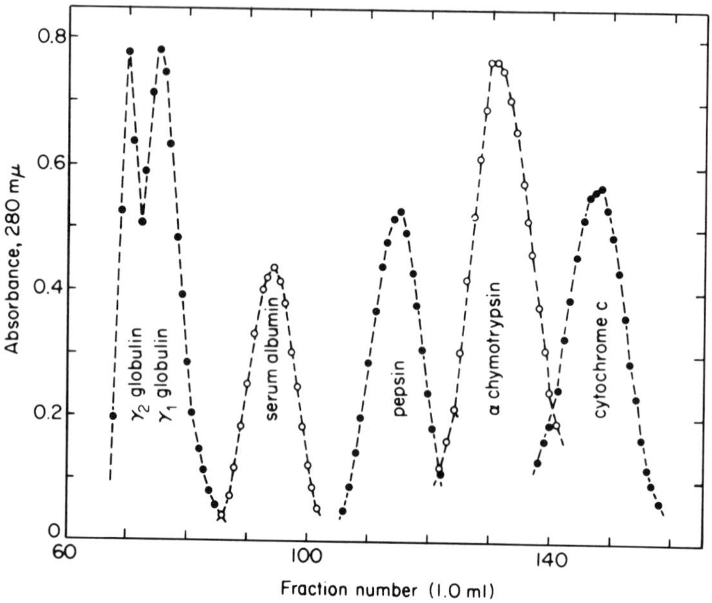

Figure 18 Elution curves for several proteins from a 1.1- by 192-cm Sephadex G-100 column. Amounts of protein applied to top of column were 7.5 mg for serum albumin and 5.0 mg for each of the other proteins. (From Ref. 25, p. 1951, by courtesy of the American Chemical Society.)

the gel structure (determined by γ_2-globulin in Figure 18 but more often by use of Blue Dextran-2000, MW $\sim 2 \times 10^6$) is plotted against the logarithm of the molecular weight, a straight-line relation is obtained (Fig. 19). The slope and position of the line are determined by the particular grade and batch of Sephadex used. By use of the same column and standard curve the molecular weight of an unknown protein can be determined. An advantage of this procedure is that the material placed on the column need not be pure if the elution volume of the protein under study can be determined by enzymatic or other biological activity. Therefore, molecular weights of specific enzymes in crude preparations can be determined—something that is not easy to accomplish with an ultracentrifuge.

The largest proteins come off the column first because the large molecules cannot diffuse into as much of the gel matrix as can the smaller molecules. Therefore, their pathway from the top to the bottom of the column is considerably shorter than that of the smaller molecules.

2. Types of Gel Filtration Material

Sephadex gels are made by cross-linking dextran molecules, which are long polymers of glucose linked 1-6 (~95%) and 1-3 (~5%), with epichlorohydrin:

$$2ROH + H_2C\overset{O}{—}CHCH_2Cl \xrightarrow{OH^-} ROCH_2CHOHCH_2OR \qquad (5)$$

Dextran   Epichlorohydrin   Sephadex

The degree of cross-linkage can be controlled by the amount of epichlorohydrin added to give different grades of Sephadex (Table 5). The lower the number of the Sephadex type, the more extensive the degree of cross-linking and the lower the molecular weight of the compounds that can be separated. For example, Sephadex G-10 is used to separate molecules up to a molecular weight of 700, while Sephadex G-200 is used to separate compounds ranging in molecular weight from 5000 to 800,000. Resolution between proteins is improved by selecting a Sephadex that covers only a limited range. For example, for a mixture of proteins with molecular weights ranging between 15,000 and 30,000, much better resolution would be obtained by the use of Sephadex G-50 than with Sephadex G-200. The fractionation range varies depending on the nature of the molecules to be fractionated. (Compare protein and dextran exclusion size, Table 5.)

The Sephadex materials in Table 5 are for use with hydrophilic compounds such

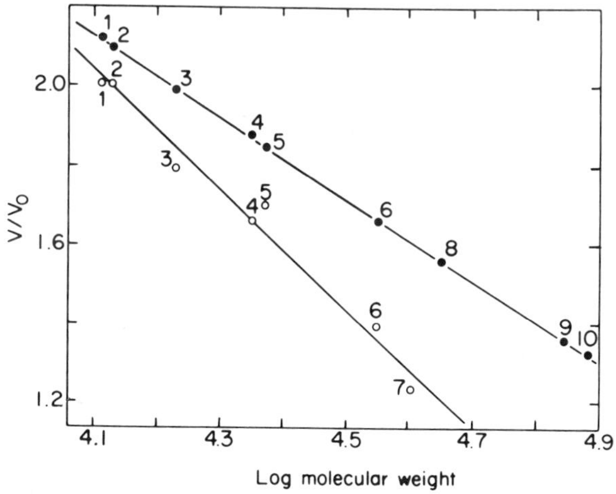

Figure 19 Relationship between elution volume from Sephadex columns and molecular weights of proteins. ○, 1.2- by 184-cm Sephadex G-75 column; ●, 1.1- by 192-cm Sephadex G-100 column. Eluting fluid was 0.1 *M* acetate buffer, pH 6.0 containing 0.4 *M* NaCl. Proteins used were (with molecular weight in parentheses): 1, cytochrome *c* (13,000); 2, ribonuclease (13,600); 3, myoglobin (16,900); 4, α-chymotrypsin (22,500); 5, trypsin (23,800); 6, pepsin (35,500); 7, peroxidase (40,200); 8, ovalbumin (45,000); 9, serum albumin (70,000); 10, iron conalbumin (76,600). (From Ref. 25, p. 1951, by courtesy of the American Chemical Society.)

Table 5 Specifications of Different Types of Sephadex

Sephadex type	Fractionation range for: Peptides and globular proteins	Dextrans
G-10	Up to 700	Up to 700
G-15	Up to 1500	Up to 1500
G-25	1000–5000	100–5000
G-50	1500–30,000	500–10,000
G-75	3000–70,000	1000–50,000
G-100	4000–150,000	1000–100,000
G-150	5000–400,000	1000–150,000
G-200	5000–800,000	1000–200,000

Source: Ref. 26, p. 10, by courtesy of Pharmacia Fine Chemicals, Inc.

as carbohydrates and proteins. For separation of hydrophobic compounds such as lipids, sterols, and so on, Sephadex materials, treated to render them hydrophobic, are available.

Gel filtration materials are also prepared by polymerization of acrylamide in the presence of low and variable concentrations of bisacrylamide as a cross-linking agent. These insoluble polymers operate on the same principle as Sephadex in separating molecules on the basis of size. Polyacrylamide gels are sold under the trade name Bio-Gel P (Table 6). As with Sephadex, the lower the number, the more

Table 6 Bio-Gel P Gel Filtration Materials

Bio-Gel type	Exclusion limit	Operating range
P-2	1,800	200–1800
P-4	4,000	800–4000
P-6	6,000	1000–6000
P-10	20,000	1500–20,000
P-30	40,000	2500–40,000
P-60	60,000	3000–60,000
P-100	100,000	5000–100,000
P-150	150,000	15,000–150,000
P-200	200,000	30,000–200,000
P-300	400,000	60,000–400,000

Source: Ref. 27, p. 40, by courtesy of Bio-Rad Laboratories.

highly cross-linked the material and the lower the molecular weight of material excluded from the gel matrix.

Materials that fractionate molecules as large as 100×10^6 are also available. These include Sepharose and Bio-Gel A, which are prepared from the neutral fraction of agar, agarose. Glass beads with pores the size of molecules are also available for separation of compounds on the basis of size. Separation of proteins on the basis of molecular size by gel filtration should always be considered among the steps used for purification of a protein.

G. Ultracentrifugation

Ultracentrifugation is used more as a method for determining molecular weight and homogeneity of an enzyme preparation than for purification of proteins because the capacity of the method is rather limited. However, density gradient centrifugation does offer some advantages in purification of proteins.

Separation by ultracentrifugation depends on differences in density, size, and, to some degree, shape of molecules. Proteins do not settle from solution on standing under the gravitational pull of the earth. However, when a protein-containing solution is spun at a high velocity so as to generate tremendous gravitational forces (on the order of 100,000 to 500,000 *g*), the large, dense molecules are forced outward from the center of the rotor at a faster rate than those of smaller size or those that are less dense. Lipoproteins, with densities below 1, "float" to the top of the centrifuge tube.

The rate of movement of molecules in a centrifugal field is related to molecular weight, M, of the molecule and its partial specific volume, $\bar{v}$ (reciprocal of density; $\bar{v} = 0.70$ to 0.76 for most proteins) by the equation

$$M = \frac{RTs}{D(1 - \bar{v}\rho)} \tag{6}$$

where D is the diffusion coefficient of the protein, R is the gas constant (8.317×10^7 ergs mol^{-1} deg^{-1}), T the absolute temperature (K), s the sedimentation velocity, and ρ the specific gravity of the protein solution. The sedimentation velocity, s, is dependent on the distance (cm) of the molecules from axis of rotation, x, and the square of the angular velocity, ω^2 (rad/sec), of the rotor:

$$s = \frac{dx/dt}{\omega^2 x} \tag{7}$$

A unit of 1×10^{-13} sec rad^{-2} has been defined as 1 Svedberg unit in honor of The Svedberg, the father of the ultracentrifuge. The sedimentation velocity of most proteins ranges between 1 and 200 Svedberg units. The higher the molecular weight, the larger the Svedberg units.

Purity of a protein is determined by this method from the symmetry of the sedimentation pattern (Fig. 20). The sedimenting boundary is detected by a Schlieren optical system which measures the refractive index increment, dn, along the length of the tube.

The molecular weight of a protein can also be determined by the *approach-to-equilibrium method*. When a protein is subjected to a low centrifugal force, 20,000 to 50,000g, the molecules approach a position in the tube such that the transport of molecules toward the bottom of the tube due to the centrifugal force is just counterbalanced by the rate of diffusion of molecules toward the axis of rotation. The molecular weight, M, is given by the relationship

$$M = \frac{2RT \ln (c_2/c_1)}{\omega^2(1 - \bar{v}\rho)(x_2^2 - x_1^2)} \tag{8}$$

where c_1 and c_2 are the concentrations of protein at distances x_1 and x_2 from the axis of rotation and the other symbols have the same meaning as for Eq. (6). This method does not require a knowledge of the diffusion coefficient, D. Detailed application of this method is described in Ref. 28.

Sucrose density gradient ultracentrifugation has some advantages over that performed in dilute buffer solutions. Most proteins have densities between 1.31

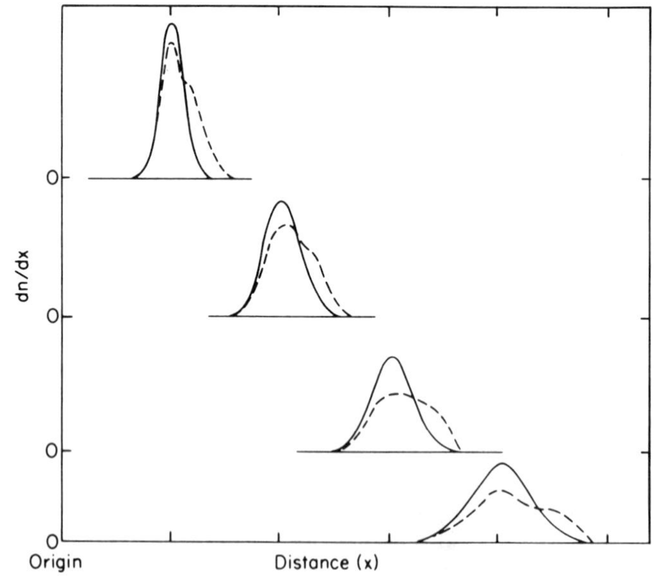

Figure 20 Schematic representation of sedimentation velocity Schlieren patterns for homogeneous (solid curve) and heterogeneous (dashed curve) protein preparations taken at various distances, x, from the origin. Distance moved is proportional to time.

and 1.44. A small amount of protein solution in sucrose of 1.30 density is layered on top of a sucrose gradient ranging from 1.30 at the top to 1.45 at the bottom of a centrifuge tube. When subjected to a high centrifugal force the proteins migrate toward the bottom of the tube until they reach a density of sucrose equal to their own density. At this point, the protein molecules accumulate in thin bands, each band being characteristic of a different type of protein; the solution can be removed in segments starting at the bottom of the tube.

H. Electrophoresis

Electrophoresis is generally used for determination of protein purity rather than for purification, although there are procedures whereby as much as 100 mg of protein may be purified in a single separation. Sometimes, 1 mg or less of protein is adequate for several assays. Separation of proteins by this method is dependent on differences in charge density, charge distribution, and shape and size (in starch gel and polyacrylamide gel electrophoresis) of the molecules. The charge density (charge per unit weight) of a protein can be changed by changing the pH of the solution (see Section IV.C.1), which offers an invaluable tool in the separation of one protein from another (Fig. 21). As the pH is raised there is a decrease in mobility until at a certain pH, characteristic for each protein, the mobility is zero. This is the isoelectric point of that protein. At pH values below this point the protein migrates as a cation. Above this pH there is an increase in rate of mobility of the protein as an anion. Importantly, the mobility profile is different for different proteins (Fig. 21).

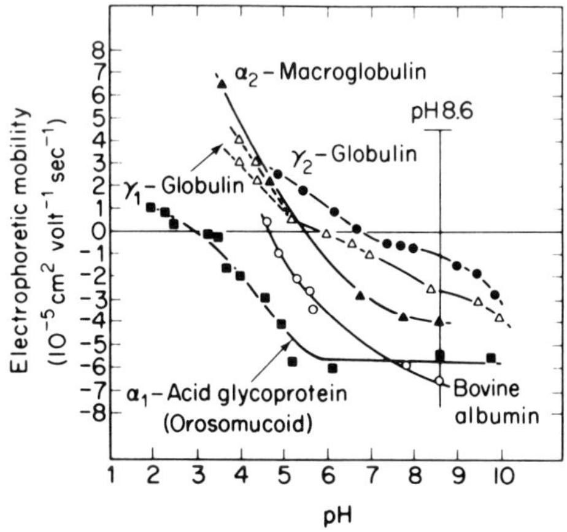

Figure 21 pH–electrophoretic mobility curves of purified human serum proteins and of bovine serum albumin. (From Ref. 29, p. 163, by courtesy of Academic Press.)

Electrophoresis may be divided into two types, *free boundary* and *zone electrophoresis*, depending on whether the support is an aqueous solution or a solid, stabilizing medium, respectively.

1. Free Boundary Electrophoresis

Figure 22 shows schematically the experimental arrangement for free boundary electrophoresis. When a source of direct current is applied to the electrodes, cations move toward the cathode and anions toward the anode. As in the analytical ultracentrifuge, the rate of migration of the molecules may be measured by following changes in refractive index by a Schlieren optical system. The ascending boundary for normal human serum proteins is shown in Fig. 23A.

In free boundary electrophoresis the electrophoretic mobility μ of a protein at constant pH is given by

$$\mu = \frac{K_p A d}{It} \tag{9}$$

where K_p is the specific conductance of the protein solution, A the cross-sectional cell area (Fig. 22), d the distance moved (cm), I the current applied in amperes, and t the time in seconds.

2. Zone Electrophoresis [22]

Zone electrophoresis uses a stabilizing medium through which protein molecules migrate under the influence of an electrical current. The most commonly used support materials are cellulose paper, cellulose triacetate, starch granules, starch

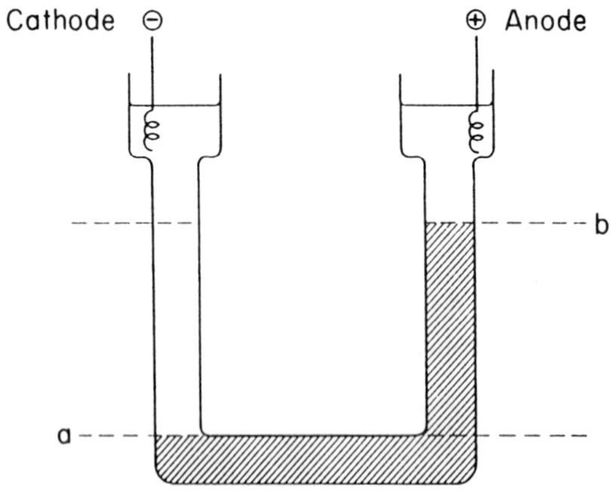

Figure 22 Schematic arrangement of cell for free boundary electrophoresis. The silica cell is made in three segments to permit formation of sharp boundaries between buffer (white) and protein solution (shaded) at points a and b. The arrangement of proteins in the cell is shown before any current has been applied.

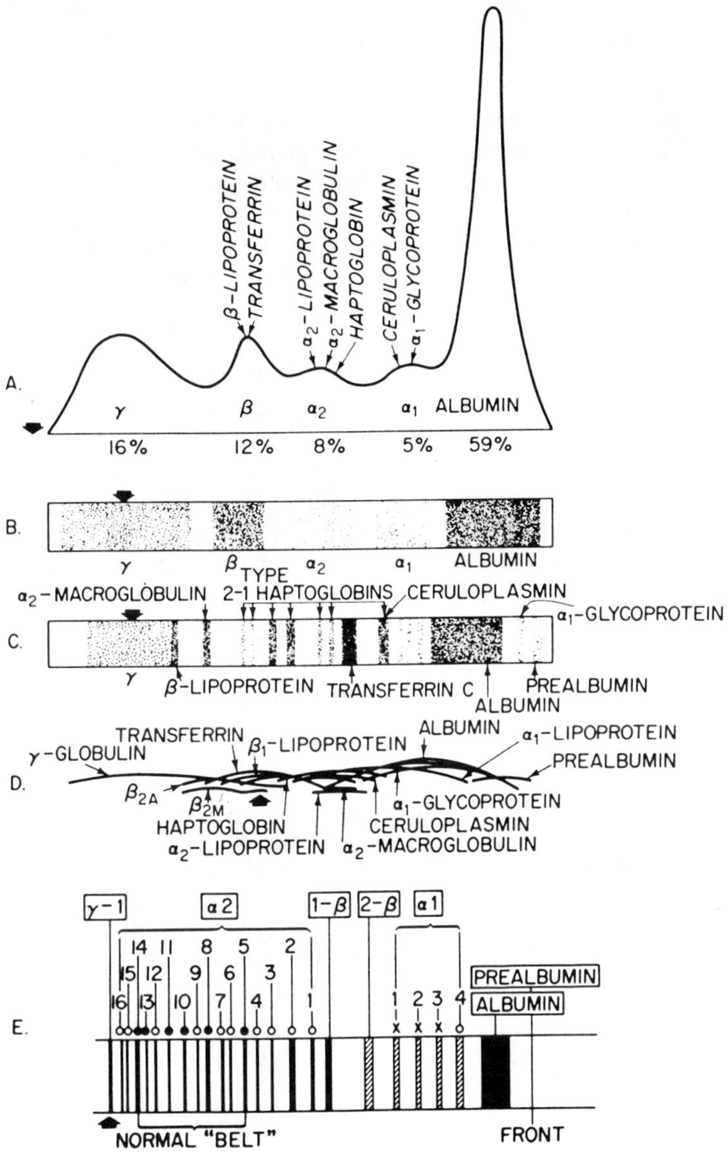

Figure 23 Schematic representation of patterns of normal human serum proteins in pH 8.6 buffer by electrophoresis on different support media: A, free boundary electrophoresis; B, paper electrophoresis; C, starch gel electrophoresis; D, agar gel electrophoresis as visualized by immunochemical analysis; and E, disk polyacrylamide gel electrophoresis. The large vertical arrows indicate the starting point in each case. In E, the following symbols are used: □, found in all sera examined; •, found in 85 to 95% of sera examined; ○, found in about 50% of sera examined; and X, found in about 25% of sera examined. (Diagrams A to D reprinted from Ref. 30, p. 164, by courtesy of Academic Press; diagram E redrawn from Ref. 31, p. 432, by courtesy of the New York Academy of Sciences.)

gel, polyacrylamide gel, and Sepharose. Cellulose paper is most useful for separation of low-molecular-weight compounds such as amino acids and peptides, but it is not useful for proteins because of their adsorption to cellulose. Because of adsorption there is streaking and 16 hours at 200 to 400 V is required to achieve adequate separation. The same separation can be achieved on cellulose triacetate strips in 1 to 2 hours. Gels prepared from starch and polyacrylamide are more useful because they permit separation on the basis of molecular size as well as charge.

The resolving capabilities of several support media for blood serum proteins are shown in Fig. 23. Each band in zone electrophoresis is due to a separated protein. Polyacrylamide gel offers the best resolving capability.

After separation by zone electrophoresis the positions of the protein bands are visualized by protein stains (Coomassie Brilliant Blue dye is preferred), by enzyme activity stains (Fig. 24), and by immunology (antigen–antibody reaction) as shown in Fig. 23D. The last two are the most discriminating since Coomassie Brilliant Blue dye stains all proteins. Several types of gel electrophoresis are possible: regular polyacrylamide gel electrophoresis (PAGE), gradient gel electro-

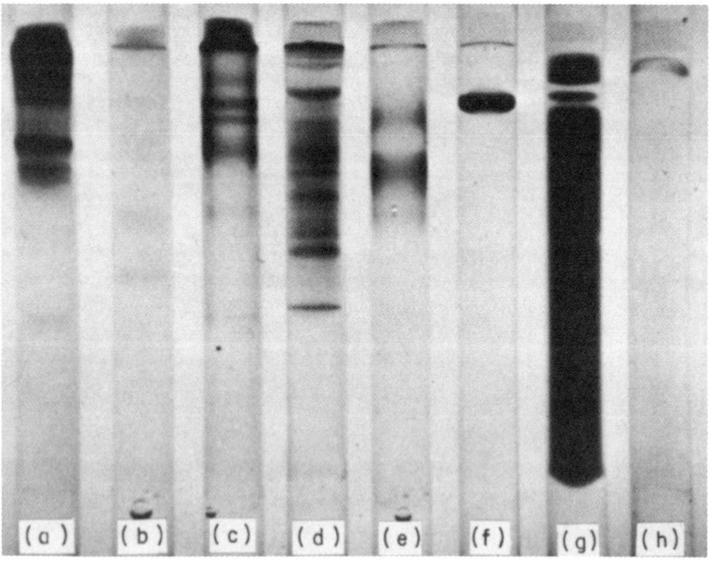

Figure 24 Use of specific enzyme staining techniques for the location of enzymes after separation by polyacrylamide gel electrophoresis. Shown are (a) alkaline phosphatase; (b) dehydrogenases; (c) esterases; (d) proteins; (e) malic dehydrogenase; (f) malic enzyme; (g) amylase; (h) phosphorylase. (Courtesy of Dr. Robert Clements, Ohio Experiment Station, Wooster, Ohio.)

phoresis, sodium dodecylsulfate gel electrophoresis (SDS-PAGE), and isoelectric focusing electrophoresis.

Isoelectric focusing electrophoresis is a very sensitive method for separation and determination of purity of proteins. Electrophoresis is performed in a pH gradient produced by an aqueous solution of an ampholite. Proteins migrate in an electrical field until they reach a point in the solution where the pH is equal to their isoelectric point. Since at the isoelectric point the net charge on the protein is zero, the protein cannot move but concentrates in a narrow band at that position. By measuring the pH along the gel, the pI (isoelectric point) of each protein can be determined.

Regular gel electrophoresis is run at a constant fixed pH along the gel. All the cationic proteins move toward the cathode and all the anionic proteins move toward the anode. At low gel concentrations separation is based on the charge density of the proteins. At high gel concentrations that sterically hinder movement, separation of proteins is based on a combination of charge density and size.

Gradient gel electrophoresis is run at constant pH but through gels that increase continuously in concentration from one end to the other. The proteins move based on their charge but at a rate dependent upon size.

Sodium dodecylsulfate polyacrylamide gel electrophoresis (SDS-PAGE) is run in a buffer containing sodium dodecylsulfate (a detergent; 1 to 2%) and β-mercaptoethanol. Before placing on the gel, the protein solution is treated with β-mercaptoethanol to reduce disulfide bonds and denatured by heating. Subunit proteins separate into the subunits. The highly negative charged proteins and subunits (because of the SDS^-) migrate in the electrical field at a rate dependent on size, with the smallest molecular weight units moving the fastest.

I. Hydrophobic Chromatography [22]

While most of the hydrophobic amino acids are folded into the interior of a protein, some will be on the surface. Therefore, the surface of proteins will have different degrees of hydrophobicity, which can be used to separate them by hydrophobic chromatography. Octyl-Sepharose or phenyl-Sepharose, which are hydrophobic, is used to pack the column using a buffer containing 0.5 to 1 *M* phosphate. The protein sample is added to the top of the column, where the proteins are bound due to the high phosphate concentration. The proteins then are selectively eluted, in increasing order of lowest to highest hydrophobicity, by a linearly decreasing phosphate concentration and with a linearly increasing concentration of ethylene glycol. The eluant is collected in fractions from the bottom of the column. A typical separation is shown in Fig. 25.

J. Other Separation Methods [22]

Special methods are available for some proteins based on functional groups of the protein, cofactors of an enzyme, or "tails" added in biosynthesis by genetic

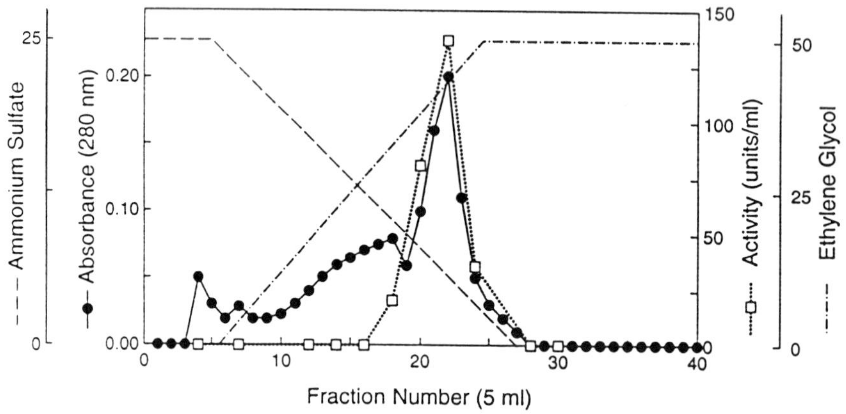

Figure 25 Chromatogram of purification of white kidney bean 858A α-amylase inhibitor on a phenyl-Sepharose CL-4B column (2.0 × 9.5 cm). Sample: 70 mg of 30 to 60% $(NH_4)_2SO_4$ precipitated proteins. (Unpublished data of Ho, F. M. and Whitaker, J. R.)

engineering. Proteins with sulfhydryl groups bind specifically to metal ions bound to an insoluble support. The bound proteins then are eluted selectively with an increasing gradient of a thiol compound. The converse reaction can be used for metal-containing proteins using an insoluble support containing sulfhydryl groups. The proteins are eluted selectively by use of an increasing gradient of metal ions. NAD^+- and $NADP^+$-containing proteins are bound selectively to Cibacron Blue 3GA bound to an insoluble support. Elution is with NAD^+ or soluble Cibacron Blue 3GA. "Tails" or "heads" can be added to genetically engineered proteins by increasing the gene with codons for acidic, basic, cysteine, or metal-binding groups. The tail is used to separate the protein selectively and then is removed from the protein.

V. ORDER OF APPLICATION OF PURIFICATION STEPS

It is a rare exception when an enzyme can be obtained in pure form by the application of only one purification procedure. In view of the hundreds of proteins in a tissue and the limited number of ways in which proteins differ, it is not surprising that a number of the proteins will be found identical by one or several techniques. The question then arises as to criteria to use in proper selection of order of purification steps. There is no single criterion for making such a selection, and the same order will not work in all cases. In terms of time and overall yield of enzyme it is best to apply those procedures that can handle large quantities of material first and reserve the low-capacity methods for the later stages of purification. Based on this, and experience, we give our preferred order of steps in the purification of an enzyme.

After the enzyme has been extracted and centrifuged to remove suspended material, it is best to apply pH adjustment and addition of protamine sulfate or streptomycin to remove nucleic acids, or heat treatment. At this point, since the volume of solution is usually large and enzyme concentration low, a concentration step is in order. This can be performed by fractional precipitation with ammonium sulfate or with organic solvents (acetone or ethanol), by adsorption on an ion-exchange column, by perevaporation, by selective removal of water by partial freezing and/or thawing of the solution, or by absorption onto dry Sephadex or Cryogel. Alternatively, the solution can be concentrated by ultrafiltration. By use of membranes of selective porosity small molecules (water, salts, etc.) can be removed while the protein molecules are retained. The rate of removal of water by ultrafiltration depends on the pressure used, protein concentration, and the size of membrane. Figure 26 shows some characteristics of cellulose acetate membranes used in ultrafiltration.

Following a precipitation step the material should be dissolved in the smallest volume of water or buffer possible. This is done by adding small quantities of solvent and stirring until no more of the precipitate goes into solution. This is repeated until it appears that the remainder of the precipitate is insoluble. There is often some denatured protein at this point which may be removed, because of lower solubility with an increase in the purity of the preparation. It is well to use gel filtration at this point since the material can be placed directly on a column without the need to remove salts. This eliminates the necessity to dialyze the material.

The fractions from gel filtration containing the majority of enzyme activity are pooled and concentrated by precipitation with $(NH_4)_2SO_4$ or acetone or by ultrafiltration. The solution is then dialyzed overnight against the buffer chosen for chromatography of the material on an ion-exchange column (see Section IV.E.3.a). The preparation is chromatographed on an ion-exchange column using a linear NaCl gradient. Fractions from the column that contain the majority of enzyme activity are pooled, concentrated, dialyzed, and put back on an ion-exchange column that has been equilibrated against a buffer containing 20% less NaCl concentration than that at which the peak of the enzyme activity was eluted by the linear gradient. The fractions containing the majority of the enzyme are pooled, concentrated, and tested for homogeneity by one or more of the methods described below. If the enzyme is still heterogeneous at this point, it is well to consider further purification by use of affinity chromatography, hydrophobic chromatography, chromatofocusing chromatography, or one of the other specialty methods. Each step in the purification should be evaluated by regular polyacrylamide gel electrophoresis to assess progress toward a single protein band (see later for determination of purity).

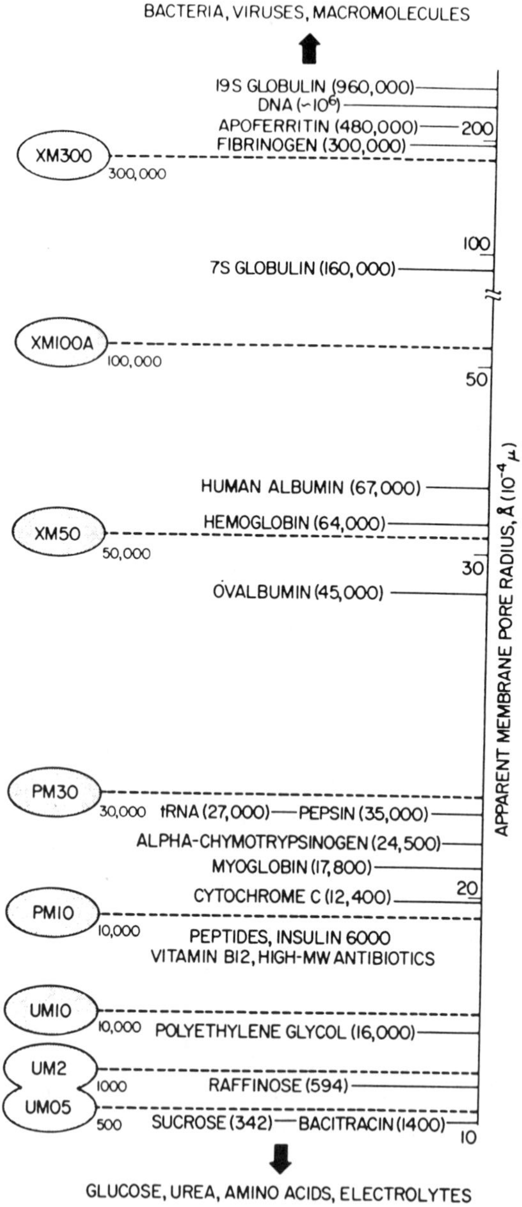

Figure 26 Retention characteristics of Diaflo membranes. A particular membrane will retain 90% or more of all species with molecular weights above the horizontal dashed line associated with that membrane. An indication of the pore size and therefore the size of the molecule is given along the right-hand edge of the diagram. (From Ref. 32, p. 7, by courtesy of Amicon Corporation.)

When the enzyme is obtained in pure form it can be stored at 0°C, frozen and stored at −20°C, or lyophilized and stored at −20°C. Some thought and experimentation should be given to the method of storage; otherwise, a tremendous amount of effort can be lost. If freezing and storage of the solution is selected, it is best to divide the solution into small aliquots before freezing. This way only the amount of enzyme needed is thawed at one time. Repeated thawing and freezing of an enzyme preparation usually results in appreciable loss of activity.

VI. QUANTITATION OF PURIFICATION STEPS

A. Activity Assays

The success of every step in the purification procedure must be ascertained by determining enzyme activity and protein concentration and comparing the two. The amount of enzyme is expressed in units of activity per milliliter, which may be defined as a change in some experimental parameter associated with conversion of substrate to product (absorbance, optical rotation, equivalents of acid produced, micromoles substrate converted or product formed) per unit time under a standardized set of conditions. Any unit of enzyme activity can be used; however, it should be expressed in units that can be reproduced by other workers. The Commission on Enzymes of the International Union of Biochemistry has recommended that a unit of enzyme be defined as that amount which will catalyze the conversion of 1 micromole of substrate per minute under defined conditions.

Protein concentration should be expressed as milligrams protein (or nitrogen) per milliliter. When units of enzyme per milliliter are divided by milligrams of protein per milliliter, the result is *specific activity*. This is the number of units of enzyme per milligram of protein. It is easy to follow the degree of purification by comparing the specific activity at that step with the specific activity of the original extract. The ratio of the two gives the fold purification.

Purification data should be kept in a balance sheet as shown in Table 7. From this table it is easy to follow the success of the purification. Notice that in purification of alkaline phosphatase the original volume was 1.3×10^6 mL, which contained 1.81×10^7 units of enzyme and 4.14×10^4 g of protein (5.10×10^{-3} g N/mL $\times 6.25 \times 1.3 \times 10^6$ mL). The final purified enzyme was in 10 mL and contained 2.68×10^5 units of enzyme (1.5% yield) and 109 mg of protein (0.0016% of original). It was necessary to lose 98.5% of the original enzyme activity and 99.998% of the original protein to attain an enzyme that was 5660-fold purified with respect to the original crude extract.

Percentage loss of protein is not of concern; it is the percentage loss of activity that is important. To achieve purification, it is necessary to lose enzyme. How much loss of enzyme can be tolerated will depend on how much starting material is available and how much enzyme is needed at the end. The 109 mg of enzyme recovered in the final step (Table 7) would permit a large number of assays to be

Table 7 Purification of Alkaline Phosphatase from Milk

Step	Procedure	Vol. (mL)	Activity (units/mL)	Total (units × 10^{-3})	Protein N (mg/mL)	Specific activity (units/mg N)	Yield (%)	Purification (fold)
1	Whole milk	1.3×10^6	13.9	18,070	5.10	2.7	100.0	1
2	Buttermilk	32,500	80	2,600	2.80	28.6	14.4	11
3	Filtrate, butanol treatment	27,120	66	1,790	0.21	314.3	9.9	116
4	Filtrate, treatment at pH 8.5, 12 hours, 5°C	27,000	65	1,755	0.20	325.0	9.7	121
5	Redissolved ppt, acetone pptn (45%)	1,480	1,092	1,616	1.20	910.0	8.9	337
6	Redissolved ppt, 63–85% $(NH_4)_2SO_4$	127	9,230	1,172	2.25	4,102	6.5	1,520
7	Supernatant, charcoal treatment	280	3,340	935	0.46	7,261	5.2	2,690
8	Redissolved ppt, 40–49% acetone	75	9,800	735	1.04	9,423	4.1	3,490
9	Supernatant, charcoal treatment	89	5,730	510	0.49	11,690	2.8	4,330
10	Redissolved ppt, 69–82% $(NH_4)_2SO_4$	12	29,000	348	2.04	14,200	1.9	5,270
11	Redissolved ppt, 42–48% acetone	10	26,800	268	1.75	15,314	1.5	5,660

Source: Ref. 33, p. 797, by courtesy of the Biochemical Society.

done on parameters that affect activity of the enzyme as well as the chemical and physical nature of the protein (based on available micromethods).

Fold purification is not an index of purity of an enzyme. It only reflects the increase in specific activity with respect to starting material. If the percentage of enzyme in a crude preparation is 20%, a fivefold purified material will be pure; on the other hand, if the enzyme is present as 0.1% of the total protein in a crude preparation, an 861-fold purified material will be only 86.1% pure.

In determining the units of enzyme per milliliter, standardized conditions should be used where the substrate concentration, pH, temperature, and any other parameters are kept constant. Any convenient assay can be used for following the rate of conversion of substrate to product. Because of its sensitivity, a continuous spectrophotometric method or a pH Stat method is preferred. Any procedure chosen must be precise, specific, rapid, and inexpensive since a large number of activity determinations will be needed, particularly in monitoring eluates from column fractions.

B. Protein Determination

There are a number of methods that are used to determine protein content of a preparation. The method chosen should be relatively precise, rapid, and if possible not use up too much of the sample for the assay.

1. Kjeldahl Method

There are many modifications of this method ranging from macrotechniques requiring several hundred milligrams to microtechniques involving milligram quantities of protein. They all have in common oxidation of organic material with concentrated H_2SO_4, H_2O_2, $HClO_4$, and so on, to CO_2, H_2O, and NH_3. In the macrotechniques the amount of ammonia is determined by distillation into standard acid. In the microtechniques the amount of ammonia is determined by Nessler's reagent or with hypochlorite. The amount of protein is calculated by multiplying the amount of nitrogen (calculated from ammonia content) found in the sample by the factor 6.25. Proteins vary in their nitrogen content from 13 to 19%, but the average nitrogen content of proteins is 16% (100%/16% = 6.25). The disadvantages of the Kjeldahl method are the length of time required for digestion and analyses, destruction of sample, other nitrogen-containing compounds (such as nucleic acids) are detected, and all ammonium sulfate added to the preparation during purification must be rigorously removed.

2. Biuret Method

The biuret method [34] is the single most accurate procedure for protein determination. It depends on reaction of Cu^{2+}, in alkaline solution, with peptide linkages of the protein to produce a red color that is read at 540 nm. Since all proteins contain approximately the same number of peptide bonds per unit weight of protein, the color value is essentially independent of the nature of the protein. The usual biuret method requires 1 to 10 mg protein per analysis, although there are micromethods that require about one-tenth this amount of material. When there is plenty of material available, it is an excellent method. The method only determines protein and is standardized against a solution of known protein content. Crystallized bovine serum albumin is often used as a standard protein. Reducing sugars interfere with the assay.

3. Lowry Method

The Lowry method [35] depends on the reaction of phosphomolybdate in alkaline solution with the tyrosyl residues of proteins. Cu^{2+} is included to increase sensitivity of the reaction. The blue color formed is quite stable and has a maximum absorbance at 750 nm, although absorbance is sometimes measured at 600 nm. As little as 5 to 40 μg of protein is needed for the assay. Therefore, the method is 200 to 1000 times more sensitive than the standard biuret method. It is the method of choice when the protein concentration or amount is low, although the method suffers in that the results are quite dependent on nature of the protein since the tyrosine content of proteins varies markedly. Bovine serum albumin is used as a standard. A number of other compounds, especially phenols, interfere in the assay.

4. Bradford Method

The Bradford method [36] is based on the binding of Coomassie Brilliant Blue G-250 (dye) to proteins in phosphoric acid. The principle is the binding of a negatively charged dye to a positively charged protein, which results in a shift in λ_{max} from 465 nm for free dye to 595 nm for bound dye. Some hydrophobic interaction between dye and protein may also occur. The increase in absorbance at 595 nm is related to protein concentration with a standard curve prepared with bovine serum albumin. The range of determination is 5 to 40 μg of protein. Since not all proteins have equal numbers of positive charges per unit weight, the observed absorbance is somewhat protein specific.

5. Bicinchoninic Acid Method

The bicinchoninic acid (BCA) method [37] is based on the reduction of Cu(II) to Cu(I) in alkaline solution by proteins. The Cu(I) reacts with BCA to give an intense purple color with λ_{max} of 562 nm. A standard curve is prepared with bovine serum albumin. Range of protein needed is 0.5 to 10 μg/mL. This method is particularly useful for detergent-containing protein solutions.

6. Spectrophotometric Method

All proteins have at least two regions of maximum absorbance, one near 280 nm due to the aromatic amino acid residues of tyrosine and tryptophan and the second at 180 nm due to peptide bonds of the protein (Fig. 27). Other compounds also

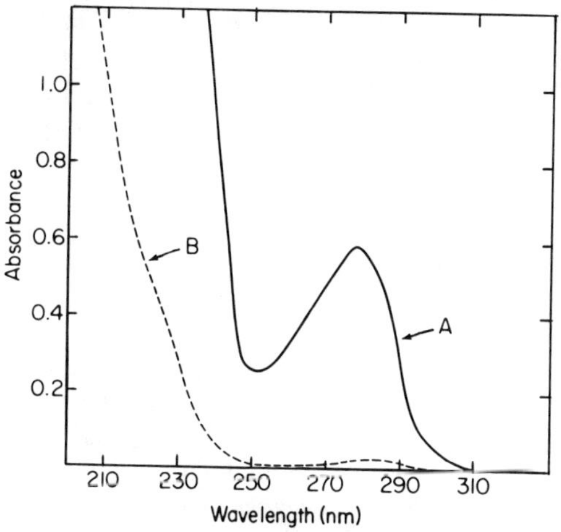

Figure 27 Typical absorbance spectrum of a protein as illustrated for bovine serum albumin. Curve A, 1 mg/mL protein, is shown as a solid line; curve B, for 0.0667 mg/mL protein, is shown as a dashed line. (From Ref. 38.)

absorb within these regions and may interfere with the spectrophotometric assay. For example, nucleic acids have a maximum absorbance at 260 nm. The following equation has been suggested for use in determining the amount of protein in a solution in the presence of nucleic acids [39]:

$$\text{protein concentration (mg/mL)} = 1.55A_{280} - 0.76A_{260} \quad (10)$$

It must be clear that only an approximate value will be obtained for most preparations by use of this equation, since it was established for enolase in the presence of yeast nucleic acid. (For pure yeast nucleic acid, $A_{280}/A_{260} = 0.49$; for pure enolase, $A_{280}/A_{260} = 1.75$.)

Determination of protein content by absorbance at 280 nm has the advantages that it does not destroy the enzyme, it is rapid, and it is reasonably sensitive (0.1 to 0.5 mg/mL). However, it has the disadvantage that proteins vary quite widely in their absorbance at 280 nm. Most proteins have $E_{280}^{0.1\%}$ of 0.5 to 2.5, where $E_{280}^{0.1\%}$ is the absorbance of a 0.1% solution of protein at 280 nm in a 1-cm cuvette.

Sensitivity can be improved by determining protein concentration in the region of 215 to 225 nm. For example, $E^{0.1\%}$ is 0.58, 5.0, and 11.7 at 280, 225, and 215 nm, respectively, for the solution of bovine serum albumin shown in Fig. 27. Another advantage is that the extinction coefficient is nearly independent of the nature of the protein at these lower wavelengths. A disadvantage is that many other substances absorb in the region 210 to 230 nm, including certain buffer salts, and the wavelength must be set with precision because of the rapid change in absorbance in this region. Because of this, Whitaker and Granum [40] recommended use of 235 and 280 nm for protein concentration according to Eq. (11). It turns out that aromatic residues have the same absorbance at 280 and 235 nm, thereby correcting the 235 nm absorbance, and other compounds do not interfere so much at 235 nm.

$$\text{Protein concentration (mg/mL)} = \frac{A_{235} - A_{280}}{2.51} \tag{11}$$

7. Drying and Weighing

With a good analytical balance the dry weight content of a sample can be accurately determined to the nearest 0.1 mg and with a micro electrobalance to 0.01 mg. This method consists of placing measured aliquots of the preparation on tared planchets, drying to constant weight, and determining total solid material of the sample by weight change. The method has the advantage that large numbers of samples can be done simultaneously and that all the solid material of the sample, including any salts, and so on, are determined. It has the disadvantage of requiring several hours before the results are known. Since salts contribute to the dry weight, it is necessary to dialyze the preparation exhaustively before analysis by this method.

8. Chromic Acid Oxidation

In 1949, Johnson [41] proposed an excellent method for determining the total organic material present in a preparation. The method is fast, precise, and sensitive. The range of organic material needed is about 5 to 40 μg. A fixed amount of chromic acid is added to an aliquot containing organic material. On heating, chromic acid is reduced to Cr_2O_3, the color changes from yellow to green and decreases in absorbance are read at 440 nm. The method may be standardized against a known concentration of sucrose or bovine serum albumin. The amount of Cr_2O_3 produced is independent of the chemical nature of the organic material (carbohydrate, lipid, or protein).

VII. CRITERIA OF PURITY

When is an enzyme solution pure and further application of purification techniques futile? This is often a difficult question to answer, but there are some techniques which, when properly applied, give definitive information about purity of an enzyme, both as a protein and as active enzyme.

A. Electrophoresis

Two of the most definitive determinations of purity are regular polyacrylamide gel electrophoresis and isoelectric focusing (see Section H.2). If a protein migrates as a single band at two or more pH values and over a range of gel concentrations sufficient to cause separation on the basis of size as well as charge, the protein is considered to be pure. A single band by isoelectric focusing electrophoresis indicates purity.

B. Operational Normality of an Enzyme Solution

In Chapter 7, Section III, methods are described that permit determination of the purity of an enzyme based on titration of the active site of the enzyme. The rationale of the method is to detect how much enzyme intermediate (EA′) or product (P_1) is formed when the enzyme is permitted to go through the sequence of converting substrate to product only one time, which means that $k_3 \ll k_2$ or zero.

$$\mathrm{E} + \mathrm{A} \underset{k_{-1}}{\overset{k_1}{\rightleftharpoons}} \mathrm{E{\cdot}A} \xrightarrow{k_2} \underset{\substack{+\\ \mathrm{P_1}}}{\mathrm{EA'}} \xrightarrow{k_3} \mathrm{E} + \mathrm{P_2} \tag{12}$$

Methods for accomplishing this are described in Chapter 7. Of all the methods of purity listed in this section, this and a special procedure described in Section VII.C are the only ones that specifically determine the true purity of the enzyme as a catalyst. The other methods determine purity of the enzyme as a protein or, if based on specific activity, on a relative basis.

C. Chromatographic Behavior

Important information about purity of an enzyme preparation can be obtained from the symmetry and specific activity of the protein elution peak (Figs. 28 and 29). In the pure preparation, the protein peak is quite symmetrical and the specific activity is constant over the entire peak, which is not true of the impure preparation (Fig. 29), where the impurity is coming off the column slightly later than the enzyme.

Special affinity chromatographic procedures on columns containing covalently bound competitive enzyme inhibitors permit a determination of purity of enzyme preparations on the basis of their active site properties (see Refs. 10 to 12). Only active enzyme binds to the column. The peak that elutes in the void volume is inactive protein, while that eluted by washing the column with soluble competitive inhibitor is active enzyme.

D. Specific Activity

Specific activity determinations under rigorously controlled reaction conditions can be used to determine the percent active enzyme, provided that the specific activity of 100% active enzyme is known. For example, if 100% active enzyme is known to have 1000 units/mg and another preparation has 650 units/mg, the latter preparation is 65% active enzyme.

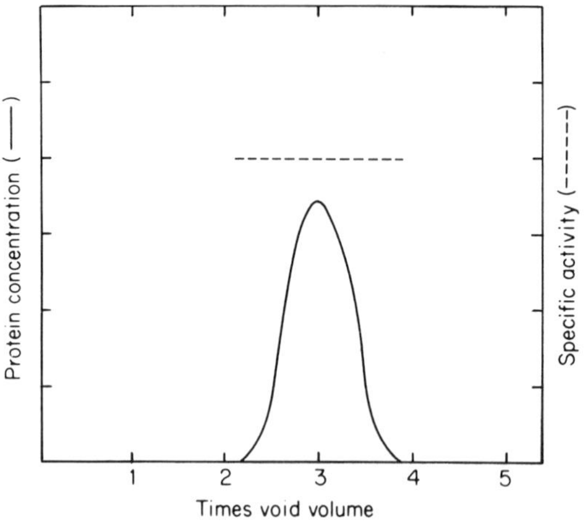

Figure 28 Schematic representation of results expected from chromatography of a pure enzyme on an ion-exchange column by starting buffer elution. Criteria of purity include symmetry of protein peak and constancy of specific activity.

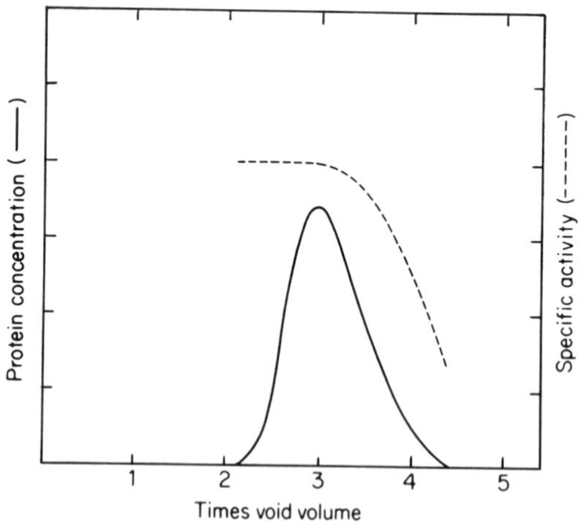

Figure 29 Schematic representation of results expected from chromatography of an impure enzyme preparation on an ion-exchange column by starting buffer elution. Heterogeneity is indicated by the asymmetric protein peak and the decrease in specific activity on the right side of the curve.

E. Presence of Other Enzymes

An enzyme is purified from a crude preparation containing large numbers of other enzymes. It is often overlooked that one excellent method of establishing enzymatic purity, but not necessarily protein purity, is to show that all other enzyme activities are absent. Since enzyme activities can be detected at a very low concentration of protein this is an excellent method.

F. Crystallinity

Crystallinity is an excellent criterion of purity for small molecules but with proteins that have high molecular weights this is not true. Proteins, on crystallizing, still contain considerable amounts of impurities. Many isozymes have very similar or identical crystal structures, so that they cocrystallize. An enzyme can crystallize in several forms, depending on experimental conditions.

VIII. LOSS OF ENZYME ACTIVITY DURING PURIFICATION

At some step in purification all or a large part of the desired enzymatic activity may be lost. This may be due to denaturation of protein. If so, the procedure must be modified to prevent denaturation either by eliminating this step or by changes in pH, temperature, and so on. Enzyme activity may also be lost because of removal of an essential cofactor (see Chapter 12) during purification. This may happen after a dialysis step, for example. Or the measured activity may depend on the action of more than one type of protein. In either case, combining the two fractions obtained at this stage of purification should restore part or all of the enzymatic activity. This situation is frequently encountered in enzyme purification and the investigator should be forewarned and prepared to take appropriate action when appreciable loss of activity is observed.

REVIEW QUESTIONS

1. In separation of subcellular organelles by differential centrifugation it is essential to have methods for determining which organelles are present in a fraction. How would you accomplish this experimentally?

2. During ripening of an apple there are a number of enzyme activities which increase quite markedly. List some of these. Devise an experiment or experiments that would permit you to distinguish whether this increased activity is due to *de novo* synthesis of enzyme, conversion of inactive protein to active enzyme, or to increased accessibility of enzyme to substrate.

3. A solution is known to contain a mixture of three proteins with isoelectric points of 5.0, 7.0, and 8.1. Design an experiment using ion-exchange chromatography that would be expected to separate the three.

4. Could dialysis be used to separate proteins? How would you design a dialysis experiment that would be effective in such a separation? What would be the most important difference among the proteins that you would use?

5. What are some advantages and disadvantages of the use of sodium chloride in place of ammonium sulfate in fractional precipitation of proteins?

6. Preparations of several times crystallized trypsin which contain approximately 50% $MgCl_2$ are often sold by biochemical supply companies. Would you classify these preparations as pure, 50% pure, or otherwise? What is the basis for your answer?

7. A number of proteins contain, in addition to amino acids, considerable amounts of carbohydrates. How would you determine whether the carbohydrate found in a protein preparation is the result of contamination of the preparation or an integral part of the molecule?

8. What would be the criteria you would use to decide whether two isolated enzymes are isozymes or two different enzymes?

9. Under what conditions do you think it would be preferable to use isopropanol rather than ammonium sulfate for fractionation of proteins?

10. In column chromatography it is necessary to determine protein and enzyme activity in a large number of fractions. Specifically, what assay for protein would you be inclined to select? What criteria would you set in selection of a suitable activity assay under these conditions?

11. A solution at pH 7.0 contains three different proteins: A, with molecular weight of 100,000 and a net positive charge of 5; B, with a molecular weight of 40,000 and a net positive charge of 1; and C, with a molecular weight of 20,000 and a net positive charge of 2. On the basis of these properties depict graphically the separation expected at pH 7.0 by

a. Zone electrophoresis on 5% polyacrylamide gel
b. Chromatography on a DEAE-cellulose column
c. Chromatography on a CM-cellulose column
d. Chromatography on a Sephadex G-100 column

12. In a partial purification of pectin methylesterase from tomatoes at 0°C the following data were obtained. Ammonium sulfate solution was added to each tube over a period of several minutes with continuous stirring and the tubes were allowed to stand for 1 hour. They were centrifuged, the supernatant liquid removed and analyzed. For activity, 1.0 mL of solution was used. Absorbance at 280 nm was measured on the supernatant liquid without dilution.

(a) Calculate the expected activity and absorbance of the precipitate when dissolved in 10 mL of water. Is this the best way to get these data?

(b) How would you obtain the purest enzyme preparation by $(NH_4)_2SO_4$ precipitation?

(c) What is the specific activity of this purest enzyme fraction?

(d) What is the percent recovery of activity in the purest enzyme fraction?

(e) What is the degree of purification of this purest enzyme fraction?

Tube	Enzyme solution (mL)	Water (mL)	Sat'd $(NH_4)_2SO_4$ (mL)	Supernatant activity (mEq acid/min)	Absorbance (280 nm)
1	3.0	7.0	0.0	0.500	0.650
2	3.0	5.0	2.0	0.486	0.450
3	3.0	4.0	3.0	0.396	0.400
4	3.0	3.0	4.0	0.152	0.350
5	3.0	2.0	5.0	0.128	0.308
6	3.0	1.0	6.0	0.120	0.256

REFERENCES

General References

Altgelt, K. H., Theory and mechanism of gel permeation chromatography, in *Advances in Chromatography* (J. C. Giddings and R. A. Keller, eds.), Vol. 7, Marcel Dekker, New York, 1968.

Angeletti, R. H., ed., *Techniques in Protein Chemistry III*, Academic Press, San Diego, 1992.

Bruno, T. J., *Chromatography and Electrophoretic Methods*, Prentice Hall, Engelwood Cliffs, N.J., 1991.

Deutscher, M. P., ed., Guide to protein purification, *Methods Enzymol.*, **182** (1990).

Everson Pearse, A. G., *Histochemistry: Theoretical and Applied*, Vol. I, Little, Brown, Boston, 1968.

Fischer, L., *An Introduction to Gel Chromatography*, American Elsevier, New York, 1969.

Grossman, P. D., and J. C. Colburn, eds., *Capillary Electrophoresis: Theory and Practice*, Academic Press, San Diego, 1991.

Harris, E. L. V., and S. Angal, eds., *Protein Purification Methods: A Practical Approach*, IRI Press, Oxford, 1989.

Kennedy, A., and S. Fowell, eds., *Practical Protein Chromatography*, Humana Press, Totowa, N.J., 1992.

Scopes, R. K., *Protein Purification: Principles and Practice*, 2nd Ed., Springer-Verlag, New York, 1987.

Shaw, D. J., *Electrophoresis*, Academic Press, New York, 1969.

Villafranca, J. J., ed., *Techniques in Protein Chemistry II*, Academic Press, San Diego, 1991.

Cited References

1. G. E. Hobson, *Nature*, **195**, 804 (1962).
2. A. M. Siddiqi and A. L. Tappel, *J. Am. Oil Chemists' Soc.*, **34**, 529 (1957).
3. M. A. Joslyn and E. Zuegg, *J. Assoc. Offic. Agr. Chemists*, **39**, 267 (1956).

4. B. A. Dave, Studies on polygalacturonic acid *trans*-eliminase produced by *Bacillus pumilus*, Ph.D. thesis, University of California, Davis, Calif., 1968.
5. J. Brachet, *Sci. Am.*, **205**(3), 50 (1961).
6. C. C. Cooper, R. G. Cassens, and E. J. Briskey, *J. Food Sci.*, **34**, 299 (1969).
7. L. Boscan, W. D. Powrie, and O. Fennema, *J. Food Sci.*, **27**, 574 (1962).
8. R. K. Morton, *Nature*, **166**, 1092 (1950); *Biochem. J.*, **55**, 795 (1953); *Biochem. J.*, **57**, 595 (1954).
9. H. C. Isliker, *Ann. N.Y. Acad. Sci.*, **57**, 225 (1953); *Advan. Protein Chem.*, **12**, 387 (1957); L. H. Kent and J. H. R. Slade, *Biochem. J.*, **77**, 12 (1960).
10. J. C. Beeley and H. Neurath, *Biochemistry*, **7**, 1239 (1968).
11. S. Blumberg, I. Schechter, and A. Berger, *European J. Biochem.*, **15**, 97 (1970).
12. P. Cuatrecasas, *J. Biol. Chem.*, **245**, 3059 (1970).
13. A. A. Green, *J. Biol. Chem.*, **95**, 47 (1932).
14. E. J. Cohn and J. D. Ferry, in *Proteins, Amino Acids and Peptides as Ions and Dipolar Ions* (E. J. Cohn and J. T. Edsall, major contributors), Reinhold, New York, 1943, p. 586.
15. A. Grönwall, *Compt. Rend. Trav. Lab. Carlsberg*, **24**(8), 185 (1942).
16. R. Willstätter and H. Kraut, *Ber.*, **56**, 1117 (1923).
17. D. Keilin and E. F. Hartree, *Proc. Roy. Soc. (London)*, **B124**, 397 (1938).
18. S. Paleus and J. B. Nielands, *Acta Chem. Scand.*, **4**, 1024 (1950).
19. E. A. Peterson and H. A. Sober, *J. Am. Chem. Soc.*, **78**, 751 (1956).
20. M. B. Rhodes, P. R. Azari, and R. E. Feeney, *J. Biol. Chem.*, **230**, 399 (1958).
21. V. C. Sgarbieri, S. M. Gupte, D. E. Kramer, and J. R. Whitaker, *J. Biol. Chem.*, **239**, 2170 (1964).
22. M. P. Deutscher, ed., Guide to protein purification, *Methods Enzymol.*, **182** (1990).
23. R. L. M. Synge and A. Tiselius, *Biochem. J.*, **46**, XLI (1950).
24. P. Flodin, *Dextran Gels and Their Application in Gel Filtration*, Pharmacia, Uppsala, Sweden, 1962.
25. J. R. Whitaker, *Anal. Chem.*, **35**, 1950 (1963).
26. *Sephadex-Gel Filtration in Theory and Practice*, Pharmacia Fine Chemicals, Inc., Piscataway, N.J., Dec. 1966-1.
27. Price List U, Bio-Rad Laboratories, Richmond, Calif., Sept. 1969.
28. H. K. Schachman, in *Methods in Enzymology* (S. P. Colowick and N. O. Kaplan, eds.), Vol. 4, Academic Press, New York, 1957, p. 32.
29. R. A. Phelps and F. W. Putnam, in *The Plasma Proteins* (F. W. Putnam, ed.), Vol. 1, Academic Press, New York, 1960, p. 143.
30. F. W. Putnam, in *The Proteins* (H. Neurath, ed.), Vol. 3, Academic Press, New York, 1965, p. 153.
31. J. T. Clark, *Ann. N.Y. Acad. Sci.*, **121**(2), 428 (1964).
32. *Ultrafiltration for Laboratory and Clinical Uses*, Publication 403, Amicon Corporation, Lexington, Mass., 1970.
33. R. K. Morton, *Biochem. J.*, **55**, 795 (1953).
34. A. G. Gornall, C. J. Bardawill, and M. M. David, *J. Biol. Chem.*, **177**, 751 (1949).
35. O. H. Lowry, N. J. Rosebrough, A. L. Farr, and R. J. Randall, *J. Biol. Chem.*, **193**, 265 (1951).

36. M. Bradford, *Anal. Biochem.*, **72**, 248 (1976).
37. P. K. Smith, R. I. Krohn, G. T. Hermanson, A. K. Mallia, F. H. Gartner, M. D. Provenzano, F. K. Fujimoto, N. M. Goeke, B. J. Olson, and D. C. Klenk, *Anal. Biochem.*, **150**, 76 (1985).
38. J. R. Whitaker, unpublished data, 1970.
39. O. Warburg and W. Christian, *Biochem. Z.*, **310**, 384 (1941).
40. J. R. Whitaker and P. E. Granum, *Anal. Biochem.*, **109**, 156 (1980).
41. M. J. Johnson, *J. Biol. Chem*, **181**, 707 (1949).

4
Active Sites and Factors Responsible for Enzyme Catalysis

I. ACTIVE SITES OF ENZYMES

Before discussing the parameters that influence rates of enzyme-catalyzed reactions it is quite important to achieve a visual picture of the active site of an enzyme. We shall depict the active site as being composed of a binding locus and a transforming locus. Binding may take place without transformation but transformation cannot take place catalytically without binding. Separating the active site into binding and transforming loci does not exclude binding from contributing to the activation energy required for substrate to be converted to product.

A. Binding Locus

Enzymes are quite specific for the compounds that serve as substrates. Small changes in structure of a substrate molecule may lead to inability of an enzyme to convert the compound to products. These findings led Fischer to propose a "lock-and-key" model to explain enzyme specificity [1]. The substrate is analogous to the key, the lock to the enzyme. According to this model, there is on the surface of the protein a region (a template) that is complementary to the structure of the substrate. This complementary nature may involve size, shape, and/or charge. A molecule whose structure is exactly complementary to this template area will bind to the enzyme. When the susceptible bond in the substrate is positioned properly with respect to the transforming site, the substrate may be converted to product.

Exercise 1. Sketch an illustration depicting what the lock-and-key analogy of enzyme–substrate combination means to you.

1. Stereospecific Binding to Active Site

A substance must meet the very rigid restrictions of structure as imposed by the enzyme before it is bound into the binding locus. There are distinct recognition areas within the binding locus. Consider, for example, the molecule CH_3CH_2OH. Ethanol has no asymmetric carbon atoms, yet alcohol dehydrogenase must recognize a difference between the two methylene hydrogens since it stereospecifically removes the same hydrogen atoms (one methylene hydrogen and the hydroxyl hydrogen) from the molecule to produce acetaldehyde. It does this repeatedly when the reaction is reversed several times in each direction. The same hydrogen of the methylene group is always removed or added. The enzyme must recognize and bind to more than one part of the ethanol molecule simultaneously (Fig. 1). When ethanol is so attached to the enzyme surface, the position of the second methylene hydrogen is fixed. Orientation of this hydrogen is quite critical as it must be positioned in proximity to the C(4) of the nicotinamide moiety of

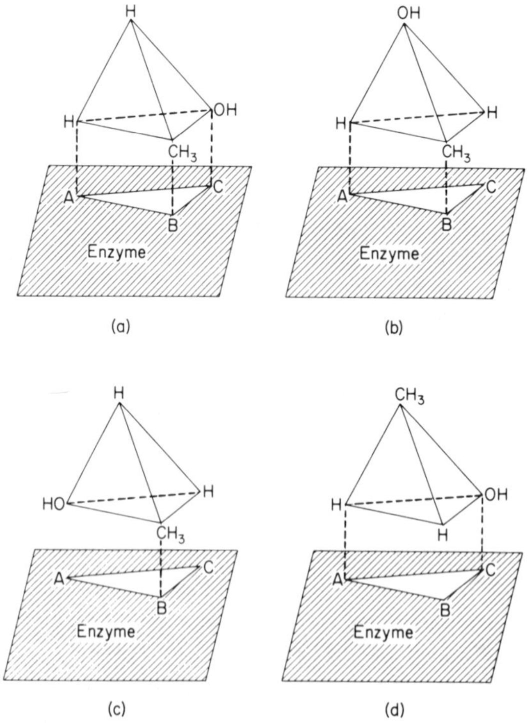

Figure 1 Schematic representation of several possible modes of orientation of ethanol on the surface of alcohol dehydrogenase.

NAD^+ (Fig. 2). Other possible modes of combination of ethanol with the enzyme are depicted in Fig. 1. Only (a) leads to correct alignment for transformation to product and the complex is also the most stable thermodynamically. Both isomers of a compound, such as a D- and an L-amino acid, may bind at the binding locus of the active site but one of these will lead to improper positioning with respect to the transforming locus. The improperly oriented compound is not a substrate but a competitive inhibitor.

2. Active Site of Carboxypeptidase A

Figure 3 shows some proposed binding modes of substrate in the active site of carboxypeptidase A. In this model based on experimental data, the active site contains a hydrophobic "pocket," Zn(II), arginines 71 and 145, phenylalanine 279, and tyrosines 198 and 248. While Zn(II) may be involved in binding substrate, it is certainly involved in catalytic hydrolysis of substrate.

One of the best substrates for carboxypeptidase A is the compound benzyloxycarbonylglycyl-L-phenylalanine, shown in Fig. 3. For the substrate to be hydrolyzed it must be bound correctly so that the carbonyl group supplied by glycine interacts with Zn(II). The substrate is bound to the enzyme in Fig. 3(a) and (b), but this requirement for correct positioning of the carbonyl group is not met. In Fig. 3(a) the carboxyl group has interacted with Zn(II), while in 3(b) the substrate is oriented backward in the active site with the wrong carbonyl group interacting with Zn(II). Proper binding of substrate is shown in Fig. 3(c) and (d). In 3(c) two molecules of substrate are attempting to occupy the same space, one in a correct position, the other incorrectly. The molecule showing incorrect binding is a competitive inhibitor of the other molecule since if it is bound, the other substrate molecule cannot be bound, and vice versa. Only binding as shown in Fig. 3(d) is productive binding.

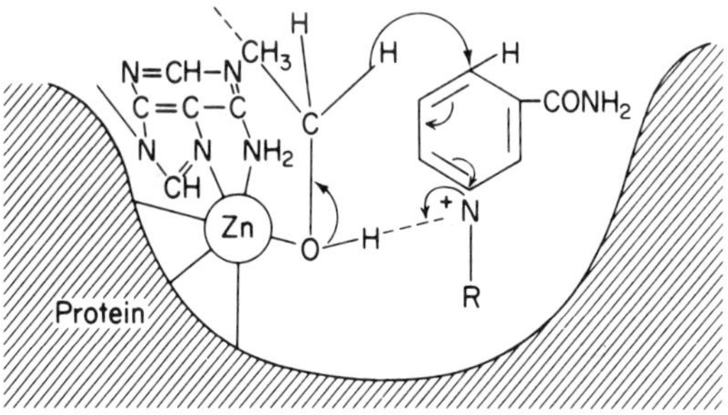

Figure 2 Proposed transition state in the oxidation of ethanol to acetaldehyde by alcohol dehydrogenase. Notice the stereospecific orientation of ethanol, NAD^+, and Zn(II) so that a hydrogen atom is transferred stereospecifically to only one position on C(4) of the nicotinamide ring. The R group is the remainder of NAD^+ with only adenine and nicotinamide shown. (From Ref. 2, p. 1863, by courtesy of Acta Chemica Scandinavica.)

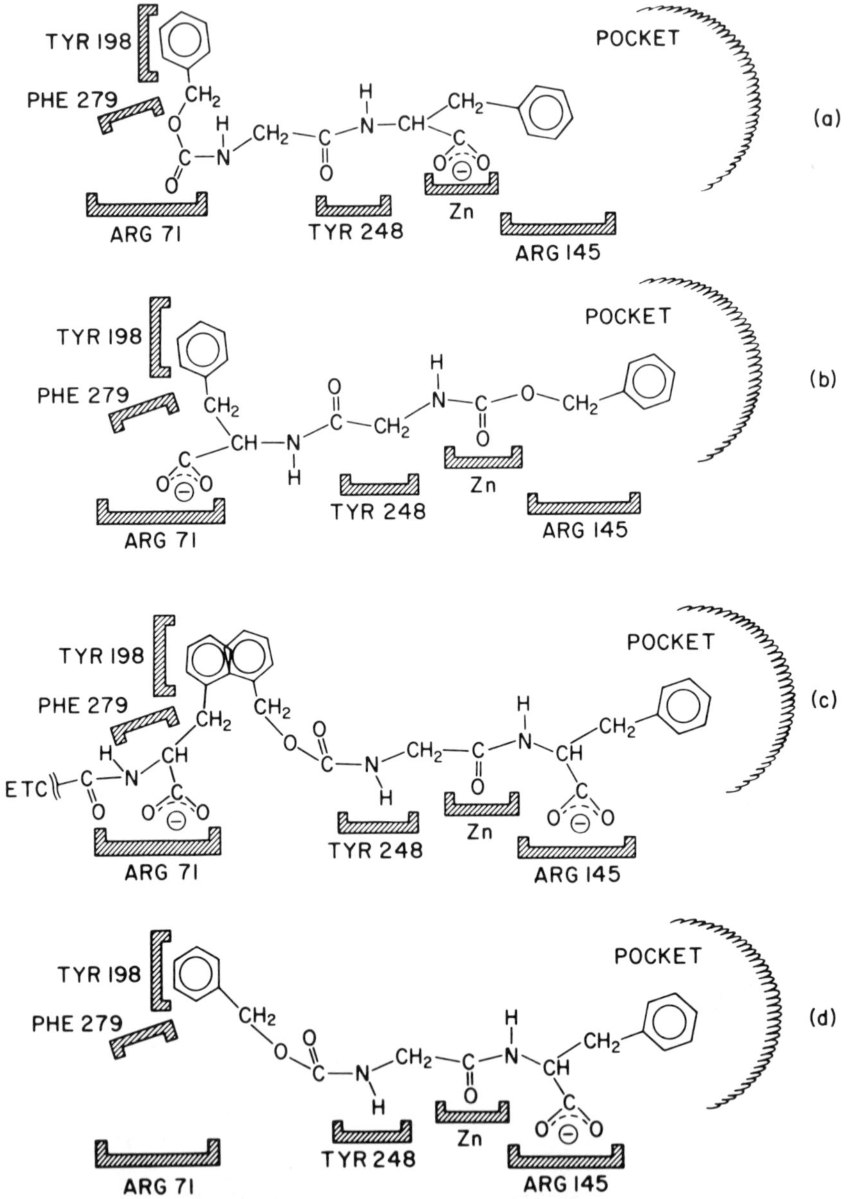

Figure 3 Proposed modes of binding of benzyloxycarbonylglycyl-L-phenylalanine into the active site of carboxypeptidase A. (From Ref. 3, p. 80, by courtesy of Brookhaven National Laboratory.)

While only Zn(II), Arg-71 and Arg-145, Phe-279, and Tyr-198 and Tyr-248 appear to be involved in binding of benzyloxycarbonylglycyl-L-phenylalanine in the active site of carboxypeptidase A, additional binding positions are utilized with a larger substrate. Systematically increasing the length of the substrate by adding amino acid residues to the glycyl residue of glycyl-L-phenylalanine indicates that as many as five amino acid residues may be involved in binding of substrate to enzyme. The generalization may be drawn from these data that the measured size of the binding locus of an enzyme may be dependent on the size and nature of the substrate.

3. Active Site of Lysozyme

Binding of a hexa-*N*-acetylchitohexaose in the active site of lysozyme is shown in Fig. 4. The substrate (Fig. 5) lies in a long cleft on the lysozyme surface with the reducing end of the molecule pointing downward. The susceptible glycosidic bond is between residues D and E and lies approximately 3 Å from the carboxyl groups of Asp-52 and Glu-35 of the enzyme, which are involved in transforming the substrate to product. The 12 groups on the substrate that may interact with enzyme are shown in Fig. 5 and the number of possible contact points between substrate and enzyme are given in Table 1. The number of possible interactions is impressive. With the number of points of contact between the enzyme and substrate so large, it is not surprising that stereospecific requirements are so critical.

The major types of bonds involved in substrate binding to enzyme are hydrogen bonds, hydrophobic bonds, and electrostatic bonds. The strengths of these bonds are indicated in Table 2, where the covalent bond is included for comparison. Covalent bonds *are not involved* in adsorptive complex formation between enzyme and substrate.

B. Flexibility of the Active Site

There are a number of examples in which the classical theory of the lock-and-key model of Fischer are in conflict with experimental data. Koshland developed an alternative theory which explains not only results that agree with the lock-and-key theory but also those data that conflict with this theory [6–8]. The postulates of Koshland's "induced fit" model are: (a) there is a considerable change in the geometry of the protein as the substrate binds into the active site; (b) a delicate orientation of catalytic groups is required for transformation to take place; and (c) the substrate "induces" this proper orientation by the change in geometry it causes on binding into the active site.

A model for the induced fit theory is shown in Fig. 6. In (a) the enzyme and substrate are shown separated before formation of the adsorptive complex. There is little in the conformation of the active site that is complementary to the substrate structure. In (b) the substrate is shown adsorbed to the enzyme. As the substrate molecule approaches the enzyme surface it causes the peptide chains to change their position so that they now conform more closely to the shape of the substrate.

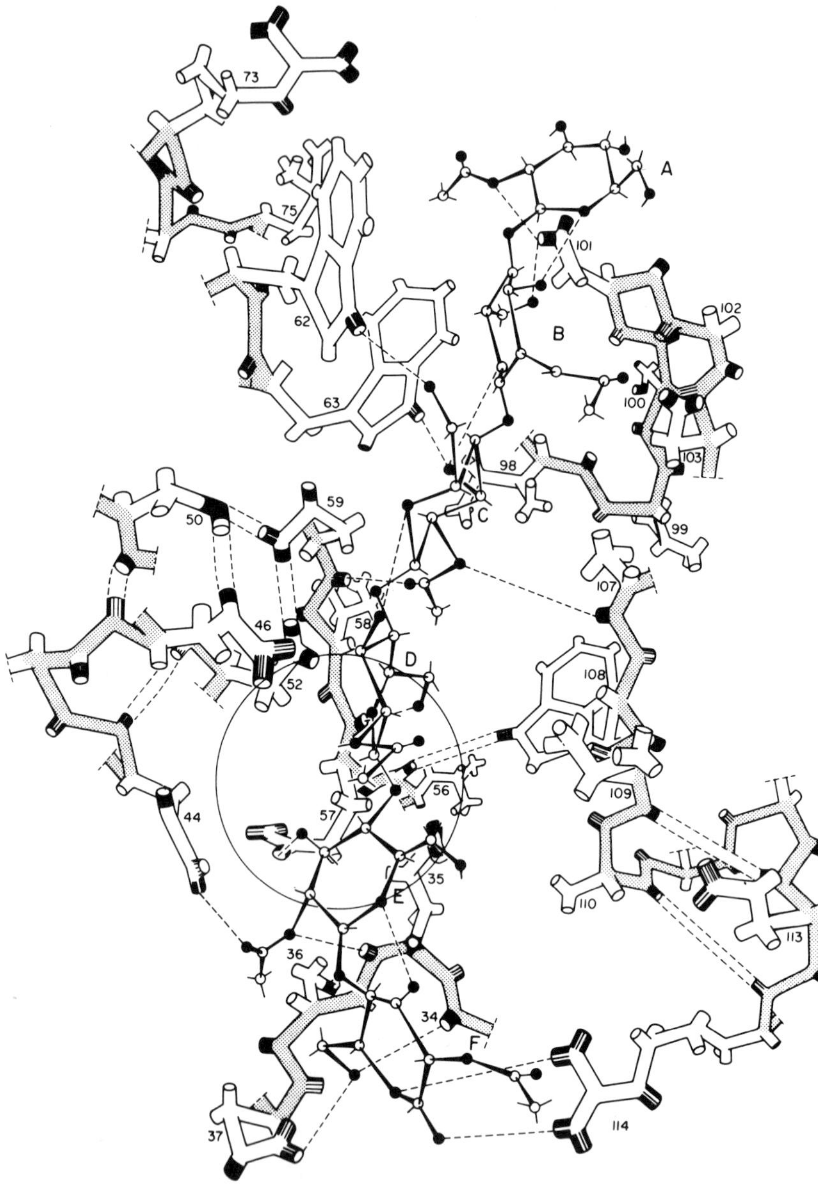

Figure 4 Binding of a hexa-*N*-acetylchitohexaose molecule in the active site of lysozyme. The substrate molecule is oriented vertically in the active site, with the terminal reducing residue, F, at the bottom. The susceptible glycosidic bond is between monosaccharide residues D and E and is in proximity to carboxyl groups of Asp-52 (~3 Å) and Glu-35 (~3 Å). Residues of the enzyme near the substrate are numbered, and proposed interactions between substrate and enzyme are shown by single dashed lines. The double dashed lines show hydrogen bonding between NH and CO on the enzyme. (Reprinted from Ref. 4, p. 122, by courtesy of Brookhaven National Laboratory.)

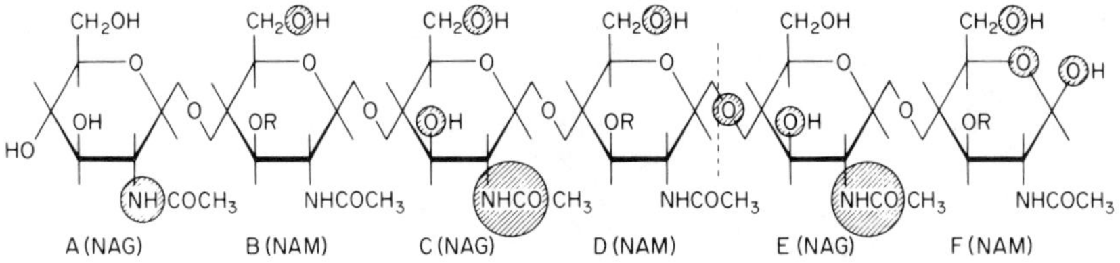

Figure 5 Structure of a hexa-*N*-acetylchitohexaose with repeating NAG:NAM units. The 12 groups on the substrate postulated to interact with groups in the active site of lysozyme are shaded. The susceptible glycosidic bond is between residues D and E (dashed vertical line).

In particular, the A, B, and C groups are brought into close proximity to the substrate molecule. The group C functions in binding the substrate to the enzyme while groups A and B are involved in transforming the susceptible bond in the substrate (shown as the neck of the substrate) to the product. The lower two diagrams show enzyme·compound complexes that are inactive. In both cases the compounds are positioned correctly with respect to the binding group, C, so that

Table 1 Principal Interatomic Contacts Between a Hexa-*N*-acetylchitohexaose Substrate and the Lysozyme Molecule

Monomer unit	Polar contacts[a]	Total number of van der Waals contacts <4 Å
A	NH—Asp-101	7
B	O(6)—Asp-101	11
C	O(6)—Try-62	30
	O(3)—Try-63	
	NH—CO-107	
	CO—NH-59	
D	O(6)—CO-57	35
	O(1)—Glu-35	
E	O(3)—Gln-57	45
	NH—CO-35	
	CO—Asn-44	
F	O(6)—CO-34	13
	O(6)—Asn-37	
	O(5)—Arg-114	
	O(1)—Arg-114	

Source: Ref. 4, p. 124, by courtesy of Brookhaven National Laboratory.
[a]The first listed group is on the substrate; the second group with amino acid residue number is on the enzyme.

Table 2 Energies and Bond Distances of Several Types of Bonds Between Enzyme and Substrate in Enzyme·Substrate Complexes

Type of bond	Energy (kcal/mol)	Distance of interaction (Å)
Hydrogen	2–10	2–3
Electrostatic	10–20	2–3
Hydrophobic	1–3	3–5
Covalent[a]	30–100	1–2

Source: Ref. 5.
[a]Included for comparison only.

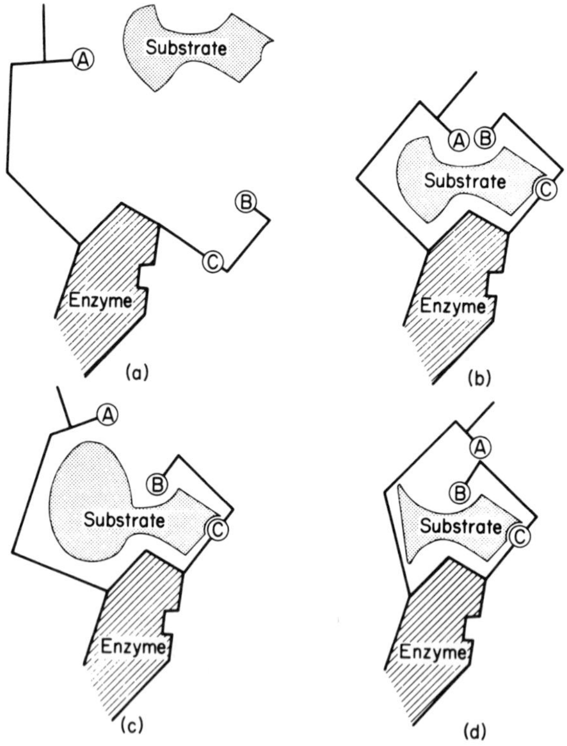

Figure 6 Schematic model depicting the induced fit hypothesis. Black lines indicate protein chain containing catalytic groups A and B and substrate binding group C. See the text for an explanation. (From Ref. 7, p. 724, by courtesy of the Federation of American Societies of Experimental Biology.)

adsorptive complexes are formed. However, the shapes of the compounds prevent the catalytic groups A and B from being properly positioned with respect to the susceptible bond of the substrate. These compounds would compete with the substrate for binding at the active site.

There are many data to support the induced fit theory of Koshland, which involves a change in conformation of the active site on binding of substrate. The evidence includes changes in reactivity of groups within the active site, changes in fluorescent and absorbance properties of groups, and changes demonstrable by thermodynamic measurements on binding of substrate. Perturbation methods provide additional evidence for conformational changes. For example, several data indicate that aspartate transcarbamylase undergoes one or more conformational changes on binding of substrate and during transformation of substrate to product [9].

C. Efficiency of Enzyme–Substrate Combination

Although a molecule has all the prerequisites to be a substrate, it is unreasonable to expect that all adsorptive complexes formed between enzyme and substrate will be productive (i.e., lead to product). As shown in Figs. 1 and 3, the substrate may bind in the active site in incorrect positions. Only those complexes that correctly place the susceptible bond in proximity to the catalytic groups will be productive.

D. Transforming Locus of Active Site

We have alluded several times to the requirement that the correct mode of binding bring the susceptible bond adjacent to groups involved in transformation of substrate to product. The groups involved in the transforming locus of enzymes are those contributed by the side chains of amino acid residues such as the hydroxyl group of serine, the sulfhydryl group of cysteine, the imidazole group of histidine, the carboxyl groups of aspartic and glutamic acids, and the ϵ-amino group of lysine. In some enzymes there are specific requirements for cofactors. A cofactor may be involved in binding of substrate to the enzyme, in transformation of substrate(s) to product(s), or it may play a role in both capacities.

1. Formation of Covalent Intermediates

Participation of groups of the active site in transformation of substrate to product may result in formation of a covalently bound intermediate between enzyme and the substrate. This has not been demonstrated in all enzyme-catalyzed reactions; therefore, enzymes may be divided into two groups, those in which a covalent intermediate has been demonstrated (or is expected) and those in which a covalent intermediate has not been found. The hydrolytic enzymes all appear to form covalent intermediates which are then hydrolyzed by participation of the nucleophile, water. A general rule is that nucleophilic–electrophilic mechanisms of catalysis involve covalent intermediates, while general acid–general base mechanisms of catalysis do not involve covalent intermediates.

Evidence for covalent bond formation is substantial. In the case of several of the proteolytic enzymes, for example, chymotrypsin, trypsin, and papain, the intermediates have been isolated and their properties studied (see Chapter 19). In other cases, as for example the aldolases, a stable covalent intermediate can be isolated following reduction with $NaBH_4$. The covalent intermediate is a Schiff's base that is not stable enough to permit isolation [Eq. (1)]. However, on reduction a stable bond is formed [Eq. (2)], the intermediate can be isolated and the position of the substrate residue located on an ϵ-amino group of a particular lysine in aldolase.

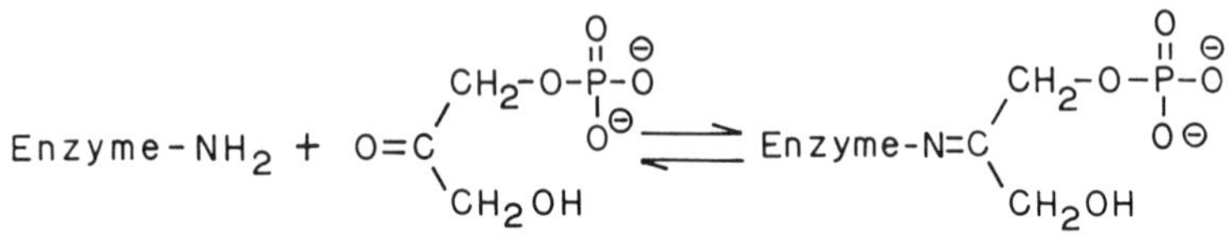
(1)

Dihydroxyacetone phosphate

Schiff's base

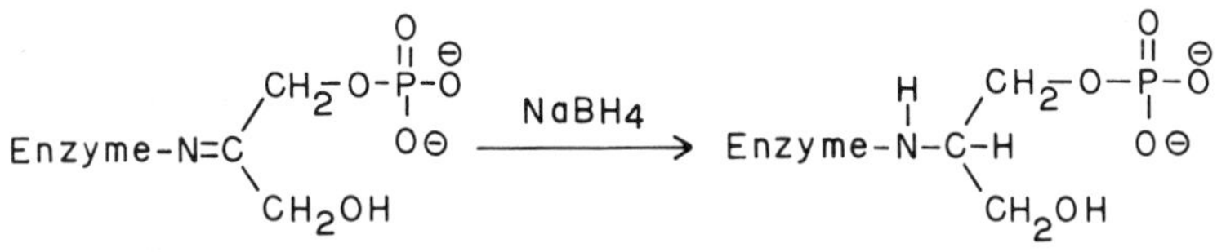
(2)

In some cases, binding of substrate stereospecifically on the enzyme surface may produce strain in the susceptible bond (such as in the rack mechanism; see Section II.B), while at the same time there may be bonding between the substrate and the groups involved in transformation. However, in visualizing the mechanism of action of an enzyme it is often advantageous to talk about the binding and transforming loci separately.

E. Size of the Active Site

Definition of size of the active site is somewhat arbitrary. The number of interacting positions between substrate and enzyme on binding is often a function of substrate size. In a large polymer such as a starch, protein, or nucleic acid molecule a very large number of positions may be interacting with the enzyme. Granted this, there is still a large portion of the peptide chain of the enzyme that cannot be in direct contact with substrate. How does the remainder of the protein molecule function?

The groups involved in the active site are usually not in proximity in terms of the primary sequence of the protein chain. This is shown in Figs. 3 and 4. In carboxypeptidase A, the groups thought to be involved in the active site are His-69, Glu-72, and His-196 [which function in binding the Zn(II)], Zn(II), Arg-71, Arg-145, Tyr-198, Tyr-248, and Phe-279. These groups are well separated in terms of primary sequence but must be brought in proximity through folding of protein into the tertiary structure. Therefore, a large part of the remainder of the peptide chain is involved in maintaining the conformational features of the active site (i.e., in keeping all these groups properly positioned in the active site). A change in the relative position of one of the groups by as little as a fraction of 1 Å may be sufficient to cause loss of activity. However, there is reason to believe that a part of the protein chain in many enzymes is not essential for enzyme activity and that when proper methods are used, this portion can be removed without loss of activity.

In lysozyme Glu-35, Asn-37, Asn-44, Asp-52, Asp-59, Try-62, Try-63, Asp-101, Arg-114, and the carbonyl oxygen of peptide bonds between Phe-34–Glu-35, Gln-57–Leu-58, and Ala-107–Try-108 all appear to be in proximity in the active site. In the enzyme ribonuclease the following groups are known to be involved in the active site: Lys-7, Arg-10, His-12, Lys-41, and His-119. Again, the amino acid residues are well separated in the primary sequence but they are brought into proximity with each other by the tertiary structure.

The number of active sites per enzyme molecule is small. In general, there does not appear to be more than one active site per polypeptide chain. Many enzymes have only one polypeptide chain and these all apparently have one active site. In multichain enzymes there may be as many active sites as there are polypeptide chains, or some of the chains may perform regulatory or other functions, so that the number of active sites is smaller than the number of chains.

II. FACTORS RESPONSIBLE FOR EFFICIENCY OF ENZYME-CATALYZED REACTIONS

The most remarkable thing about an enzyme is its ability to increase the rate of a reaction several orders of magnitude over that of the same reaction in the absence of enzyme. In Chapter 11 we describe this catalytic efficiency of an enzyme as being the result of its ability to lower the activation energy needed for transformation of substrate to products. This statement is true, but it does not help us to see how the enzyme is such an efficient catalyst.

In this section we do not attempt to catalog the many types of specific reactions catalyzed by enzymes with a proposed mechanism for each. Such are well documented in the references given at the end of this chapter; in Chapters 15 to 27 we present specific mechanisms wherever it seems appropriate. In this section we deal with the factors that may account for the efficiency of enzyme-catalyzed reactions.

A. Proximity and Orientation Effects

An enzyme is very specific in selection of substrate. The specificity extends to both the binding and the transforming loci of the active site, as indicated above. How much is this proper positioning of the substrate with respect to the catalytic groups worth in increasing the efficiency of the enzyme? There is no general agreement on the absolute value of the gain in efficiency, but some reasonable approximations can be made.

1. Intra- Versus Intermolecular Catalysis

By binding of substrate the enzyme converts an intermolecular reaction (second- or higher-order reaction) to an intramolecular reaction (first-order reaction). Certainly, it must gain something in entropy by doing this. Attempts to measure the effect of converting an intermolecular reaction to an intramolecular reaction have used models of one kind or another. One type of model is to build into a single molecule both a catalytic group and a susceptible bond. This is shown in the following two examples.

a. Intra- versus intermolecular-catalyzed hydrolysis of a thiol ester

(i)

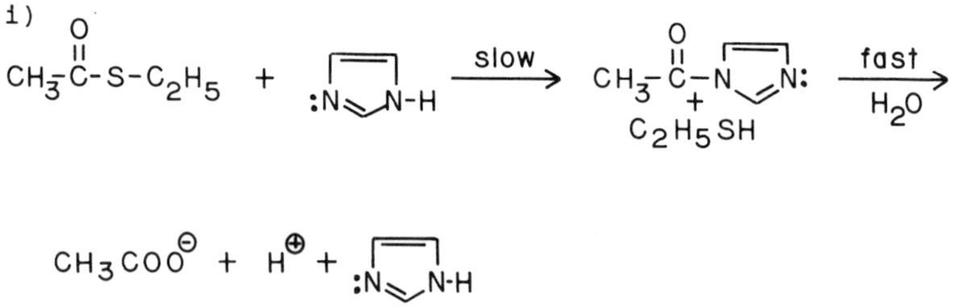

(3)

(ii)

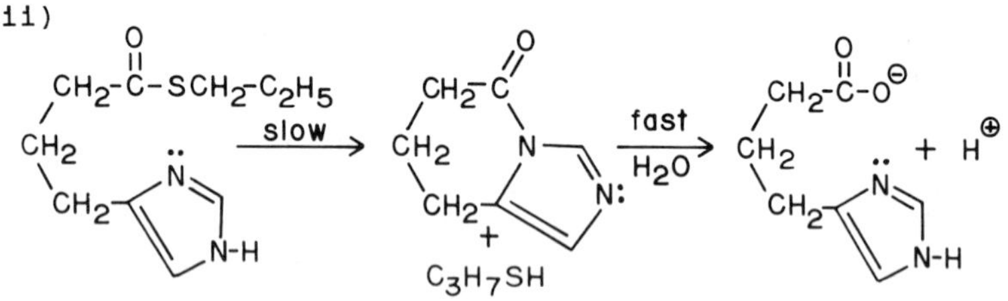

(4)

b. Intra- versus intermolecular-catalyzed hydrolysis of phenylacetate

(i)

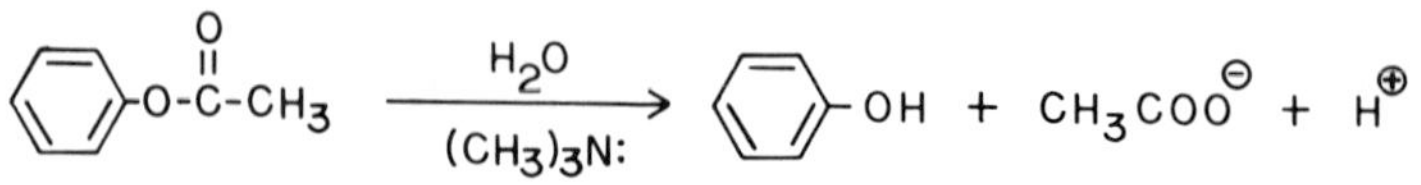

(5)

(ii)

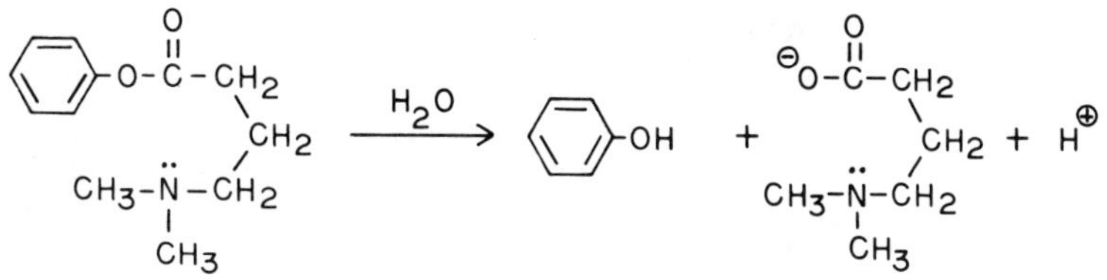

(6)

In example a an imidazole group is the catalyst, whereas in example b a tertiary amino group is the catalyst. There are 100- and 1000-fold enhancements in rate in going from an inter- to an intramolecular reaction in examples a and b, respectively.

Calculations of the effect of orientation and proximity factors on rates of reactions made on theoretical grounds [8] indicate an enhancement of 5.5×10^4 for a bimolecular reaction and 1×10^{22} for a termolecular reaction based on 10^{-3} *M* concentrations of each of the reactants.

2. Effect on Entropy of Reaction

The models presented in Section II.A.1 are not adequate for an enzyme-catalyzed reaction since only a fraction of the total number of molecules will at any time have the catalytic group positioned properly with respect to the carbonyl group, which is the center of attack. The importance of the enzyme holding the substrate in a fixed position can be seen from considering the relative rates of hydrolysis of acetylchymotrypsin and *N*-acetyl-L-tyrosylchymotrypsin (Table 3). These are intermediates formed in chymotrypsin-catalyzed reactions involving the respective substrates, *p*-nitrophenylacetate and *N*-acetyl-L-tyrosine ethyl ester.

N-Acetyl-L-tyrosylchymotrypsin is deacylated 3540 times faster than is acetylchymotrypsin. This marked difference in rate is due entirely to the difference in $\Delta S^{\ddagger}$ values for the two reactions since the $\Delta H^{\ddagger}$ values are essentially the same. It should be recalled that $\Delta G^{\ddagger} = \Delta H^{\ddagger} - T\,\Delta S^{\ddagger}$. The effect on $\Delta S^{\ddagger}$ may be

Table 3 Activation Parameters of Deacylation of Two Acylchymotrypsins

Acylenzyme	Relative k_3	$\Delta G^{\ddagger}$ (kcal/mol)	$\Delta H^{\ddagger}$ (kcal/mol)	$\Delta S^{\ddagger}$ (eu)[a]
N-Acetyl-L-tryrosyl-	3540	14.3	10.3	−13.4
Acetyl-	1	20.4	9.7	−35.9

Source: Ref. 10.
[a]Cal mol^{-1} deg^{-1}.

pictured as shown in Fig. 7, where the bulky *N*-acetyl-L-tyrosyl group has lost its freedom to move due to its immobilization on binding into the binding locus. The small acetyl group is not restricted much in its ability to rotate because of its ill-fit in the binding locus. That this is an essentially correct picture of what is involved can be shown by observing the effect of adding indole to a solution of acetylchymotrypsin. The rate of deacylation is markedly increased because of an increase in $\Delta S^{\ddagger}$ in the presence of indole. Because of the small size of the acetyl group, indole can also bind in the binding locus of acetylchymotrypsin. In so doing, it cuts down on the rotational freedom of the acetyl group and increases the rate of deacylation. The closer the ground state resembles the transition state, the more positive will be $\Delta S^{\ddagger}$ and the faster the rate, all other things being equal.

3. Effective Concentration of Reactive Groups

The *effective* concentration of an intramolecular catalytic group is equivalent to ~10 *M* of the corresponding intermolecular catalyst [11]. This is because the attacking species is positioned close to the scissile bond, where it is most effective. As shown in Fig. 7, the attacking nucleophilic species on the carbonyl group of the acylenzyme is the hydroxide ion, which has been produced from water through participation of the imidazole group of the enzyme in proximity to the carbonyl group it is to attack. There is no effective dilution by the bulk water of the system.

B. Catalysis by Distortion

The concept of substrate activation by distortion on binding to the active site of an enzyme has been proposed repeatedly. This has often been called the "rack mechanism." This theory should not be confused with that in which the electronic configuration of the susceptible bond is distorted because of the participation of acidic and basic groups in the active site (concerted acid-base catalysis; see below) following binding. In the rack mechanism the binding forces between substrate and enzyme are used directly to supply part of the energy to reach the transition state.

The binding locus of the active site of an enzyme and its substrate is shown schematically and hypothetically in Fig. 8. The binding locus is longer than the substrate by 0.5 Å. When the substrate is bound to the enzyme there is lengthening of the susceptible bond between groups A and B by 0.5 Å, leading to the bound substrate more closely resembling the transition state. The process of binding supplies part of the energy required to achieve the transition-state intermediate.

Figure 7 Effect of structure of acyl group on specificity of deacylation of acylchymotrypsin and resemblance of the ground state to the transition-state structure: (a) acyl group is *N*-acetyl-L-tyrosyl; (b) acyl group is acetyl; and (c) acyl group is acetyl with indole also bound in the active site. The acyl group is covalently bound to Ser-195 of the enzyme. (Modified from Ref. 10, p. 3718, by courtesy of the American Chemical Society.)

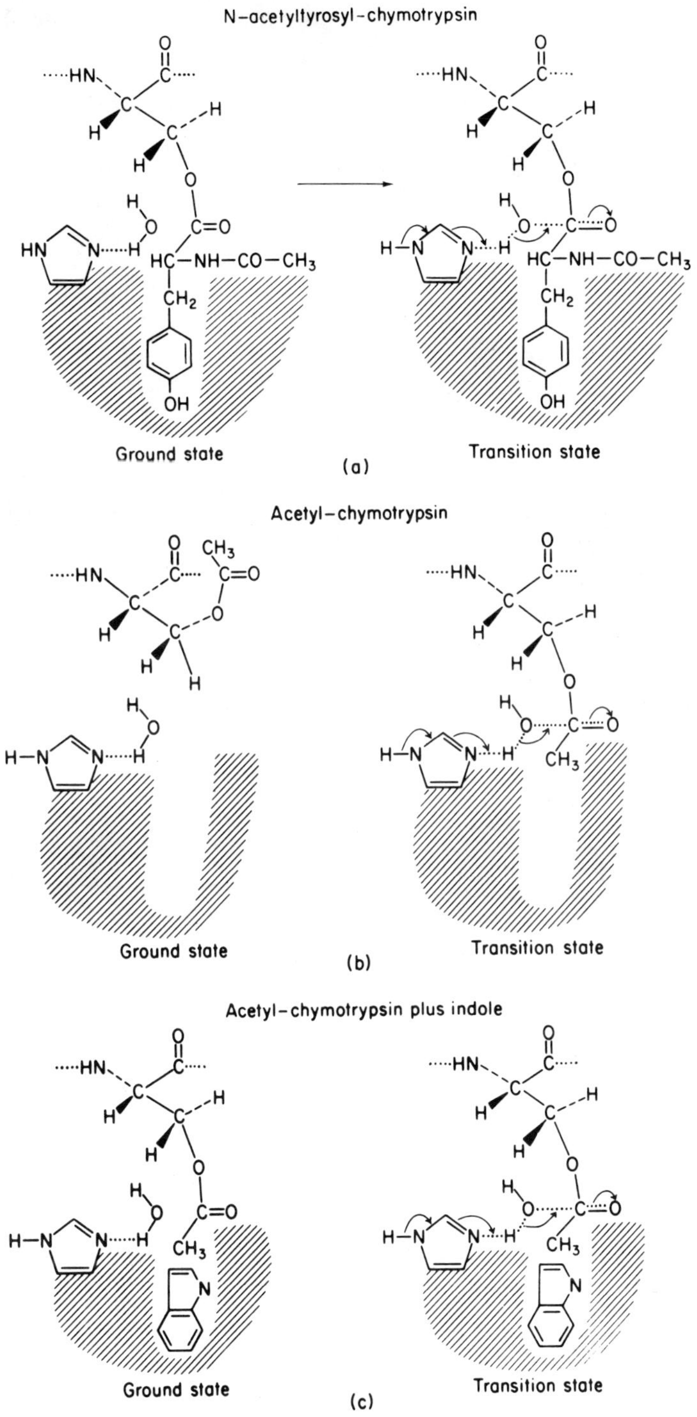
N-acetyltyrosyl-chymotrypsin
Ground state
Transition state
(a)
Acetyl-chymotrypsin
Ground state
Transition state
(b)
Acetyl-chymotrypsin plus indole
Ground state
Transition state
(c)

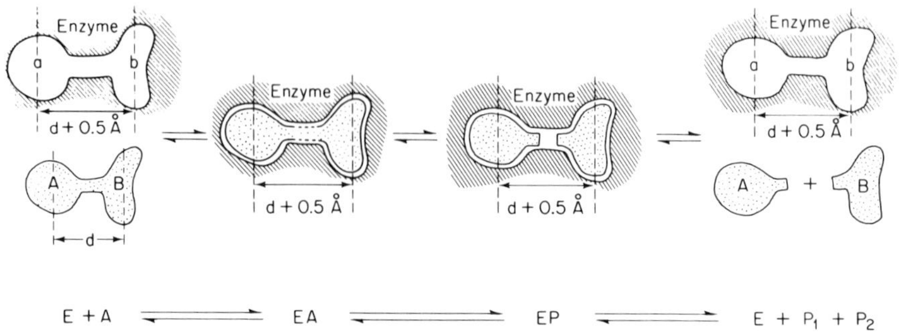

Figure 8 Schematic representation of the rack mechanism.

If the dimensions of the binding locus are intermediate between the substrate and product, both will be distorted when bound and the rate of the reaction will be accelerated from both directions.

Many examples from chemical reactions can be used to illustrate the effect of strain in enhancing reaction rates. One of the best examples is that of the alkaline hydrolysis of ethylene phosphate (A) compared with the structurally similar compound, dimethyl phosphate (B). Hydrolysis of ethylene phosphate at the O-P

A B

bond occurs at least 1×10^8 times faster than the same reaction in dimethyl phosphate. This difference in rate must reflect distortion in ethylene phosphate which forces its structure to approach that of the transition state. Postulations concerning the importance of strain in active sites of enzymes such as α-chymotrypsin, trypsin, and lysozyme have been made. Transition-state substrate analogs are very potent inhibitors, with K_i values in the n*M* range. Substrates designed to mimic the transition state have high turnover numbers.

C. General Acid/General Base-Catalyzed Reactions

The transforming locus of the active site of a number of enzymes has been shown to involve prototropic groups. In enzymes that are composed only of amino acid residues, the prototropic groups are contributed by the side chains of the acidic and basic amino acids. These prototropic groups are the α-carboxyl group of the

C-terminal amino acid residue and the carboxyl groups of aspartyl and glutamyl residues, respectively; the imidazole group of histidyl residues; the α- and ϵ-amino groups of the N-terminal amino acid residue and of lysine, respectively; the sulfhydryl group of cysteine; the phenolic hydroxyl group of tyrosine; and the guanidino group of arginine. Analyses of the mechanisms of action of the cofactors of enzymes (Chapter 12) show that with the exception of cations and possibly the metal-coordinated prosthetic groups (such as heme or hemin), these components of the active site do not act by general acid-general base catalysis. This does not preclude rate-determining proton transfer reactions.

By definition an acid is a substance that can donate a proton and a base is a substance that can accept a proton. Therefore, it should be clear that each of the prototropic groups listed above has the potential to act either as a general acid or as a general base, depending on its state of ionization. This is illustrated for the imidazole group in Eq. (7). The imidazole group is ideally suited to participate in general acid/general base reactions because it can perform both functions simultaneously. Many enzyme-catalyzed reactions involve the transfer of a proton from one position to another in the substrate, between two substrates, or between a prototropic group on the enzyme and the substrate.

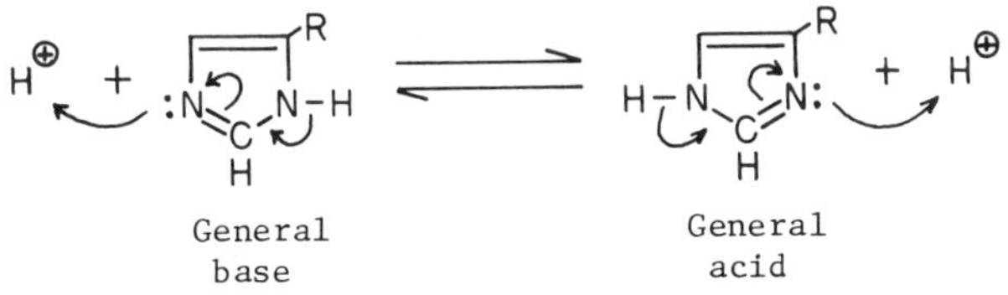

(7)

General acid or *general base catalysis* is used to describe a reaction when the rate is affected by transfer of protons from one or more of the prototropic groups on the side chains of amino acid residues. If, on the other hand, the rate of a reaction is affected by the concentration of hydrogen or hydroxide ions from water because of their direct participation, the reaction is said to be *specific acid* or *specific base catalyzed*. The effect of pH on ionization of prototropic groups on an enzyme and therefore rate of reaction is due to an indirect effect of hydrogen or hydroxide ions from water.

1. Mutarotation of Tetramethylglucose

Mutarotation of glucose has been studied extensively as an example of a general acid/general base-catalyzed reaction. In water, the rate of mutarotation is catalyzed both by general acids and general bases as well as by the specific acid, hydrogen ion, and specific base, hydroxyl ion. To eliminate the contribution of specific acid/specific base catalysis, the rate of mutarotation of tetramethylglucose has been studied in benzene [12]. The four hydroxyl groups were methylated so that the compound would be soluble in benzene; however, methylation does not affect the essential features of the mutarotation process. In benzene alone, there is a very slow rate of mutarotation. If pyridine or *p*-cresol is added alone, the rate of mutarotation is still very slow. If the two are added together, there is a 10-fold increase in rate of mutarotation over that expected from the individual effects (Table 4).

An explanation for the effect of these compounds on rate of mutarotation is that the slow step in the reaction is the rate of proton transfer involved in opening of the ring so that mutarotation can take place [Eq. (8)]. In formation of the open-ring

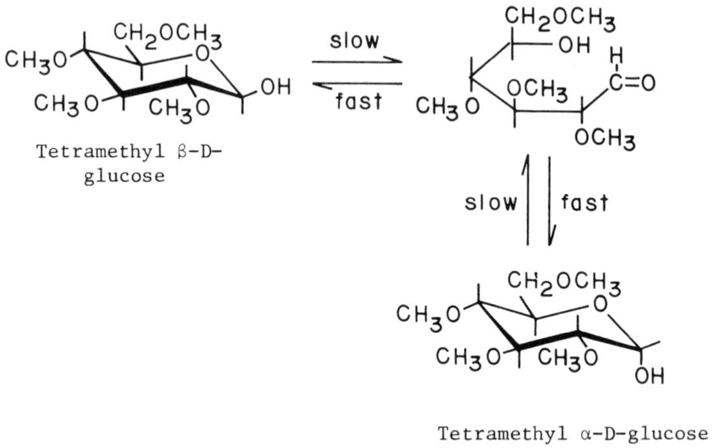

(8)

Table 4 Catalysis of Mutarotation of Tetramethylglucose

Catalyst	Catalyst concentration[a] (*M*)	*k* (min^{-1})
None	0	0.000078
Pyridine	0.1	0.00037
p-Cresol	0.1	0.00042
Pyridine + *p*-cresol	0.1 each	0.0079
2-Methoxypyridine	0.1	0.00018
N-methyl-α-pyridone	0.1	0.00054
α-Pyridone	0.1	0.18
Benzoic acid	0.1	2.20
Trichloroacetic acid	0.1	27.00

Source: Ref. 12.

[a]Concentration reported as total concentration. In some cases values calculated to concentrations given. Concentration of tetramethylglucose was 0.09 *M* in all cases.

intermediate there is removal of a proton from the OH of C(1) and addition of a proton to the acyclic O(6). Pyridine alone can remove the proton from the OH of C(1) and *p*-cresol alone can donate a proton to O(6) (Fig. 9). Alone, the two are rather ineffective, but when both are present in the solution, removal and addition of protons can occur simultaneously with a marked increase in rate. This would be an example of a *concerted general acid/general base* catalysis, but it is a crude model for an enzyme because there is need for three individual molecules to be involved in the transition state:

$$\text{rate} = k[\text{tetramethylglucose}][\text{pyridine}][p\text{-cresol}] \tag{9}$$

A test of the hypothesis would be to put both an acidic and a basic group on the catalyst. This is the situation with the compound α-pyridone (Fig. 10). In benzene, carboxylic acids are also very effective catalysts for mutarotation for the same reason as that shown for α-pyridone (Table 4).

The effectiveness of α-pyridone as a catalyst is appreciated when it is realized that it is only 1×10^{-4} as strong a base as pyridine and 1×10^{-2} as strong an acid as *p*-cresol. On this basis alone, it would be expected to be 1×10^{-3} as effective as a mixture of pyridine and *p*-cresol. However, with a reaction 1×10^{-3} *M* in α-pyridone and 0.1 *M* in tetramethylglucose, the rate is 7000 times that calculated for the combined catalytic effect of 1×10^{-3} *M* pyridine and 1×10^{-3} *p*-cresol.

When either the basic or acidic group of α-pyridone is eliminated as in 2-methoxypyridine or *N*-methyl-α-pyridone the rate of mutarotation is decreased markedly (Table 4).

D. Nucleophilic and Electrophilic Catalysis

A *nucleophile* is a group that has a strong tendency to donate an electron pair, while an *electrophile* is a group that has a strong tendency to accept an electron pair. A catalyzed reaction whose rate is controlled (totally or partially) by a step involving the donation of an electron pair from the catalyst to the substrate is

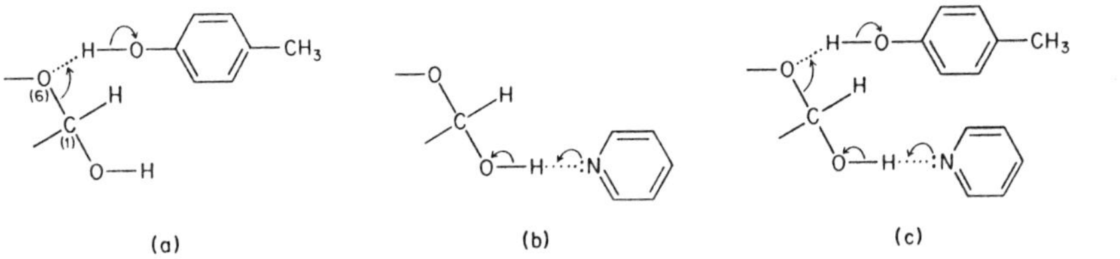

Figure 9 Proposed transition states in the pyridine- and *p*-cresol-catalyzed mutarotation of tetramethylglucose. (a) *p*-cresol alone; (b) pyridine alone; (c) concerted effect of the two combined.

Figure 10 Proposed transition state (left) in the α-pyridone-catalyzed mutarotation of tetramethylglucose.

termed a *nucleophilic catalysis*; a catalyzed reaction whose rate is controlled (totally or in part) by a step involving the acceptance of an electron pair by the catalyst from the substrate is termed an *electrophilic catalysis*.

In a discussion of general acid/general base-catalyzed reactions there is no need to postulate the formation of a covalent intermediate (Fig. 10). With nucleophilic or electrophilic catalysis, however, there should be a transient covalent intermediate formed as shown for several model reactions [Eqs. (10) to (12)]. In all cases there is a transient intermediate formed.

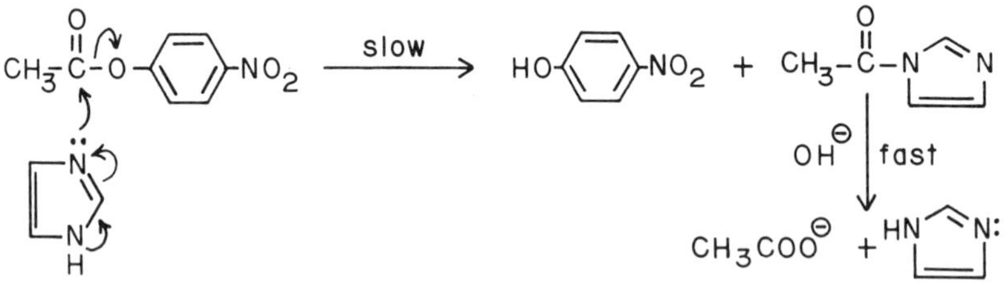

(10)

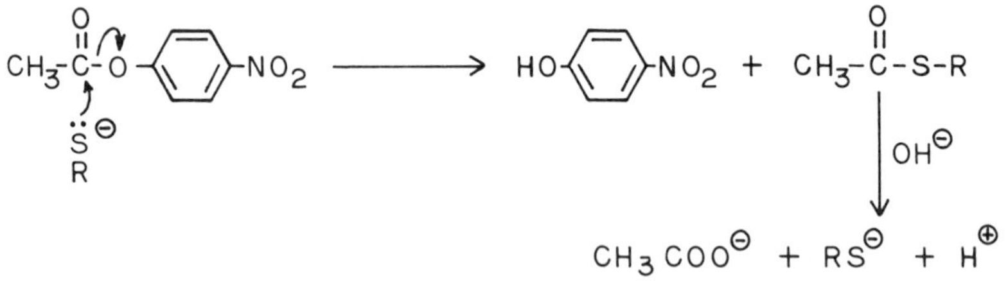

(11)

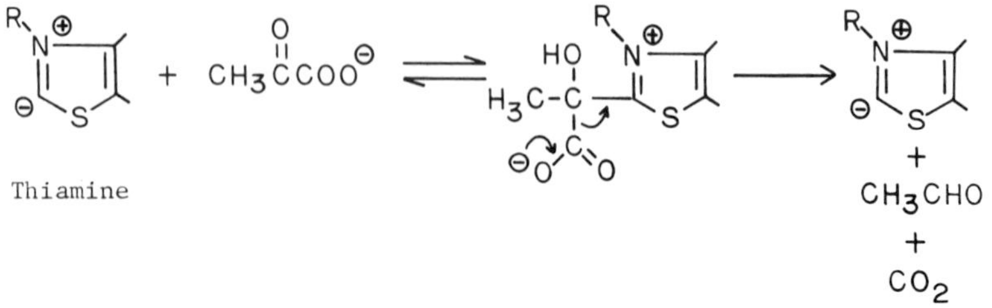

(12)

We conclude our discussion of nucleophilic and electrophilic catalysis with two postulated mechanisms, one for chymotrypsin-catalyzed reactions and the other for ribonuclease-catalyzed reactions (Figs. 11 and 12). Nucleophilic catalysis is aided by concerted general acid/general base catalysis in both cases.

In the proposed ribonuclease-catalyzed reaction the substrate, cytidine 2′,3′-phosphate, is bound at the binding locus of the active site so as to position the susceptible 2′,3′-phosphate bonds between two imidazole groups of the enzyme. As shown, the imidazole group on the left serves as a general base to remove a proton from a water molecule. The hydroxyl ion, formed in proximity to the electrophilic phosphorus, then attacks this atom in a nucleophilic attack. It is helped in this attack by the concerted action of the imidazole group on the right serving as a general acid. The reaction can then be designated as a *concerted general acid/general base* catalysis. It will be noted that in the proposed mechanism there are no covalent bonds formed between the substrate (or a substrate component) and the enzyme since the nucleophile is hydroxide ion.

In the proposed α-chymotrypsin mechanism a single imidazole serves as both a general acid and a general base in a concerted mechanism (Fig. 12). The extensive hydrogen bonding network involving Asp-102, His-57, and Ser-195 is an essential feature of the mechanism proposed. In formation of the first transition-state intermediate (A), N(3) of His-57 acts as a general base to abstract a proton from the hydroxyl group of Ser-195, thus facilitating the nucleophilic attack of the oxygen on the carbonyl carbon of the peptide bond of bound substrate. This is facilitated further by the carboxylate group of Asp-102 acting as a general base to abstract the proton from N(1) of His-57. This leads to formation of a tetrahedral intermediate whose transition state is shown (B). N(3) serves as a general acid to protonate the nitrogen of the peptide bond of substrate in a facilitated reaction that

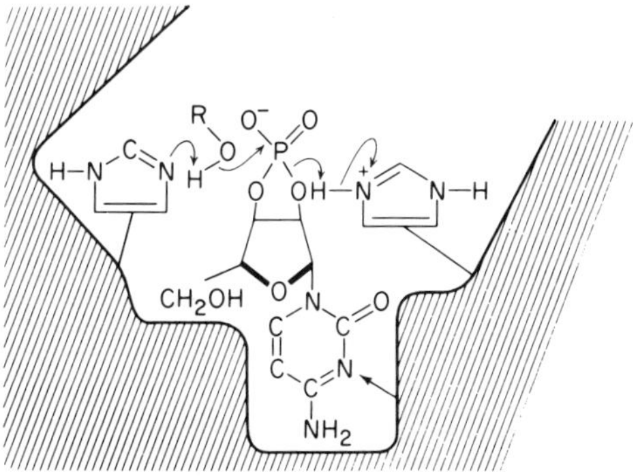

Figure 11 Proposed transition state in the ribonuclease-catalyzed hydrolysis of cytidine 2′,3′-phosphate. Two imidazole groups of histidine residues are involved in the transforming locus. (From Ref. 13, p. 27, by courtesy of Academic Press.)

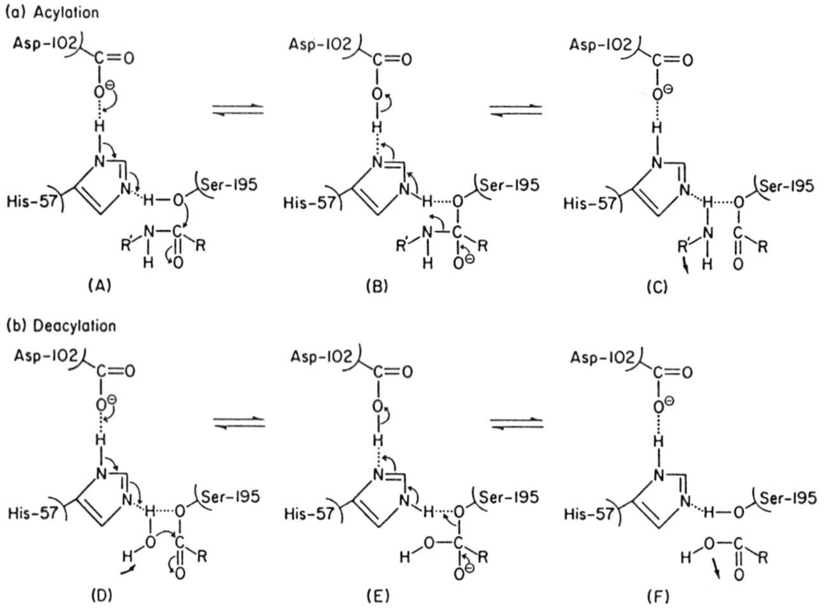

Figure 12 Proposed mechanism for α-chymotrypsin-catalyzed reactions. The top reactions show acylation of the seryl OH of the enzyme by the substrate, the bottom reactions show removal of the acyl group by hydrolysis. (Modified from Ref. 14.)

leads to the acylenzyme (C) plus the first product, an amine. The position formerly occupied by the amine is taken by a water molecule in the first step of the deacylation pathway (D). In the transition state shown (D), N(3) of His-57 serves as a general base to abstract a proton from water facilitating the nucleophilic attack of the OH^- on the acylenzyme. This proceeds to a tetrahedral intermediate that goes through a transition state (E) to give the free enzyme and the second product.

Unlike the case of ribonuclease where no covalent intermediate is formed, there is a covalent acylenzyme intermediate formed in the chymotrypsin-catalyzed hydrolysis because the first nucleophile is the hydroxyl group of Ser-195.

E. Factors Responsible for α-Chymotrypsin Catalysis

Based on proximity and orientation effects, general acid/general base catalysis, and on nucleophilic (or electrophilic) catalysis, can one account for the tremendous effectiveness of an enzyme? Unfortunately, there are insufficient experimental data to permit this to be done for most enzyme-catalyzed reactions. Bender et al. [10] attempted this for the α-chymotrypsin-catalyzed hydrolysis of *N*-acetyl-L-tryptophanamide (Table 5). The calculated rate constant is very close

Table 5 Kinetic Factors Responsible for the Difference between the Hydroxide Ion- and α-Chymotrypsin-Catalyzed Hydrolysis of *N*-Acetyl-L-tryptophanamide

Rate constant of hydroxide ion catalysis	$3 \times 10^{-4}\ M^{-1}\ \text{sec}^{-1}$
1. Conversion to an intermolecular general base-catalyzed reaction involving imidazole ($1.6 \times 10^{-6}\ \text{M}^{-1}\ \text{s}^{-1}$)	$4.8 \times 10^{-10}\ M^{-1}\ \text{sec}^{-1}$
2. Conversion to an intramolecular general base-catalyzed reaction involving imidazole (10 *M*)	$4.8 \times 10^{-9}\ \text{sec}^{-1}$
3. Nucleophilic catalysis: change in rate-determining step (10^2)	$4.8 \times 10^{-7}\ \text{sec}^{-1}$
4. Proximity and orientation effects: freezing of the substrate (specificity factor, reflected by increase in $\Delta S^{\ddagger}$; 10^3)	$4.8 \times 10^{-4}\ \text{sec}^{-1}$
5. General acid catalysis by imidazole (10^2)	$4.8 \times 10^{-2}\ \text{sec}^{-1}$
Calculated enzymatic rate based on five factors above	$4.8 \times 10^{-2}\ \text{sec}^{-1}$
Experimentally observed rate constant	$4.4 \times 10^{-2}\ \text{sec}^{-1}$

Source: Ref. 10, p. 3721, by courtesy of the American Chemical Society.

to the experimentally observed rate constant. Some approximations have been used in the calculation, but it is encouraging to see that at least in one case, the marked efficiency of an enzyme can be accounted for by experimentally sound factors.

REVIEW QUESTIONS

1. How might one experimentally distinguish between a general acid–catalyzed reaction and a specific acid (H^+)–catalyzed reaction?

2. For the reactions shown in Eqs. (1), (3), (4), (5), and (6) indicate whether they are examples of nucleophilic, electrophilic, general acid, or general base catalysis.

3. Outline the essential differences between the lock-and-key and induced fit models for enzyme–substrate interaction.

4. Aconitase is an enzyme that converts citrate to α-ketoglutarate via isocitrate. When citrate, labeled at only one of the primary carboxyl groups, was converted to α-ketoglutarate the labeled carboxyl group was found to be 100% adjacent to the carbonyl group of α-ketoglutarate. Draw some models showing binding between citrate and aconitase that can explain the observed results.

5. Predict how many incorrectly oriented enzyme·substrate complexes might be formed between a hexa-*N*-acetylchitohexaose and lysozyme.

6. In your own words, give an interpretation of the meaning of $\Delta S^{\ddagger}$ and its importance in relation to enzyme-catalyzed reactions.

7. Several types of proposals are discussed in the chapter with respect to the

factors that may account for the efficiency of enzyme-catalyzed reactions. Are these proposals mutually exclusive or do each of them contribute to the overall efficiency of an enzyme? Explain.

8. How close must a group on a substrate be to a group on an enzyme for there to be a possibility of interaction between the two?

9. Draw the structures of the side chains of the amino acid residues found in an enzyme. For each, indicate whether you would expect it to be involved primarily in binding the substrate or in transforming it to product. For each group involved in transformation indicate whether you would expect it to act as a general base, a general acid, a nucleophile, or an electrophile.

10. In Section II.A.3 it was indicated that the effective concentration of an intramolecular catalytic group is equivalent to $\sim$10 *M* of the corresponding intermolecular catalyst. Elaborate on the meaning of this.

11. In Table 4, trichloracetic acid is shown to be 13 times as effective a catalyst as benzoic acid in catalyzing the mutarotation of tetramethylglucose. Draw a proposed mechanism involving the carboxylic group of each. Explain the difference in effectiveness of the two.

12. Based on the discussion in Section II.D with respect to formation of a covalent intermediate in chymotrypsin-catalyzed reactions versus lack of formation of such an intermediate in ribonuclease-catalyzed reactions, state some general conditions that might serve to indicate when a covalent intermediate would be expected to be formed or not formed.

REFERENCES

General References

Bender, M. L., and R. Breslow, Mechanisms of organic reactions, in *Comprehensive Biochemistry* (M. Florkin and E. H. Stotz, eds.), Vol. 2, Elsevier, Amsterdam, 1962, p. 1.

Bruice, T. C., and S. Benkovic, *Bioorganic Mechanisms*, Vols. 1 and 2, W.A. Benjamin, New York, 1966.

Hammes, G. G., *Enzyme Catalysis and Regulation*, Academic Press, San Diego, 1982.

Jencks, W. P., *Catalysis in Chemistry and Enzymology*, McGraw-Hill, New York, 1969.

Koshland, D. E., Jr., The active site and enzyme action, *Advan. Enzymol.*, **22**, 45 (1960).

Koshland, D. E., Jr., and K. E. Neet, The catalytic and regulatory properties of enzymes, *Ann. Rev. Biochem.*, **37**, 359 (1968).

Koshland, D. E., Jr., Conformational changes at the active site during enzyme action, *Federation Proc.*, **23**, 719 (1964).

Koshland, D. E., Jr., Mechanisms of transfer enzymes, in *The Enzymes* (P. D. Boyer, H. Lardy, and K. Myrbäck, eds.), Vol. 1, Academic Press, New York, 1959, p. 305.

Silverman, R. B., *Mechanism-Based Enzyme Inactivation: Chemistry and Enzymology*, Vols. 1 and 2, CRC Press, Boca Raton, Fla., 1988.

Walsh, C., *Enzymatic Reaction Mechanisms*, W.H. Freeman, San Francisco, 1979.
Page, M. I., and A. Williams, eds., *Enzyme Mechanisms*, Royal Society of Chemistry, London, 1987.

Cited References

1. E. Fischer, *Ber.*, **27**, 2985 (1894); *Z. Physiol. Chem.*, **26**, 60 (1898).
2. H. Theorell and J. S. McKinley-McKee, *Acta Chem. Scand.*, **15**, 1834 (1961).
3. W. N. Lipscomb, J. A. Hartsuck, G. N. Reeke, Jr., F. A. Quiocho, P. H. Bethage, M. L. Ludwig, T. A. Steitz, H. Muirhead, and J. C. Coppola, *Brookhaven Symp. Biol.*, **21**, 24 (1969).
4. L. N. Johnson, D. C. Phillips, and J. A. Rupley, *Brookhaven Symp. Biol.*, **21**, 120 (1969).
5. R. T. Jones, in *Symposium on Foods: Proteins and Their Interactions* (H. W. Schultz and A. F. Anglemeir, eds.), AVI, Westport, Conn., 1964, p. 33.
6. D. E. Koshland, Jr., in *The Enzymes* (P. D. Boyer, H. Lardy, and K. Myrbäck, eds.), Vol. 1, Academic Press, New York, 1959, p. 305.
7. D. E. Koshland, Jr., *Federation Proc.*, **23**, 719 (1964).
8. D. E. Koshland, Jr., *Advan. Enzymol.*, **22**, 45 (1960).
9. K. D. Collins and G. R. Stark, *J. Biol. Chem.*, **244**, 1869 (1969).
10. M. L. Bender, F. J. Kézdy, and C. R. Gunter, *J. Am. Chem. Soc.*, **86**, 3714 (1964).
11. M. L. Bender, *Chem. Rev.*, **60**, 53 (1960); D. E. Koshland, Jr., *J. Theoret. Biol.*, **2**, 75 (1962).
12. C. G. Swain and J. F. Brown, Jr., *J. Am. Chem. Soc.*, **74**, 2534, 2538 (1952).
13. A. P. Mathias, A. Deavin, and B. R. Rabin, in *Structure and Activity of Enzymes* (T. W. Goodwin, J. I. Harris, and B. S. Hartley, eds.), Academic Press, New York, 1964, p. 19.
14. D. M. Blow, J. J. Birktoft, and B. S. Hartley, *Nature*, **221**, 337 (1969).

5
Rates of Reactions

To study the catalytic properties of enzymes, one must observe them in action. To do this the enzyme is mixed with its substrate(s) under the proper conditions of pH, temperature, and cofactors (if needed). The effect of this mixing of reactants is observed by measuring the rate of disappearance of substrate or the rate of appearance of product. It is the rate of change that is usually most important in understanding the mechanism of action of an enzyme, not just an observation that a change takes place. Experimentally, the amount of substrate left (or product formed) is determined at timed intervals (Fig. 1). The time axis may be expressed in any convenient units (seconds, minutes, hours, days).

I. WRITING RATE EXPRESSIONS

To express a rate change in a quantitative fashion, it is of great advantage to derive proper mathematical equations that express the changes in concentrations of reactant and product during the reaction. We begin this process by writing generalized rate equations in their differential forms.

A. Irreversible Reactions

Consider the reaction

$$A \xrightarrow{k_1} P \qquad (1)$$

where A is the reactant, P the product, and k_1 the specific reaction rate constant

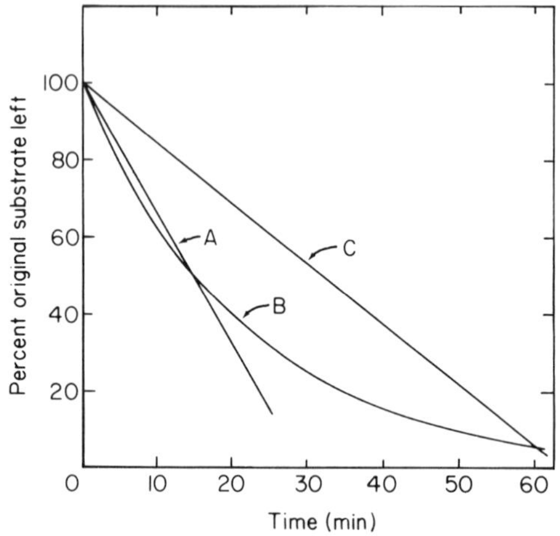

Figure 1 Disappearance of substrate as a function of time. With the data indicated by lines A and C the amount of substrate lost is a linear function of time with $t_{1/2}$ of 15 and 32 min, respectively. In the data indicated by line B the amount of substrate lost is not a linear function of time but rather it decreases with time. In B, $t_{1/2}$ is 15 min.

for the irreversible conversion of A to P. The rate of conversion is expressed mathematically as

$$-\frac{dA}{dt} = k_1(A) \tag{2}$$

which indicates that the rate of disappearance of A, expressed as $-dA/dt$, is proportional to concentration of reactant at any time. The negative sign indicates that the reactant, A, is disappearing from the reaction. Equation (2) is the mathematical expression of a first-order process.

Reaction (1) is described equally well as

$$\frac{dP}{dt} = k_1(A) \tag{3}$$

which states that the rate of appearance of product (indicated by positive sign, dP/dt) is proportional to the concentration of the reactant, A. Therefore, $-dA/dt = dP/dt = k_1(A)$.

B. Reversible Reactions

Equation (1) describes an irreversible process. Consider the reversible reaction represented by

$$\mathrm{A} \underset{k_{-1}}{\overset{k_1}{\rightleftharpoons}} \mathrm{P} \tag{4}$$

In this reaction the reactant is disappearing by one process but is being formed by another. In studying the rate of change we are interested only in the net change. Therefore, the overall rate of disappearance of A (which is equal to the overall rate of formation of P) is given as

$$-\frac{d\mathrm{A}}{dt} = k_1(\mathrm{A}) - k_{-1}(\mathrm{P}) = \frac{d\mathrm{P}}{dt} \tag{5}$$

When the reaction described by Eq. (4) comes to equilibrium (since it is a reversible process), the net rate of substrate disappearance and the net rate of product formation are equal to zero.

$$-\frac{d\mathrm{A}}{dt} = 0 \qquad \frac{d\mathrm{P}}{dt} = 0 \tag{6}$$

Therefore,

$$k_1(\mathrm{A_e}) - k_{-1}(\mathrm{P_e}) = 0 \tag{7}$$

and

$$k_1(\mathrm{A_e}) = k_{-1}(\mathrm{P_e}) \tag{8}$$

Rearranging yields

$$\frac{k_1}{k_{-1}} = \frac{(\mathrm{P_e})}{(\mathrm{A_e})} = K_{\mathrm{eq}} \tag{9}$$

Equation (9) indicates that the ratio of the two specific reaction rate constants, k_1 and k_{-1}, can be defined as a new constant called the equilibrium constant, K_{eq}. This quantity is also defined as the ratio of concentration of product at equilibrium ($\mathrm{P_e}$) to concentration of reactant at equilibrium ($\mathrm{A_e}$). In these equations it is really the activity of the compound that is most important. Since the concentrations involved in enzyme-catalyzed reactions are small (10^{-8} to 10^{-1} *M*), we shall use the symbol of the compound enclosed in parentheses to indicate the molar concentration of that compound.

A rate expression may be written for each of the molar species that occur in a reaction. Consider the following equation, which is the simplest description of an enzyme-catalyzed reaction.

$$\mathrm{E} + \mathrm{A} \underset{k_{-1}}{\overset{k_1}{\rightleftharpoons}} \mathrm{EA} \overset{k_2}{\rightarrow} \mathrm{E} + \mathrm{P} \tag{10}$$

The rates of change of the various molecular species are expressed by the following rate equations:

$$-\frac{dE}{dt} = k_1(E)(A) - k_{-1}(EA) - k_2(EA) \tag{11}$$

$$-\frac{dA}{dt} = k_1(E)(A) - k_{-1}(EA) \tag{12}$$

$$-\frac{dEA}{dt} = k_{-1}(EA) - k_2(EA) - k_1(E)(A) \tag{13}$$

$$\frac{dP}{dt} = k_2(EA) \tag{14}$$

Equations may also be written which account for the total material of the system. These are called conservation equations. For the enzyme involved in Eq. (10) this is

$$(E_0) = (E) + (EA) \tag{15}$$

This equation indicates that the total enzyme concentration (E_0) of the system is accounted for by the sum of the free enzyme concentration (E) and the enzyme·substrate complex concentration (EA).

In the reactions considered so far, the coefficients in front of the reactants and products of the stoichiometric equations [Eqs. (1), (4), and (10)] are unity. Consider the case of the reversible reaction described by

$$2A + B \underset{k_{-1}}{\overset{k_1}{\rightleftharpoons}} 3P \tag{16}$$

The rates of change of the various molecular species are described by the following rate equations:

$$\tfrac{1}{2}\left(-\frac{dA}{dt}\right) = k_1(A)^2(B) - k_{-1}(P)^3 \tag{17}$$

$$-\frac{dB}{dt} = k_1(A)^2(B) - k_{-1}(P)^3 \tag{18}$$

$$\tfrac{1}{3}\left(\frac{dP}{dt}\right) = k_1(A)^2(B) - k_{-1}(P)^3 \tag{19}$$

It will be noted that

$$\tfrac{1}{2}\left(-\frac{dA}{dt}\right) = -\frac{dB}{dt} = \tfrac{1}{3}\left(\frac{dP}{dt}\right) = k_1(A)^2(B) - k_{-1}(P)^3 \tag{20}$$

From the specific equations (16) and (20), one may then deduce that the corresponding general equations can be written as

$$aA + bB \underset{k_{-1}}{\overset{k_1}{\rightleftharpoons}} pP + qQ \tag{21}$$

and

$$\frac{1}{a}\left(-\frac{dA}{dt}\right) = \frac{1}{b}\left(-\frac{dB}{dt}\right) = \frac{1}{p}\left(\frac{dP}{dt}\right) = \frac{1}{q}\left(\frac{dQ}{dt}\right) \tag{22}$$

Therefore, when the rate of a chemical reaction may be described by a simple stoichiometric equation, it is possible to define a rate of reaction for the system uniquely. The rate of change in the concentration of A for the forward reaction can then be written in the general form

$$-\frac{dA}{dt} = kA^aB^bC^c \cdots \tag{23}$$

where A, B, and C are concentrations of reactants at any time and the exponents, a, b, and c are coefficients of the balanced equation. This may or may not be true for reactions that go by alternative pathways.

The molecular order of the reaction depicted in Eq. (23) is a with respect to A, b with respect to B, c with respect to C, and so on, while the overall (or total order) is the sum of the exponents $a + b + c \cdots$.

C. Consecutive Reactions

Equation (10) describes a case of *consecutive reactions*. This equation indicates that enzyme and substrate must combine to form an enzyme·substrate complex, which in turn reacts to form enzyme and product. The enzyme·substrate complex, EA, is an obligatory intermediate in the reaction as written. In the case of consecutive reactions it is valid to write the equation in parts as

$$E + A \underset{k_{-1}}{\overset{k_1}{\rightleftharpoons}} EA \tag{24}$$

$$EA \xrightarrow{k_2} E + P \tag{25}$$

D. Simultaneous Reactions

The case of consecutive reactions is to be contrasted to the case of *simultaneous reactions*. If several reactions occur simultaneously, they are independent of each other as follows:

$$\begin{aligned} &A \xrightarrow{k_1} P \\ &A \xrightarrow{k_2} Q \\ &A \xrightarrow{k_3} R \end{aligned} \tag{26}$$

The rates of change of the various molecular species in Eq. (26) are described by the following rate equations:

$$-\frac{dA}{dt} = k_1(A) + k_2(A) + k_3(A) \tag{27}$$

$$\frac{dP}{dt} = k_1(A) \tag{28}$$

$$\frac{dQ}{dt} = k_2(A) \tag{29}$$

$$\frac{dR}{dt} = k_3(A) \tag{30}$$

II. RATE LAWS

The rate expressions above indicate the relationship between the rate of disappearance of reactant (or rate of appearance of product) and the reactant concentration at that time. Integration of these differential equations leads to equations that express this relationship at any time during the reaction. The following discussion will deal with those rate laws encountered most frequently in the study of enzyme systems.

A. Zero-Order Rate Law

A zero-order rate process is characterized by the rate of disappearance of reactant (or rate of appearance of product) being independent of concentration of reactant. For the reaction

$$A \xrightarrow{k} P \tag{31}$$

$$-\frac{dA}{dt} = k = \frac{dP}{dt} \tag{32}$$

To obtain the integrated form, this equation is integrated between the limits of substrate concentration at zero time (A_0) and at time t (A):

$$-\int_{A_0}^{A} dA = k\int_{t_0}^{t} dt \tag{33}$$

$$-[(A) - (A_0)] = kt - kt_0 \tag{34}$$

When $t_0 = 0$, $kt_0 = 0$ and

$$(A_0) - (A) = kt = (P) \tag{35}$$

[The difference $(A_0) - (A) = (P)$, so that Eq. (35) can be written as $kt = (P)$, an expression also arrived at by integration of $dP/dt = k$.] Graphically, Eq. (35) is presented as in Fig. 2, where (P) [or $(A_0) - (A)$] is plotted against time. The plot gives a straight line with an intercept of zero and a slope of k.

The specific reaction rate constant, k, can be obtained from the experimental

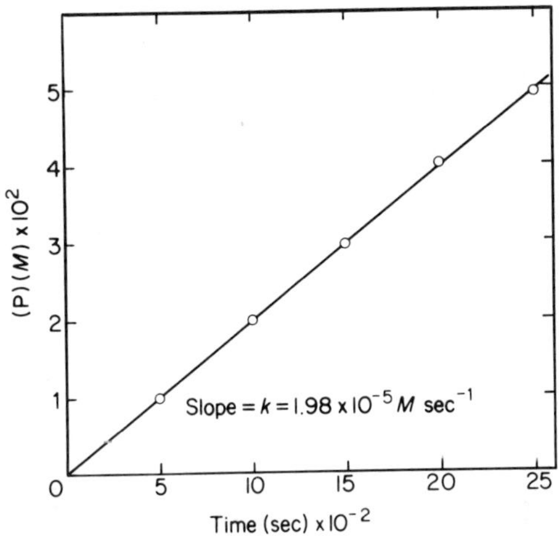

Figure 2 Zero-order rate process. The data of Table 1 are plotted according to Eq. (35). The slope of the line is the specific reaction rate constant.

data by calculating the rate constant for each data point using Eq. (35) or by plotting the data as shown in Fig. 2. This is illustrated by the data of Table 1, which are plotted in Fig. 2.

The specific reaction rate constant, k, is found to be 1.98×10^{-5} M sec^{-1} by either method. In terms of time, the graphical method is preferred. Although there is some loss in precision by use of the graphical method as compared with point-by-point calculation of the rate constant, the precision of the graphical

Table 1 Experimental Data and Calculations of k for Zero-Order Reaction

Time (sec) $\times 10^{-2}$	(A)[a] (M) $\times 10^2$	(P) (M) $\times 10^2$	$k = (P)/t$ (M sec^{-1}) $\times 10^5$
0	9.00	0	—
5.0	8.01	0.99	1.98
10.0	7.02	1.98	1.98
15.0	6.04	2.96	1.97
20.0	4.99	4.01	2.00
25.0	4.08	4.92	1.97
			average = 1.98

[a](A_0) is 9.00×10^{-2} M. The designation (M) $\times 10^2$ is to be read as the actual concentration multiplied by 10^2 to give the number 9.00.

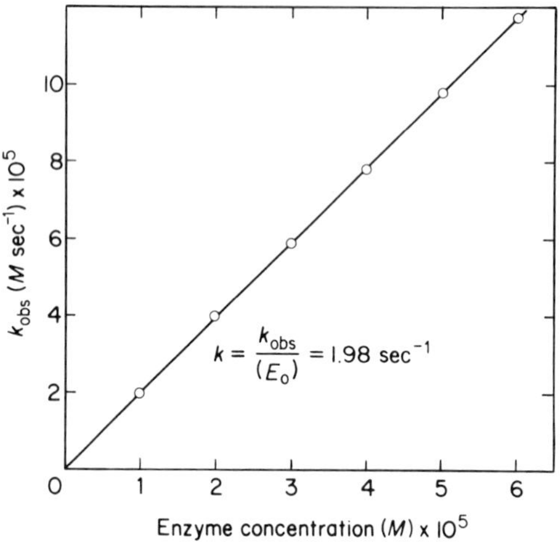

Figure 3 Effect of enzyme concentration on observed rate constant of a zero-order rate process with respect to substrate concentration. Data plotted according to Eq. (36).

method usually exceeds the precision of the actual experimental data and it clearly demonstrates the linearity of the data. In either case, all available data should be utilized in determining k.

From the data of Table 1 it should be noted that the zero-order rate process is characterized by the amount of reactant removed per unit time being a constant.

Zero-order reactions, with respect to substrate, are encountered whenever the enzyme is saturated with substrate (see Chapter 6). For enzyme-catalyzed reactions the amount of product formed per unit time is dependent on the enzyme concentration:

$$(E_0)k = \frac{(P)}{t} = k_{obs} \tag{36}$$

k_{obs} changes with concentration of enzyme, while k [$= k_{obs}/(E_0)$] is a true constant and independent of enzyme concentration. The rate is independent of substrate concentration. The effect of enzyme concentration on k_{obs} is shown in Fig. 3.

B. First-Order Rate Law

A first-order rate process is characterized by a rate of disappearance of reactant that is dependent on reactant concentration according to Eq. (38) as derived from Eq. (37):

$$\mathrm{A} \xrightarrow{k} \mathrm{P} \tag{37}$$

$$-\frac{d\mathrm{A}}{dt} = k(\mathrm{A}) = \frac{d\mathrm{P}}{dt} \tag{38}$$

On integration of Eq. (38) between the limits of reactant concentration of (A_0) at time t_0 and (A) at time t, Eq. (40) is obtained.

$$-\int_{\mathrm{A}_0}^{\mathrm{A}} \frac{d\mathrm{A}}{(\mathrm{A})} = k\int_{t_0}^{t} dt \tag{39}$$

$$-[\ln(\mathrm{A}) - \ln(\mathrm{A}_0)] = kt = kt_0 \tag{40}$$

Rearranging and setting $t_0 = 0$ yields

$$\ln\frac{(\mathrm{A}_0)}{(\mathrm{A})} = kt = 2.3\log\frac{(\mathrm{A}_0)}{(\mathrm{A})} \tag{41}$$

Equation (41) can be written in terms of initial reactant concentration (A_0) and product concentration (P) at time t as shown in

$$2.3\log\frac{(\mathrm{A}_0)}{(\mathrm{A}_0) - (\mathrm{P})} = kt \tag{42}$$

The specific reaction rate constant, k, is evaluated graphically or by point-by-point calculation by use of Eq. (41) or (42). This is shown in Table 2 and Fig. 4.

Inspection of Eq. (42) and the data of Table 2 indicate the following unique features of a first-order process.

1. Concentration units do not occur in the rate constant because the units cancel in taking the ratio $(\mathrm{A}_0)/(\mathrm{A})$. This means that any parameter, such as

Table 2 Experimental Data and Calculations of Rate Constant k for a First-Order Reaction

Time (sec)	(A) $(M) \times 10^4$	(P) $(M) \times 10^4$	$k = \frac{2.3}{t}\log\frac{(\mathrm{A}_0)}{(\mathrm{A})}$ $(\mathrm{sec}^{-1}) \times 10^4$
0	9.00	0	—
300	8.55	0.45	1.69
600	8.12	0.88	1.69
900	7.71	1.29	1.72
1200	7.32	1.68	1.73
1500	6.95	2.05	1.72
1800	6.60	2.40	1.72
		average =	1.71

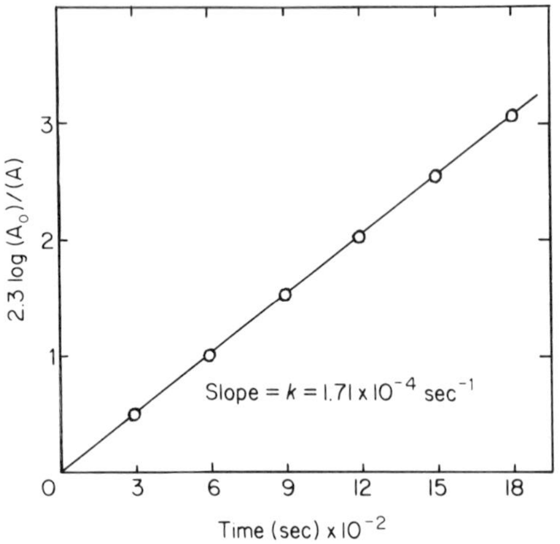

Figure 4 First-order rate process. The data of Table 2 are plotted according to Eq. (41). The slope of the line is the specific reaction rate constant.

absorbance, optical rotation, milliliters of titrant used, and so on, can be used as long as it is proportional to concentration.

2. The time required to deplete a fixed percentage of the reactant is independent of the initial reactant concentration. For the same reaction under the same conditions, except for initial reactant concentration, it will take the same time to convert 50% of a reactant at 1.0 M to product as it takes for 0.0001 M reactant. Equation (41) can be rewritten to illustrate this.

$$t_{1/2} = \frac{2.3}{k} \log \frac{(A_0)}{(A)} = \frac{2.3}{k} \log \frac{(A_0)}{0.5(A_0)} = \frac{0.693}{k} \tag{43}$$

Therefore, one criterion that may be used to determine if a reaction is first order is to measure $t_{1/2}$ at two different initial reactant concentrations. If $t_{1/2}$ is a constant, the reaction is first order with respect to that reactant.

3. In a first-order process the *percentage* of reactant that disappears is constant between two fixed intervals of time. In Table 2, 5% of the reactant disappears between each 300-sec interval. This permits a rapid calculation of how much reactant is left. For example, if one-half of the reactant is left in one half-life, $t_{1/2}$, then $\frac{1}{4}$, $\frac{1}{8}$, and $\frac{1}{16}$ of the reactant will be left after 2, 3, and 4 half-lives, respectively. For most purposes all of the reactant is considered to have disappeared after 10 half-lives (reactant concentration left = $(A_0/2^{10})$).

4. As a result of feature 3, if one takes the logarithm of the reactant (or product) concentration at fixed intervals of time, the increments between the successive values will be a constant if the reaction is first order.

In enzyme-catalyzed reactions, the rate of the reaction is first order with respect to substrate concentration when (A) $\ll K_m$ (i.e., when the enzyme is less than 1 to 5% saturated with substrate). In enzyme-catalyzed reactions the observed rate of reaction is dependent on enzyme concentration:

$$\frac{(E_0)k}{K_m} = \frac{2.3}{t}\log\frac{(A_0)}{(A)} = k_{obs} \tag{44}$$

Therefore, the rate is dependent on both the substrate concentration and the enzyme concentration. The enzyme concentration, although usually much lower than the substrate concentration, remains constant throughout the reaction, unless it is denatured or inhibited as a result of the reaction.

The discussion above has dealt with a first-order irreversible process. Often, first-order reversible processes are encountered as indicated by

$$A \underset{k_{-1}}{\overset{k_1}{\rightleftharpoons}} P \tag{45}$$

The differential equation is

$$-\frac{dA}{dt} = k_1(A) - k_{-1}(P) = \frac{dP}{dt} \tag{46}$$

Assuming that only reactant is present at zero time, we can write

$$(P) = (A_0) - (A) \tag{47}$$

On substituting these values into Eq. (46), we obtain

$$-\frac{dA}{dt} = k_1(A) + k_{-1}(A) - k_{-1}(A_0) = (k_1 + k_{-1})(A) - k_{-1}(A_0) \tag{48}$$

Integration of Eq. (48) gives

$$2.3\log\frac{k_1(A_0)}{(k_1 + k_{-1})(A) - k_{-1}(A_0)} = (k_1 + k_{-1})t \tag{49}$$

To solve Eq. (49), we need prior knowledge of one of the constants k_1 or k_{-1} or of their ratios, k_1/k_{-1}. K_{eq} ($= k_1/k_{-1}$) can be determined by letting the reaction come to equilibrium and measuring the concentration of reactant left or of product formed. This relationship is

$$k_1(A_e) = k_{-1}(P_e) \tag{50}$$

where (A_e) and P_e) represent the concentrations of A and P present at equilibrium. Rearranging Eq. (50) gives

$$\frac{k_1}{k_{-1}} = \frac{(P_e)}{(A_e)} = \frac{(A_0) - (A_e)}{(A_e)} \tag{51}$$

which on solving for (A_e) gives

$$(A_e) = \frac{k_{-1}}{k_1 + k_{-1}}(A_0) \tag{52}$$

By substitution of Eq. (52) into Eq. (49), it may then be written in the form

$$2.3 \log \left(\frac{(A_0) - (A_e)}{(A) - (A_e)} \right) = (k_1 + k_{-1})t \tag{53}$$

which expresses all variables in readily measurable quantities. Equation (53) can be expressed in terms of concentrations of (P) and (P_e) at any time since $(A_0) - (A_e) = (P_e)$ and $(A) = (A_0) - (P)$. The following equation shows this relationship:

$$2.3 \log \frac{(P_e)}{(P_e) - (P)} = (k_1 + k_{-1})t \tag{54}$$

A plot of data according to Eq. 54 [or (53)] gives a straight line with intercept of zero and slope of $k_1 + k_{-1}$ if the reaction is first order, reversible (Fig. 5). The individual rate constants are calculated by recourse to Eq. (51).

C. Second-Order Rate Laws

A second-order rate process is characterized by the rate being dependent on two molecules either of the same compound or of two different compounds. Consider the reaction described by

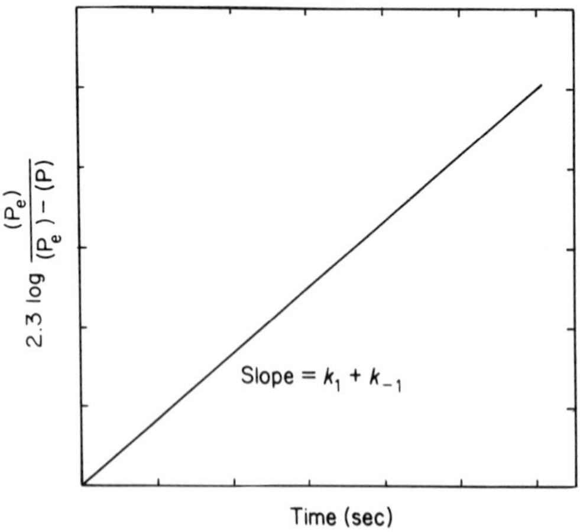

Figure 5 First-order reversible rate process. The line is drawn according to Eq. (54). The slope of the line is the sum of the forward and reverse specific reaction rate constants.

$$\mathrm{A} + \mathrm{B} \xrightarrow{k} \mathrm{P} \tag{55}$$

The rate of the reaction may be written as

$$-\frac{d\mathrm{A}}{dt} = -\frac{d\mathrm{B}}{dt} = \frac{d\mathrm{P}}{dt} = k(\mathrm{A})(\mathrm{B}) \tag{56}$$

Experimentally, we can distinguish three types of second-order reactions and the forms of the integrated rate expressions will be dependent on the type.

1. Type I

Type I second-order reactions will be designated as those in which the initial concentrations of A and B are identical and in which the two react in equimolar amounts. In this case Eq. (56) may be written as

$$-\frac{d\mathrm{A}}{dt} = k(\mathrm{A})^2 \tag{57}$$

Integration between the limits of A_0 and A at times t_0 and t,

$$-\int_{\mathrm{A}_0}^{\mathrm{A}} \frac{d\mathrm{A}}{(\mathrm{A})^2} = k\int_{t_0}^{t} dt \tag{58}$$

gives

$$-\left[-\frac{1}{(\mathrm{A})} + \frac{1}{(\mathrm{A}_0)}\right] = kt - kt_0 = kt \qquad (\text{when } t_0 = 0) \tag{59}$$

which rearranges to the form

$$\frac{(\mathrm{A}_0) - (\mathrm{A})}{(\mathrm{A})(\mathrm{A}_0)} = kt = \frac{(\mathrm{P})}{(\mathrm{A}_0)[(\mathrm{A}_0) - (\mathrm{P})]} \tag{60}$$

A plot of data according to Eq. (60) gives a straight line with intercept of zero and slope of k provided that the reaction is type I second order (Fig. 6).

The half-life, $t_{1/2}$, of the reaction is calculated by using the relation (A) = $0.5(A_0)$, and thus Eq. (60) becomes

$$t_{1/2} = \frac{1}{k(\mathrm{A}_0)} \tag{61}$$

The half-life of a second-order reaction (type I) depends on the initial concentration of the reactant, but the half-life of a first-order reaction does not. This provides a simple experimental method of distinguishing between the two.

2. Type II

Type II second-order reactions will be designated as those in which the initial concentrations of A and B are not equal but are similar. In actual practice this

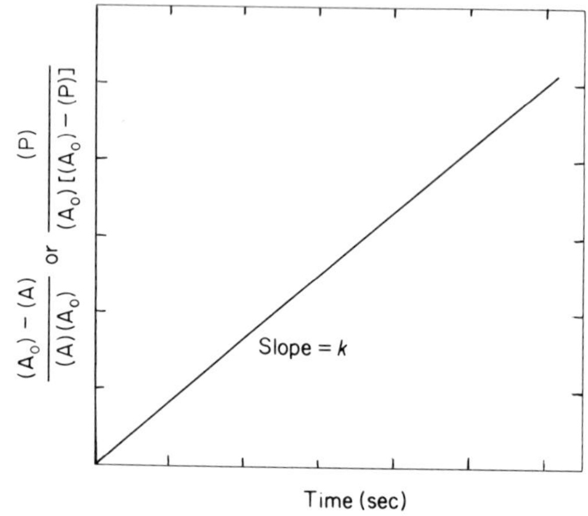

Figure 6 Type I second-order rate process. The line is drawn according to Eq. (60). The slope of the line is the specific reaction rate constant.

is when $(B_0)/(A_0) > 1 < 20$. The reaction rate is first order with respect to each of the reactants.

Equation (56) may be integrated after the variable B is replaced by the relationship $(B) = (B_0) - (A_0) + (A)$ as derived from rearrangement of $(A_0) - (A) = (B_0) - (B)$, which is correct as shown for the stoichiometric equation (62).

$$-\frac{dA}{dt} = k(A)(B) = k(A)[(B_0) - (A_0) + (A)] \tag{62}$$

which for integration rearranges to

$$-\int_{A_0}^{A} \frac{dA}{(A)[(B_0) - (A_0) + (A)]} = k\int_{t_0}^{t} dt \tag{63}$$

The left-hand side of Eq. (63) may be integrated by consulting a table of integrals or by use of the method of partial fractions to give, finally,

$$\frac{2.3}{(B_0) - (A_0)} \log \frac{(A_0)(B)}{(B_0)(A)} = kt \tag{64}$$

Since $(B) = (B_0) - (P)$ and $(A) = (A_0) - (P)$, Eq. (64) may be written in terms of initial reactant and product concentrations as

$$\frac{2.3}{(B_0) - (A_0)} \log \frac{(A_0)[(B_0) - (P)]}{(B_0)[(A_0) - (P)]} = kt \tag{65}$$

If a reaction follows the type II second-order rate law, a plot of the left-hand side of Eq. (64) [or (65)] versus t will give a straight line with an intercept at zero and a slope of k.

The precision with which the rate constant may be determined is dependent on the difference $(B_0) - (A_0)$. When these are quite similar, the precision with which k can be evaluated will be low. Experimentally, type II second-order reactions may be avoided by arranging conditions so that the reactions will follow type I or type III second-order kinetics.

3. Type III

Type III second-order reactions will be designated as those in which $(B_0)/(A_0) > 20$. This means that for all practical purposes the concentration of B remains constant throughout the reaction and the reaction then is first order with respect to A. In this case the data are analyzed according to Eq. (41) [or (42)] to give the first-order rate constant k_{obs}, which is then divided by the concentration (B_0) to give k. The rate constant k is a true constant, whereas the rate constant k_{obs} is dependent on (B_0). Therefore, type III second-order rate reactions are covered by the special situation described by Eq. (44), where (E_0) is replaced by the more general term (B_0).

4. Second-Order Autocatalytic Reactions

Because of its importance in enzymology, special mention will be made of second-order autocatalytic reactions. These are encountered wherever the product of a reaction becomes one of the reactants. Consider as a specific example of this the conversion of the proenzyme trypsinogen to trypsin as catalyzed by trypsin (Fig. 7; see Chapter 20, Section II.C.1):

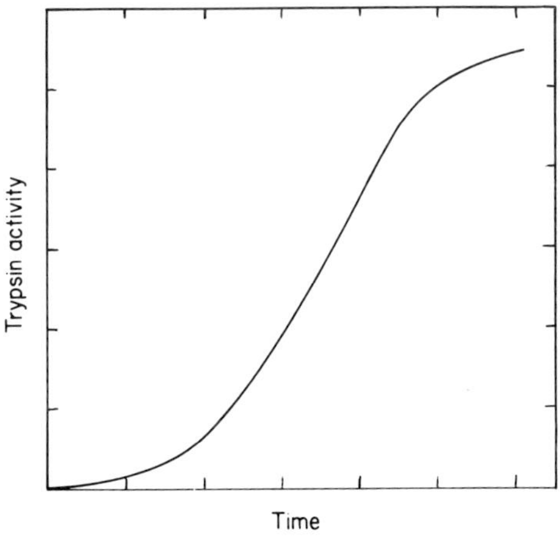

Figure 7 Typical autocatalytic progress curve using the trypsin-trypsinogen system as an example.

$$\text{trypsinogen} \xrightarrow{\text{trypsin}} \text{trypsin} \tag{66}$$

The rate of reaction is described by the rate expression $-dA/dt = k(A)(B)$, where A is the concentration of trypsinogen and B the concentration of trypsin at any time. But unlike the case of type II second-order reactions discussed above, the concentration of B increases with time, so that the relationship $(A_0) - (A) = (B) - (B_0)$ holds. The expression

$$-\frac{dA}{(A)[(A_0) + (B_0) - (A)]} = k\,dt \tag{67}$$

then is integrated to give

$$\frac{2.3}{(B_0) + (A_0)} \log \frac{(A_0)(B)}{(B_0)(A)} = kt \tag{68}$$

III. MOLECULARITY VERSUS ORDER OF REACTION

In the discussion so far we have assumed that the stoichiometric equation as written also indicates the experimentally determined dependence of rate of reaction on the individual reactants. This is not necessarily so. If one reactant is in large excess (e.g., water), the rate of reaction will appear to be independent of that reactant. In this case the reaction is an example of type III second order. In cases where the stoichiometric equation does not adequately reflect the individual rate-determining steps the experimentally determined order will not correspond to the rate predicted from the balanced equation. In summary, molecularity reflects the number of reactant species involved in the stoichiometric equation, while order is the experimentally determined dependence of rate on concentrations of the species. For example, take the general reaction

$$2A + B \rightarrow C + D \tag{69}$$

which experimentally might be found to obey the rate law $-dA/dt = k(A)^2$. This is a termolecular reaction but experimentally might be found to be second order as a result of the slow step involving a combination of two molecules of A before B reacts with A_2 in a fast step. Fractional orders of reactions are rarely observed in enzyme-catalyzed reactions, unlike those observed for noncatalyzed reactions.

IV. RATES OF ENZYME-CATALYZED REACTIONS

In the discussion that follows we consider an enzyme-catalyzed reaction to be described by

$$E + A \underset{k_{-1}}{\overset{k_1}{\rightleftharpoons}} EA \xrightarrow{k_2} E + P \tag{70}$$

The rate of the reaction will always depend on the concentration of the enzyme even though it is a catalyst that should remain constant throughout the reaction.

Enzyme-catalyzed reactions follow the progress curve shown in Fig. 8. The initial part of the curve, where the rate dP/dt is increasing with time, is called the *pre-steady-state* part of the curve and is a result of the equilibration of the reaction shown in Eq. (70). It is the time required for $-dEA/dt = dEA/dt$. Since EA is an adsorptive complex and its rate of formation is generally diffusion controlled, the pre-steady-state is over in a few *milliseconds*. The rate of equilibration must be studied by special techniques such as stopped-flow or temperature, pH, or pressure perturbation methods. With the techniques used in the majority of laboratories, evidence for this pre-steady state is not obtained. When the experiment is not performed properly, what looks like a pre-steady state may be observed. However, this is a result of failure to equilibrate the solutions at the temperature used, to have the enzyme in the fully active form, to mix properly, hysteresis, and so on.

The pre-steady state is followed by the constant rate portion of the progress curve. If (A_0) is much larger than K_m (i.e., the enzyme is saturated with substrate), there will be a period in which $dP/dt = k$ [zero order with respect to (A)]. The length of this period will depend, among other things, on (A_0) in relation to K_m.

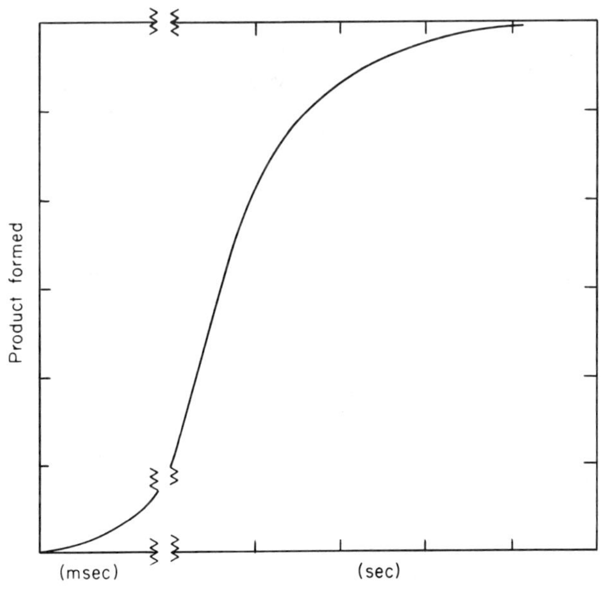

Figure 8 Typical enzyme-catalyzed reaction showing the pre-steady-state (millisecond scale), the constant rate, and the declining rate portions of the progress curve.

Following the linear portion, the rate of reaction will decrease. This decrease may be the result of one or more of the following: (a) decrease in (A) until the enzyme is no longer saturated with A; (b) instability of the enzyme so that (E_0), assumed to be constant, is not constant; (c) inhibition of the reaction by accumulation of products; and (d) in a reversible reaction the back reaction will become more and more important as product concentration increases.

A. Initial Rates

Rates of enzyme-catalyzed reactions are most frequently determined by a method referred to as the *initial rate* (or *initial velocity*) *method*. Essentially, this method involves determination of rate of the reaction as close to time zero as possible. The method has the advantages of (a) not being greatly affected by instability of the enzyme (if reaction is initiated by addition of enzyme and the rate of activity loss of enzyme is low compared to rate of reaction), (b) not being influenced by product since this is zero initially, and (c) the substrate concentration can be taken as that initially added to the reaction provided that no more than about 5% is converted to product. In the case where K_P [K_P = (E)(P)/(EP)] is equal to or smaller than K_m, initial rates will be too low, because P is a competitive inhibitor with the substrate, competing for E. This is not found with most of the enzymes that we deal with. The disadvantage of the initial rate method is that it does not tell one anything about the effect of products, which can be quite important for understanding the mechanism of action of the enzyme.

Determination of initial rates is shown schematically in Fig. 9, where five reactions have been run under identical conditions except for the variable (A_0). (In experiments designed for different purposes, another variable, such as temperature or pH, could be used.) The method most frequently used is to plot the data as shown and to draw a line tangent to the initial part of the curve with the requirement also that the line go through zero if the proper controls are included. The slope of the line then gives the initial rate of that reaction. An even better determination of initial rates can be obtained by determining dP/dt versus t as t approaches zero. However, when the data are properly obtained, the first method is usually adequate.

Several features of the data presented in Fig. 9 should be noted. The initial rates may differ appreciably from the rates calculated at point t. Frequently, experiments are performed where the rates of a reaction are determined by analysis of a single aliquot at time t (Fig. 9). This procedure is valid only if the relation between P and t is linear over that time. The method is hazardous at best. It should also be noted that, in general, the faster the rate, the more difficulty there is in obtaining consistent initial rates. To offset this, more data points are needed close to beginning of the reaction. Most of the problems of obtaining good initial rates can be alleviated if the reaction can be monitored continuously (in a spectrophotometer, pH Stat, etc.) with the aid of a variable-speed recorder. In this way

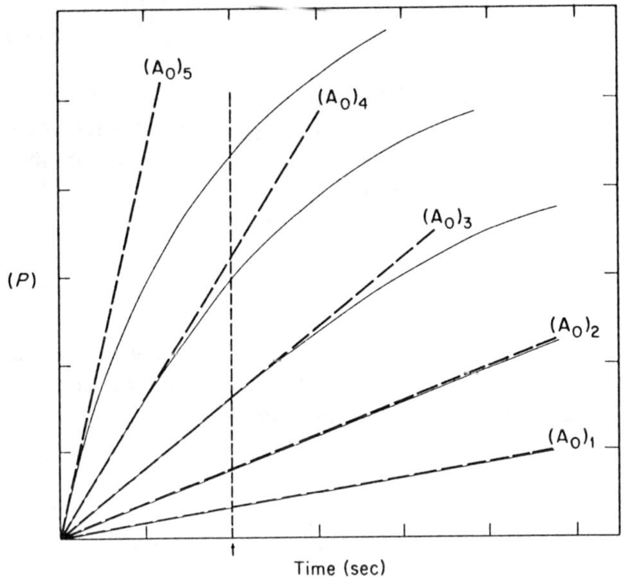

Figure 9 Method of determining initial rates of reactions. The solid lines are the experimentally determined data, the dashed lines are tangents drawn to the initial slope of the experimental data. Note the marked difference between the actual amount of product formed at time *t* in reactions $(A_0)_4$ and $(A_0)_5$ compared with that predicted from initial rates.

the time scale can be stretched out for fast reactions. Because of mixing rates of 5 to 10 sec, initial rates should be based on data over at least 60 to 120 sec. In general, our discussions in the following sections of this book will deal only with initial rates.

In a few cases, initial rates are not linear, as shown in Fig. 9. There is a lag period prior to reaching a constant rate. The lag period, which may be a few seconds to 120 min in one case [1], is called *hysteresis* [2]. The lag period may result from slow responses caused by ligand-induced conformational changes of the enzyme, displacement of a tightly bound ligand by substrate or by enzyme polymerization–depolymerization. This can be visualized as $E_{inact} \rightleftharpoons E_{act}$. The equilibrium is shifted to the right on adding substrate, which binds only to E_{act}. Initial velocities are measured after the rate becomes constant.

B. Rates and Rate Constants from Complete Reaction Curves

The velocity of a reaction can be determined at any point on a total progress curve. Best values are obtained by use of a computer and the integrated rate equation

$$(E_0)kt = 2.3K_m \log \frac{(A_0)}{(A_0) - (P)} + (P) \tag{71}$$

However, good values can be obtained by the method illustrated in Fig. 10 using a differential approach. The data are plotted, preferably by continuous recording of the reaction progress curve, on as large a scale as practicable. Lines are then drawn tangent to the curve as shown. Slopes of the straight lines give a measure of the rate of reaction at the substrate concentration calculated from

$$(A_i) = (A_0) - \frac{(P_{i+1}) + (P_i)}{2} \tag{72}$$

where the meaning of A_i, P_{i+1}, and P_i are indicated in Fig. 10. This method of determining velocities and thus V_{max} and K_m, although used infrequently, can be extremely useful. If there are no problems with enzyme instability, product inhibition, or reversible reaction, the results obtained by this method should agree with those obtained by the initial rate method with a fraction of the total experimental effect and cost. Starting the reaction at $(A_0) \gg K_m$ so that the initial rate is the maximum rate (V_{max}) provides an internal check on the system since the

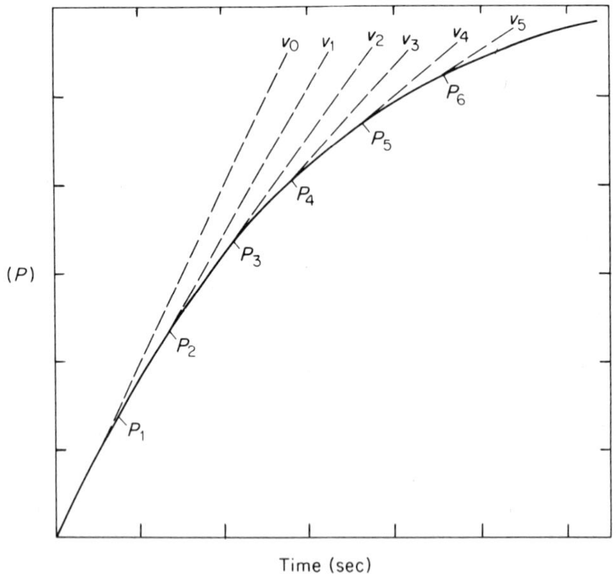

Figure 10 Method of determining the velocities, dP/dt, as a function of substrate concentration at points along a complete reaction progress curve. The closer together the points P_i are taken, the better the fit to the experimental line.

same V_{max} should be obtained by analysis of the complete data by methods described in Chapter 6.

REVIEW QUESTIONS

1. Write all possible differential rate expressions for the reaction 2A + 3B → 2P. What is the molecularity of this reaction?

2. What does the rate expression $-dB/dt = 3k_1(A)(B)^3 - 3k_{-1}(P)^2$ indicate in terms of a stoichiometric equation?

3. Experimentally, how might you distinguish between a consecutive reaction in which A is converted to P and two simultaneous reactions that lead to the conversion of A to P and Q? How would you determine the rate constant for each of these processes?

4. The half-life of a first-order reaction was found to be 300 sec at pH 6.0 and 30.0°C. Calculate the fraction of original substrate left after 30 min.

5. Substrate A was converted to product by a first-order reaction in which $t_{1/2}$ was 200 sec at pH 5 and 25.0°C. In a different reaction, substrate B was converted to product by a zero-order reaction with $t_{1/2}$ = 500 sec at pH 5 and 25.0°C. Calculate the time it will take for 99% of the original substrates to be converted to product in each case.

6. Consider the consecutive reactions

$$A \xrightarrow{k_1} B \xrightarrow{k_2} C$$

Calculate the concentrations of A, B, and C at 100-sec intervals between 0 and 900 sec when

(a) $(A_0) = 1 \times 10^{-2}\ M$, $(B_0) = (C_0) = 0$, $k_1 = 1 \times 10^{-3}\ \text{sec}^{-1}$, and $k_2 = 1 \times 10^{-2}\ \text{sec}^{-1}$.

(b) $(A_0) = 1 \times 10^{-2}\ M$, $(B_0) = (C_0) = 0$, $k_1 = 1 \times 10^{-2}\ \text{sec}^{-1}$, and $k_2 = 1 \times 10^{-3}\ \text{sec}^{-1}$.

7. A reaction was run at 25.0°C and pH 7.0. Aliquots removed at 0, 95, 195, 250, 310, 380, 450, 510, 600, and 685 sec contained ($\times 10^2$) 5.63, 5.15, 4.61, 4.33, 4.07, 3.76, 3.50, 3.30, 3.00, and 2.76 *M* substrate, respectively. What is the order of the reaction? Calculate the specific reaction rate constant.

8. In two separate experiments, (a) and (b), reactants A and B were mixed at equimolar concentrations of $3.60 \times 10^{-2}\ M$ and $4.50 \times 10^{-3}\ M$, respectively, at pH 5.5 and 28°C. In (a), $t_{1/2}$ was 135 sec, while in (b) it was $1.08 \times 10^3\ \text{sec}^{-1}$. What order would you predict for the reactions? Prove that the reaction follows the predicted order by calculating the specific reaction rate constants for the two reactions. What conclusions do you draw from data of this nature?

9. Many reactions, such as hydrolysis of esters, involve the participation of

water as one of the substrates. What is the usual molecularity and order of these reactions?

REFERENCES

General References

Amdur, I., and G. G. Hammes, *Chemical Kinetics: Principles and Selected Topics*, McGraw-Hill, New York, 1966.

Benson, S. W., *The Foundations of Chemical Kinetics*, McGraw-Hill, New York, 1960.

Frost, A. A., and R. G. Pearson, *Kinetics and Mechanism*, 2nd Ed., Wiley, New York, 1961.

Glasstone, S., K. J. Laidler, and H. Eyring, *The Theory of Rate Processes*, McGraw-Hill, New York, 1941.

Gutfreund, H., *Enzymes: Physical Principles*, Wiley-Interscience, London, 1972.

Laidler, K. J., *Chemical Kinetics*, 2nd Ed., McGraw-Hill, New York, 1965.

Laidler, K. J., *Reaction Kinetics*, Pergamon Press, Elmsford, N.Y., 1963.

Cited References

1. J. R. Whitaker and P. V. Caldwell, *Arch. Biochem. Biophys.*, **159**, 188 (1973).
2. C. Frieden, *J. Biol. Chem.*, **245**, 5788 (1970).

6

Effect of Substrate Concentration on Rates of Enzyme-Catalyzed Reactions

Rates of enzyme-catalyzed reactions are influenced by a number of experimental parameters. Among these parameters are substrate concentration, enzyme concentration, pH, temperature, presence of activators and inhibitors, dielectric constant, and ionic strength of solvent. In the next several chapters we systematically discuss several of these parameters.

I. GENERAL CONCEPT OF ENZYME–SUBSTRATE COMBINATION

When all other parameters, including enzyme concentration, are kept constant in an experiment and the initial substrate concentration is varied between wide limits, changes in initial rates (same as initial velocities) are described by Fig. 1, if the reaction follows Michaelis–Menten kinetics. When more than one substrate is involved in the reaction, the concentration of the second substrate is kept constant.

The biphasic nature of this curve was first studied independently by Henri [1] and Brown [2] in 1902 for the hydrolysis of sucrose by the enzyme invertase (sucrase):

$$\underset{\text{sucrose}}{C_{12}H_{22}O_{11}} + H_2O \xrightarrow{\text{invertase}} \underset{\text{glucose}}{C_6H_{12}O_6} + \underset{\text{fructose}}{C_6H_{12}O_6} \qquad (1)$$

Henri explained the biphasic nature of the curve by assuming that substrate and enzyme combine during the reaction and that the substrate undergoes a chemical change only after undergoing this combination:

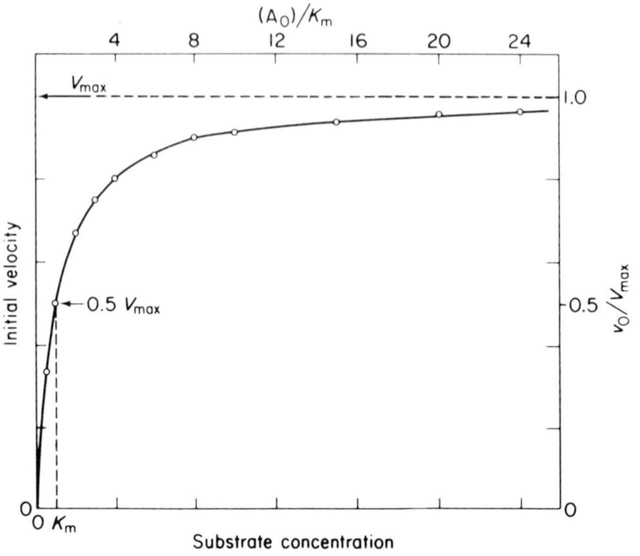

Figure 1 Variation of observed velocity with substrate concentration for an enzyme-catalyzed reaction. The data are plotted according to Eq. (17) of this chapter.

$$\text{sucrose} + \text{invertase} \underset{k_{-1}}{\overset{k_1}{\rightleftharpoons}} \text{sucrose}\cdot\text{invertase} \underset{H_2O}{\overset{k_2}{\rightleftharpoons}}$$
$$\text{invertase} + \text{glucose} + \text{fructose} \tag{2}$$

Equation (2) may be written in a more general form as

$$E + A \underset{k_{-1}}{\overset{k_1}{\rightleftharpoons}} EA \underset{H_2O}{\overset{k_2}{\rightleftharpoons}} E + P_1 + P_2 \tag{3}$$

It follows from Henri's assumption that at low substrate concentrations not all molecules of enzyme are combined with substrate. As the substrate concentration is increased, more and more enzyme molecules are combined with substrate until finally, at high concentration, all enzyme molecules are so combined (Fig. 2). Once all the enzyme molecules have combined with substrate (enzyme saturated with substrate), further increases in substrate concentration will not increase the rate of reaction. The velocity at this point is the maximum velocity, V_{max}.

It is difficult to verify experimentally the assumption that an enzyme·substrate complex is an obligatory intermediate in the pathway for conversion of substrate to product. Spectroscopic evidence for an enzyme·substrate complex in the reaction of D-amino acid oxidase with a D-amino acid in the absence of the sec-

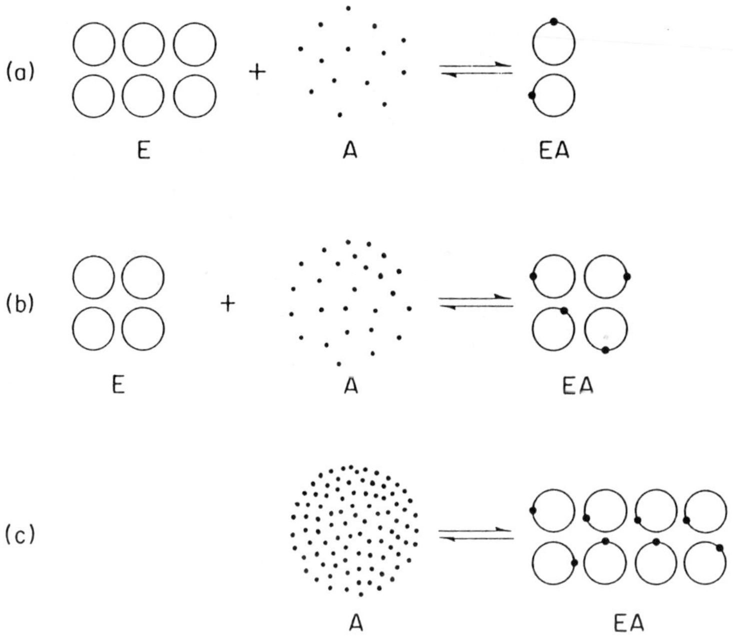

Figure 2 Schematic representation of effect of substrate A concentration on the amount of enzyme·substrate complex, EA, formed.

ond substrate, O_2, has been reported [3]. Direct evidence for existence of enzyme·substrate complexes comes from x-ray crystallographic work on enzymes such as ribonuclease, chymotrypsin, lysozyme, and carboxypeptidase A. More indirect evidence is obtained from kinetic studies in which binding constants of substrate (or competitive inhibitor) to the enzyme can be measured and from studies in which modifications are made to the substrate or groups on the enzyme.

The first substantial mathematical treatment of effect of substrate concentration on rate of enzyme-catalyzed reactions was by Michaelis and Menten in 1913 [4] and was based on the earlier hypotheses of Brown and of Henri that an adsorption complex is an obligatory intermediate on the pathway to formation of products. In their treatment according to Eq. (3), Michaelis and Menten assumed that equilibrium among the enzyme E, substrate A, and enzyme·substrate complex EA was reached very rapidly and that k_2 was small compared to k_{-1} (and k_1). Based on this assumption, $K_m = k_{-1}/k_1 = K_s$, which is an equilibrium constant. Later, Briggs and Haldane [5] suggested that this assumption may not be correct for all enzymes. They derived an expression relating the observed velocity with the maximum velocity and K_m based on a steady-state assumption. We shall derive the

Michaelis–Menten equation by the Briggs–Haldane method, but later we shall show that this is only a special case of a more general situation.

Under steady-state conditions it is assumed that the concentration of enzyme·substrate intermediate is constant over the period of point measurement (absorbance, for example, requiring 1 to 5 sec) or at least its rate of change is small compared with the rate of attainment of the steady state. Constant concentrations of intermediate are maintained over a long time (many seconds or minutes) only under conditions where $(A_0) \gg K_m$, so that the enzyme is completely saturated with substrate. Some of these relationships are shown in Fig. 3. When $(E_0) = (A_0)$ [Fig. 3(a)], there is a considerable buildup of (EA) with a corresponding decrease in (E) when the rate k_1(E)(A) is equal to or faster than the rate k_2(EA). Notice the short lag period (msec scale) in the production of product P, which reflects the time

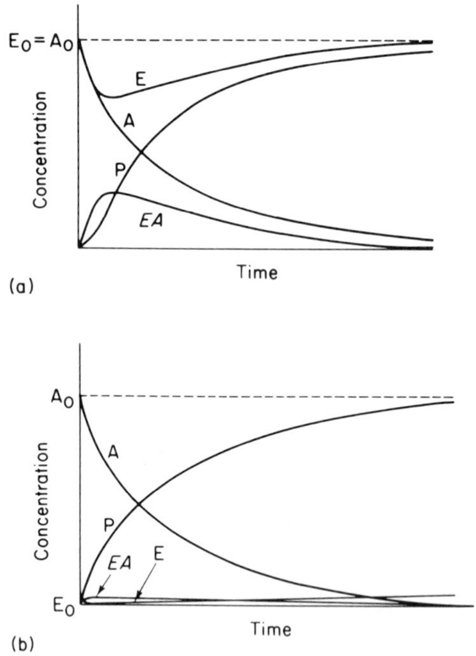

Figure 3 Hypothetical time course of enzyme-catalyzed reactions. In (a) the reaction is irreversible, initial starting concentrations of enzyme and substrate are equal, and the rate k_1(E)(A) is comparable with or faster than the rate k_2(EA). In (b) the reaction is irreversible but initial substrate concentration is much higher than initial enzyme concentration. The symbols E, A, EA, and P represent free enzyme, free substrate, enzyme·substrate complex, and product concentrations, respectively at any time. (From Ref. 6, pp. 25 and 26, by courtesy of J. Westley.)

for the system to reach equilibrium. In most cases substrate concentration is much higher than enzyme concentration [(A)/(E) ratio = 10^5 to 10^6:1] and the reaction is described by Fig. 3(b). Compared to (A_0), (EA) never reaches a very high level because of the small (E_0) used. This is true even though the enzyme is saturated with substrate in the example shown as indicated by the absence of free enzyme early in the reaction. The free enzyme concentration increases steadily with time as the substrate concentration decreases. This reflects insufficient substrate to keep the enzyme saturated. Only at one time is $|d(\mathrm{EA})/dt| = 0$ for a long time. This does not invalidate steady-state assumptions, however, since even at (A_0) much less than saturating, the rate of change of concentration of intermediate is small compared to the rate of attainment of the steady state and the time required for a point measurement. So steady-state assumptions are valid particularly for initial rate data.

II. ONE–SUBSTRATE REACTIONS

A. One Kinetically Observable Intermediate; Initial Rate Conditions

The situation treated is defined kinetically by Eq. (3). The rate expression for Eq. (3) is derived based on the following definitions and assumptions. Let

(E_0) = total enzyme concentration
(A_0) = total substrate concentration, which is assumed to be much larger than (E_0); therefore, (A) = (A_0) during initial measurement (<5% A converted to P)
(EA) = enzyme·substrate complex concentration
(P) = product concentration ~ 0 [i.e., initial rates are being observed (see Chapter 5, Section IV.A]; in Eq. (3), (P_1) = (P_2) (we assume that only P is formed in the general derivation)
(E) = free enzyme concentration

The rate of formation of EA is

$$\frac{d\mathrm{EA}}{dt} = k_1(\mathrm{E})(\mathrm{A}_0) \tag{4}$$

while the rate of disappearance of EA is

$$-\frac{d\mathrm{EA}}{dt} = k_{-1}(\mathrm{EA}) + k_2(\mathrm{EA}) \tag{5}$$

When steady-state conditions are reached, $d\mathrm{EA}/dt = -d\mathrm{EA}/dt$, so that

$$k_1(\mathrm{E})(\mathrm{A}_0) = k_{-1}(\mathrm{EA}) + k_2(\mathrm{EA}) = (k_{-1} + k_2)(\mathrm{EA}) \tag{6}$$

The free enzyme concentration (E) cannot be measured experimentally but it is related to (E_0) by the conservation equation

$$(E) = (E_0) - (EA) \tag{7}$$

Substitution of relationship (7) into Eq. (6) gives

$$k_1[(E_0) - (EA)](A_0) = k_1(E_0)(A_0) - k_1(EA)(A_0) = (k_{-1} + k_2)(EA) \tag{8}$$

Rearrangement of Eq. (8) results in

$$\begin{aligned} k_1(E_0)(A_0) &= (k_{-1} + k_2)(EA) + k_1(EA)(A_0) \\ &= (EA)[k_{-1} + k_2 + k_1(A_0)] \end{aligned} \tag{9}$$

and on solving for (EA), we obtain

$$(EA) = \frac{k_1(E_0)(A_0)}{k_{-1} + k_2 + k_1(A_0)} \tag{10}$$

Dividing both numerator and denominator by k_1 gives

$$(EA) = \frac{(E_0)(A_0)}{(k_{-1} + k_2)/k_1 + (A_0)} \tag{11}$$

In the denominator is a collection of rate constants. These are defined as

$$K_m = \frac{k_{-1} + k_2}{k_1} \tag{12}$$

K_m is called the Michaelis constant. Substitution of Eq. (12) into Eq. (11) yields

$$(EA) = \frac{(E_0)(A_0)}{K_m + (A_0)} \tag{13}$$

(EA) cannot be determined experimentally in most cases but we can substitute for it since the rate of formation of product is

$$\frac{dP}{dt} = k_2(EA) = v_0 \tag{14}$$

where v_0 is the observed velocity under initial rate conditions. Combination of Eqs. (13) and (14) gives

$$v_0 = \frac{k_2(E_0)(A_0)}{K_m + (A_0)} \tag{15}$$

The maximum velocity, V_{max}, is attained when all active sites of the enzyme are combined with substrate so that (E_0) = (EA). Under these conditions

$$k_2(EA) = k_2(E_0) = V_{max} \tag{16}$$

and

$$v_0 = \frac{V_{max}(A_0)}{K_m + (A_0)} \tag{17}$$

Equation (17) is the equation of a right hyperbola and describes the data as plotted in Fig. 1. The initial velocity of the reaction is one-half the maximum velocity when the substrate concentration is equal to K_m. However, even at $(A_0) = 24K_m$, the observed velocity (v_0) is only 0.96 of the maximum velocity (V_{max}). These relationships are best seen by the plot of v_0/V_{max} versus $(A_0)/K_m$ shown (Fig. 1). This curve will be identical for all enzyme-catalyzed reactions that obey Eq. (17).

It is essential to stress that V_{max} is dependent on enzyme concentration [Eq. (16)]. Whenever enzyme concentration is changed in a reaction, V_{max} will change. On the other hand, k_2 is a true constant.

1. Specific Relations of Substrate Concentration to K_m

Consider the consequences of the following three situations.

a. $(A_0) \ll K_m$. When (A_0) is less than 0.01 of K_m the (A_0) term in the denominator of Eq. (17) may be ignored so that

$$v_0 = \frac{V_{max}(A_0)}{K_m} = \frac{k_2(E_0)(A_0)}{K_m} = k'(A_0) \tag{18}$$

where $k' = V_{max}/K_m$. Experimentally, data obtained at $(A_0) \leqslant 0.05K_m$ are treated sufficiently well by Eq. (18). As shown in Chapter 5, Eq. (18) is a description of a first-order rate reaction with respect to substrate concentration.

b. $(A_0) \gg K_m$. When $(A_0) \geqslant 100K_m$, the K_m term in the denominator of Eq. (17) may be ignored, so that

$$v_0 = V_{max} = k_2(E_0) \tag{19}$$

which describes the situation when the enzyme is saturated with substrate. Equation (19) is the description of a zero-order rate reaction with respect to substrate concentration.

c. $(A_0) = K_m$. Substitution in Eq. (17) for $K_m = A_0$ yields

$$v_0 = \frac{V_{max}}{(A_0) + (A_0)} = \frac{V_{max}(A_0)}{2(A_0)} = \frac{V_{max}}{2} \tag{20}$$

K_m is then equal to the initial substrate concentration at which $v_0 = 0.5\ V_{max}$. This, then, is the operational definition of K_m. K_m is the apparent dynamic dissociation constant under steady-state conditions and approaches K_s, the true dissociation constant, only when $k_2 \ll k_{-1}$ [Eq. (12)] in this particular case.

Some typical values for k_{-1}, k_1, and k_2 are given in Table 1. In general, k_1 is on the order of 10^6 to $10^8\ M^{-1}\ \text{sec}^{-1}$, which approaches the value of $10^9\ M^{-1}\ \text{sec}^{-1}$ for the maximum rate of diffusion of a small molecule from solution to the active

Table 1 Rate Constants for Some Selected Enzymes

Enzyme	Substrate	k_1[a] (M^{-1} sec^{-1})	k_{-1}[b] (sec^{-1})	k_0[c] (sec^{-1})
Fumarase	Fumarate	$>10^9$	$>4.5 \times 10^4$	10^3
Acetylcholinesterase	Acetylcholine	10^9	—	10^3
Liver alcohol	NAD^+	5.3×10^5	74	10^3
dehydrogenase	NADH	1.1×10^7	3.1	
	Ethanol	$>1.2 \times 10^4$	>74	
Catalase	H_2O_2	5×10^6	—	10^7
Peroxidase	H_2O_2	9×10^6	<1.4	10^6
Hexokinase	Glucose	3.7×10^6	1.5×10^3	10^3
Urease	Urea	$>5 \times 10^6$	—	10^4
Chymotrypsin				10^2 to 10^3
Trypsin				10^2 to 10^3
Ribonuclease				10^2
Papain				10^1

Source: Ref. 7.
[a]Rate of formation of enzyme·substrate complex.
[b]Rate of dissociation of enzyme·substrate complex.
[c]Order of magnitude of the turnover number in moles of substrate converted to product per second per mole of enzyme. k_0 ($\equiv k_{cat}$) is the observed rate constant and may or may not involve a single rate-limiting step.

site of an enzyme. k_{-1} is usually much smaller than k_1 and varies from 10^1 to 10^4 sec^{-1}. The value of k_2 ranges from 10^1 to 10^6 sec^{-1}. It is better to refer to this constant as k_{cat} generally since in many cases the experimentally observed constant is not a single constant as is k_2 defined by Eq. (3).

2. Determination of K_m and V_{max}

If experimental data were found to fit closely the theoretical curve shown in Fig. 1, K_m and V_{max} could be evaluated from such a plot. However, plots based on experimental data are often complicated by substrate inhibition or activation at high A concentration, by substrate insolubility, and so on. Therefore, it is best to use methods of plotting v_0 versus (A_0) data to display them in linear form. Three of these methods are described below.

Data for evaluation of K_m and V_{max} should be obtained within the region of K_m. For best results initial velocities should be obtained at 6 to 10 substrate concentrations, ranging from $0.1K_m$ to $10K_m$. However, good results may be obtained where (A_0) = $0.3K_m$ to $3K_m$. The data should be treated statistically, including a measurement of the standard deviations in both K_m and V_{max} [8].

a. Lineweaver–Burk Method

The Lineweaver–Burk method, first described in 1934, is the method most frequently used [9]. Taking the reciprocal of Eq. (17) gives

$$\frac{1}{v_0} = \frac{K_m}{V_{max}(A_0)} + \frac{1}{V_{max}} \tag{21}$$

Equation (21) is the equation of a straight line, $y = ax + b$, where y is $1/v_0$, a (the slope) is K_m/V_{max}, x is $1/(A_0)$, and b (the intercept) is $1/V_{max}$. Data of Table 2 plotted according to Eq. (21) are shown in Fig. 4.

The intercept of the line on the y axis is $1/V_{max}$, and the intercept on the x axis is $-1/K_m$. The slope of the line is K_m/V_{max}. Therefore, V_{max} and K_m can be obtained readily by this method.

b. Augustinsson Method

If the Lineweaver–Burk equation [Eq. (21)] is multiplied by (A_0), one obtains

$$\frac{(A_0)}{v_0} = \frac{K_m}{V_{max}} + \frac{(A_0)}{V_{max}} \tag{22}$$

Equation (22) is the equation of a straight line where y is $(A_0)/v_0$, a is $1/V_{max}$, x is (A_0), and b is K_m/V_{max}. Data of Table 2 plotted according to Eq. (22) are shown in Fig. 5. Again, both K_m and V_{max} may readily be determined.

c. Eadie–Hofstee Method

The Michaelis–Menten equation [Eq. (17)] can be algebraically rearranged to give Eq. (25). On cross-multiplying,

Table 2 Effect of Substrate Concentration on the Velocity of an Enzyme-Catalyzed Reaction[a]

(A_0) (M) $\times 10^5$	v_0 $(M\ sec^{-1})$ $\times 10^6$	$1/(A_0)$ (M^{-1}) $\times 10^{-4}$	$1/v_0$ $(M^{-1}sec^{-1})$ $\times 10^{-5}$	$(A_0)/v_0$ (sec)	$v_0/(A_0)$ (sec^{-1}) $\times 10^2$
2.00	1.67	5.00	6.00	12.0	8.35
4.00	2.86	2.50	3.50	14.0	7.15
6.00	3.70	1.67	2.70	16.2	6.17
10.0	4.95	1.00	2.02	20.2	4.95
20.0	6.50	0.50	1.54	30.8	3.25
30.0	7.40	0.33	1.35	40.5	2.47
50.0	8.14	0.20	1.23	61.5	1.63

[a]The data are theoretical.

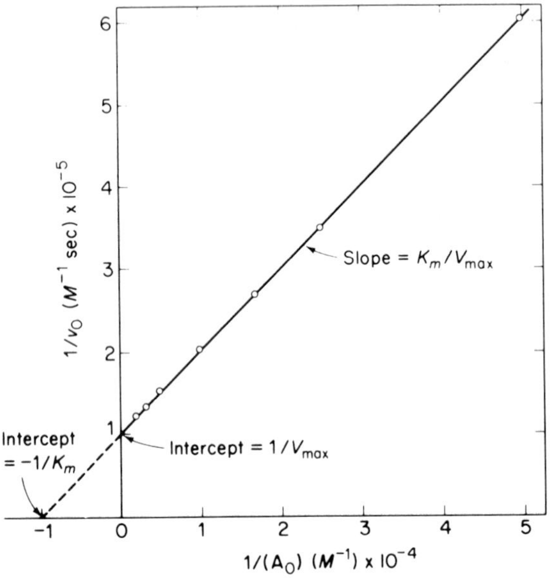

Figure 4 Plot of the substrate–velocity data of Table 2 according to the Lineweaver–Burk method as shown in Eq. (21).

$$v_0K_m + v_0(A_0) = V_{max}(A_0) \tag{23}$$

and subtracting v_0K_m from each side of Eq. (23) yields

$$v_0(A_0) = V_{max}(A_0) - v_0K_m \tag{24}$$

and dividing Eq. (24) by (A_0) gives

$$v_0 = V_{max} - \frac{v_0}{(A_0)}K_m \tag{25}$$

Equation (25) is that of a straight line where y *is* v_0, a is $-K_m$, x is $v_0/(A_0)$, and b is V_{max}. Data of Table 2 plotted according to this equation are shown in Fig. 6.

A number of advantages and disadvantages have been given for each type of plot. In deciding which plot to use for a particular set of data, one should consider such factors as (a) more even spacing of data points along the line, (b) best fit of data points to a straight line, and (c) that in the Eadie–Hofstee plot the least precise experimental parameter, v_0, appears in both the x and y terms, so that the overall precision of determination of K_m and V_{max} may be impaired. With respect to factor (a), the investigator may design the experiment so that the data will be uniformly distributed along the line by whichever method is used to plot the data. A

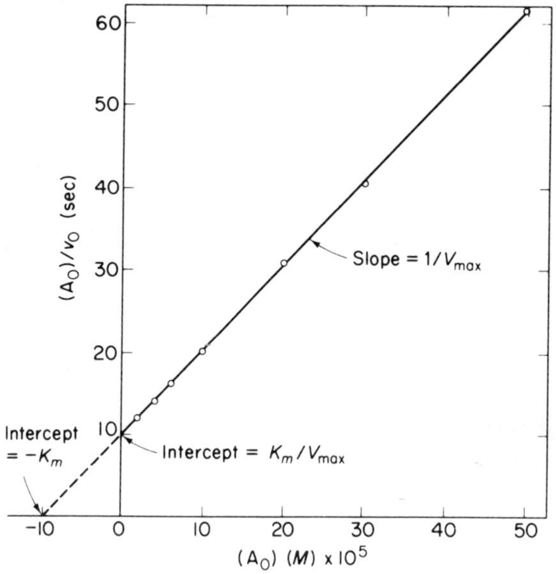

Figure 5 Plot of the substrate–velocity data of Table 2 according to the Augustinsson method, as shown in Eq. (22).

comprehensive discussion of this subject is given in Ref. 10. More workers use the Lineweaver–Burk method than the other two combined.

B. General Case for One-Substrate Reactions

We shall take the general case defined by

$$E + A \underset{k_{-1}}{\overset{k_1}{\rightleftharpoons}} EA \underset{k_{-2}}{\overset{k_2}{\rightleftharpoons}} EX \underset{k_{-3}}{\overset{k_3}{\rightleftharpoons}} E + P \tag{26}$$

There are two intermediates, EA and EX, in the reaction, EA is the Michaelis–Menten complex, but no assumptions are made as to the nature of EX. No assumptions will be made regarding initial rates, but it is assumed that $(A_0) \gg (E_0)$.

The first step in obtaining a velocity expression analogous to Eq. (17) for this reaction is to write differential equations showing rates of change of the intermediate EA and EX (the same designations used in Section II.A are used here):

$$\left|\frac{dEA}{dt}\right| = k_1(E)(A_0) + k_{-2}(EX) - k_{-1}(EA) - k_2(EA) \tag{27}$$

$$\left|\frac{d\text{EX}}{dt}\right| = k_2(\text{EA}) + k_{-3}(\text{E})(\text{P}) - k_{-2}(\text{EX}) - k_3(\text{EX}) \tag{28}$$

The observed velocity of the reaction is

$$v = \frac{d\text{P}}{dt} = k_3(\text{EX}) - k_{-3}(\text{E})(\text{P}) \tag{29}$$

The conservation equation involving enzyme is

$$(\text{E}_0) = (\text{E}) + (\text{EA}) + (\text{EX}) \tag{30}$$

The system may be solved by use of determinants or alternatively by following the procedure shown in Section II.A. In the present situation this becomes quite laborious. The best procedure is the King–Altman method [11].

The King–Altman method can be applied to multisubstrate–enzyme-catalyzed reactions as well as one-substrate reactions. The key features of this method are demonstrated by a solution of Eq. (26) for the observed velocity.

1. The first step in the solution is to write down all differential expressions and the conservation equation(s) [Eqs. (27) to (30)].

2. The second step is to draw a geometrical diagram the corners of which are the various forms of the enzyme, E, EA, and EX, and which shows the pathway of

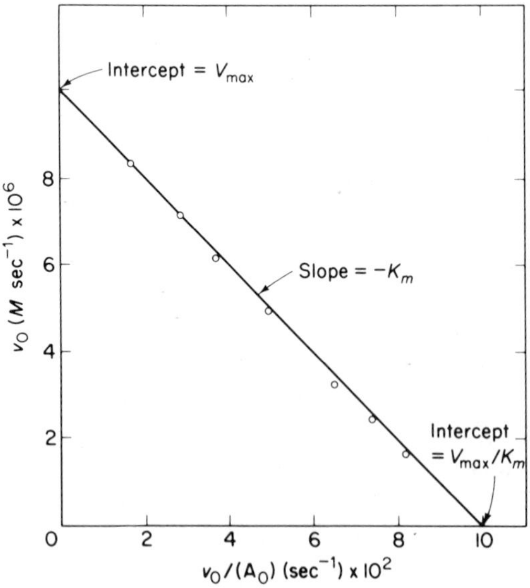

Figure 6 Plot of the substrate–velocity data of Table 2 according to the Eadie–Hofstee method as shown in Eq. (25).

interconversion among the forms. This diagram includes species up to any irreversible step.

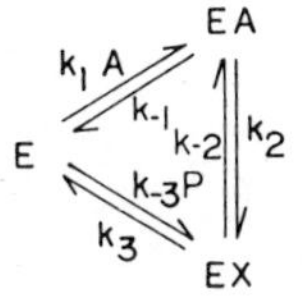

(31)

3. Then draw all possible patterns which show conversion to each of the enzyme forms. These patterns will consist of an $n - 1$ (where n is number of enzyme forms) number of converging lines. Closed loops are prohibited and, if found, are discarded. The general patterns for Eq. (31) are

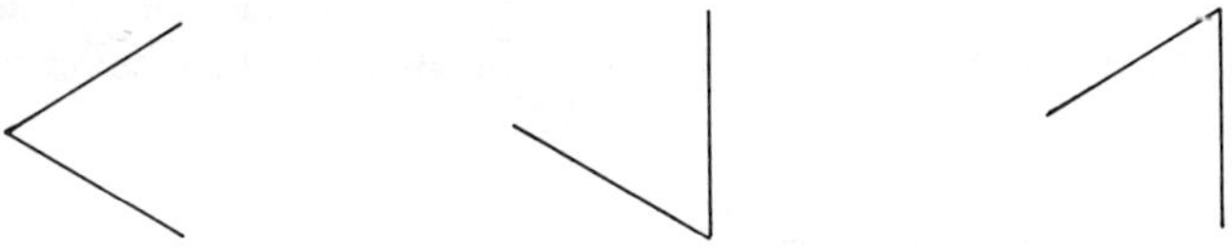

4. Next, write expressions that show the fraction of the total enzyme in each form (the partition functions), which are (E)/(E_0), (EA)/(E_0), and (EX)/(E_0) for Eq. (26). The terms for the individual numerators are obtained from application of step 3 and the denominator of each expression is the sum, Σ, of all the individual numerator terms.

a. $$\frac{(\mathrm{E})}{(\mathrm{E}_0)} = \frac{k_{-1}k_3 + k_2k_3 + k_{-1}k_{-2}}{\Sigma} \tag{32}$$

where

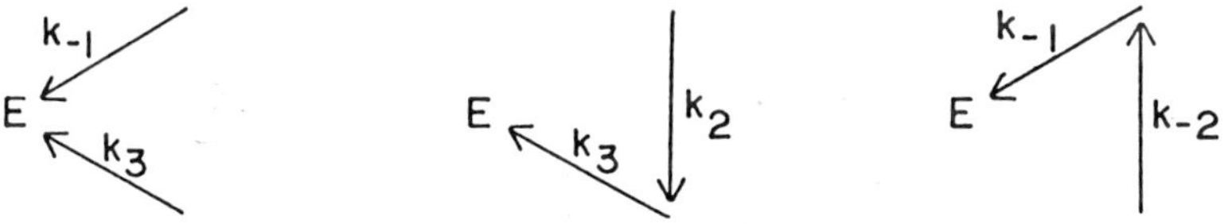

Note that only rate constants on route toward E are used, not those leading away from E.

b. $$\frac{(\mathrm{EA})}{(\mathrm{E}_0)} = \frac{k_1k_3(\mathrm{A}) + k_{-2}k_{-3}(\mathrm{P}) + k_1k_{-2}(\mathrm{A})}{\Sigma} \tag{33}$$

where

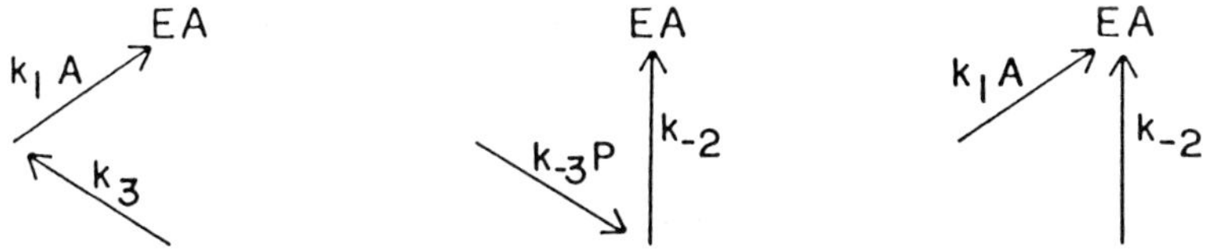

Note that only rate constants A and P on route to EA are used, not those leading away from EA.

c. $$\frac{(\mathrm{EX})}{(\mathrm{E}_0)} = \frac{k_{-1}k_{-3}(\mathrm{P}) + k_2k_{-3}(\mathrm{P}) + k_1k_2(\mathrm{A})}{\Sigma} \tag{34}$$

where

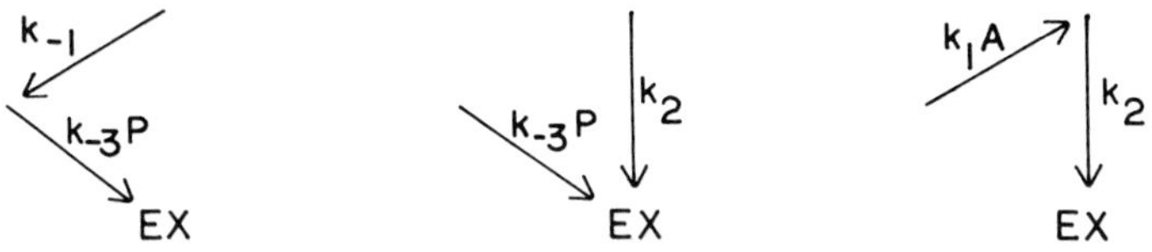

Note that only rate constants A and P on route to EX are used, not those leading away from EX. The denominator, Σ, is given by summation of all the numerator terms.

$$\Sigma = k_{-1}k_3 + k_2k_3 + k_{-1}k_{-2} + k_1k_3(\mathrm{A}) + k_{-2}k_{-3}(\mathrm{P}) + k_1k_{-2}(\mathrm{A}) + k_{-1}k_{-3}(\mathrm{P}) + k_2k_{-3}(\mathrm{P}) + k_1k_2(\mathrm{A}) \tag{35}$$

5. The partition equations (32) to (35) are substituted into Eq. (29), $v = k_3(\mathrm{EX}) - k_{-3}(\mathrm{E})(\mathrm{P})$, which on rearrangement gives an expression for the velocity of the reaction:

$$v = \frac{[k_1k_2k_3(\mathrm{A}) - k_{-1}k_{-2}k_{-3}(\mathrm{P})](\mathrm{E}_0)}{k_{-1}k_3 + k_{-1}k_{-2} + k_2k_3 + k_1(k_2 + k_{-2} + k_3)(\mathrm{A}) + k_{-3}(k_{-1} + k_2 + k_{-2})(\mathrm{P})} \tag{36}$$

This expression may be written as

$$v = \frac{[N_0(\mathrm{A}) - N_1(\mathrm{P})](\mathrm{E}_0)}{D_0 + D_1(\mathrm{A}) + D_2(\mathrm{P})} \tag{37}$$

by letting the indicated letters stand for the set of rate constants, $N_0 = k_1k_2k_3$, $N_1 = k_{-1}k_{-2}k_{-3}$, $D_0 = k_{-1}k_3 + k_{-1}k_{-2} + k_2k_3$, $D_1 = k_1(k_2 + k_{-2} + k_3)$ and $D_2 = k_{-3}(k_{-1} + k_2 + k_{-2})$.

It should be noted that regardless of the complexity of the equation, a plot of experimental data will still fit Figs. 1, 4, 5, and 6 if the reactions conform to Michaelis–Menten kinetics. However, the numerical interpretations of K_m and $V_{\max}$ are more complex.

C. Special Cases for One-Substrate Reactions

Let us consider some restrictions on the general case of Eq. (26) and note the effect that this will have on Eq. (36). Special cases 1 through 4 are not just shown as examples of mental gymnastics but represent specific cases encountered with most of the one-substrate enzymes. Only the forward initial rates will be treated. It should be clear that analogous expressions involving P can be written for the reverse reaction if it is truly reversibly converted back to A.

1. Assume that only initial rate data are being treated. Therefore, all terms involving (P) will be deleted since (P) = 0. In this case, Eq. (37) becomes

$$v_0 = \frac{N_0(\mathrm{A}_0)(\mathrm{E}_0)}{D_0 + D_1(\mathrm{A}_0)} = \frac{N_0(\mathrm{A}_0)(\mathrm{E}_0)/D_1}{(D_0/D_1) + (\mathrm{A}_0)} = \frac{V_{\max}(\mathrm{A}_0)}{K_m + (\mathrm{A}_0)} \tag{38}$$

where $V_{\max} = N_0(\mathrm{E}_0)/D_1$ and $K_m = D_0/D_1$. When N_0, D_0, and D_1 are replaced by the sets of specific rate constants,

$$v_0 = \frac{\dfrac{k_2k_3}{k_2 + k_{-2} + k_3}(\mathrm{E}_0)(\mathrm{A}_0)}{\dfrac{k_{-1}k_3 + k_{-1}k_{-2} + k_2k_3}{k_1(k_2 + k_{-2} + k_3)} + (\mathrm{A}_0)} = \frac{V_{\max}(\mathrm{A}_0)}{K_m + (\mathrm{A}_0)} \tag{39}$$

2. Assume that initial rate data are being analyzed for a system where $k_{-2} \ll k_1$ and k_{-1}. In addition to those terms involving (P), the terms involving k_{-2} will be deleted from Eq. (36) and the relationship will be

$$v_0 = \frac{k_1k_2k_3(\mathrm{A}_0)(\mathrm{E}_0)}{k_{-1}k_3 + k_2k_3 + k_1(k_2 + k_3)(\mathrm{A}_0)} \tag{40}$$

On dividing Eq. (40) by $k_1(k_2 + k_3)$, the modifier term of (A_0) in the denominator, to put it in the general form of the Michaelis–Menten equation, one obtains

$$v_0 = \frac{\dfrac{k_2k_3}{k_2 + k_3}(\mathrm{A}_0)(\mathrm{E}_0)}{\dfrac{(k_{-1} + k_2)k_3}{k_1(k_2 + k_3)} + (\mathrm{A}_0)} = \frac{V_{\max}(\mathrm{A}_0)}{K_m + (\mathrm{A}_0)} \tag{41}$$

3. Assume that initial rate data are being analyzed for a system where $k_{-2} \ll k_1$ and k_{-1}, and $k_2 \ll k_3$. The effect of this on the general equation (36) can best be seen by use of Eq. (41). When $k_2 \ll k_3$ the term $k_2k_3/(k_2 + k_3)$ becomes k_2 since k_2 may be deleted from the denominator whereupon the k_3 terms cancel. The term $k_3/(k_2 + k_3)$ in the denominator is eliminated by the same reasoning, so that

$$v_0 = \frac{k_2(\mathrm{E}_0)(\mathrm{A}_0)}{\dfrac{k_{-1} + k_2}{k_1} + (\mathrm{A}_0)} = \frac{k_2(\mathrm{E}_0)(\mathrm{A}_0)}{K_m + (\mathrm{A}_0)} \tag{42}$$

which is identical to Eq. (15).

4. Assume that initial rate data are being analyzed for a system where $k_{-2} \ll k_{-1}$ and k_1, and $k_3 \ll k_2$. By use of Eq. (41) and the same reasoning as in case 3, the general equation (36) becomes

$$v_0 = \frac{k_3(\mathrm{E}_0)(\mathrm{A}_0)}{\frac{(k_{-1} + k_2)k_3}{k_1 k_2} + (\mathrm{A}_0)} = \frac{V_{\max}(\mathrm{A}_0)}{K_m + (\mathrm{A}_0)} \tag{43}$$

In all cases the data may be plotted by the methods described in Section II.2 so as to obtain $V_{\max}$ and K_m. The specific meaning of $V_{\max}$ and K_m as the maximum velocity and substrate concentration at which $v = 0.5V_{\max}$, respectively, are unchanged. However, the complexity of $V_{\max}$ and K_m, mathematically, will depend on the mechanism in each specific case. The mechanisms cannot be distinguished by initial velocity plots.

III. TWO-SUBSTRATE REACTIONS

While the enzymes belonging to the classification *hydrolases* can be treated satisfactorily by the equations developed for one-substrate reactions, most other enzyme reactions involve two or more substrates. Some enzymes require three or more substrates with formation of multiple products [12–16]. In most cases a cofactor involved is best treated kinetically as a substrate or product of the reaction. Many enzymes of interest to workers in the area of food science require two or more substrates. In the treatment that follows we deal only with special cases of two-substrate systems.

Consider the overall reaction

$$\mathrm{A} + \mathrm{B} \rightleftharpoons \mathrm{P} + \mathrm{Q} \tag{44}$$

As suggested by Cleland [12], substrates will be designated as A, B, C, and D with the first substrate on the enzyme being A, products will be designated at P, Q, R, and S with the first product off being P, and forms of the enzyme involved will be designated as E, F, G, and H with the first form being E.

At least three general mechanisms can be considered by which an enzyme combines with two reactants to form two products.

A. Ordered Mechanism

In this mechanism it is assumed that there is an order in which the substrates are added to and the products removed from the enzyme:

$$\mathrm{E} + \mathrm{A} \underset{k_{-1}}{\overset{k_1}{\rightleftharpoons}} \mathrm{EA} \underset{k_{-2}}{\overset{k_2\mathrm{B}}{\rightleftharpoons}} \mathrm{EAB} \underset{k_{-3}}{\overset{k_3}{\rightleftharpoons}} \mathrm{EPQ} \underset{k_{-4}\mathrm{P}}{\overset{k_4}{\rightleftharpoons}} \mathrm{EQ} \underset{k_{-5}}{\overset{k_5}{\rightleftharpoons}} \mathrm{E} + \mathrm{Q} \tag{45}$$

In this reaction two binary adsorption complexes (EA and EQ) and two ternary adsorption complexes (EAB and EPQ) are involved. The covalent bond making and breaking step(s) is defined by EAB$\rightleftharpoons$EPQ without specifying how many intermediates may be involved.

Cleland [12] suggested that this case be designated as an Ordered Bi Bi mechanism, indicated schematically as

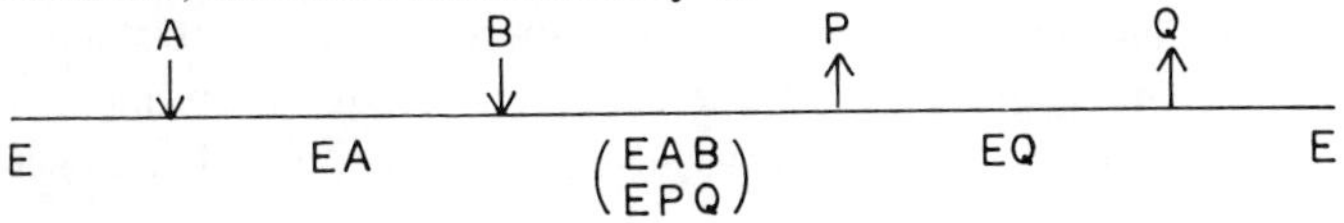

(46)

The designation Bi Bi indicates there are two substrates and two products involved.

B. Random Mechanism

In this mechanism it is assumed that the substrates A and B add to and the products P and Q are released from the enzyme in a random fashion:

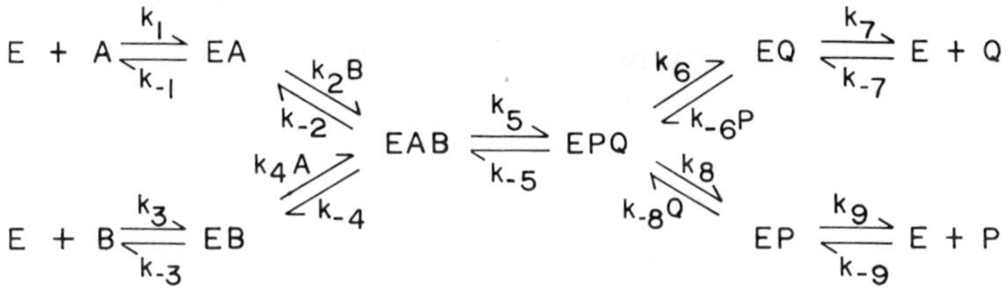

(47)

We treat a Rapid Equilibrium Random mechanism in Eqs. (51) and (52) below. In Eq. (47) there are four binary adsorption intermediates (EA, EB, EP, EQ) and two ternary adsorption intermediates (EAB, EPQ). The bond-making, bond-breaking step is defined by EAB⇌EPQ. This is designated as a Random Bi Bi mechanism (*12*), which is presented diagramatically as

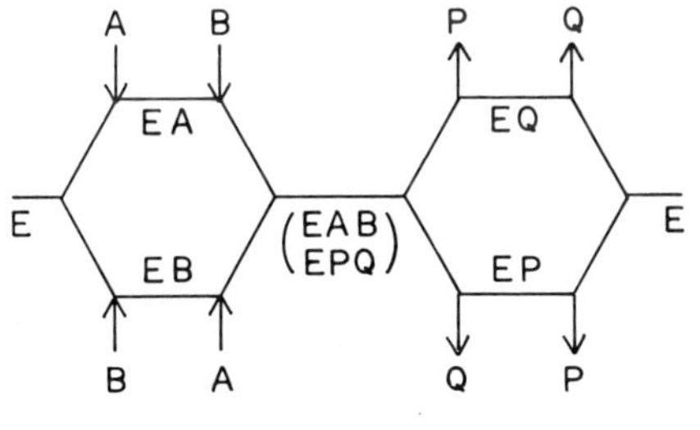

(48)

C. Ping-Pong Mechanism

The Ping-Pong mechanism is so-called because only one substrate (or one product) can be bound to the enzyme at a time in a two substrate-two product reaction:

$$\mathrm{E} + \mathrm{A} \underset{k_{-1}}{\overset{k_1}{\rightleftharpoons}} \mathrm{EA} \underset{k_{-2}}{\overset{k_2}{\rightleftharpoons}} \mathrm{FP} \underset{k_{-3}\mathrm{P}}{\overset{k_3}{\rightleftharpoons}} \mathrm{F} \underset{k_{-4}}{\overset{k_4\mathrm{B}}{\rightleftharpoons}} \mathrm{FB} \underset{k_{-5}}{\overset{k_5}{\rightleftharpoons}} \mathrm{EQ} \underset{k_{-6}}{\overset{k_6}{\rightleftharpoons}} \mathrm{E} + \mathrm{Q} \tag{49}$$

In this reaction four binary adsorption complexes (EA, FP, FB, EQ) are involved. No ternary complexes are involved. The bond-making, bond-breaking steps are indicated as EA$\rightleftharpoons$FP and FB$\rightleftharpoons$EQ. In the reaction there is a transitory change of the nature of the enzyme as indicated by use of F. This enzyme form may differ from the original enzyme in terms of oxidation state of a metal or other cofactor, conformational change, and so on.

This mechanism, referred to as a Ping-Pong Bi Bi mechanism [12], is diagramed as

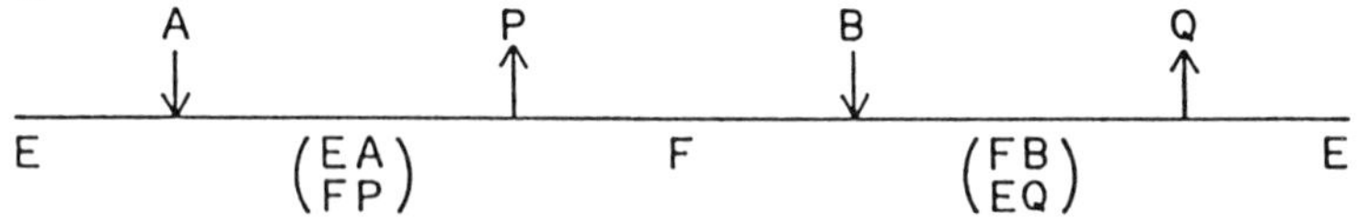

(50)

D. Rate Equations for Two-Substrate Reactions

Differential rate expressions can be written for two-substrate reactions and solved by the same procedures discussed for one-substrate reactions. Because of the complexity of the equations involved, the King–Altman procedure should be used. The rate equations will not be derived here because of the length of their presentations.

1. Sequential Mechanism

Mechanisms where all substrates must add to the enzyme before any products are released are called *sequential* [12]. Both the Ordered and Random mechanism (above) but not the Ping-Pong mechanisms are sequential. The final forms of the general equations for a two-substrate sequential mechanism, assuming initial rates, are given by

$$v_0^f = \frac{V^f_{\max}}{1 + K_a/(\mathrm{A}_0) + K_b/(\mathrm{B}_0) + K_{ia}K_b/(\mathrm{A}_0)(\mathrm{B}_0)} \tag{51}$$

and

$$v_0^r = \frac{V^r_{\max}}{1 + K_p/(\mathrm{P}_0) + K_q/(\mathrm{Q}_0) + K_{iq}K_p/(\mathrm{P}_0)(\mathrm{Q}_0)} \tag{52}$$

which describe the initial rates for the forward and reverse reactions, respectively.

It is helpful to note the similarity of Eq. (51) to Eq. (17) for a one-substrate reaction, which may be written as

$$v_0^f = \frac{V^f_{\max}}{1 + K_a/(\mathrm{A}_0)}$$

by dividing Eq. (17) by (A_0) and substituting the term K_a for K_m without change in meaning. Thus Eq. (51) contains two additional terms, $K_b/(\mathrm{B}_0)$ involving the second substrate and an expression $K_{ia}K_b/(\mathrm{A}_0)(\mathrm{B}_0)$ which involves both substrates.

In Eqs. (51) and (52), K_{ia} and K_{iq} are inhibition constants for A and Q, respectively. K_a is defined as the concentration of reactant A that gives an observed initial velocity v_0 of $0.5V_{max}$ when B is present in saturating concentration. K_b is defined as the concentration of reactant B that gives an observed initial velocity v_0 of $0.5V_{max}$ when A is present in saturating concentration. K_p and K_q are defined in an analogous fashion. K_a, K_b, K_p, and K_q each involve several rate constants and have dimensions of concentration. v_0 and V_{max} have dimensions of concentration per unit time.

Evaluation of the constants K_a, K_b, K_{ia}, and Vf_{max} is made in a straightforward manner. In gathering of experimental data, the concentration of one substrate must be kept constant (usually in less than saturating amounts), as the initial velocity is determined as a function of the concentration of the other substrate. As described for the one-substrate reactions, the experimental concentrations of substrate to be used are those around the values of K_a and K_b.

When v_0 is plotted versus the concentration of variable substrate [designated as (B_0) in the example] while the concentration of the other substrate (A_0) is kept constant plots are obtained as shown in Fig. 7. These plots are identical in form to that shown in Fig. 1. However, although the enzyme concentration is kept constant,

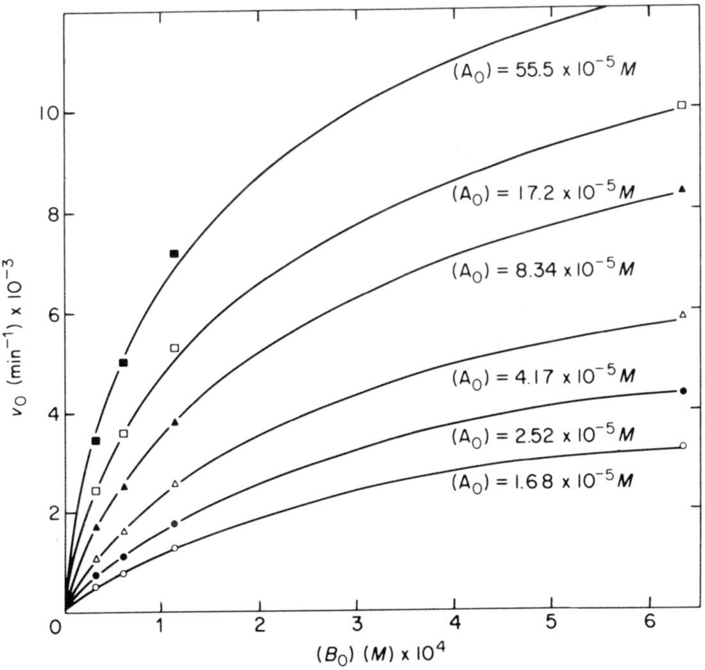

Figure 7 Effect of substrate concentration on initial velocities of an enzyme-catalyzed reaction that requires two substrates, A and B. The data plotted are from Table 4 for lactate dehydrogenase, where A is NAD^+ and B is lactate.

Table 3 Effect of Substrate Concentration (A_0) on Apparent K_b and Vf_{max} in a Two-Substrate Reaction Where (A_0) Is Not Present in Saturating Amounts[a]

(A_0) (*M*) × 10^5	Apparent K_b (*M*) × 10^3	Apparent Vf_{max} (min^{-1}) × 10^{-4}
1.68	3.73	0.544
2.52	3.26	0.688
4.17	2.72	0.862
8.34	2.06	1.08
17.2	1.56	1.23
55.5	1.16	1.36
(A_0) ≫ K_a	0.855	1.43

[a]Data are calculated from data of Tables 4 and 5.

V'_{max} and k'_b are not true constants, but rather, their values depend on the concentration of the second substrate, (A_0). Only when A is present in saturating amounts [(A_0) ≫ K_a] will V'_{max} and k'_b become true constants (Table 3).

The experimental data are analyzed best by the Lineweaver–Burk method. Taking the reciprocal of Eq. (51) yields

$$\frac{1}{vf_0} = \frac{1}{Vf_{max}} + \frac{K_a}{(A_0)Vf_{max}} + \frac{K_b}{(B_0)Vf_{max}} + \frac{K_{ia}K_b}{(A_0)(B_0)Vf_{max}} \tag{53}$$

which can then be rearranged to

$$\frac{1}{vf_0} = \frac{1}{(B_0)}\left[\frac{K_b}{Vf_{max}} + \frac{K_{ia}K_b}{(A_0)Vf_{max}}\right] + \frac{1}{Vf_{max}}\left[1 + \frac{K_a}{(A_0)}\right] \tag{54}$$

or

$$\frac{1}{vf_0} = \frac{1}{(A_0)}\left[\frac{K_a}{Vf_{max}} + \frac{K_{ia}K_b}{(B_0)Vf_{max}}\right] + \frac{1}{Vf_{max}}\left[1 + \frac{K_b}{(B_0)}\right] \tag{55}$$

where (B_0) is the variable in Eq. (54) and (A_0) the variable in Eq. (55). These are the equations of straight lines where in Eq. (54),

$$y = \frac{1}{vf_0}, \quad a = \frac{K_b}{Vf_{max}} + \frac{K_{ia}K_b}{(A_0)Vf_{max}}, \quad x = \frac{1}{(B_0)}, \quad \text{and} \quad b = \frac{1}{Vf_{max}}\left[1 + \frac{K_a}{(A_0)}\right]$$

To determine K_a, K_b, K_{ia}, and Vf_{max} from the data of Table 4, one calculates the reciprocals of the velocities and the reciprocals of (B_0). The data are plotted according to either Eq. (54) [(B_0) variable, (A_0) constant] or Eq. (55) [(A_0) variable, (B_0) constant], depending on which substrate concentration was kept

Table 4 Variation of Initial Velocity of Reaction with Variation of Initial Concentration of Two Substrates, A and B[a]

	v_0[c] at (B_0) $(M) \times 10^4$			
(A_0)[b] $(M) \times 10^5$	3.85[d] $(min^{-1}) \times 10^{-3}$	6.32[d] $(min^{-1}) \times 10^{-3}$	11.3[d] $(min^{-1}) \times 10^{-3}$	63.2[d] $(min^{-1}) \times 10^{-3}$
1.68	0.508	0.788	1.27	3.23
2.52	0.715	1.10	1.75	4.35
4.17	1.08	1.61	2.56	5.89
8.34	1.70	2.50	3.85	8.35
17.2	2.44	3.57	5.26	10.0
55.5	3.45	5.00	7.15	12.5

[a]These data are for lactate dehydrogenase where A is NAD^+ and B is lactate. Calculated from data of Ref. 17.
[b]This column contains initial concentrations of NAD^+ used: for example, 1.68×10^{-5} M.
[c]Expresed in moles of lactate (or NAD+) converted to pyruvate (or NADH) per minute per mole of enzyme concentration at 29°C and pH 6.80. Therefore, this is four times the turnover number of the enzyme since it has four active sites.
[d]These are the initial concentrations of lactate used: for example, 3.85×10^{-4} M.

constant and which was varied in the experiments. Figure 8 shows a plot of the experimental data of Table 4 according to Eq. (54). The initial velocities were determined as a function of both (A_0) and (B_0) concentrations. Within any one experiment (A_0) was kept constant and (B_0) was varied. The experiment was repeated at different (A_0). Notice the effect of both substrate concentrations on the initial velocity (see also Table 3). The initial velocity that is more nearly that of the maximum velocity is the one obtained at the highest concentrations of both substrates.

The slopes and y-intercepts from data plotted as in Figure 8 are determined (Table 5). These data are replotted according to

$$\text{slope} = \frac{K_b}{Vf_{\max}} + \frac{K_{ia}K_b}{(A_0)Vf_{\max}} \tag{56}$$

$$\text{intercept} = \frac{1}{Vf_{\max}} + \frac{K_a}{(A_0)Vf_{\max}} \tag{57}$$

which are the slope and intercept terms of Eq. (54) as shown in Figs. 9 and 10. From the data plotted in Fig. 10, $Vf_{\max}$ is the reciprocal of the y-intercept value, and K_a is the slope divided by the y-intercept. From the data plotted as in Fig. 9, K_b is the y-intercept multiplied by $Vf_{\max}$ and K_{ia} is the slope divided by the y-intercept. Therefore, all four constants can readily be determined. K_a, K_{ia}, K_b, and $Vf_{\max}$ for the data of Table 4 have values of 2.75×10^{-5} M, 1.70×10^{-4} M, 8.85×10^{-4} M, and 1.43×10^4 min^{-1}, respectively.

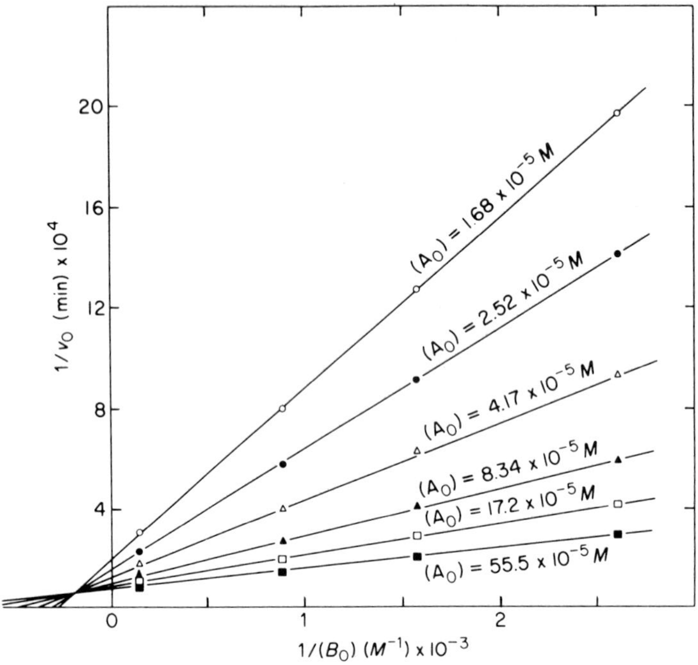

Figure 8 Plot of the substrate–velocity data of Table 4 according to the Lineweaver–Burk method as shown in Eq. (54). Within a single experiment (A_0) was kept constant while (B_0) was varied.

Table 5 Slope and y-Intercept Terms of Data Plotted in Fig. 8

(A_0) (M) $\times$ 10^5	Slope (M min) $\times$ 10^7	Intercept (min) $\times$ 10^5
1.68	6.86	18.4
2.52	4.80	14.6
4.17	3.16	11.6
8.34	1.91	9.30
17.2	1.27	8.11
55.5	0.85	7.35

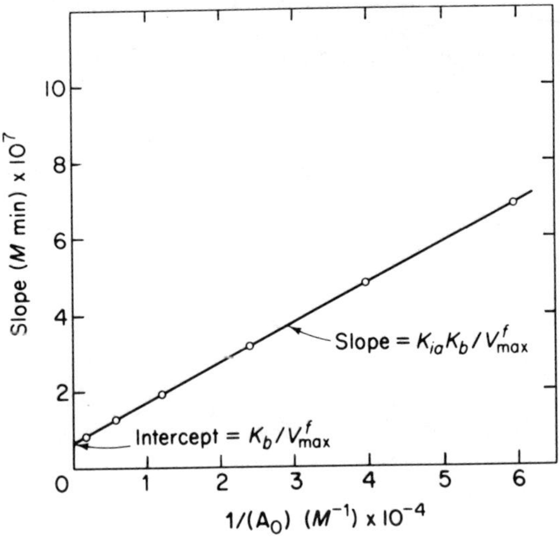

Figure 9 Plot of the slopes of the lines of Fig. 8 versus $1/(A_0)$ according to Eq. (56). The numerical values of the slopes plotted are shown in Table 5.

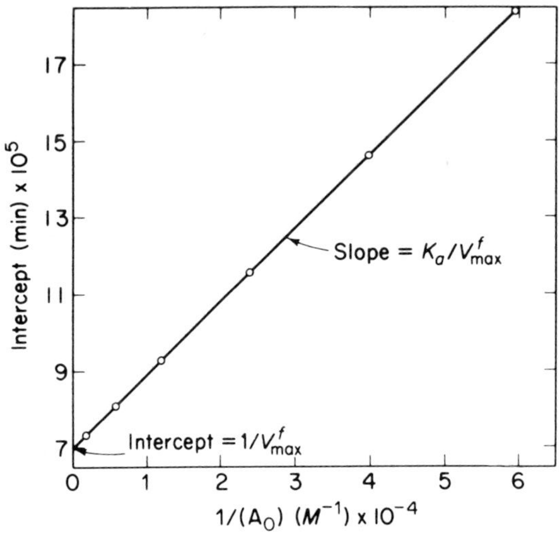

Figure 10 Plot of the y-intercepts of the lines of Fig. 8 versus $1/(A_0)$ according to Eq. (57). The numerical values of the intercepts plotted are shown in Table 5.

2. Ping-Pong Mechanism

The final form of the general equation for a two-substrate Ping-Pong mechanism in the forward direction, assuming initial rates, is given by

$$v^f_0 = \frac{V^f_{max}}{1 + K_a/(A_0) + K_b/(B_0)} \tag{58}$$

The fourth term in the denominator [see Eq. (52)] is missing since there is no ternary complex in the mechanism.

The data in Fig. 11 are plotted according to

$$\frac{1}{v^f_0} = \frac{1}{V^f_{max}} + \frac{K_a}{(A_0)V^f_{max}} + \frac{K_b}{(B_0)V^f_{max}} \tag{59}$$

[the reciprocal of Eq. (58)], where $y = 1/v^f_0$, $a = k_b/V^f_{max}$, $x = 1/(B_0)$, and $b = (1/V^f_{max}) + [K_a/(A_0)V^f_{max}]$. A replot of the y-intercepts from Fig. 11 versus $1/(A_0)$ gives Fig. 12. Therefore, the constants K_a, K_b, and V^f_{max} can be determined from Figs. 11 and 12.

E. Determination of Mechanism

Several methods are available to distinguish among the Ordered, Random, and Ping-Pong mechanisms described above. Only brief mention of these methods will be given here. References 12 to 16 should be consulted for a thorough treatment of this topic.

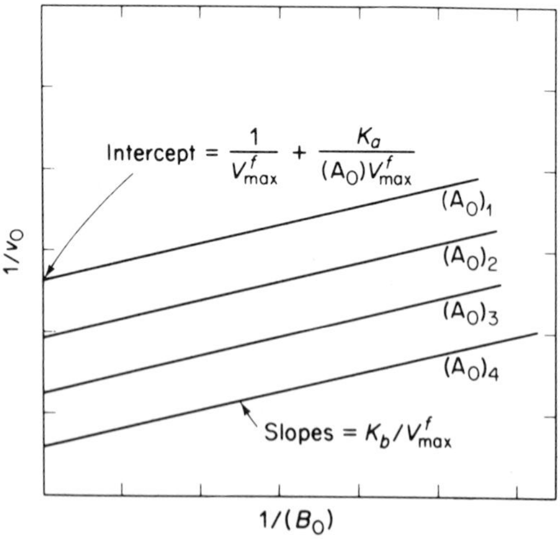

Figure 11 Hypothetical $1/v_0$ versus $1/(B_0)$ plots for a Ping-Pong Bi Bi mechanism. Data are plotted according to Eq. (59).

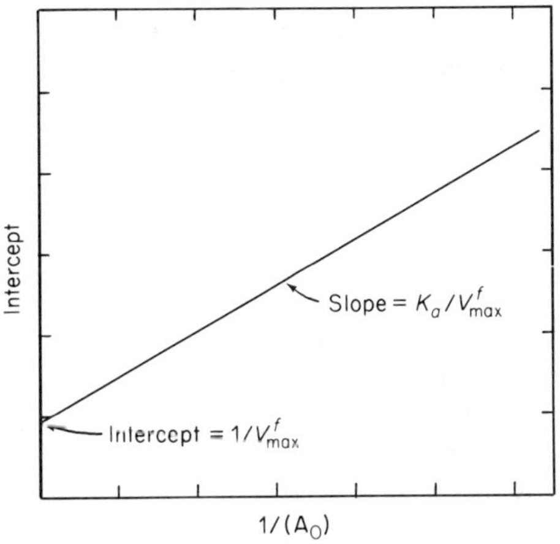

Figure 12 Plot of the *y*-intercepts of the lines of Fig. 11 versus $1/(A_0)$ according to the intercept terms of Eq. (59), $(1/V^f_{max}) + [K_a/(A_0)V^f_{max}]$.

The Ping-Pong mechanism can be distinguished from the sequential mechanism by analysis of initial rate data as described in Section III.D. The enzyme systems that follow sequential mechanisms (random or ordered) give lines that intersect to the left of the y-axis (Fig. 8), while those systems that follow the Ping-Pong mechanism give parallel lines (Fig. 11).

The Random and Ordered mechanisms are not distinguished by plots of initial rate data. The overall reaction will not take place if one substrate is missing. However, exclusion of one substrate offers an opportunity to evaluate the order of addition of substrates to the enzyme and to determine if part of the overall reaction will take place. Determination of binding of one substrate in the absence of the other can be performed by equilibrium dialysis or by spectral changes in some cases. In a Random mechanism, regardless of which substrate is added first, each should bind to the enzyme in the absence of the other. In an Ordered mechanism, when A is added to the enzyme, binding will be observed. However, when B alone is added to the enzyme, there will be no binding until A is added. In a Ping-Pong mechanism, addition of A to the enzyme not only results in a complex between A and enzyme, but the first product is formed and the nature of the enzyme is changed. The simplest way of distinguishing a Ping-Pong from a sequential mechanism is to observe isotopic exchange on adding one of the substrates but not the other.

Initial rate data can be used to determine K_a, K_b, K_{ia}, and V^f_{max} and to distinguish between sequential and Ping-Pong mechanisms. However, initial rate data for detailed mechanistic studies is of extremely limited utility. Integrated rate expressions analogous to that shown in Eq. (36) for one-substrate reactions should be developed. These equations, in conjunction with techniques involving product inhibition, deadend inhibition, and isotopic exchange rates can be used unequivocally to determine the mechanism of multisubstrate reactions. References 12 to 16 should be consulted for development and use of these concepts and equations.

IV. SUBSTRATE-INDUCED "ABNORMAL" BEHAVIOR

In the preceding discussion we have treated enzyme systems that follow the classic Michaelis–Menten pattern [i.e., when the initial rate, v_0, is observed as a function of substrate concentration, a hyperbolic relationship is found (Fig. 1)]. This treatment holds true not only for enzymes that have a single active site, it is also applicable to multiactive site enzymes as long as there is no interaction among the sites.

A. Single-Site/Single-Peptide Chain Enzymes

There is evidence for binding of a second molecule of the same substrate at a locus near the active site in a number of single-site enzymes, for example, trypsin and carboxypeptidase A. The second molecule of substrate does not undergo reaction to products, but its binding does influence behavior at the active site either in changing the binding of the first molecule of substrate, or in transforming to product, or both. If binding of the second molecule leads to a reduction in observed initial velocity, it is called an inhibitor, and if it increases the observed initial velocity, it is called an activator. The influence of this second substrate molecule is shown graphically in Figs. 13 and 14, where Fig. 13 shows the second molecule of substrate acting as an inhibitor and Fig. 14 shows it acting as an activator. One would observe behavior identical to that depicted in Fig. 14 if there are two enzymes present in a system capable of transforming a common substrate to product but which have different K_m values for the substrate.

B. Multisite–Multichain Enzymes

1. Distinguishing Features

Let us now turn to a discussion of a multiactive site enzyme. In general, it is expected that such an enzyme will be composed of more than one polypeptide chain. If there is no interaction among the sites, normal Michaelis–Menten behavior will be observed. However, consider the possibility that when a substrate is bound in one of the active sites, it influences one or more of the remaining active sites, so that the second substrate molecule is bound with a different affinity. This possibility can be extended to include an effect on the rate of transformation of reactant(s) to product(s). The result of this interaction will be seen as a cooperation among the active sites. The kinetic behavior of the enzyme system will then differ

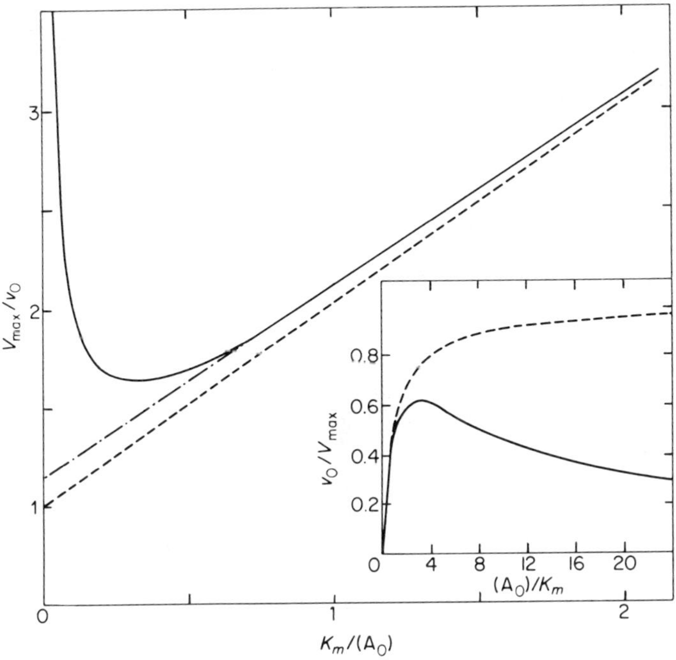

Figure 13 Effect of substrate inhibition on the initial velocity of an enzyme-catalyzed reaction. The dashed line shows the normal curve in absence of inhibition, the solid line shows inhibition by a second substrate molecule for which the dissociation constant, K'_s, is 10 times larger that K_m. The solid line is calculated according to the equation $v_0 = V_{max}/\{1 + K_m/(A_0)] + [(A_0)/K'_s]\}$ of Ref. 18. EA_2 does not form product. Large graph plotted by Lineweaver–Burk method; the insert is as shown in Fig. 1.

from classic Michaelis–Menten kinetics (Fig. 15). In Fig. 15 relative v_0 (v_0/V_{max}) values have been plotted so that the different behaviors of the two enzyme systems can be compared easily.

Experimentally, the two behaviors (Fig. 15) are most readily distinguished by comparing the ratio of substrate concentration necessary to achieve an observed velocity of $0.9V_{max}$ to that needed to achieve $0.1V_{max}$. This ratio, called R_s [19], measures the cooperativity of the interaction. For all enzymes that follow Michaelis–Menten kinetics $R_s = 81$ no matter what the values of K_m and V_{max}. For allosteric-behaving systems where binding of a prior substrate molecule enhances binding of each succeeding substrate molecule at the remaining sites, $R_s < 81$; this system is said to show positive cooperativity. In cases where binding of a prior substrate molecule decreases binding of the next substrate molecule at one of the

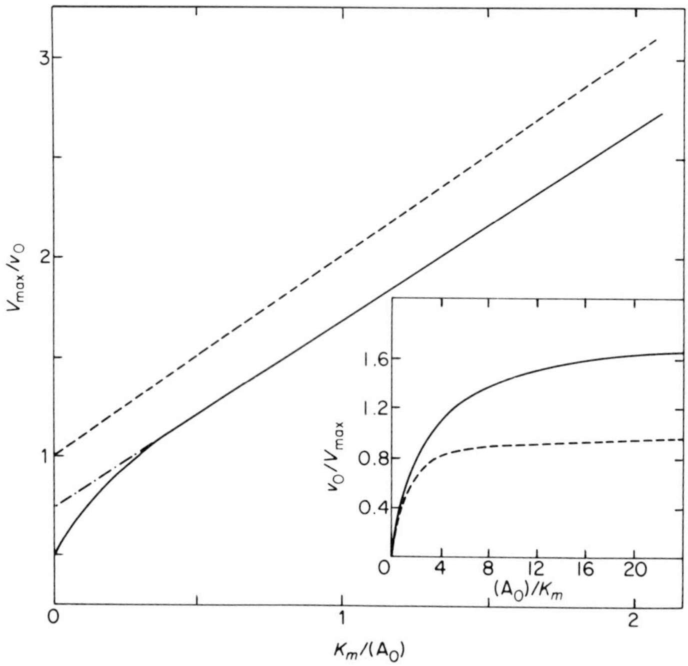

Figure 14 Effect of substrate activation on initial velocity of an enzyme-catalyzed reaction. The dashed line shows a normal reaction in absence of activation, the solid line shows a reaction in the presence of activation. The solid line is calculated on the assumptions that K_s' for the second substrate molecule is $10K_m$, that V_{max} is doubled when all the enzyme is saturated with a second substrate molecule (i.e., EA_2 goes to product twice as fast as EA), and that the second substrate molecule does not form product. The large graph is plotted by the Lineweaver–Burk method; the insert is as shown in Fig. 1.

remaining sites, $R_s > 81$; this system shows negative cooperativity. For the case shown in Fig. 15, R_s is 3.5.

At least two important features of allosteric behavior should be noted. One is the marked dependence of initial velocity on substrate concentration in the region of $v_0/V_{max} = 0.5$. The second feature is that enzymes showing allosteric behavior become saturated with substrate at lower substrate concentrations in relation to $A_{0.5}$ (or K_m) than do enzymes showing Michaelis–Menten behavior (Fig. 15).

2. Models

Mathematical models have been developed to describe these sigmoidal-behaving systems [19,20]. Although it is beyond the scope of this book to consider these

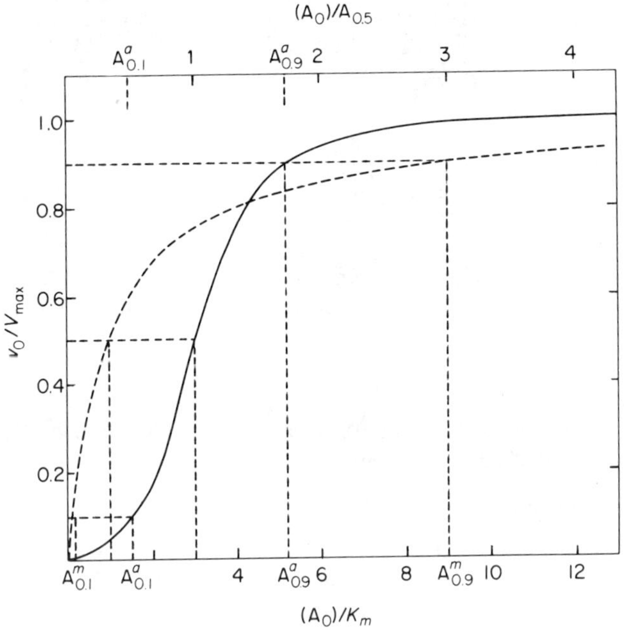

Figure 15 Comparison of effect of substrate concentration on the initial velocities of two enzyme-catalyzed reactions, one that obeys Michaelis–Menten kinetics (dashed line) and one that shows allosteric behavior (solid line). The designations $A^m_{0.1}$ and $A^m_{0.9}$ indicate the substrate concentrations at which the initial velocity is 0.1 and 0.9 of V_{max}, respectively, for Michaelis–Menten kinetics, while $A^a_{0.1}$ and $A^a_{0.9}$ have the same meaning for the allosteric behaving system. The solid line is calculated by using an n value of 4 in Eq. (61).

models in detail, it is well to consider the postulates of these models in connection with the general description given above. The model of Monod et al. [20] (Fig. 16) postulates (a) that an equilibrium exists between different conformational states of the enzyme, (b) that the conformations of all subunits of the enzyme change simultaneously and (c) that there is a differential affinity of ligands (substrate, inhibitor, cofactors) for one state of the enzyme relative to the other. The model of Koshland et al. [19] postulates that (a) a substrate molecule induces a conformational change in the enzyme that may be limited to the single subunit to which the substrate binds, (b) the conformational changes in the individual subunits occur sequentially rather than simultaneously, and (c) a differential binding affinity for ligands exists between one conformational state and another. The two models differ only in degree, the essential difference being that in the model of Monod et al. the two conformational forms are in equilibrium even in the absence of

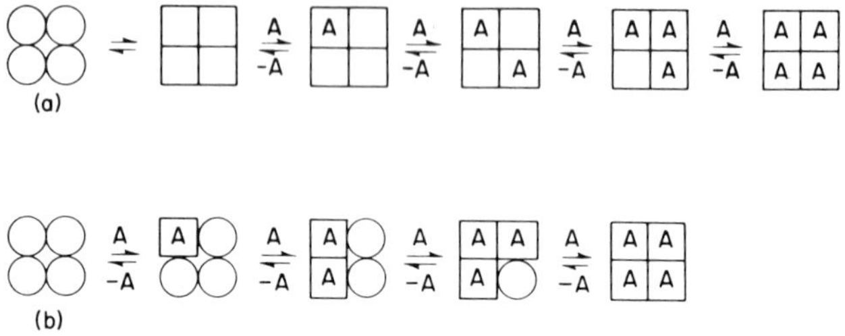

Figure 16 Schematic representation of the (a) Monod et al. and (b) Koshland et al. models for binding of substrate A to subunits of an enzyme that shows allosteric behavior.

substrate and the substrate can combine with only one form, while the model of Koshland et al. postulates that this may occur in a sequential fashion.

3. Evaluation of Constants of Rate Equations

The relationship between velocity and ligand (substrate, inhibitor, cofactors) concentration for a number of regulatory enzymes has been treated [21] over a limited ligand concentration range by the Hill equation [22]. The following equation was derived in connection with the interactions of O_2 with hemoglobin, which shows allosteric behavior:

$$y = \frac{Kx^n}{1 + Kx^n} \tag{60}$$

where y is the fraction of saturation of hemoglobin by O_2, x the partial pressure of O_2, n the number of oxygen molecules bound per hemoglobin molecule, and K is a constant.

Equation (60) resembles the Michaelis–Menten equation except for the power function of substrate concentration. When the following change is made in the Michaelis–Menten equation,

$$v_0 = \frac{V_{\max}(\mathrm{A}_0)^n}{K + (\mathrm{A}_0)^n} \tag{61}$$

the equation may be rearranged to

$$\log \frac{v_0}{V_{\max} - v_0} = n \log(\mathrm{A}_0) - \log K \tag{62}$$

where n is the number of sites and K is an average of all the individual K_m (or K_{eq}) values involved in the various steps of binding and transformation of substrate to product.

Experimental data plotted according to Eq. (62) conform to the predicted linear

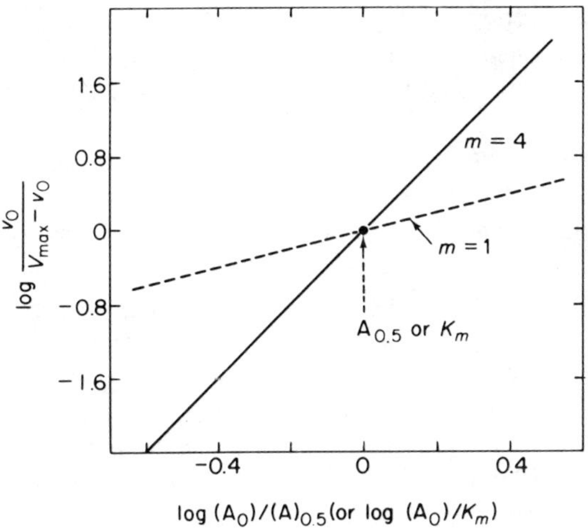

Figure 17 Plot of the data of Fig. 15 according to Eq. (62). The dashed line is for a system following Michaelis–Menten kinetics; the solid line is for a system showing allosteric behavior with $m = 4$. Inclusion of the constant term $A_{0.5}$ in the abscissa label does not change the relationship shown in Eq. (62); it only changes the absolute values.

relationship only to an extent dependent on the degree of cooperativity involved. In the case of infinite cooperativity, experimental data will fit the relationship well since n, the number of total sites involved, is equal to the number of interacting sites observed experimentally. However, as the degree of cooperativity decreases from infinite cooperativity, a more realistic situation, the fit of experimental points to a straight line will be less good.

The point where $\log[v_0/(V_{max} - v_0)] = 0$, although not a true constant, has important features, for at this point $v_0 = 0.5\ V_{max}$ and thus defines the substrate concentration necessary to give $v_0 = 0.5\ V_{max}$ as $A_{0.5}$. The slope of a line drawn tangent to the experimental data at this point gives a measure of the degree of interaction of the active sites (Fig. 17). This slope, designated as m by Atkinson [21], is a function of n, the total number of sites and the strength of interactions among the sites. m will range between the limits of n when there is infinite cooperativity among n active sites to 1 when there is no cooperativity.

REVIEW QUESTIONS

1. Given that K_m is 0.01 M, (A_0) is $1 \times 10^{-4}\ M$ and that there is 20% conversion of substrate to product in 10 min, calculate

(a) the percent conversion of substrate to product after 30 min.

(b) the half-life, $t_{1/2}$, of the reaction.
(c) the specific reaction rate constant of the reaction.
(d) the percent conversion of substrate to product after 10 min if (A_0) were 1×10^{-6} M.

2. Given that K_m is 1.0×10^{-4} M, (A_0) is 0.2 M, and that there is 20% conversion of substrate to product after 15 min, calculate
(a) the specific reaction rate constant of the reaction.
(b) the half-life, $t_{1/2}$, of the reaction.
(c) the percent conversion of substrate to product after 30 min.
(d) the percent conversion of substrate to product after 30 min when (A_0) is 0.05 M. Do you wish to reevaluate your answer to part (c) after answering part (d)?

3. When (A_0) was 0.1 M, a 1×10^{-6} M solution of an enzyme gave 1.0% substrate conversion to product in 30 min, while at (A_0) of 1.0×10^{-3} M it gave 5% conversion of substrate to product in 3 min. Calculate the approximate value of K_m.

4. By having one of the substrates present in saturating amounts, a two-substrate enzyme-catalyzed reaction appears kinetically as a one-substrate reaction. Since this situation is easier to treat mathematically, why isn't it used for determining K_a, K_b, K_{ia}, and V_{max} for the reaction?

5. When enzyme B was allowed to act upon either substrate C or D at saturating levels, V_{max} was 3×10^{-4} M min^{-1} for each substrate. However, when substrates C and D were present together at the same concentrations as used above and with the same enzyme concentration, the observed velocity was 2.60×10^{-4} M min^{-1} for C and 0.40×10^{-4} M min^{-1} for D. Explain how this can be.

6. Using Cleland's method of presentation, diagram the reaction pathway of
(a) an Ordered Uni Bi mechanism.
(b) a Random Ter Bi mechanism (more than one possibility).
(c) a Ping Pong Ter Bi mechanism (more than one possibility).

7. For an allosteric-behaving system $A_{0.5}$ was found to be 1.0×10^{-4} M while R_s was 20. Calculate the substrate concentrations at which the observed velocities would equal 0.1 and 0.9 of V_{max}. Do the same calculations for a system that follows Michaelis–Menten kinetics and has a K_m of 1.0×10^{-4} M.

8. By use of Eq. (36), write down the mathematical expression for the observed velocity under the following conditions:
(a) when the reaction is run in the reverse direction so that (P_0) is a finite value and $(A_0) = 0$.
(b) when initial velocities in the forward direction are being measured and k_1 is the rate-determining step.
(c) when initial velocities in the reverse direction are being measured and k_{-3} is the rate-determining step.
(d) when initial velocities in the reverse direction are being measured and k_{-2} is the rate-determining step.

9. What is meant by the *rate-determining step*?

REFERENCES

General References

Boyer, P. D., H. Lardy, and K. Myrbäck, eds., *The Enzymes*, 3rd Ed., Vol. 1, Academic Press, New York, 1970.

Cleland, W. W., *Biochim. Biophys. Acta*, **67**, 104, 173, 188 (1963).

Dawes, E. A., Enzyme kinetics, in *Comprehensive Biochemistry* (M. Florkin and E. G. Stotz, eds.), Vol. 12, Elsevier, New York, 1964, p. 89.

Dixon, M., and E. C. Webb, *Enzymes*, 3rd Ed., Academic Press, New York, 1980.

Eigen, M., and G. G. Hammes, *Advan. Enzymol.*, **25**, 1 (1963).

Hammes, G. G., *Enzyme Catalysis and Regulation*, Academic Press, New York, 1982.

Segel, I. H., *Biochemical Calculations*, Wiley, New York, 1976.

Wong, J. T., and C. S. Hanes, *Can. J. Biochem. Physiol.*, **40**, 763 (1962).

Cited References

1. V. Henri, *Acad. Sci., Paris*, **135**, 916 (1902); *Lois générales de l'action des diastases*, Hermann, Paris, 1903.
2. A. Brown, *J. Chem. Soc.*, **81**, 373 (1902).
3. K. Yagi, T. Ozawa, and M. Naoi, *Biochim. Biophys. Acta*, **185**, 31 (1969), and papers referenced therein.
4. L. Michaelis and M. L. Menten, *Biochem. Z.*, **49**, 333 (1913).
5. G. E. Briggs and J. B. S. Haldane, *Biochem. J.*, **19**, 338 (1925).
6. J. Westley, *Enzymic Catalysis*, Harper & Row, New York, 1969.
7. M. Eigen and G. G. Hammes, *Advan. Enzymol.*, **25**, 1 (1963).
8. G. N. Wilkerson, *Biochem. J.*, **80**, 324 (1961).
9. H. Lineweaver and D. Burk, *J. Am. Chem. Soc.*, **56**, 658 (1934).
10. W. W. Cleland, *Advan. Enzymol.*, **29**, 1 (1967).
11. E. L. King and C. Altman, *J. Phys. Chem.*, **60**, 1375 (1956).
12. W. W. Cleland, *Biochim. Biophys. Acta*, **67**, 104 (1963).
13. W. W. Cleland, *Biochim. Biophys. Acta*, **67**, 173 (1963).
14. W. W. Cleland, *Biochim. Biophys. Acta*, **67**, 188 (1963).
15. W. W. Cleland, *Ann. Rev. Biochem.*, **36**, 77 (1967).
16. W. W. Cleland, in *The Enzymes*, 3rd Ed. (P. D. Boyer, H. Lardy, and K. Myrbäck, eds.), Vol. 2, Academic Press, New York, 1970, p. 1.
17. M. T. Hakala, A. J. Glaid, and G. W. Schwert, *J. Biol. Chem.*, **221**, 191 (1956).
18. M. Dixon and E. C. Webb, *Enzymes*, Longmans, Green, London, 1958, p. 83.
19. D. E. Koshland, Jr., G. Némethy, and D. Filmer, *Biochemistry*, **5**, 365 (1966).
20. J. Monod, J. Wyman and J. P. Changeux, *J. Mol. Biol.*, **12**, 88 (1965).
21. D. E. Atkinson, *Ann. Rev. Biochem.*, **35**, 85 (1966).
22. A. V. Hill, *Biochem. J.*, **7**, 471 (1913).

7

Effect of Enzyme Concentration on Rates of Enzyme-Catalyzed Reactions

Rates of enzyme-catalyzed reactions depend directly on enzyme concentration. In general, enzyme concentration in a reaction ranges from 1×10^{-8} to 1×10^{-5} M, depending on the turnover number of the enzyme, while substrate concentration generally ranges from 1×10^{-5} to 1×10^{-2} M. Enzyme concentration is low compared to substrate concentration; however, its total concentration (present in all enzyme-containing species) is expected to remain constant throughout the reaction. Therefore, $dE_0/dt = 0$ for the overall reaction.

Many enzyme assays are done with the sole purpose of measuring the amount of enzyme in a sample. Such assays are done routinely in clinical applications, in following the blanching and pasteurization of foods, in standardizing enzyme preparations, in following purification of enzymes, and so on. In these assays, it is taken for granted that the amount of product formed in a given time is a linear function of enzyme concentration (Fig. 1). In this chapter we delineate some of the factors that influence this expected linear relationship.

I. ENZYME CONCENTRATION–VELOCITY RELATIONSHIPS

A. Influence of Substrate Concentration

The substrate concentration–velocity curve for an enzyme-catalyzed reaction is composed of three distinct regions. At $(A_0) \leqslant 0.01K_m$ the reaction is first order with respect to substrate concentration; at $(A_0) \geqslant 100K_m$ the reaction is zero order

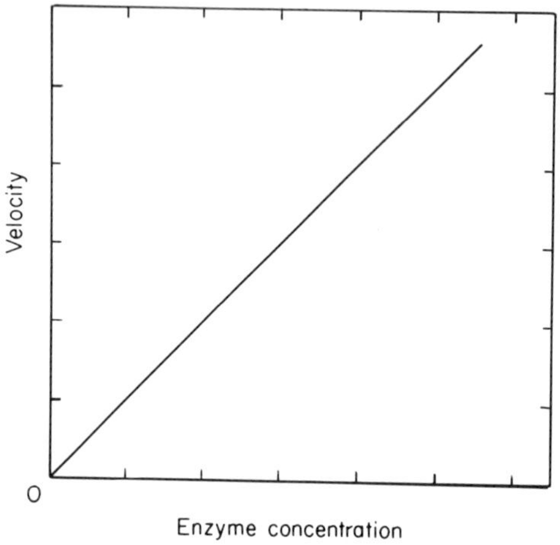

Figure 1 Expected relationship between enzyme concentration and observed velocity of reaction. Substrate concentration, pH, temperature, and buffer are kept constant.

with respect to substrate concentration, and in between $(A_0) \geqslant 0.01K_m$ and $(A_0) \leqslant 100K_m$ the reaction is mixed order with respect to substrate concentration and is best treated by the relationship $v_0 = V_{max}(A_0)/[K_m + (A_0)]$. These relationships between substrate concentration and velocity must be kept in mind throughout discussion of the effect of enzyme concentration on velocity.

Whenever accurate initial rates are obtained, in the absence of perturbing factors, the relationship between v_0 and (E_0) is linear (Fig. 1) as long as $(A_0) \gg (E_0)$, regardless of (A_0) in relation to K_m. Unfortunately, many laboratories run routine enzyme assays by taking aliquots of the solution for analysis at one time interval. Such velocity measurements should not presume to be a measure of initial rates unless the initial substrate concentration is much greater than K_m $[(A_0) \geqslant 100K_m]$ or previous analyses at intermediate times showed this to be the case. In Fig. 2 the slope of the line, dP/dt, is equal to $\bar{k}(E_0)_i$. Therefore, a plot of the slopes of the lines in Fig. 2 versus $(E_0)_i$ gives the relationship shown in Fig. 1. Since the relationship between (P) and time is linear, taking aliquots at any time would give the same velocity for a given enzyme concentration.

An accurate measure of enzyme concentration will be obtained at low substrate concentrations $[(A_0) \leqslant 0.01K_m]$ when initial rates are measured. This procedure is particularly valuable whenever the substrate is expensive, solubility of substrate is low, or when K_m is large (10^{-2} to 10^0 M). An accurate measure of enzyme concentration can also be obtained when the reaction is monitored continuously (or at frequent time intervals) over at least 10 to 25% of the total reaction. The reaction follows first-order rate laws with respect to substrate concentration (Fig. 3). The

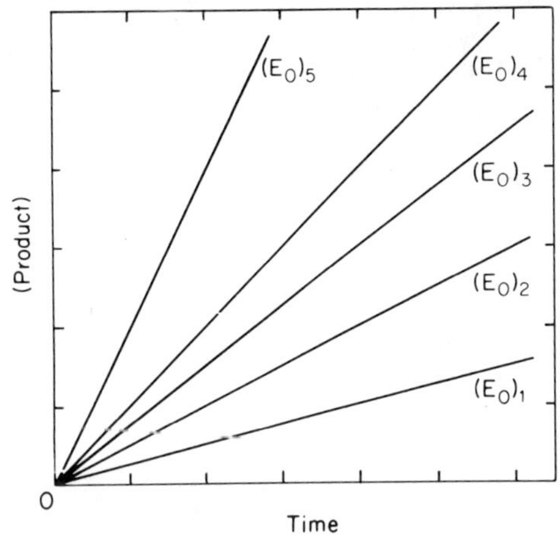

Figure 2 Relationship between concentration of product formed and time at various enzyme concentrations. $(E_0)_5$, $(E_0)_4$, $(E_0)_3$, $(E_0)_2$, and $(E_0)_1$ are in the relationship of 8, 4, 3, 2, and 1, respectively.

slopes of the straight lines are $k(E_0)_i$, so that a plot of slope versus $(E_0)_i$ gives the relationship shown in Fig. 1.

At substrate concentrations near K_m, the only accurate way to obtain a linear relationship between velocity and (E_0) is to determine initial rates. Therefore, initial rate data are invaluable in accurate determinations of active enzyme concentration.

B. Determination of K_m

If a knowledge of K_m is important in designing an experiment for determining enzyme concentration, is it worth the time required to determine K_m by the usual method (see Chapter 6) when one is not really interested in that aspect? The order of magnitude of K_m can be determined by just two sets of data. Each set is performed in an identical fashion except that one contains a 5- to 20-fold greater concentration of substrate than the other. The highest concentration of substrate used should not saturate the enzyme. Initial rates are determined. For set 1 the observed initial velocity is

$$v_{01} = \frac{V_{\max}(A_0)_1}{K_m + (A_0)_1} \tag{1}$$

while for set 2 it is

$$v_{01} = \frac{V_{\max}(A_0)_2}{K_m + (A_0)_2} \tag{2}$$

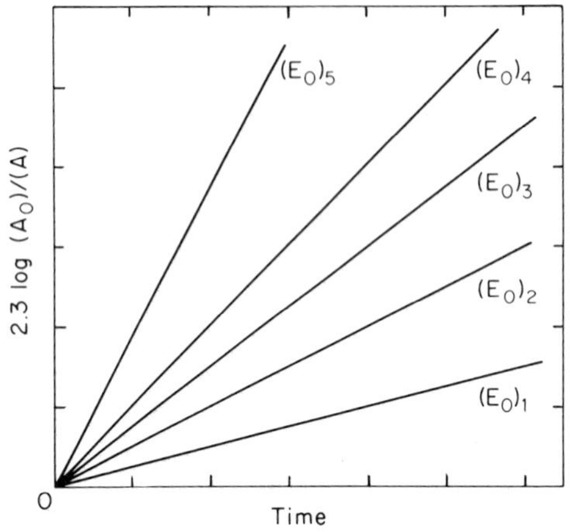

Figure 3 Relationship between disappearance of substrate and time at several enzyme concentrations where $(A_0) \leq 0.01K_m$. $(E_0)_5$, $(E_0)_4$, $(E_0)_3$, $(E_0)_2$, and $(E_0)_1$ are in the relationship 8, 4, 3, 2, and 1, respectively.

Taking the ratio of v_{02}/v_{01} and solving for K_m gives

$$K_m = \frac{(A_0)_1(A_0)_2(v_{01} - v_{02})}{v_{02}(A_0)_1 - v_{01}(A_0)_2} \tag{3}$$

The greater the difference between $(A_0)_1$ and $(A_0)_2$, the greater the precision with which K_m can be determined. Equation (3) was derived assuming that the reaction follows Michaelis–Menten kinetics.

II. PERTURBING FACTORS IN ENZYME CONCENTRATION–VELOCITY RELATIONSHIPS

The expected relationship between velocity and enzyme concentration is that shown in Fig. 1; however, there are a number of situations where such a relationship is not observed.

A. Limitations Due to Assay Method

1. Limitations on Substrate Concentration

Consider the following reaction, which involves two substrates, one being a gas (O_2):

$$AH_2 + \text{E-FAD} \xrightarrow{k_1} A + \text{E-FADH}_2 \xrightarrow{k_2, O_2} \text{E-FAD} + H_2O_2 \tag{4}$$

In this example, AH_2 is the reduced form of substrate A and E-FAD is an enzyme that requires flavin adenine dinucleotide (FAD) as a cofactor to accept the

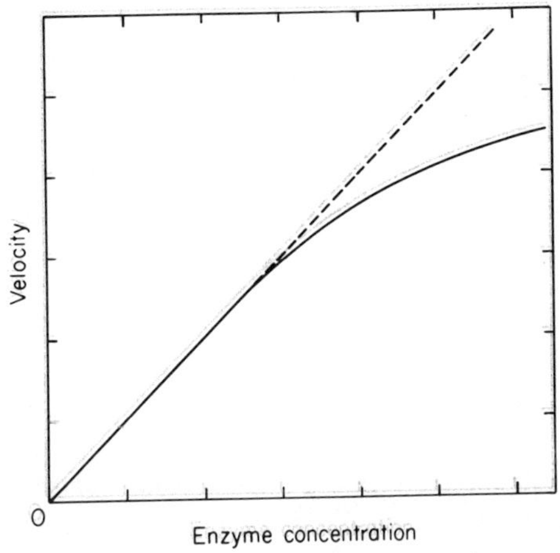

Figure 4 Enzyme concentration–velocity relationship when there is a limitation imposed by inadequate substrate concentration at high enzyme concentrations. The dashed line is the expected relationship; the solid line is the observed relationship.

hydrogens from AH_2. For E-FAD to act in a catalytic manner (i.e., to be repetitively cycled through the system), it must be continuously regenerated from E-$FADH_2$ by involvement of the second substrate, O_2. The rate of formation of E-$FADH_2$ is $d(\text{E-FADH}_2)/dt = k_1(\text{AH}_2)(\text{E-FAD})$, while its rate of disappearance is $-d(\text{E-FADH}_2)/dt = k_2(\text{O}_2)(\text{E-FADH}_2)$. To obtain a linear relationship between velocity and enzyme concentration the oxygen concentration must be high enough so that its rate of diffusion into the reaction solution, and thus $-d(\text{E-FADH}_2)/dt$, does not become the limiting factor. Otherwise, the results will be those shown in Fig. 4, where the relation is no longer linear at high enzyme concentration. This situation can be checked and corrected by increasing the rate of diffusion of O_2 into the system (e.g., by increasing shaking rate, stirring rate or by bubbling 100% O_2 through the system).

2. Limitations Due to Other Enzyme Reactions

As noted in Chapter 1, it is often convenient to use coupled enzyme assays in following the reaction of certain enzymes or in determining the amount of a compound in a system. A coupled system is described by

$$\underset{}{\text{A}} \underset{\text{E}_1}{\overset{k_1}{\rightarrow}} \text{B} \underset{\text{E}_2}{\overset{k_2}{\rightarrow}} \text{C} \underset{\text{E}_3}{\overset{k_3}{\rightarrow}} \text{P} \tag{5}$$

where the product of one enzyme-catalyzed reaction is the substrate for another enzyme-catalyzed reaction. The rate of formation of P is

$$\frac{dP}{dt} = v_{03}\frac{k_3(E_0)_3(C)}{K_c + (C)} \tag{6}$$

However, the rate of formation of C is

$$\frac{dC}{dt} = v_{02}\frac{k_2(E_0)_2(B)}{K_b + (B)} \tag{7}$$

which in turn depends on the rate of formation of B:

$$\frac{dB}{dt} = v_{01}\frac{k_1(E_0)_1(A)}{K_a + (A)} \tag{8}$$

To obtain a linear relationship between dP/dt and $(E_0)_3$ at all levels of $(E_0)_3$, then, dB/dt and dC/dt must be $\geqslant$10 times faster than dP/dt. The variables that are controlled to assure this condition are (A_0), $(E_0)_1$, and $(E_0)_2$, taking into account k_1 and k_2. An excellent discussion of the kinetics of coupled assays is available [1].

B. Coupled Reactions Involving a Common Form of Coenzyme in Two-Enzyme Systems

Consider the following overall reactions:

$$AH_2 + NAD^+ \underset{k_{-1}}{\overset{k_1,E_1}{\rightleftharpoons}} P + NADH + H^+ \tag{9}$$

$$B + NADH + H^+ \underset{k_{-2}}{\overset{k_1,E_2}{\rightleftharpoons}} QH_2 + NAD^+ \tag{10}$$

where AH_2 and B are substrates and P and QH_2 are products of E_1 and E_2, respectively, E_1 and E_2 are different enzymes, and NADH is the common form of the coenzyme in the two reactions. Assume that the experiment involves a determination of the velocity, dP/dt, as a function of concentration of enzyme 1 and that we are depending on the reaction catalyzed by enzyme 2 to replenish the supply of NAD^+. Two important factors are involved.

1. The rate of formation of NAD^+ should equal its rate of disappearance $[-d(NAD^+)/dt = d(NAD^+)/dt]$ as shown in Eqs. (11) and (12) if the concentration of NAD^+ is not to diminish.

$$-d(NAD^+)/dt = k_1(E_0)_1(AH_2)(NAD^+) + k_{-2}(E_0)_2(QH_2)(NAD^+) \tag{11}$$

$$d(NAD^+)/dt = k_{-1}(E_0)_1(P)(NADH)(H^+) + k_2(E_0)_2(NADH)(B)(H^+) \tag{12}$$

2. Both E_1 and E_2 have affinities for NAD^+ and NADH. This can be expressed best in terms of binding constants.

$$E_1 + NADH \underset{k_{-1}}{\overset{k_1}{\rightleftharpoons}} E_1\cdot NADH;\ K_{(NADH\cdot E_1)} = \frac{(E_1\cdot NADH)}{(E)_1(NADH)} \tag{13}$$

and

$$E_1 + NAD^+ \underset{k_{-2}}{\overset{k_2}{\rightleftharpoons}} E_1\cdot NAD^+;\ K_{(NAD^+\cdot E_1)} = \frac{(E_1\cdot NAD^+)}{(E)_1(NAD^+)} \quad (14)$$

$$E_2 + NADH \underset{k_{-3}}{\overset{k_3}{\rightleftharpoons}} E_1\cdot NADH;\ K_{(NADH\cdot E_2)} = \frac{(E_2\cdot NADH)}{(E)_2(NADH)} \quad (15)$$

and

$$E_2 + NAD^+ \underset{k_{-4}}{\overset{k_4}{\rightleftharpoons}} E_2\cdot NAD^+;\ K_{(NAD^+\cdot E_2)} = \frac{(E_2\cdot NAD^+)}{(E)_2(NAD^+)} \quad (16)$$

The relative magnitudes of these binding constants are very important in determining the concentration of NAD^+ available for the first reaction, as well as $(E_0)_1$, $(E_0)_2$, and the rate constants. The analysis is further complicated in that NAD^+ and NADH are competitive inhibitors of each other, since only one form can be bound to E_1 and E_2 at any time.

The net result of factors 1 and/or 2 is that the velocity of the reaction will be decreased at high concentrations of E_1 (Fig. 5). This situation can be avoided in practice by adding sufficient coenzyme to the reaction to maintain enzyme 1 saturated with coenzyme at all enzyme concentrations, and delete enzyme 2, if possible (purified system).

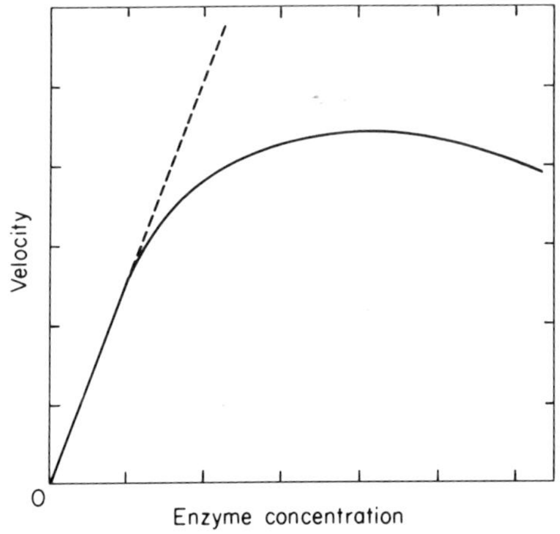

Figure 5 Enzyme concentration–velocity relationship where there is competition of two enzymes for a common cofactor. The dashed line is the expected relationship, the solid line the observed relationship. The activity of the first enzyme, E_1, is monitored.

C. Presence of Irreversible Inhibitor in Reaction System

Consider that the irreversible inhibitor is present in the substrate, buffer, or water but not in the enzyme preparation. Assume also that the reaction is initiated by addition of enzyme (the usual procedure). As a consequence, the total enzyme concentration, (E_0), of the system is distributed as

$$(E_0) = (E) + (EA) + (E\text{-}I) \tag{17}$$

The net result of the presence of an irreversible inhibitor will be a reduction in the amount of enzyme available for combination with substrate (Fig. 6). The extent of displacement, x, along the x axis is a function of the amount of inhibitor present. Once sufficient enzyme has been added to react with all the inhibitor, the relation between velocity and enzyme concentration is a normal one, except that the extrapolated line does not go through zero and the slope of the line is smaller by x amount than the expected line.

Two specific examples of this situation are given. In the case of an enzyme whose activity depends on a sulfhydryl group in the active site, the presence of heavy metal ions (Cu^{2+}, Pb^{2+}, etc.) in the reaction will give the results indicated, as the enzyme forms ES-M^+ and/or ES-M-SE. Addition of EDTA and/or a thiol compound (dithiothreitol, mercaptoethanol, cysteine) prevents this situation. A second example is the presence of a contaminant in the substrate which is an irreversible inhibitor of the enzyme.

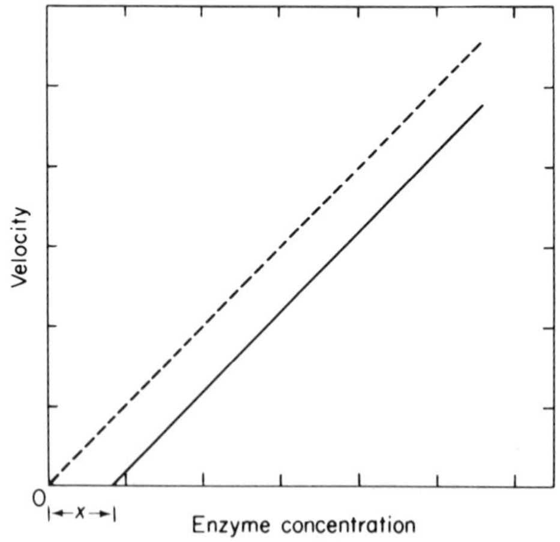

Figure 6 Enzyme concentration–velocity relationship in the presence of irreversible inhibitor. The dashed and solid lines are for the expected and observed relationships, respectively.

D. Competitive Inhibitor in Enzyme Preparation

A competitive inhibitor might be added to the reaction from any of the constituents. However, only if it is added by the enzyme preparation so that it increases as the enzyme concentration increases will an effect on the velocity–enzyme relationship be observed, as shown in Fig. 7. The reactions involved are described by

$$E + A \underset{k_{-1}}{\overset{k_1}{\rightleftharpoons}} EA \overset{k_2}{\rightarrow} E + P \tag{18}$$

$$E + I \underset{k_{-3}}{\overset{k_3}{\rightleftharpoons}} EI \qquad K_i = \frac{(E)(I)}{(EI)} \tag{19}$$

The total enzyme in the system will be distributed as

$$(E_0) = (E) + (EA) + (EI) \tag{20}$$

As long as substrate concentration remains high enough to saturate the enzyme, the presence of a competitive inhibitor will not affect the results. However, in reactions where substrate concentration is less than saturating, an effect will be seen (Fig. 7). The higher the enzyme concentration, the greater the deviation from expected results.

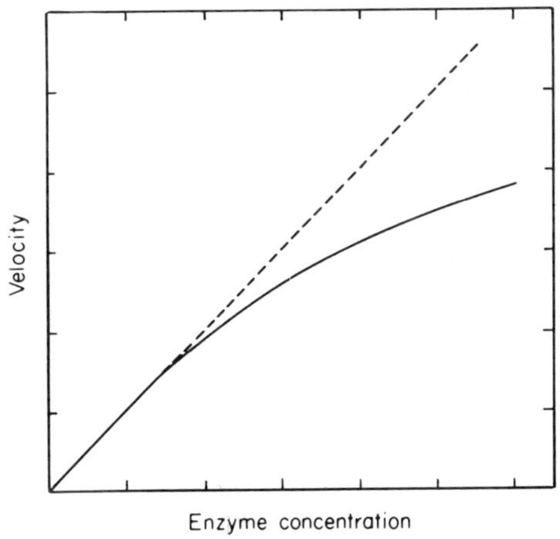

Figure 7 Enzyme concentration–velocity relationship in the presence of competitive inhibitor. The competitive inhibitor is increased as the enzyme concentration is increased by being present in the enzyme preparation. The dashed and solid lines show expected and observed results, respectively. (A_0) is not saturating.

E. Noncompetitive Inhibitor in Enzyme Preparation

Experimentally, the same effect as shown in Fig. 7 would be found. However, the two situations are different in that the observed results in the latter case are independent of substrate concentration. Since inhibitor and substrate are noncompetitive, the effect due to inhibition cannot be overcome by saturating the enzyme with substrate.

F. Uncompetitive Inhibitor in the Enzyme Preparation

Experimentally, the same qualitative effect as for the presence of a competitive or noncompetitive inhibitor in the enzyme preparation would be seen (Fig. 7). However, the effect of the inhibitor would increase as substrate concentration is increased since the inhibitor combines with a species of the enzyme which has combined with substrate.

G. Essential Cofactor Requirement of Enzyme

When the cofactor is attached firmly to the enzyme so that the dissociation constant is very small (0 to 10^{-9} *M*), the enzyme will be fully saturated with cofactor at all enzyme concentrations [when (cofactor) > (E_0)]. However, when K_d is large, an effect on the velocity versus (E_0) curve will be seen. Consider the following reaction,

$$\mathrm{E{\cdot}CoE} + \mathrm{A} \underset{k_{-1}}{\overset{k_1}{\rightleftharpoons}} \mathrm{E{\cdot}CoE{\cdot}A} \xrightarrow{k_2} \mathrm{E{\cdot}CoE} + \mathrm{P} \tag{21}$$

$$\mathrm{E{\cdot}CoE} \underset{k_{-3}}{\overset{k_3}{\rightleftharpoons}} \mathrm{E} + \mathrm{CoE} \qquad K_d = \frac{(\mathrm{E})(\mathrm{CoE})}{(\mathrm{E{\cdot}CoE})} \tag{22}$$

where CoE is the cofactor involved. Only the enzyme form E·CoE is active. Two different results can be seen, depending on the experimental conditions. The situation is different depending on the relative concentrations of enzyme, E_0, and coenzyme, CoE_0. We treat two extreme cases to show the differences between the two conditions.

1. (E_0) and (CoE_0) Approximately Equal

Consider a situation where (E_0) and (CoE_0) are approximately equal in the enzyme preparation and small aliquots of this preparation are added to the substrate–buffer system to initiate a reaction. At low concentrations of (E_0) the enzyme will not be fully saturated with cofactor even though the substrate concentration is much greater than K_m. Under these conditions less than the expected activity will be observed at low enzyme concentrations (Fig. 8).

This observed behavior has been called a dilution effect. This same experimental behavior can also be observed because of instability of enzyme at low

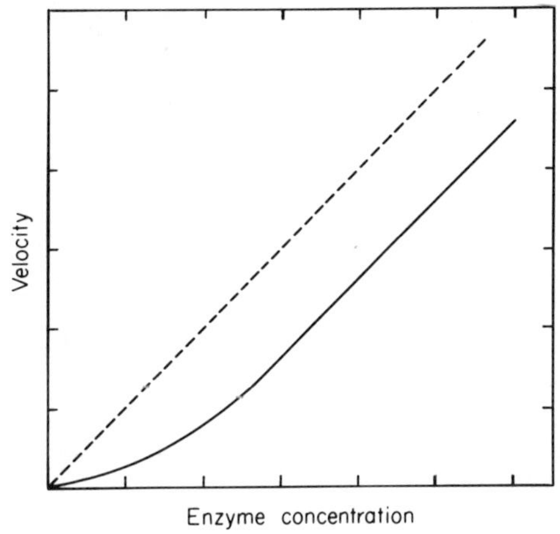

Figure 8 Enzyme concentration–velocity relationship when dissociable cofactor required. (E_0) and (CoE_0) are approximately equal in concentration. The dashed and solid lines show expected and observed results, respectively.

concentrations or by adsorption of enzyme to the surface of the reaction container. The dilution effect can be eliminated by adding sufficient cofactor so that all the enzyme will be in the active form at all concentrations. The instability of an enzyme at low concentrations can often be prevented by adding gelatin or bovine serum albumin (~0.1%) to the reaction mixture.

2. (CoE_0) Maintained Constant

A different situation will be seen if the cofactor concentration is maintained constant by addition to the reaction mixture but at less than the concentration needed to saturate the enzyme at all enzyme concentrations. Consider, for example, the results when $K_d = 1 \times 10^{-6}$ *M*, (CoE_0) is maintained constant at 2.5×10^{-7} *M*, (A_0) $\gg K_m$, and the enzyme concentration is varied from 1×10^{-7} to 5×10^{-5} *M* (Fig. 9).

H. Polymeric Substrate with Several Susceptible Bonds

The natural substrates for the amylases, pectic enzymes, proteolytic enzymes, and nucleases are polymers. These polymers may consist of repeating identical units (homopolymers) or of different repeating units (heteropolymers). Consider the example of α-amylase acting on the substrate amylose where all the glucosidic bonds are identical (α1-4 glucose bonds). The intermediate-size products serve as substrates for the enzyme until all material has been hydrolyzed to maltose. However, the relative affinity between the various fragments and α-amylase and

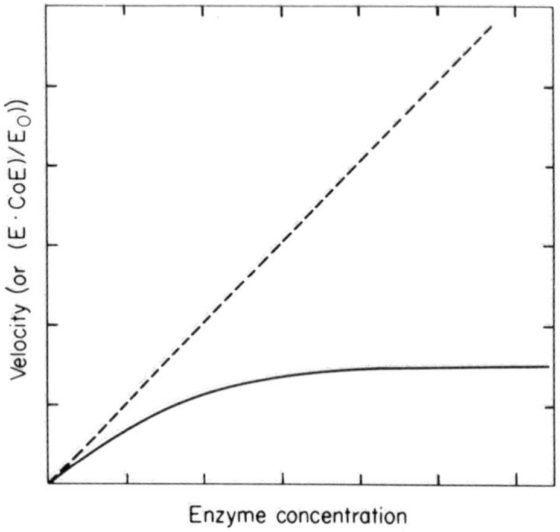

Figure 9 Enzyme concentration–velocity relationship when dissociable cofactor required. The concentration of cofactor is kept constant and in insufficient amount to saturate enzyme at all (E_0). The dashed and solid lines show expected and observed results, respectively.

the relative rates of hydrolysis of the glucosidic bonds may be quite different, particularly for the lower-molecular-weight fragments.

Consider the example of a proteolytic enzyme, trypsin, in the hydrolysis of a denatured protein (heteropolymer). Trypsin has specificity for the peptide bonds in which arginine and lysine residues provide the carbonyl group. The rate at which trypsin hydrolyzes a peptide bond will be determined not only by whether the bond involves lysine or arginine but also by which of the 20 amino acids furnishes the amino portion of the peptide bond. Therefore, hydrolyses of the susceptible peptide bonds proceed at a large number of different rates, so that the observed rate is an average of all of these.

The kinetic result of both examples described is that the most susceptible bonds will be hydrolyzed at the fastest rate. Therefore, as the reaction proceeds, the rate will become slower and slower. In these cases, the relationship between velocity and enzyme concentration will be linear only as long as there is a sufficient concentration of the most susceptible bonds so that an initial rate may be properly obtained. The plot of number of peptide bonds hydrolyzed versus (E) would resemble Fig. 4.

III. ACTIVE ENZYME CONCENTRATION

As indicated earlier in this chapter, often the primary purpose of determining the activity of an enzyme is to measure how much of the enzyme is present in a preparation. The slope of the line in Fig. 1 gives a measure of this since the slope is specific activity. Specific activity gives a measure of relative and not absolute concentration of an enzyme preparation. By knowing the specific activity it is possible to say that one preparation is more concentrated than another. If there were a method of determining the specific activity of a 100% pure enzyme preparation, there would be an absolute basis for comparison. The crux of the problem here is to determine what is a 100% pure enzyme.

In Chapter 3 methods for purification of enzymes and analyses designed to determine homogeneity of both protein and enzyme are described. As noted, a preparation could be 100% homogeneous as a protein but 0 to 100% homogeneous as an enzyme. We now want to expand on that discussion.

A. Determining Absolute Enzyme Concentration

Methods have been designed for determining purity of a number of enzymes. The rationale of all these methods is the same. The only true measure of whether a protein molecule is active as a catalyst is to observe it in action. This is done in normal assays but the enzyme is permitted to cycle through the system repeatedly so that each molecule of enzyme is responsible for the breakdown of several hundred substrate molecules during the course of the assay, and an averaging effect is seen. A true measure of concentration of active enzyme molecules could be obtained if there were a way to limit each active enzyme molecule to a single passage through the reaction while preventing the inactive molecules from reacting. A compound is needed that meets all the specificity requirements of the enzyme as a substrate so that it is stereospecifically bound into the active site to test for proper conformation in the binding locus of the active site. To test for proper conformation in the transforming locus of the active site, once bound stereospecifically, the compound must form a covalent linkage with a group on the enzyme that is a part of the transforming locus of the active site. To meet the requirement that each enzyme molecule be permitted to go through the cycle only once, the covalent linkage formed between the enzyme and compound should be stable (or relatively so). For experimental convenience the product(s) formed should have unique absorption and a high molar extinction coefficient. This idea is expressed more clearly by

$$\mathrm{E} + \mathrm{A} \underset{k_{-1}}{\overset{k_1}{\rightleftharpoons}} \mathrm{EA} \xrightarrow{k_2} \underset{\substack{+\\ \mathrm{P}_1}}{\mathrm{E\text{-}A}} \xrightarrow{k_3} \mathrm{E} + \mathrm{P}_2 \tag{23}$$

where E-A involves covalent bond formation between a reactive group in the active site of the enzyme and a part of the substrate. If $k_3 = 0$ or $k_3 \ll k_2$, all the enzyme will be trapped in the form E-A [where $(\mathrm{A}_0) \gg (\mathrm{E}_0)$]. Therefore, by determining the concentration of E-A (or of P_1) the amount of active enzyme in a preparation will be known. Let us take some specific examples.

B. Organophosphates

There are a number of enzymes that have a seryl group involved in the transforming locus of the active site. Among these enzymes are chymotrypsin, trypsin, thrombin, elastase, subtilisin, phosphoglucomutase, cholinesterase, and various esterases. These enzymes are inhibited by the organophosphorus compound diisopropylphosphorofluoridate (DFP), which will react only with the active enzyme.

$$((CH_3)_2CHO)_2POF + HOCH_2\text{-E} \rightarrow ((CH_3)_2CHO)_2POOCH_2\text{-E} + HF \qquad (24)$$

Although there are many other seryl groups in the enzyme, only the one in the active site reacts. For convenience and sensitivity in determining concentration of active enzyme, the enzyme preparation is treated with ^{32}P-labeled DFP. The amount of DFP incorporated into the protein is determined by measuring the amount of radioactivity and protein present.

Exercise 1. Five milliliters of a chymotrypsin solution that was $1.00 \times 10^{-4}\,M$ in protein was treated with a slight excess of ^{32}P-labeled DFP, which had a radioactivity value of 300,000 cpm per milligram of phosphorus. The protein isolated had a radioactivity value of 3600 cpm/mg. Using a molecular weight of 23,500 for chymotrypsin, calculate the active enzyme concentration and purity of the chymotrypsin preparation.

C. Chloromethyl Ketones

Chymotrypsin and trypsin react specifically with *N*-tosyl-L-phenylalanine chloromethyl ketone (TPCK) and *N*-tosyl-L-lysine chloromethyl ketone (TLCK), respectively. In these reactions the specific histidyl residue involved in the transforming locus of the active site of each of these enzymes is alkylated [Eqs. (25) and (26)].

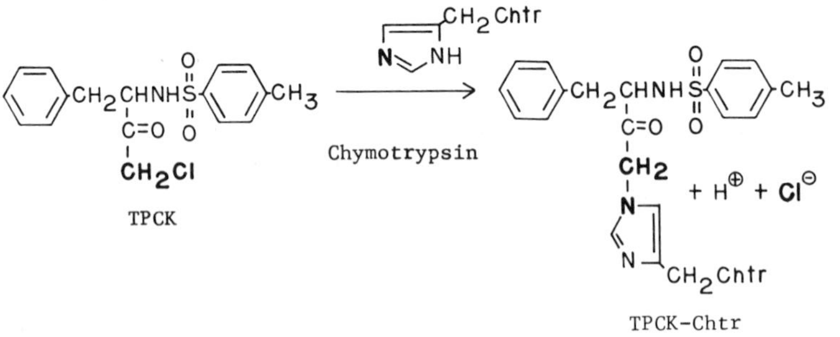

(25)

The amino acid residues phenylalanyl and lysyl fit the specificity requirements of chymotrypsin and trypsin, respectively. The alkyl derivative formed is stable. By use of radioactively labeled TPCK or TLCK and the general procedure described for DFP, the amount of active enzyme can be determined. Inactive enzyme will not react.

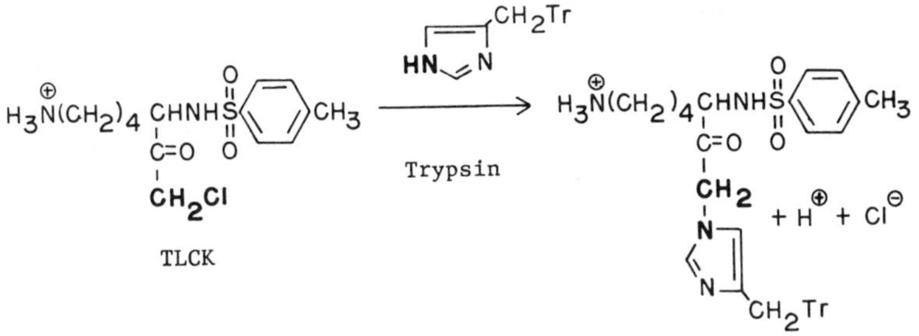 (26)

TLCK

TLCK-Tr

The same procedure may be used for the sulfhydryl proteolytic enzymes papain and ficin, which react with both TPCK and TLCK. In these cases it is the essential sulfhydryl group of the cysteine residue in the catalytic locus of the active site that is alkylated, and only the active enzyme will react with these reagents.

D. Substrates with Good Leaving Groups

The concentration of active enzyme may be determined by use of substrates that have good leaving groups, so that $k_2 \gg k_3$ [Eq. (23)].

1. Cinnamoylimidazole

The active enzyme concentration of chymotrypsin can be determined by use of *N-trans*-cinnamoylimidazole [2].

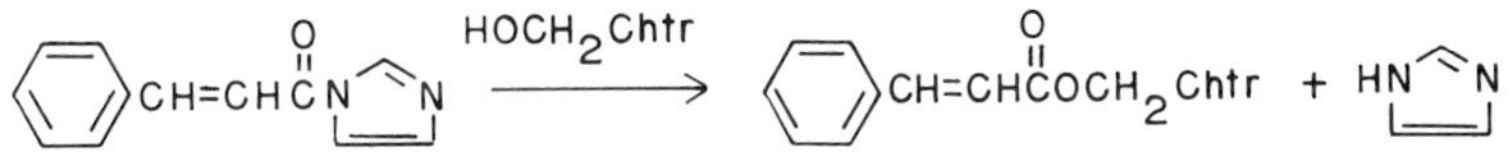 (27)

N-*trans*-cinnamoyl-imidazole

N-*trans*-cinnamoyl-chymotrypsin

The cinnamoyl group confers on the substrate the specificity needed for chymotrypsin, and imidazole is a good leaving group. Therefore, the rate of acylation of the enzyme is rapid, whereas the rate of deacylation is quite slow. The reaction is followed spectrophotometrically at 310 nm (Fig. 10) which measures the cinnamoylchymotrypsin concentration. The concentration of active enzyme is calculated as

$$\frac{A_1 - A_2 - A_3}{\epsilon l} = \text{concentration of active enzyme} \tag{28}$$

where A_1, A_2, and A_3 are absorbances of chymotrypsin plus cinnamoylimidazole, chymotrypsin alone, and cinnamoylimidazole alone, respectively; ϵ is the molar extinction coefficient of cinnamoylchymotrypsin; and l is the light path, in centimeters, through the solution in the cuvet.

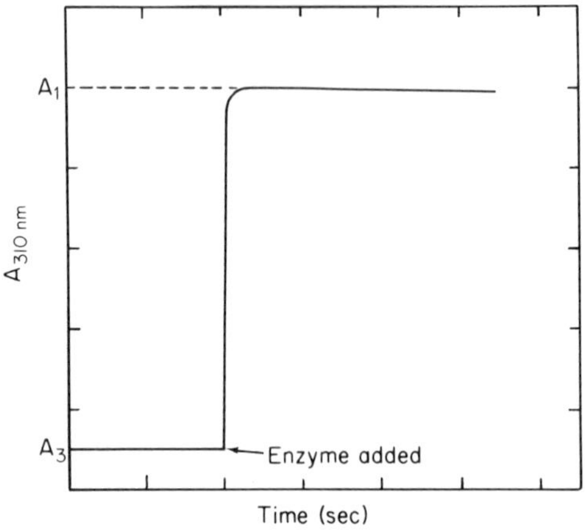

Figure 10 Schematic representation for determination of concentration of active chymotrypsin by use of *N-trans*-cinnamoylimidazole. The rapid increase in absorbance on adding enzyme is due to formation of *N-trans*-cinnamoylchymotrypsin.

2. Nitrophenyl Esters

Nitrophenyl ester derivatives of a specific substrate of an enzyme are used for determination of active enzyme concentration. Use of nitrophenyl esters achieves two purposes, in that the nitrophenyl group is a good leaving group and the concentration of nitrophenol [P_1 in Eq. (23)] liberated is easily determined spectrophotometrically at 403 nm above pH 7 or at 340 nm below pH 7. In order to be generally useful k_2 must be much larger than k_3 (Fig. 11). Table 1 lists several enzymes whose concentrations may be determined by the methods described.

E. Separation of Active and Inactive Enzyme Molecules

Inactive enzyme molecules that have lost a large part of their secondary and/or tertiary structure can be readily separated from active enzyme molecules because they differ in solubility, electrophoretic, and possibly chromatographic properties. However, subtle changes in conformation of the active site resulting in loss of catalytic activity will not be detected by these techniques. Remarkable success has been achieved in separating active and inactive enzyme molecules by affinity chromatography [10–12]. As examples, active trypsin was separated from inactive trypsin by the use of a column packed with guanidinocellulose [13], while active papain was separated from inactive papain by use of a column of Sepharose to which glycylglycyl-L-tyrosyl(*bzl*)-L-arginine had been covalently linked through the amino group of glycine [14]. An adsorptive complex is formed

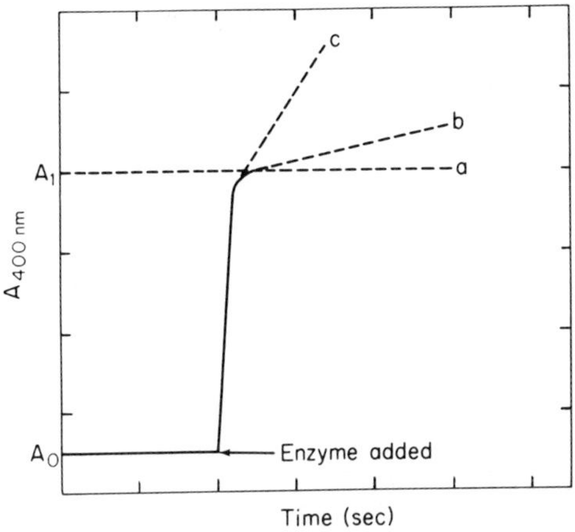

Figure 11 Schematic representation of determination of concentration of active enzyme by use of *p*-nitrophenyl ester as substrate. The rapid increase in absorbance on adding enzyme is due to *p*-nitrophenol production. In case a the acylenzyme formed is stable, in case b the acylenzyme is slowly hydrolyzed to enzyme and acid and in case c the acylenzyme is hydrolyzed to enzyme and acid almost as rapidly as it is formed. Case c would not be useful for determination of active enzyme concentration.

Table 1 Determination of Active Enzyme Concentration

Enzyme	Reagent	Reference
Chymotrypsin	*N-trans*-Cinnamoylimidazole	2
	o-Nitro-p-carboxyphenyl-*N,N*-diphenylcarbamate	3
	p-Nitrophenyl N^2-acetyl-N^1-benzylcarbazate	4
	p-Nitrophenyl acetate	5
	p-Nitrophenyl *p*-(*N,N,N*-trimethylammonium) cinnamate	6
Trypsin	*p*-Nitrophenyl N^2-benzyloxycarbonyl-L-lysinate	5
	p'-Nitrophenyl-*p*-guanidinobenzoate	7
	p-Nitrophenyl N^2-acetyl-N^1-benzylcarbazate	4
	α-*N*-Methyl-α-*N*-toluene-*p*-sulfonyl-L-lysine β-naphthyl ester	8
	p-Nitrophenyl *p*-(*N,N,N*-trimethylammonium) cinnamate	6
Thrombin	*p*-Nitrophenyl *N*-benzyloxycarbonyl-L-tyrosinate	9
Papain	*p*-Nitrophenyl *N*-benzyloxycarbonyl-L-tyrosinate	5
Subtilisin	*N-trans*-Cinnamoylimidazole	5
Elastase	Diethyl *p*-nitrophenyl phosphate	5
Acetylcholinesterase	*o*-Nitrophenyl dimethylcarbamate	5

between these groupings and the binding locus of the active site of the enzyme. This reversible process is shown as follows for trypsin (Tr) and guanidino-cellulose:

$$\mathrm{Tr} + \underset{\overset{\|}{\overset{\oplus}{N}H_2}}{H_2NCNH\text{-cellulose}} \underset{k_{-1}}{\overset{k_1}{\rightleftharpoons}} \underset{\overset{\|}{\overset{\oplus}{N}H_2}}{\mathrm{Tr}\cdot H_2NCNH\text{-cellulose}} \qquad (29)$$

where

$$K_d = \frac{k_{-1}}{k_1} = \frac{(\mathrm{Tr})(\mathrm{H_2N}(\overset{\oplus}{\mathrm{N}}\mathrm{H_2})\mathrm{CNH\text{-}cellulose})}{(\mathrm{Tr}\cdot\mathrm{H_2N}(\overset{\oplus}{\mathrm{N}}\mathrm{H_2})\mathrm{CNH\text{-}cellulose})} \qquad (30)$$

The active molecules that are complexed will be delayed in passage through the column, while those inactive molecules not complexed will pass directly through the column. The degree of delay will depend on the magnitude of K_d as well as the concentration of the compounds involved [Eq. (30)]. In general, the smaller the K_d, the longer it will take for the active enzyme molecules to pass through the column. In some cases soluble competitive inhibitor or change in pH may be required to elute the bound enzyme. Figure 12 demonstrates the expected results.

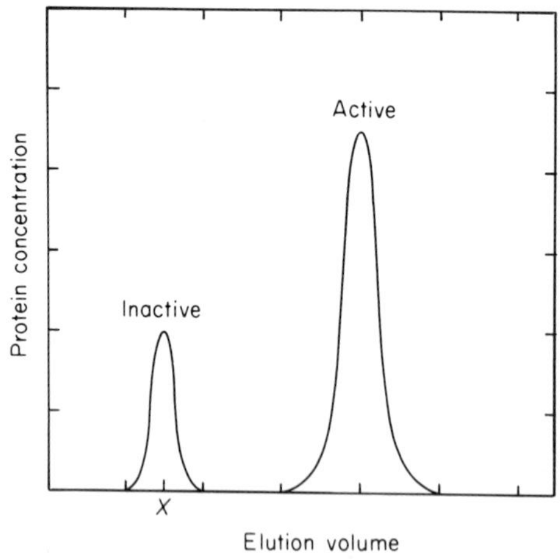

Figure 12 Schematic representation of separation of active and inactive enzyme by chromatography on a column packed with a material to which the active enzyme is adsorbed. Inactive enzyme is eluted at the void volume, X, of the column, while the active enzyme is retarded in relation to its degree of interaction with the column material.

Procedures for separating active from inactive enzyme molecules may have far-reaching consequences. Very importantly, for the first time 100% active enzyme will be available for x-ray crystallographic work.

REVIEW QUESTIONS

1. Two milliliters of enzyme solution in 10 mL of reaction mixture gave 25% conversion of substrate to product in 15 min. Four milliliters of the same enzyme solution in 10 mL of reaction mixture gave 50% conversion of substrate to product in 15 min. No inhibitors were present and no coenzymes or activators were necessary. The pH and temperature were constant. Was K_m larger or smaller than (A_0)? Explain your answer.

2. One milliliter of enzyme solution in a reaction volume of 10 mL gave 20% conversion of substrate to product in 1 hour. Two milliliters of the same enzyme solution in a reaction volume of 10 mL gave 30% substrate conversion in 30 min. The pH and temperature were constant. What conclusions can be drawn from these data?

3. Enzyme solution X gave 20% conversion of substrate in 30 min, while enzyme solution Y gave 30% conversion of the same substrate in 30 min. What can you say about the relative concentrations of the two solutions?

4. By use of proper equations, show that V_{max} is linearly related to the enzyme concentration when $(A_0) \geqslant 100K_m$ but that k_{cat}, the rate constant observed for product formation, is independent of enzyme concentration.

5. Explain the following experimental observations.

(a) Doubling the time of a reaction doubled the amount of product formed.

(b) Doubling the amount of substrate caused no increase in the amount of product formed per unit time.

(c) Doubling the enzyme concentration caused a 60% increase in the amount of product formed per unit time.

6. Why is knowledge of the specific activity of an enzyme preparation not sufficient to indicate the absolute purity of an enzyme preparation? How might you design an experiment in which you could determine the enzyme purity based on specific activity?

7. Explain what the experimental conditions must be to observe that the time for disappearance of one-half the substrate is independent of initial substrate concentration but dependent on the enzyme concentration.

8. Calculate the effect doubling the enzyme concentration will have on K_m.

9. In determining the relationship between enzyme concentration and velocity it was observed that the velocity at low enzyme concentrations was lower than expected and not proportional to enzyme concentration. Offer several explanations for this and devise experimental procedures for distinguishing among the possibilities.

10. Graphically depict the results you would expect to obtain in the relationship between enzyme concentration and velocity when a polymeric substrate is used and (a) the substrate concentration is much larger than K_m; (b) the substrate concentration is approximately K_m.

REFERENCES

General References

Dixon, M., and E. C. Webb, *Enzymes*, 3rd Ed., Academic Press, New York, 1980.

Ostrove, S., Affinity chromatography: general methods, *Methods Enzymol.*, **182**, 357 (1990).

Ostrove, S., and S. Weiss, Affinity chromatography: specialized techniques, *Methods Enzymol.*, **182**, 371 (1990).

Segel, I. H., *Biochemical Calculations*, 2nd Ed., Wiley, New York, 1976.

Cited References

1. W. R. McClure, *Biochemistry*, **8**, 2782 (1969).
2. G. R. Schonbaum, B. Zerner, and M. L. Bender, *J. Biol. Chem.*, **236**, 2930 (1961).
3. B. F. Erlanger, H. Castleman, and A. G. Cooper, *J. Am. Chem. Soc.*, **85**, 1872 (1963).
4. D. T. Elmore and J. J. Smyth, *Biochem. J.*, **107**, 103 (1968).
5. M. L. Bender, M. L. Begué-Canton, R. L. Blakeley, L. J. Brubacher, J. Feder, C. R. Gunter, F. J. Kézdy, J. V. Killheffer, Jr., T. H. Marshall, C. G. Miller, R. W. Roeske, and J. K. Stoops, *J. Am. Chem. Soc.*, **88**, 5890 (1966).
6. J. R. Knowles and J. M. Preston, *Biochim. Biophys. Acta*, **151**, 290 (1968).
7. T. Chase, Jr., and E. Shaw, *Biochem. Biophys. Res. Commun.*, **29**, 508 (1967).
8. D. T. Elmore and J. J. Smyth, *Biochem. J.*, **107**, 97 (1968).
9. F. J. Kézdy, L. Lorand, and K. D. Miller, *Biochemistry*, **4**, 2302 (1965).
10. P. Cuatrecasas, M. Wilchek, and C. B. Anfinsen, *Proc. Natl. Acad. Sci. USA*, **61**, 636 (1968).
11. S. Ostrove, *Methods Enzymol.*, **182**, 357 (1990).
12. S. Ostrove and S. Weiss, *Methods Enzymol.*, **182**, 371 (1990).
13. J. C. Beeley and H. Neurath, *Biochemistry*, **7**, 1239 (1968).
14. S. Blumberg, I. Schechter, and A. Berger, *European J. Biochem.*, **15**, 97 (1970).

8

Kinetic Consequences of Enzyme Inhibition

Any compound that reduces the velocity of an enzyme-catalyzed reaction, by whatever mechanism, is an inhibitor. Enzyme inhibitors have practical importance in fields as widely diverse as toxicology, pharmacology, food science, and nutrition. Inhibitors are of great value in elucidation of the mechanism of enzyme-catalyzed reactions, determination of groups in active sites of enzymes, and in establishing the sequences of metabolic pathways. They are also used in the control of pests, including insects and weeds.

I. TYPES OF INHIBITORS

For convenience we divide the inhibitors into two types, the irreversible and the reversible inhibitors, based on kinetic considerations.

A. Irreversible Inhibitors

The group of irreversible inhibitors includes all compounds that react with an enzyme to form kinetically stable covalent bonds. While the covalent bond formed between enzyme and inhibitor may, in some cases, be reversed by chemical treatment, from a kinetic point of view the reaction is irreversible

$$E + I \xrightarrow{k_1} EI \quad (1)$$

The rate of inhibition will depend, among other things, on k_1, (E), and (I)

$$\frac{d\mathrm{EI}}{dt} = k_1(\mathrm{E})(\mathrm{I}) \tag{2}$$

When reaction of an irreversible inhibitor with enzyme involves a group in or near the active site of the enzyme, the presence of substrate (or a competitive inhibitor) may change the rate of reaction of inhibitor with enzyme. In most cases the rate of reaction will decrease as substrate concentration is increased to saturation, although there are some examples where the rate increases as substrate concentration is increased. This effect of substrate concentration on rate of activity loss has frequently led to treatment of experimental data obtained with irreversible inhibitors by the methods discussed below for reversible inhibitors. This mistake, which must be avoided, is readily made when the effect of substrate concentration on amount of enzyme activity left is observed at one fixed interval of time because the data will approximate that expected for the case of a reversible noncompetitive inhibitor (see below). However, irreversible inhibition can readily be distinguished from true reversible inhibition by observing the activity left at different time intervals, by preincubating the enzyme with inhibitor before adding substrate or by determining the effect of dilution on activity. Dialysis and Sephadex gel chromatography are also useful in distinguishing between reversible and irreversible inhibition.

Rate of reaction of inhibitor with enzyme can be treated by the kinetic equations (Chapter 5) that apply to an irreversible process. Irreversible inhibitors are important in determining reactive groups in or near the active site of enzymes, in measuring changes in reactivity of these groups in the presence or absence of ligands (substrates, cofactors, reversible inhibitors), and in the control of enzyme activity in food processing, pharmacology, and toxicology.

B. Reversible Inhibitors

The remainder of this chapter will deal with the subject of reversible inhibition. Defined kinetically, a reversible inhibitor is one that reacts with an enzyme form, E_x, in a reversible manner:

$$\mathrm{E}_x + \mathrm{I} \rightleftharpoons \mathrm{E}_x\mathrm{I} \tag{3}$$

Reversible inhibitors can be divided into three distinct types based on their effect on the slope and intercept of a reciprocal plot of observed initial rates versus initial substrate concentrations compared to the same reaction in the absence of inhibitors:

$$\frac{1}{v_0} = \text{slope}\,\frac{1}{(\mathrm{A}_0)} + \text{intercept} \tag{4}$$

The types of inhibitors are competitive, noncompetitive, and uncompetitive.

A *competitive inhibitor* affects only the slope term of Eq. (4) (Fig. 1). Experiments are run at various concentrations of one substrate (A_0) with fixed concentrations of all other substrates in the presence and absence of inhibitor. pH, temperature, buffer concentration, cofactors, and so on, are all kept constant. Within any one experiment, the inhibitor concentration is constant.

A *noncompetitive inhibitor* affects both the vertical intercept and slope terms of Eq. (4) (Fig. 2). The lines converge at a point to the left of the vertical axis and above, below, or on the horizontal axis (see also Figs. 8, 9, and 10).

An *uncompetitive inhibitor* affects only the vertical intercept term of Eq. (4) (Fig. 3). Therefore, a series of parallel lines are obtained when different inhibitor concentrations are used in different experiments.

In understanding the mechanism of action of an inhibitor it is important to determine, in addition to whether the inhibitor changes the slope, vertical intercept, or both, whether the relationship between the slopes (and/or intercepts) obtained (Figs. 1 to 3) is a linear function of inhibitor concentration. Three relationships that have been observed in enzyme-catalyzed reactions are shown in Fig. 4, where it is seen that the relationship may be linear, hyperbolic, or parabolic. When a compound combines with only one enzyme form to decrease activity, the effect is always linear inhibition. When a compound combines with two different enzyme forms (i.e., at two points connected by a reversible pathway in the reaction sequence) to produce a multiple effect on the slope and/or intercept, and if both combinations result in a decrease in activity, parabolic inhibition is observed. When a compound combines with two different enzyme forms so as to

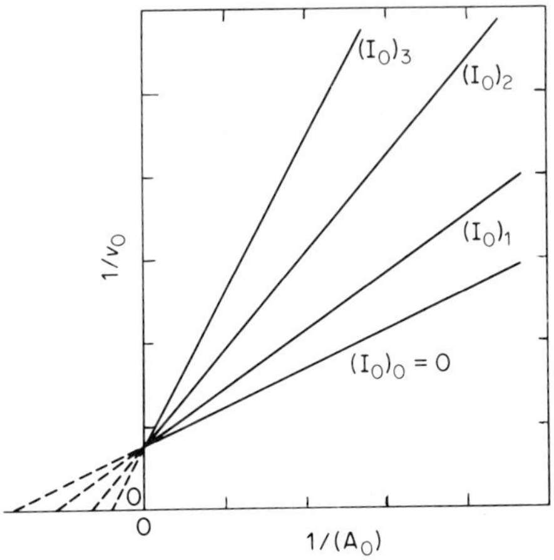

Figure 1 Effect of three different concentrations of a *competitive inhibitor* on $1/v_0$ versus $1/(A_0)$ plot. $(I_0)_0$, $(I_0)_1$, $(I_0)_2$, and $(I_0)_3$ are 0, 0.5, 1.5, and 3.0 times K_i, respectively.

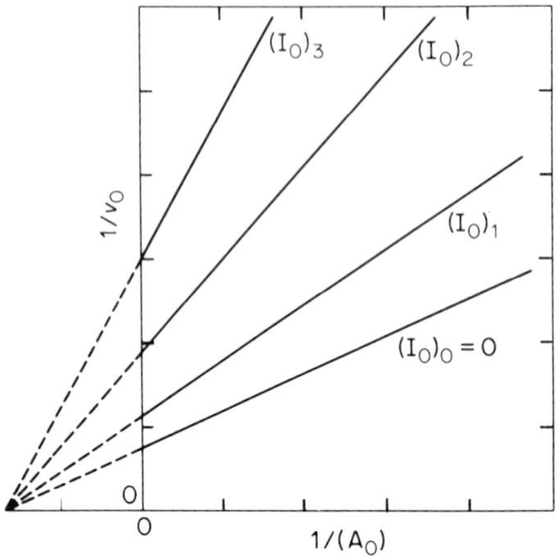

Figure 2 Effect of three different concentrations of a *noncompetitive* inhibitor on $1/v_0$ versus $1/(A_0)$ plot. $(I_0)_0$, $(I_0)_1$, $(I_0)_2$, and $(I_0)_3$ are 0, 0.5, 1.5, and 3.0 times K_i, respectively. The example shown is specifically for simple linear noncompetitive inhibition where $K_{i\,\text{slope}} = K_{i\,\text{intercept}}$ on x-axis.

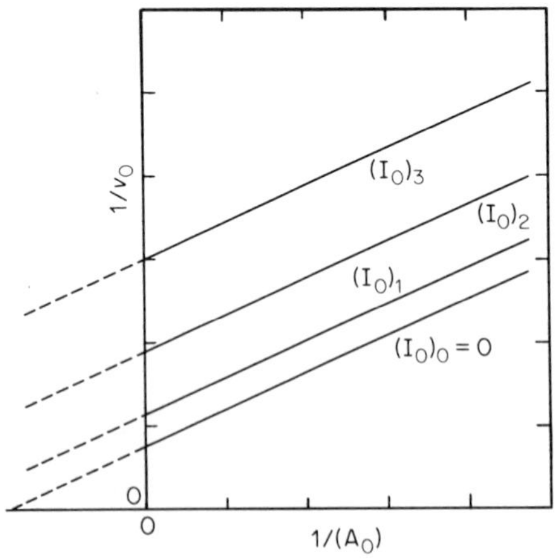

Figure 3 Effect of three different concentrations of an *uncompetitive inhibitor* on $1/v_0$ versus $1/(A_0)$ plot. $(I_0)_0$, $(I_0)_1$, $(I_0)_2$, and $(I_0)_3$ are 0, 0.5, 1.5, and 3.0 times K_i, respectively.

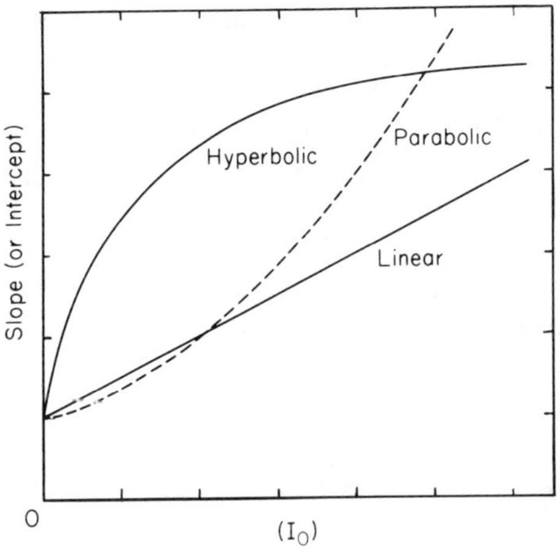

Figure 4 Linear, hyperbolic, and parabolic inhibition patterns. The slopes (or intercepts) obtained from data plotted as in Figs. 1 to 3 are plotted against (I_0).

cause a decrease at one point and activation at the other such that there is an effect on slope and/or intercept, hyperbolic inhibition is observed. Reference 1 should be consulted for discussion of more complicated cases.

In the case of competitive inhibition where only the slope is changed, one has the possibility of linear competitive, hyperbolic competitive, or parabolic competitive inhibition. While the situation is the same with uncompetitive inhibition, it is much more complex with noncompetitive inhibition. There are nine possible combinations since the shape of the replot of slope versus (I) does not restrict the shape of the replot of vertical intercept versus (I), and vice versa.

II. LINEAR INHIBITION PATTERNS

In this section special attention is given to examples where a replot of slopes (and/or intercepts) gives a linear relation with (I). Part but not all of the conclusions drawn in this section may be applicable to more complex situations.

In the discussion that follows we shall use the designations v_0, V, K, slope$_0$, and intercept$_0$ to indicate initial velocity, maximum velocity, substrate concentration at $0.5V$, slope, and y-intercept, respectively, in the absence of inhibitor and v_0', V', K', slope$_\mathrm{I}$, and y-intercept$_\mathrm{I}$ to indicate the same parameters in the presence of inhibitor, respectively. All substrates except one, all cofactors, pH, temperature, and (E_0) are kept constant within a series of experiments. It is further assumed that (I) $\gg$ (E_0) and K_i is $\geqslant 10^{-7}$ M so that (I_0) can be used in the equations. If $K_i < 10^{-7}$ M, (I_0) cannot be used and other equations must be used [2].

K' and V' will be true constants only when (a) there is only one substrate involved in the reaction, or (b) all other substrates are present in saturating amounts, which is usually not the case, or (c) when the inhibitor does not affect that particular parameter. Therefore, to avoid the implication that these are true constants, we shall use V' and K' for these parameters in the presence of inhibitors. Similarly, we shall use V and K to designate these parameters in the absence of inhibitors when describing a general case or when discussing a multisubstrate reaction. Use of A for the variable substrate in no way should be construed to imply that this equation applies only for the first substrate in a multisubstrate reaction. K_i is a constant and represents the concentration of inhibitor that doubles the slope or intercept term. This is not equal to the concentration of inhibitor that halves the initial velocity.

Under these conditions initial velocity will vary as concentration of variable substrate is changed according to

$$\frac{1}{v'_0} = \frac{K'}{V'}\frac{1}{(A_0)} + \frac{1}{V'} \tag{5}$$

A. Linear Competitive Inhibition

The consequences of the presence of a competitive inhibitor in an enzyme-catalyzed reaction is shown in Fig. 5. At high substrate concentrations where $(A_0) \gg K_m$, the influence of inhibitor is eliminated and maximum velocity in the presence and absence of inhibitor is the same. However, the substrate concentration required to give an observed velocity of 0.5 V is increased in the presence of inhibitor (K versus K'). The relationship between these two values of K is

$$K' = K\left[1 + \frac{(I_0)}{K_i}\right] \tag{6}$$

As shown by the reciprocal plot in Fig. 6, only the slope is changed in the presence of a competitive inhibitor. In the case of linear competitive inhibition this change is a constant:

$$\frac{1}{v'_0} = \left[1 + \frac{(I_0)}{K_i}\right]\frac{K}{V}\frac{1}{(A_0)} + \frac{1}{V} \tag{7}$$

K_i may be determined readily since

$$\frac{\text{slope}_1}{\text{slope}_0} = 1 + \frac{(I_0)}{K_i} \tag{8}$$

K_i may also be determined from the ratio of intercepts on the X-axis where $\text{intercept}_0/\text{intercept}_I = 1 + [(I_0)/K_i]$. However, best values of K_i are determined from the y-intercept of a replot of slope versus (I_0) from the primary plot.

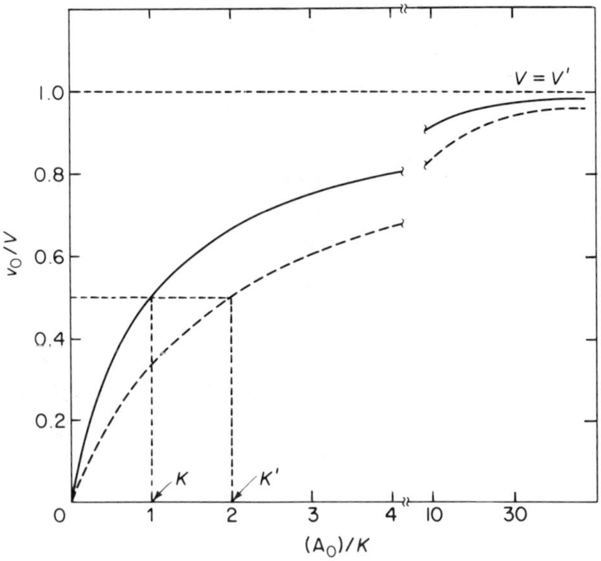

Figure 5 Linear competitive inhibition. V and K are constants in the absence of inhibitor; V' and K' are apparent constants in the presence of inhibitor. The solid line is in the absence of inhibitor; the dashed line is in the presence of $(I_0)_1 = K_i$.

B. Linear Noncompetitive Inhibition

In the presence of a noncompetitive inhibitor the maximum velocity obtained is lower than that in the absence of inhibitor (Fig. 7). Since substrate and inhibitor do not compete with each other for the same binding site on enzyme or in some cases for the same form of enzyme, inhibition cannot be eliminated by adding more substrate, as is the case with a competitive inhibitor. As a consequence of this, the difference between the maximum velocity observed in the absence and presence of a noncompetitive inhibitor is a function of the concentration of inhibitor and $K_{i\,\text{intercept}}$:

$$V = V'\left[1 + \frac{(I_0)}{K_{i\,\text{intercept}}}\right] \tag{9}$$

In noncompetitive inhibition there is a change in both the vertical intercept and slope of a reciprocal plot in the presence of inhibitor (Fig. 8). The relationship between inhibitor concentration and effect on slope and intercept is given by

$$\frac{1}{v_0'} = \left[1 + \frac{(I_0)}{K_{i\,\text{slope}}}\right]\frac{K}{V}\frac{1}{(A_0)} + \left[1 + \frac{(I_0)}{K_{i\,\text{intercept}}}\right]\frac{1}{V} \tag{10}$$

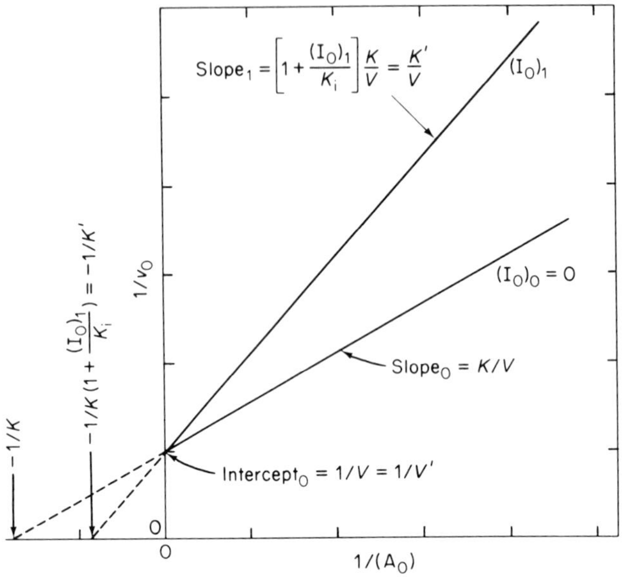

Figure 6 Linear competitive inhibition. Data of Fig. 5 plotted in reciprocal form according to Eq. (7). $(I_0)_1 = K_i$.

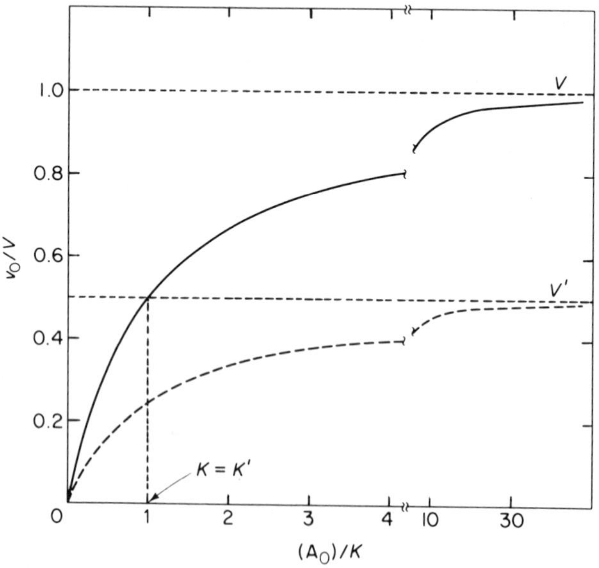

Figure 7 Simple linear noncompetitive inhibition. V and K are constants in absence of inhibitor; V' and K' are apparent constants in the presence of inhibitor. The solid line is in the absence of inhibitor; the dashed line is in the presence of $(I_0)_1 = K_i$.

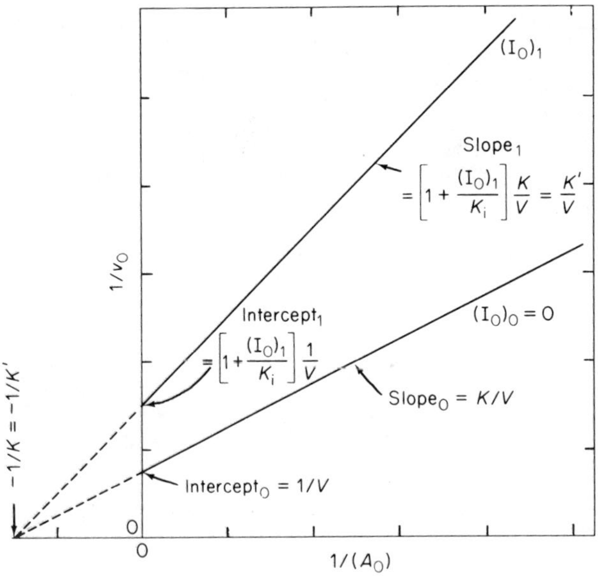

Figure 8 Simple linear noncompetitive inhibition. Data of Fig. 7 plotted in reciprocal form according to Eq. (10). $(I_0)_1 = K_i$, where $K_{i\,\text{intercept}} = K_{i\,\text{slope}}$.

The slope and vertical intercept may not change by the same extent in the presence of inhibitor. $K_{i\,\text{intercept}}$ may be greater than, equal to, or less than $K_{i\,\text{slope}}$, with the result that a family of reciprocal lines obtained at different (I_0) will converge to the left of the y-axis and above (Fig. 9), on (Figs. 2 and 8), or below the x-axis (Fig. 10), respectively. The control $[(I_0)_0 = 0]$ is the same in Figs. 2, 9, and 10, which permits a ready comparison of the data. The case where $K_{i\,\text{slope}}$ is equal to $K_{i\,\text{intercept}}$ is called *simple linear noncompetitive inhibition* and K_i may be calculated from either the change in the slope or the intercept (Fig. 8). In this case there is no change in K as a function of inhibitor concentration and $K' = K$. In the other two situations there is a change in K, so that the relationship between K in the absence and presence of inhibitor is

$$K' = K\frac{1 + (I_0)/K_{i\,\text{slope}}}{1 + (I_0)/K_{i\,\text{intercept}}} \tag{11}$$

K' may be greater than, equal to (simple linear noncompetitive inhibition), or less than K, depending on the relative values of $K_{i\,\text{slope}}$ and $K_{i\,\text{intercept}}$.

C. Linear Uncompetitive Inhibition

As in the case of linear noncompetitive inhibition, uncompetitive inhibition cannot be eliminated by increasing the substrate concentration to saturating levels

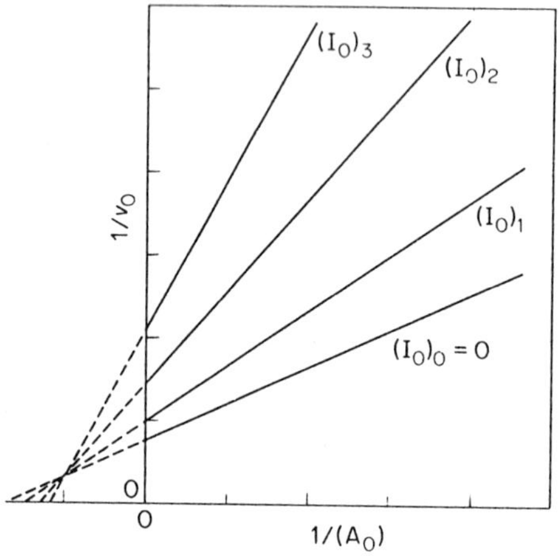

Figure 9 Linear noncompetitive inhibition where $K_{i\,\text{intercept}} = 1.67\ K_{i\,\text{slope}}$. $(I_0)_0$, $(I_0)_1$, $(I_0)_2$, and $(I_0)_3 = 0, 0.5, 1.5$, and 3.0 times $K_{i\,\text{slope}}$, respectively. The line for $(I_0)_0 = 0$ is same as in Figs. 2 and 10.

(Fig. 11). V' and K' are changed by the same extent and in the same direction, as shown by the relationships

$$V = V'\left[1 + \frac{(I_0)}{K_i}\right] \tag{12}$$

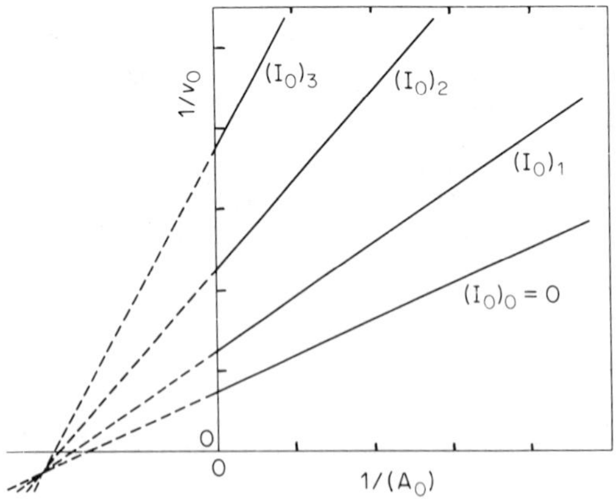

Figure 10 Linear noncompetitive inhibition where $K_{i\,\text{intercept}} = 0.75\ K_{i\,\text{slope}}$. $(I_0)_0$, $(I_0)_1$, $(I_0)_2$, and $(I_0)_3 = 0, 0.5, 1.5$, and 3.0 times $K_{i\,\text{slope}}$, respectively. The line for $(I_0)_0 = 0$ is same as in Figs. 2 and 9.

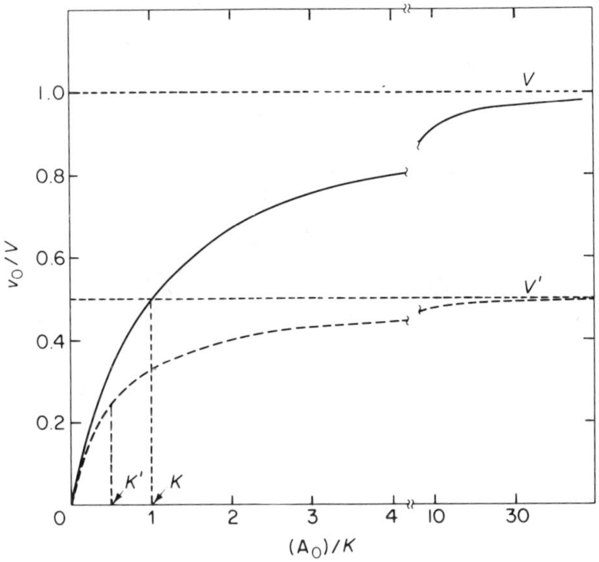

Figure 11 Linear uncompetitive inhibition. V and K are constants in absence of inhibitor; V' and K' are apparent constants in the presence of inhibitor. The solid line is in the absence of inhibitor; the dashed line is in the presence of $(I_0)_1 = K_i$.

$$K = K'\left[1 + \frac{(I_0)}{K_i}\right] \tag{13}$$

In the presence of an uncompetitive inhibitor, K' and V' will always be smaller than K and V, but the ratio K'/V' will be a constant and equal to K/V. As a consequence, a family of reciprocal lines are parallel (Fig. 12). Each line has the same slope but an intercept that differs from the intercept in the absence of inhibitor by the constant factor $1 + [(I_0)/K_i]$:

$$\frac{1}{v_0'} = \frac{K}{V}\frac{1}{(A_0)} + \left[1 + \frac{(I_0)}{K_i}\right]\frac{1}{V} \tag{14}$$

III. SIGMOIDAL BEHAVIOR AS A RESULT OF INHIBITION

Whenever binding of inhibitor at one site makes it easier for subsequent inhibitor molecules to bind at other sites in multisite enzymes (i.e., there is an increase in binding constants), positive cooperativity will be observed. Conversely, when binding of inhibitor at one site makes it more difficult to bind subsequent inhibitor molecules, negative cooperativity will be observed. When there is no effect of

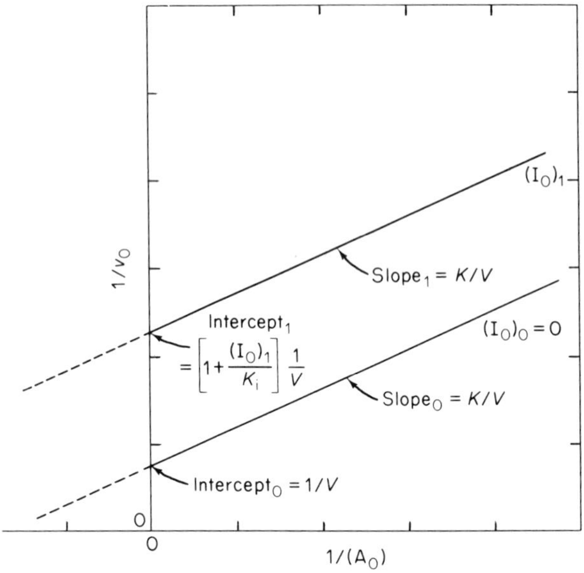

Figure 12 Linear uncompetitive inhibition. Data plotted in reciprocal form according to Eq. (14). $(I_0)_1 = 2K_i$.

bound inhibitor on subsequent binding by other inhibitor molecules, no cooperative effect will be seen and inhibition is treated by the methods described previously. The effect of inhibitor concentration on systems that follow Michaelis–Menten kinetics versus those that show allosteric behavior are shown in Fig. 13.

Quantitative description of an enzyme system showing allosteric behavior in the presence of inhibitor is possible by use of the models proposed by Monod et al. [3] and Koshland et al. [4] or by use of the modified Hill equation [Refs. 5 and 6; Eq. (62), Chapter 6]. A plot of data according to

$$\log \frac{v_0'}{V - v_0'} = n \log(\mathrm{I}_0) - \log K_{ij} \tag{15}$$

is shown in Fig. 14, where v_0' is the observed velocity in presence of inhibitor and K_{ij} is the product of all individual inhibitor constants. Substrate concentration is kept constant in less than saturating amounts (preferably at or near the value required to give $v_0 = 0.5V$ in absence of inhibitor). The slope of the line, m, at $\log v_0'/(V - v_0') = 0$ is a function of the number of interacting sites, n, and the strength of interaction among these sites. When the interaction is infinite, $m = n$. When there is no interaction among sites, $m = 1$. The concentration of inhibitor

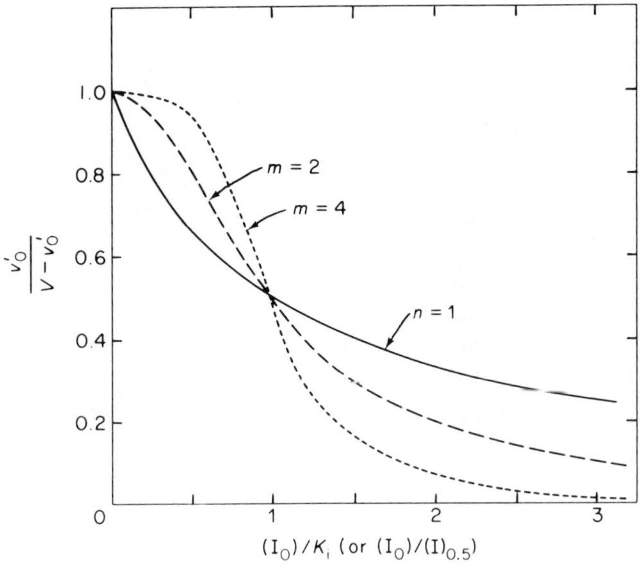

Figure 13 Comparison of effect of inhibitor concentration on an enzyme system that follows Michaelis–Menten kinetics ($n = 1$) and two systems that show allosteric behavior ($m = 2$ and 4). $(I_0)/K_i$ and $(I_0)/I_{0.5}$ are for Michaelis–Menten and allosteric-behaving systems, respectively. $(A_0) = K$ and $A_{0.5}$ for the Michaelis–Menten and the allosteric-behaving systems, respectively.

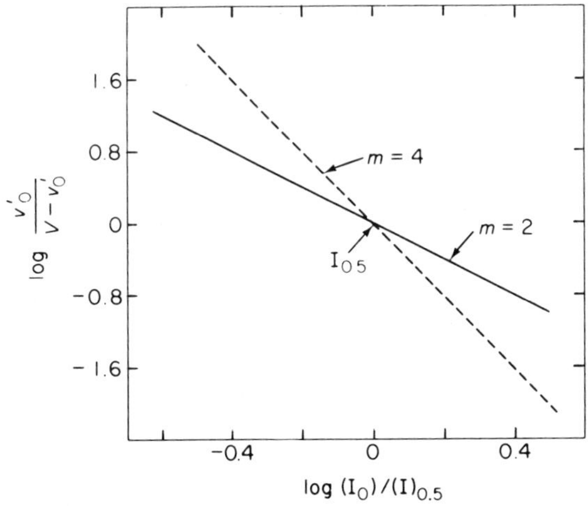

Figure 14 Effect of inhibitor concentration on activity of an allosteric-behaving enzyme system as plotted according to Eq. (15).

needed to reduce v_0' to $0.5V$, called $I_{0.5}$, is the concentration of inhibitor at $\log v_0'/(V - v_0') = 0$.

IV. INTERPRETATION OF RESULTS

From the type of inhibition pattern found, one should be able to make reasonable predictions as to what form of enzyme the inhibitor reacts with and the mechanism of addition of substrate to and release of products from the enzyme. This rationalization can be worked in reverse, in that given a particular mechanism of substrate addition and product release and the form of enzyme to which the inhibitor combines, it should be possible to predict the type of inhibition expected. Some models are presented below with full realization that they may fall short of the actual situation. Nevertheless, they are useful when appreciated only as cartoon models.

A. One-Substrate Reactions

In discussion of inhibition of those enzyme systems in which only one substrate is involved (or the second substrate, for example water, is present in such large amounts that its effect is not seen in the usual experiments), we shall deal only with dead-end inhibition. Dead-end inhibition is defined by

$$E_x + I \rightleftharpoons E_xI \tag{16}$$

where E_xI does not lead to products. E_x is an unspecified enzyme form that may be free enzyme, enzyme·substrate complex, or acylenzyme. (Other covalent enzyme·substrate forms may be formed; we shall use the acylenzyme as a typical example.)

Most cases of inhibition result from combination of a compound with unoccupied portions of the active site while substrates or products are not present. In the case of the hydrolytic enzymes, which best fit one-substrate reactions, a part of the substrate molecule is split off during formation of an acylenzyme intermediate, leaving a position that may be occupied by an inhibitor. Therefore, unoccupied areas are available in the free enzyme and in the acylenzyme.

When a compound (C) combines only with free enzyme such as to compete with binding of substrate, that compound is a *competitive* inhibitor (Fig. 15). The competitive inhibitor (C) can combine only with free enzyme (E). Effect of the inhibitor can be eliminated when the substrate is present in saturating amounts. What is a saturating amount of substrate in the absence of inhibitor [$(A_0) \geqslant 100K_m$] may not be saturating in the presence of inhibitor since $K' = K\{1 + [(I_0)/K_i]\}$.

When a compound (N) combines with free enzyme (E), enzyme·substrate complex (EA) or acylenzyme (EA′) without an effect on binding of substrate but prevents transformation of substrate to products by affecting one or more of the

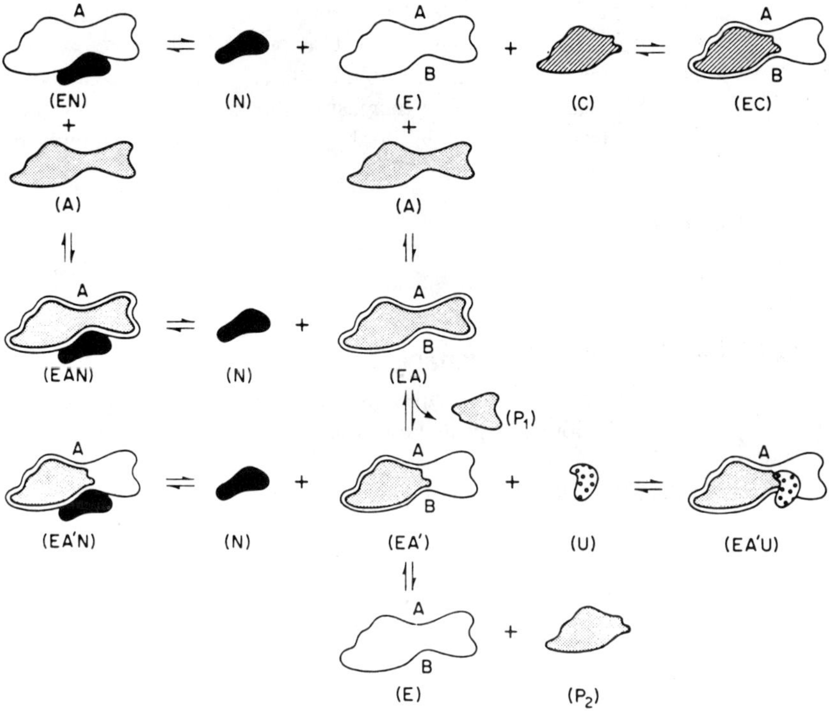

Figure 15 Schematic representation of inhibition of enzyme activity by various types of inhibitors. The model for E shows only the active site with the binding locus (inner void) and the transforming locus, with catalytic groups A and B. The symbols for the species involved (in parentheses) are: E, free enzyme; EA, enzyme·substrate complex; EA′, acylenzyme intermediate; C, competitive inhibitor; N, noncompetitive inhibitor; U, uncompetitive inhibitor; P_1 and P_2 are products formed from the substrate, A, and EN, EAN, EA′N, EC, and EA′U are complexes with respective enzyme species and inhibitors. All complexes formed involve noncovalent bonds except EA′, where a covalent bond is formed with catalytic group A.

groups involved in the transforming step, *linear noncompetitive inhibition* will be observed in that only *V* will be affected. In the model shown in Fig. 15, N combines so as to mask the catalytic group B of the enzyme. The effect of a noncompetitive inhibitor cannot be overcome by saturating amounts of substrate since the two do not compete with each other in binding.

A compound (U) that combines with acylenzyme (EA′) but not with free enzyme (E) or enzyme·substrate complex (EA) so as to prevent transformation of substrate to product results in *uncompetitive* inhibition. A locus for binding of the inhibitor must be provided by the acylenzyme which is not present on the free enzyme. In Fig. 15, part of the binding site for U is provided by the remainder of the substrate in the acylenzyme (EA′). U binds such as to mask the catalytic group B so that EA′U cannot form P_2. Increasing substrate concentration increases the effect of an uncompetitive inhibitor by forming more EA′ with which the inhibitor may combine.

B. Multisubstrate Reactions

We deal here only with the case of two-substrate reactions. Reference 1 should be consulted for more complex cases.

1. Rules for Prediction of Mechanisms

Cleland [7] has given the following rules as guidelines for determining, by inspection, the type of inhibition expected for different mechanisms. (These rules also apply to one-substrate reactions.)

1. An inhibitor affects the slope of a reciprocal plot [$1/v$ versus 1/(A)] when it and the variable substrate compete directly for the same form of the enzyme or when the inhibitor and substrate react with different forms of the enzyme which are separated from each other in the reaction pathway only by reversible steps, so that each influences the concentration of enzyme form available to the other. The effect of inhibitor can be eliminated by saturating amounts of substrate.

2. An inhibitor affects the intercept of a reciprocal plot [$1/v$ versus 1/(A)] when it reacts with a form of enzyme to which the substrate does not combine in such a way that its effect is not overcome by saturation with that substrate. The inhibitor then affects the observed velocity by lowering the total enzyme available that can participate in product formation.

3. If the effects described above occur separately, rule 1 gives competitive inhibition and rule 2 gives uncompetitive inhibition. If an inhibitor produces both effects (i.e., change in slope and intercept) by combining with more than one form of enzyme, it is a noncompetitive inhibitor.

2. Specific Examples

Let us take two examples to see how the rules above apply to both product inhibition and dead-end inhibition.

a. Ordered Bi Bi Mechanism

Consider the case of a specific ordered reaction in which NAD^+ and XH_2 are involved [in the general designation $NAD^+ = A$, $XH_2 = B$, $X = P$, and $NADH = Q$; Eq. (17)]:

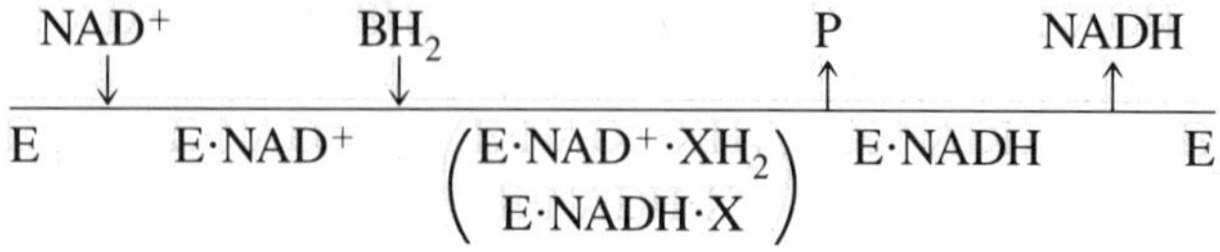

(17)

The experiment is performed by keeping the concentration of one substrate constant (and less than saturating) and varying the other substrate concentration while keeping the added inhibitor (either a product or a dead-end inhibitor, I) concentration constant. *Initial velocities* are determined in all cases. Consider each of the following as separate experiments.

1. A fixed concentration of NADH is added to reaction tubes in which (BH_2) is constant and (NAD^+) is varied.

In this case NADH and NAD^+ compete for the same enzyme form, E. Addition of NAD^+ in saturating amounts eliminates inhibition by NADH. So NADH gives *competitive* inhibition with respect to NAD^+, which on further analysis is found to be linear competitive since NAD^+ combines with only one enzyme form.

2. A fixed concentration of NADH is added to reaction tubes in which (NAD^+) is constant and (BH_2) is varied.

BH_2 combines with the enzyme form $E{\cdot}NAD^+$ while NADH combines with the enzyme form E. Therefore, the two do not compete for the same enzyme form. However, the two forms E and $E{\cdot}NAD^+$ are in equilibrium with each other. In this situation increasing (BH_2) to saturation will not displace NADH; however, since the two forms of enzyme are in equilibrium, the presence of NADH affects the amount of enzyme available to BH_2. Thus NADH gives *noncompetitive* inhibition with respect to BH_2.

3. A fixed concentration of P is added to reaction tubes in which (BH_2) is constant and (NAD^+) is varied.

Since steady-state conditions have been reached there will be a small amount of E·NADH present, and thus P will show *noncompetitive* inhibition with respect to NAD^+.

4. A fixed concentration of P is added to reaction tubes in which (NAD^+) is constant and (BH_2) is varied.

Since steady-state conditions have been reached there will be a small amount of E·NADH and thus P will show *competitive* inhibition with respect to BH_2.

5. I reacts with E (dead end inhibition)

   a. A fixed concentration of I is added to reaction tubes in which (NAD^+) varies and (BH_2) is constant.
   I and NAD^+ react with the same enzyme form, E, so that the slope is changed. The effect of I can be overcome by saturation with NAD^+. The inhibition is *competitive* with respect to NAD^+.

   b. A fixed amount of I is added to reaction tubes in which (BH_2) is varied and (NAD^+) is constant.
   I and BH_2 react with different enzyme forms (E versus $E{\cdot}NAD^+$) so that the vertical intercept is changed in the presence of I. The sequence $EI \rightleftharpoons E \rightleftharpoons E{\cdot}NAD^+$ is reversible, so the slope also varies, giving *noncompetitive* inhibition with respect to BH_2. Inhibition can be eliminated by saturation with NAD^+ but not with BH_2.

6. I reacts with E·NAD^+

a. A fixed amount of I is added to reaction tubes in which (NAD^+) varies and (BH_2) is kept constant.
NAD$^+$ and I do not combine with the same form of enzyme (E versus E·NAD^+), nor can inhibition by I be changed by the concentration of NAD^+. Therefore, only the intercept is changed and *uncompetitive* inhibition is seen with respect to NAD^+.

b. A fixed amount of I is added to reaction tubes in which (BH_2) is varied and (NAD^+) is kept constant.
Both I and BH_2 react with the same enzyme form (E·NAD^+) so that the slope is changed. The inhibition is *competitive* with respect to BH_2 and can be eliminated by saturating amounts of BH_2.

b. Ping-Pong Bi Bi Mechanism

Consider as an example the involvement of an FAD-containing enzyme. In this specific reaction, AH_2 = A, P = P, O_2 = B, and H_2O_2 = Q of a generalized equation.

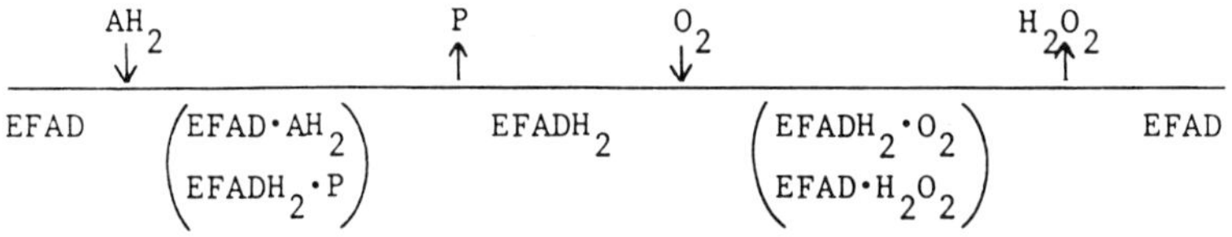

(18)

We consider only the situation with dead-end inhibition. *Initial velocities* are measured in all cases.

1. I reacts with EFAD.

a. (AH_2) is varied and (O_2) is kept constant.
Since I and AH_2 react with the same enzyme form EFAD and the effect of I can be eliminated by saturation with AH_2, only the slope of a reciprocal plot [$1/v_0$ versus $1/(AH_2)$] will be affected, so *competitive* inhibition with respect to AH_2 and I will be observed.

b. (O_2) is varied and (AH_2) is kept constant.
I and O_2 combine with different forms of the enzyme (EFAD versus $EFADH_2$), so that I will affect the intercept of a reciprocal plot. Inhibition cannot be eliminated by saturation with O_2, but only by saturation with AH_2. Therefore, *uncompetitive* inhibition with respect to O_2 and I is observed.

2. I reacts with $EFADH_2$.

a. (AH_2) is varied and (O_2) is kept constant.
I and AH_2 combine with different forms of the enzyme ($EFADH_2$ versus EFAD) and the inhibition cannot be eliminated by saturation with AH_2. *Uncompetitive* inhibition with respect to AH_2 and I will be observed.

b. (O_2) is varied and (AH_2) is kept constant.

I and O_2 combine with the same enzyme form, $EFADH_2$, and the inhibition can be eliminated by saturation with O_2. *Competitive* inhibition with respect to O_2 and I will be observed.

REVIEW QUESTIONS

1. Design an experiment or experiments that would permit you unequivocally to distinguish between reversible and irreversible inhibition.

2. Design a set of experiments that would permit a determination of whether an inhibitor was linear competitive, simple linear noncompetitive, or linear uncompetitive. Be sure to state all experimental conditions used, including the relationship of (A_0) and (I_0) to K and K_i, respectively, and indicate how the data could best be analyzed.

3. Consider a reaction that follows an Ordered Bi Uni mechanism.

(a) Write a generalized reaction for the mechanism using the method of Cleland.

(b) Predict the type of inhibition expected when

(1) I reacts with E and A is the variable substrate.
(2) I reacts with E and B is the variable substrate.
(3) I reacts with EA and A is the variable substrate.
(4) I reacts with EA and B is the variable substrate.
(5) I reacts with EAB and A is the variable substrate.
(6) I reacts with EAB and B is the variable substrate.

4. Answer Question 3 when a reaction follows a Random Bi Uni mechanism.

5. An inhibitor at 0.1 *M* was found to give 50% inhibition at all substrate concentrations. Would you classify the inhibitor as competitive, noncompetitive, or uncompetitive? (Assume linear inhibition.) Calculate K_i.

6. Inhibitor X, at a concentration of 2.5×10^{-4} *M*, gave 25% inhibition of the activity of a 1.0×10^{-7} *M* enzyme solution when the variable substrate concentration was either 0.001 *M* or 0.1 *M*. *K* had been determined previously to be 0.01 *M*.

(a) Was the inhibitor competitive or noncompetitive with respect to substrate?

(b) Calculate the dissociation constant of the enzyme·inhibitor complex.

(c) What would be the percent inhibition if the enzyme concentration were doubled?

(d) What effect would the presence of the inhibitor have on *K*?

(e) What would be the percent inhibition if the inhibitor concentration were doubled?

7. Given that *K* and *V* were 1.0 m*M* and 100 μmol/L/s, respectively, in the absence of inhibitor and 1.0 m*M* and 75 μmol/L/s in the presence of 5 m*M* inhibitor, determine the type of inhibitor involved and calculate K_i. (Assume linear and simple linear inhibition.)

8. Given that $m = n = 3$ and that $(A_0) = A_{0.5}$, calculate the expected relationship between $v_0'/(V - v_0')$ and (I_0), where (I_0) ranges from 0.33 to 6 times $I_{0.5}$.

9. What is the relationship between K and substrate concentration under each of the following situations?

(a) Addition of 1×10^{-3} *M* inhibitor, which gave linear competitive inhibition, caused a 20% reduction in rate compared to a control reaction. K_i was 5×10^{-3} *M*.

(b) Addition of 4×10^{-4} *M* of inhibitor, which gave linear uncompetitive inhibition, caused a 35% reduction in rate compared to a control reaction. K_i was 2×10^{-4} *M*.

REFERENCES

General References

Alberty, R. A., *J. Am. Chem. Soc.*, **80**, 947 (1960).
Cleland, W. W., *Biochim. Biophys. Acta*, **67**, 173 (1963).
Cleland, W. W., *Biochim. Biophys. Acta*, **67**, 188 (1963).
Fromm, J., and D. R. Nelson, *J. Biol. Chem.*, **237**, 215 (1962).
Segel, H. L., in *The Enzymes* (P. D. Boyer, H. Lardy, and K. Myrbäck, eds.), Vol. 1, Academic Press, New York, 1959, p. 1.

Cited References

1. W. W. Cleland, *Biochim. Biophys. Acta*, **67**, 173 (1963).
2. J. Bieth, in *Proteinase Inhibitors*, Bayer Symposium V (H. Fritz, H. Techesche, L. J. Greene, and E. Truscheit, eds.), Springer-Verlag, New York, 1974, p. 463.
3. J. Monod, J. Wyman, and J. P. Changeux, *J. Mol. Biol.*, **12**, 88 (1965).
4. D. E. Koshland, Jr., G. Némethy, and D. Filmer, *Biochemistry*, **5**, 365 (1966).
5. D. E. Atkinson, *Ann. Rev. Biochem.*, **35**, 85 (1966).
6. A. V. Hill, *Biochem. J.*, **7**, 471 (1913).
7. W. W. Cleland, *Biochim. Biophys. Acta*, **67**, 188 (1963).

9

Enzyme Inhibitors

In Chapter 8 the kinetic consequences of different types of general enzyme inhibitors were discussed. Most of the examples chosen involved reversible inhibition (i.e., noncovalent bonds). In this chapter we discuss specific compounds as inhibitors and how these inhibitors are used to obtain information about the nature of enzymes. Most of the inhibitors discussed in this chapter form covalent bonds with reactive groups on the enzyme. In some examples, the reaction is reversible by adding another compound.

I. IMPORTANCE OF ENZYME INHIBITORS

The effect of specific inhibitors on an enzyme system is a powerful tool to the enzymologist and protein chemist investigating the nature of the active site of an enzyme, particularly the binding and transforming loci. Conclusions regarding requirements for binding based on inhibition studies are less subject to error than are those obtained from substrate studies since K_i will reflect the equilibrium between enzyme·inhibitor complex and the free species:

$$E + I \underset{k_{-1}}{\overset{k_1}{\rightleftharpoons}} EI \qquad K_i = \frac{k_{-1}}{k_1} = \frac{(E)(I)}{(EI)} \tag{1}$$

while K_m values are complicated by including rate constants for transformation of substrate to product. An inhibitor can also be useful in providing information about the nature of functional groups at the active site.

Inhibitors have been particularly important in elucidating pathways in in vivo systems. In this connection one should mention the use of inhibitors for elucidating the glycolytic pathway by Lohmann, Embden, and Meyerhof, elucidating the sequence of enzymes of the oxidative pathway by Keilin, and mapping sites of oxidative phosphorylation by Chance.

Inhibitors are of great use to physiologists in studying relationships between the fundamental aspects of cellular activity and molecular transformation. Of particular importance are the inhibitors that uncouple oxidative phosphorylation, active transport, and cell division. It is difficult, if not impossible, to separate the action of drugs (pharmacology) from that of enzyme inhibition. Although the molecular actions of all drugs have not yet been elucidated, a large number of drugs are known to exert their function by inhibiting enzyme systems.

The pathologist has found enzyme inhibitors to be invaluable tools for blocking certain pathways in order to produce symptoms identical at the organism level with those of pathological states. Many of the agricultural chemicals used to control insect populations—such as DDT [1,1,1-trichloro-2,2-bis(*p*-chlorophenyl)ethane], the carbamates, and the organic phosphates—exert their effects through inhibition of specific enzymes of the insect, particularly acetylcholinesterase which is essential for nerve transmission.

Enzyme inhibitors and stimulators have been used extensively by the plant physiologist in the control of growth, reproduction, and maturation of plants. In this connection one can mention 2,4-D (2,4-dichlorophenoxyacetic acid), 2,4,5-T (2,4,5-trichlorophenoxyacetic acid), indoleacetic acid, and ethylene.

Addition of inhibitors to food materials to control enzyme activity is not widely used. Ascorbic acid, sodium sulfite, and thiol compounds are added to certain foods to prevent enzymatic browning. Ascorbic acid inhibits polyphenol oxidase directly as well as reducing the initial product (a benzoquinone) formed by the enzyme before the benzoquinone can undergo secondary reactions to produce brown products characteristic of the browning reaction.

II. MODE OF ACTION OF INHIBITORS

Any compound that decreases the observed activity of an enzyme is an inhibitor. Decreased enzyme activity is brought about in a number of ways. At the cellular level, the inhibitor may function by (a) altering the permeability of a cell or depressing transport mechanisms; (b) interfering with the formation of high-energy phosphate compounds (ATP, etc.); (c) interfering with synthetic reactions involved in the formation of cellular constituents; (d) interfering, at the level of DNA, RNA, or ribosomes, with biosynthesis of the enzyme; or (e) inhibiting the enzyme directly.

At the molecular level, both in vivo and in vitro, inhibition of enzyme activity by a compound can result from its reaction with apoenzyme, with necessary cofactors, with activators, or with intermediates in the pathway. An inhibitor can react directly with essential groups of the active site of the enzymes or with specific groups on the enzyme not involved in the active site per se but which are important in maintaining conformation of the active site or in holding a multi-subunit enzyme together. In vivo and in crude systems, a substance that stimulates the action of proteolytic enzymes would appear to be an inhibitor of those enzymes destroyed by proteolysis. The most useful type of inhibitor in elucidation of reaction mechanisms is one that reacts with the active site on an enzyme where substrate, cofactor, and/or activator are bound. A compound could also be inhibitory because it reacts with substrate, a cofactor, or an activator needed by the enzyme.

III. SOME FACTORS THAT INFLUENCE DEGREE OF INHIBITION

In reporting the effect of a compound on an enzyme system, full experimental details including pH, temperature, nature and concentration of buffer, ionic strength, exact source and concentration of enzyme, substrate and cofactor concentrations, concentration of inhibitor, and time of reaction must be given. Failure to record all experimental conditions have led workers to report apparently opposite effects of compounds on a system.

Just as pH affects combination of enzyme and substrate (Chapter 10), it also affects combination of enzyme with inhibitor. For example, mushroom polyphenol oxidase was 90 and 27% inhibited by 4 m*M* benzoate at pH 5.3 and 7.0, respectively [1]. pH affects the rate of reaction of the essential sulfhydryl group of ficin with chloroacetamide ([2]; Fig. 1). Such data are useful in deciding what prototropic form of inhibitor and/or enzyme is important in the reaction. With mushroom polyphenol oxidase it appears that there is a negative charge in the active site of the enzyme which prevents binding of an anionic inhibitor and that this group has a pK in the region of pH 6.

Frequently, there are differences in the effect of an inhibitor on the same enzyme from different sources (Table 1). There is more than a 100-fold difference between pig heart and *Lupinus albus* succinic dehydrogenases in their response to malonate as measured by K_i values. For this and other reasons, it is necessary to indicate the exact source of an enzyme in order that comparison among experimental values reported by different workers can be compared.

IV. SPECIFIC INHIBITORS

A. Reaction With Essential Cations

Any substance that complexes with and/or removes an essential cation from an apoenzyme will be an inhibitor of that enzyme. If removal of the cation does not

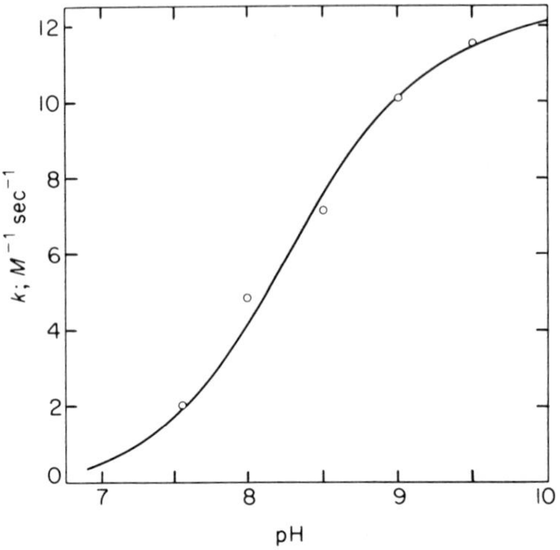

Figure 1 Effect of pH on rate of reaction of the essential sulfhydryl group of ficin with chloroacetamide [2]. The solid line is calculated for a pK value of 8.3 and a maximum rate of 12.3 M^{-1} sec^{-1}.

Table 1 Inhibition of Different Succinic Dehydrogenases by Malonate

Source of enzyme	K_i (mM)
Pig heart	0.0076
Rat heart	0.01
Yeast	0.0105
Rhizobium japonicum (nitrogen-fixing bacteria)	0.017
Claviceps purpurea (a fungus, ergot)	0.03
Corynebacterium diphtheriae (a bacterium)	0.037
Beef heart	0.041
Mytilus edulis (mussel)	0.06
Micrococcus lactilyticus (a bacterium)	0.23
Phaseolus vulgarius (bean)	0.24
Lupinus albus (white lupine)	0.91

Source: Ref. 3, p. 33, by courtesy of Academic Press.

cause irreversible conformational changes in the apoenzyme, addition of cation back to the reaction will restore activity.

1. Cyanide, Sulfide, Azide, and Carbon Monoxide

Cyanide, sulfide, azide, and carbon monoxide inhibit many of the enzymes that contain essential iron, copper, zinc, molybdenum, or other di- and trivalent cations. These compounds have been useful in establishing the importance of the cytochrome system in biological oxidation. Table 2 illustrates a variety of enzymes inhibited by cyanide. The iron-containing enzymes are more sensitive to cyanide, followed in order by those containing Cu^{2+} and Zn^{2+}. Cyanide may also inhibit enzymes that contain pyridoxal phosphate. Sulfide, at somewhat higher concentrations, also adds to the carbonyl group of pyridoxal phosphate. Both cyanide and sulfide are reducing agents and can reduce disulfide bonds in proteins, which may lead to inhibition. Generally, the concentration of cyanide and sulfide needed for reduction of disulfide bonds is in the range 10^{-3} to 10^{-1} *M*.

Carbon monoxide is perhaps the most specific of the compounds listed in that it inhibits only enzymes containing essential heavy metal cations in the active site. However, carbon monoxide does not react with all enzymes that contain heavy metal cations.

2. Ethylenediaminetetraacetic Acid

Ethylenediaminetetraacetic acid (EDTA, Versene; Fig. 2) forms strong complexes with divalent and higher-oxidation-state cations such as Fe^{2+}, Zn^{2+}, Cu^{2+}, Co^{2+}, Mn^{2+}, Mg^{2+}, Ca^{2+}, Fe^{3+}, and Mo^{6+}. Whether EDTA is inhibitory will depend on the K_b (binding constant) value for the complex formed between metal ion and EDTA compared to the K_b value for the complex of metal ion with enzyme. EDTA does not remove Ca(II) from α-amylase, even on prolonged dialysis

Table 2 Inhibition of Enzymes with Cyanide

Enzyme	Essential metal
Cytochrome oxidase	Fe^{2+}, Cu^{2+}
Catalase	Fe^{3+}
Peroxidase	Fe^{3+}
Ascorbic acid oxidase	Cu^{2+}
Carbonic anhydrase (blood)	Zn^{2+}
Nitrate reductase	Mo^{6+}
Alkaline phosphatase	Zn^{2+}
Aldolase (yeast)	Zn^{2+}
Alcohol dehydrogenase (liver)	Zn^{2+}

Source: Ref. 4.

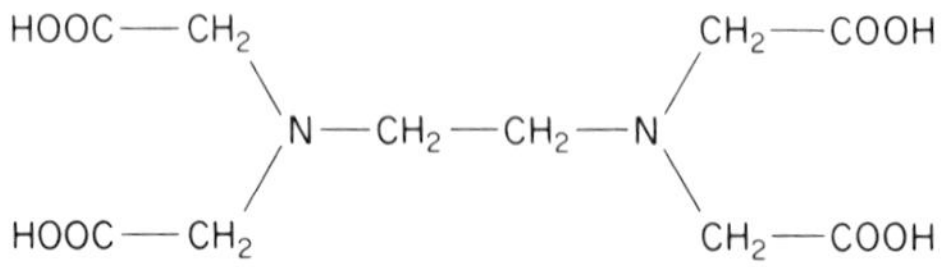

Figure 2 Ethylenediaminetetraacetic acid.

against EDTA, at pH 7 to 9. However, at pH 4 or 11, there is rapid removal of Ca(II) by EDTA [5]. In catalase and peroxidase, Fe(III) is held tightly in the protoporphyrin IX nucleus at pH 7. At pH 4 the Fe(III) may readily be removed. The same is true of Mg(II) in chlorophyll (not an enzyme).

3. 1,10-Phenanthroline

1,10-Phenanthroline (Fig. 3) forms quite stable complexes with certain metal ions, particularly Zn^{2+} and Fe^{2+}, through coordination of the metal ion with the two nitrogen atoms. The Zn(II) of carboxypeptidase A can be removed readily by dialysis against 1,10-phenanthroline. Whether or not 1,10-phenanthroline inhibits an enzyme containing Zn(II) or Fe(II) depends on the pH and relative K_b values of the complex of metal with 1,10-phenanthroline and with enzyme [6].

4. Diethyldithiocarbamate

Diethyldithiocarbamate, $(CH_3CH_2)_2NCSS^-$, forms strong chelates with Cu^{2+}. Although the compound is quite selective for enzymes that contain Cu(II), it does not inhibit all Cu(II)-containing enzymes. Table 3 lists a number of metal-complexing agents used to inhibit metalloenzymes.

B. Reaction with Thiol Groups

Inhibition of enzymes by compounds that react with sulfhydryl groups has been investigated extensively. Many enzymes contain a number of sulfhydryl groups with which these compounds may react. Inhibition may be the result of reaction of an essential sulfhydryl group or a general reaction with sulfhydryl groups so

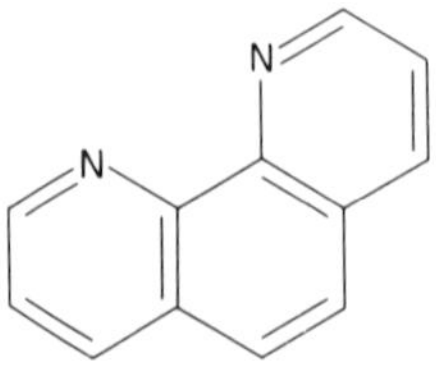

Figure 3 1,10-Phenanthroline.

Table 3 Some Metal-Complexing Agents Used to Inhibit Metalloenzymes

8-Aminoquinoline	Ethylenediaminetetraacetic acid (EDTA)
Azide	8-Hydroxyquinoline
BAL (2,3-dimercaptopropanol)	Imidazole
Carbon monoxide	Mercaptoethylamine
Cyanide	Mercaptoethanol
Diethyldithiocarbamate (DDC)	1,10-Phenanthroline (OP)
α,α′-Bipyridyl	Thiourea

Source: Ref. 7.

as to cause steric effects or denaturation of enzyme. A sulfhydryl group in the active site often shows quite different reactivity compared with those outside the active site. The essential sulfhydryl group of papain is some 30,000 times more reactive with chloroacetate than the sulfhydryl group of cysteine at the same pH [8]. On the other hand, the sulfhydryl group of carboxypeptidase A is completely unreactive in the presence of the cofactor Zn^{2+}. Restoration of activity on addition of a thiol compound is useful in determining the involvement of sulfhydryl groups in the active site. Reaction of nonessential sulfhydryl groups of an enzyme does not necessarily lead to inactivation.

Sulfhydryl groups are involved in the active site of a great many enzymes. Webb, in a partial listing, reported 161 enzymes that are inhibited by mercurials [9]. Compounds that react with sulfhydryl groups may be divided into four groups.

1. Alkylation of Sulfhydryl Groups

Examples of these alkylating reagents include the haloacids and their derivatives, the *S*- and *N*-mustards, chloroacetophenone, chloropicrin, bromobenzylcyanide, and fluoropyruvate. All of these reagents have in common a good leaving group because of the electron-withdrawing ability of both the halogen and the group attached to the methylene carbon.

$$\mathrm{I\text{-}CH_2\overset{\overset{\displaystyle O}{\|}}{C}NH_2 + {}^{-}SCH_2\text{-}E \rightarrow H_2N\overset{\overset{\displaystyle O}{\|}}{C}CH_2SCH_2\text{-}E + I^{-}} \qquad (2)$$

The thiolether formed is quite stable, so that inhibition is irreversible. The bond is stable to hydrolysis, so modification by C^{14}-labeled inhibitor, followed by partial hydrolysis, separation of peptides, and amino acid sequence determination permits location of the specific cysteine residue in the primary sequence that reacted with iodoacetamide.

The rate of reaction of these compounds with sulfhydryl groups is pH dependent since the sulfhydryl anion of the enzyme is the reactive species involved in the reaction [Eq. (2), Fig. 1]. This method may be used to determine the pK value of the reactive sulfhydryl group of the enzyme (Fig. 1).

Other groups in enzymes may also react with these alkylating compounds. For example, the two histidine residues involved in the active site of ribonuclease react with bromoacetate with loss in enzyme activity. However, the concentration of bromoacetate and the time needed to obtain inhibition are much larger than that for inhibition of enzymes with essential sulfhydryl groups.

2. Reaction of Sulfhydryl Groups at Double Bonds

Typical reactions are

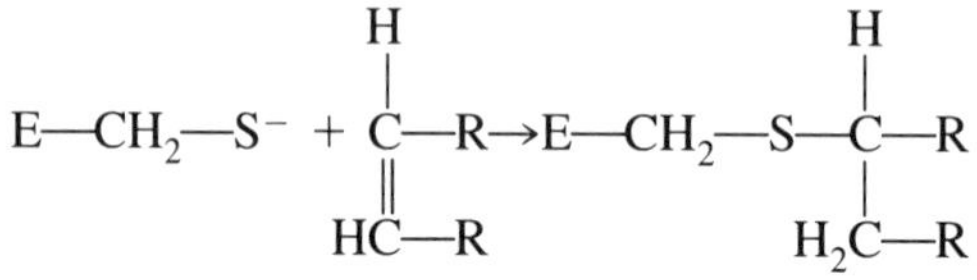

(3)

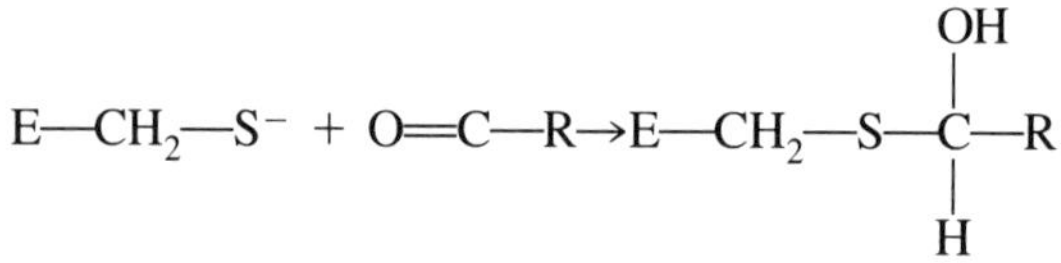

(4)

Examples of this type of compounds include maleic acid, *N*-ethylmaleimide, quinones, acrolein, sorbic acid, acetoacetate, and methylgloxal.

As in the case of alkylating reagents, the reaction is pH dependent since the reactive species is the sulfhydryl anion. In addition, there may be an effect of pH on the compound containing the double bond, depending on the nature of R or R′, which affect the inherent polarizability of carbons at the double bond. The importance of polarizability can be seen from a comparison of the reactivity of fumaric and maleic acids (Fig. 4) with sulfhydryl groups. In fumaric acid there is very little interaction of the two carboxyl groups, so that pK_1 and pK_2 are 2.85 and 4.00 (37°C, $\mu = 0.15$), respectively. The ratio of K_1/K_2 is 14.1. In maleic acid there is a strong influence of the carboxyl groups on each other. pK_1 and pK_2 values for maleic acid are 1.67 and 5.75 (37°C, $\mu = 0.15$), respectively, and the ratio K_1/K_2 is 12,000. As a net result of this difference in polarizability of the ethylenic carbons, only maleic acid reacts with sulfhydryl compounds. Sorbic acid, used as

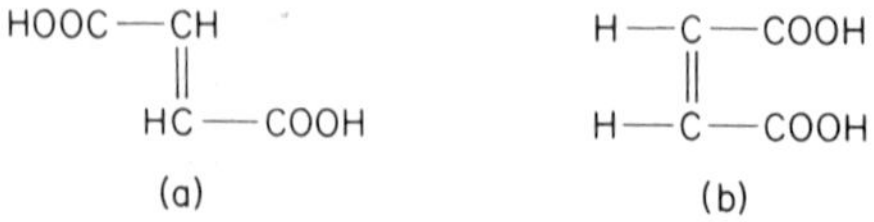

Figure 4 Structures of the geometric isomers of (a) fumaric and (b) maleic acids.

microbial inhibitor in some foods, also reacts with sulfhydryl groups. Addition of sulfhydryl groups to double bonds gives a stable thiol ether.

3. Mercaptide Formation

Examples of compounds that form mercaptides with sulfhydryl groups are $HgCl_2$, organic mercurials, arsenite, organic arsenicals, and various heavy metals such as Cu^{2+}, Pb^{2+}, Cd^{2+}, and Ag^+. Depending on the oxidation state of the mercaptide-forming reagent X^+ and its concentration relative to sulfhydryl compound, the general reactions may be written as

$$\text{E—SH} + X^+ \rightleftharpoons \text{E—S—X} + H^+ \quad (5)$$

$$2\ \text{E—SH} + X^{2+} \rightleftharpoons \text{E—S—X—S—E} + 2H^+ \quad (6)$$

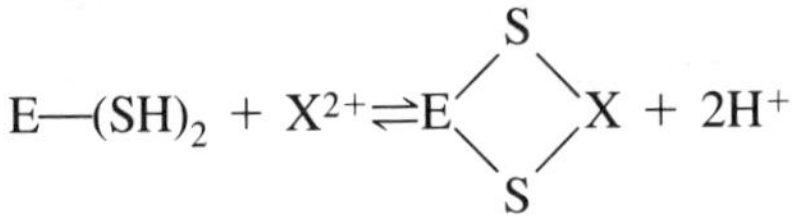

(7)

The most commonly used mercaptide-forming reagent is sodium *p*-chloromercuribenzoate:

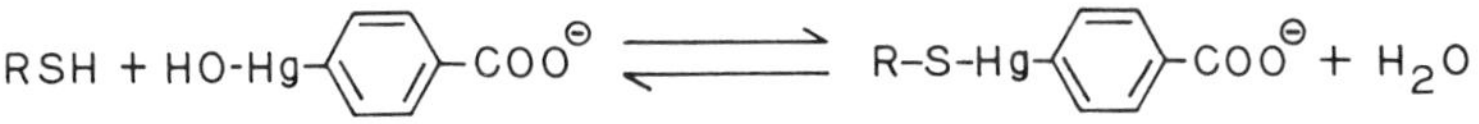

(8)

p-Hydroxymercuribenzoate is usually purchased as *p*-chloromercuribenzoate; the Cl is replaced by OH when added to an aqueous solution.

Reaction of sulfhydryl groups with mercaptide-forming compounds is readily reversible. For reversal, an excess (usually several hundredfold) of another sulfhydryl compound (e.g., cysteine, mercaptoethanol, 1,4-dithiothreitol) is added. These mercaptide-forming reagents are very useful in determining whether an enzyme has a sulfhydryl group in its active site but are of no use in further attempts to determine the location of this site by methods that involve hydrolysis of protein. Rate of reaction of sulfhydryl groups with mercaptide-forming compounds are pH independent.

Use of sodium *p*-chloromercuribenzoate has an added advantage in that there is a spectral change associated with its reaction with a sulfhydryl group (Fig. 5). The change in extinction coefficient, $\Delta\epsilon_m$, at 253 nm and pH 7.0 is 7.6×10^3 $M^{-1}\,cm^{-1}$ [10]. Therefore, this method can be used to determine quantitatively the number of sulfhydryl groups that react with *p*-hydroxymercuribenzoate.

4. Oxidation of Sulfhydryl Groups

Examples of compounds that oxidize sulfhydryl groups of an enzyme include oxidized glutathione, sulfite, tetrathionate, performic acid, hydrogen peroxide,

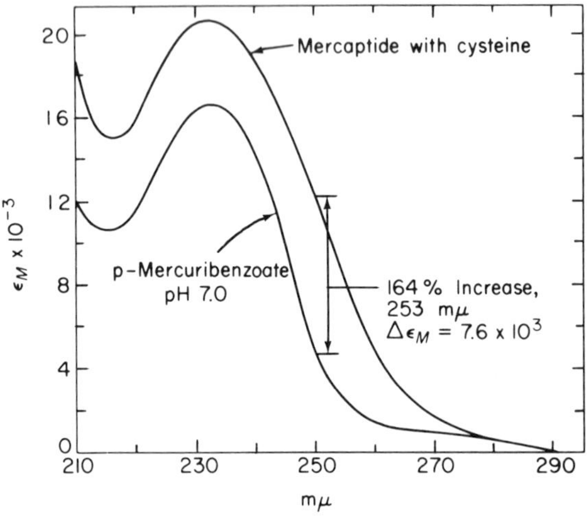

Figure 5 Absorbance of *p*-mercuribenzoate and its mercaptide with cysteine in 0.05 *M* phosphate buffer, pH 7.0. The vertical line in the graph shows the absorbance difference between mercaptide and *p*-mercuribenzoate at 253 nm, the usual wavelength used for assay. Reprinted (From Ref. 10, p. 4332, by courtesy of the American Chemical Society.)

iodine, alloxan, porphyrexide, porphyrindin, *o*-iodosobenzoate, ferricyanide, 5,5′-dithiobis(2-nitrobenzoic acid), 2-pyridine disulfide, and 4-pyridine disulfide.

The general reactions involved in oxidation of sulfhydryl groups may be written as

$$2E{-}SH + X \rightleftharpoons E{-}S{-}S{-}E + XH_2 \tag{9}$$

$$HS{-}E{-}SH + X \rightleftharpoons E\langle{}^{S}_{S}| + XH_2 \tag{10}$$

The use of performic acid, peroxides, ferricyanide, and sulfite may lead to oxidation beyond the disulfide stage, splitting of disulfide bonds, and reaction with other groups. While iodine reacts rapidly with sulfhydryl groups, it also reacts with other groups of enzymes, including the phenyl ring of tyrosine. The most specific reagents are the disulfide compounds, such as oxidized glutathione, 5,5′-dithiobis(2-nitrobenzoic acid), 2-pyridine disulfide, and 4-pyridine disulfide.

5,5′-Dithiobis(2-nitrobenzoic acid) is a particularly useful reagent in that it is used to determine quantitatively the number of sulfhydryl groups that react [11]. The reaction

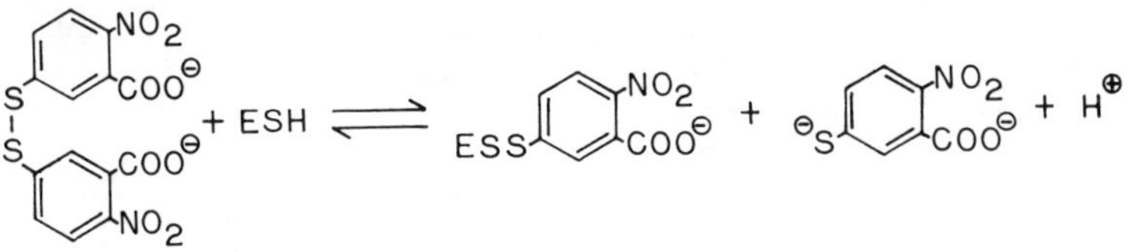

(11)

goes quite rapidly at pH 8 and there is maximum increase in absorbance (ϵ_m = 1.36×10^4 M^{-1} cm^{-1}) at 412 nm due to formation of the thionitrobenzoate ion [Eq. (11)].

C. Inhibition by Heavy Metal Ions

In Section IV.B.3 we included such metal ions as Ag^+, Pb^{2+}, Cu^{2+}, and Cd^{2+}. These heavy metal ions react rapidly, and at low concentrations, with sulfhydryl groups; however, they also complex with other groups of enzymes, including the imidazole, carboxyl, and peptide groups. At higher concentrations (>0.001 *M*) they are general protein precipitants. Therefore, one must be careful in interpretation of inhibition obtained with these heavy metal ions. If inhibition is due to reaction with essential sulfhydryl groups of an enzyme, it should be reversed by addition of an excess of a thiol compound and should occur at $\leqslant$0.001 *M*.

D. Chemical Modification and Selectivity of Chemical Modifying Reagents

A number of side-chain groups of amino acid residues in addition to sulfhydryl groups may be essential for activity of some enzymes. Among these groups are (a) the carboxyl groups of glutamic and aspartic acids, (b) the α-carboxyl group at the end of the polypeptide chain, (c) the imidazole group of histidine, (d) the aliphatic hydroxyl groups of serine and threonine, (e) the phenolic hydroxyl group of tyrosine, (f) the ϵ-amino group of lysine and the terminal α-amino group, (g) the guanidino group of arginine, and (h) the indole nucleus of tryptophan. The hydrophobic side chains of some amino acids are important not only in maintaining conformation of the active site but may also be involved in binding and catalysis of the substrate.

There have been important advances in chemical modifications of groups on enzymes [12–17]. The greatest advances have been in elimination of much of the denaturation formerly caused by experimental conditions. Suitable controls must be run to determine how much activity loss is due to chemical modification and how much is due to denaturation.

Table 4 gives a number of reagents presently used to modify groups on enzymes. Reference 12 should be consulted for a more complete list of reagents. There are no suitable chemical methods for modification of aliphatic side chains and the aromatic side chain of phenylalanine.

Table 4 Some Reagents Used in Chemical Modification of Enzymes

Reagent	Group modified							
	Amino	Carboxyl	Sulfhydryl	Ser Thr	His	Trp	Tyr	Met
Acetic anhydride	+		+	+			+	
Acetyl imidazole	±		+				+	
Aryl sulfonyl chlorides	+		+	±	+		+	
N-Bromosuccinimide			+		+	+	+	
Carbodiimides		+						
Diazomethane	+	+	+				+	
2,4-Dinitrofluorobenzene	+		+		+		+	
N-Ethyl maleimide	±		+		±			
Haloacetate, haloamide	+	+	+		+			
Hydrogen peroxide			±			±		+
Hydroxylamine		+						
Hydroxynitrobenzyl bromide						+		
Mercuribenzoate	±		+		±			
Photooxidation			+		+	+	+	±
Tetranitromethane			+				+	±
Trinitrobenzene sulfonate	+							

Source: Ref. 12.

One disadvantage of most chemical modification techniques is that there may be little or no selectivity in the reaction of the modifying agent with all of a particular group on a protein. For example, when a protein with 20 ϵ-amino groups of lysine, one α-amino group at the terminal end, and 10 tyrosine phenolic groups is treated with an excess of acetic anhydride, all may be acetylated. The question then is to decide whether loss of activity is due to (a) modification of one particular amino or phenolic group, (b) change in charge of the protein, (c) steric effects of modification, (d) conformational changes induced by extensive modification and/or change in charge, (e) modification of an essential group in the binding locus of the active site, or (f) modification of an essential group in the transforming locus of the active site.

Several methods are used to improve selectivity of the modifying reagent for groups in the active site. These methods include (a) control of pH, (b) modification of zymogen followed by activation to enzyme, (c) selective removal of modifying groups from active site, (d) correlation of rate of activity loss with rate of modification of individual types of groups, (e) blockage of active site with competitive inhibitor during modification, (f) selection of conditions where active-site groups are differentially reactive, and (g) use of site-specific reagents. Each of these will be discussed briefly.

1. Control of pH

2,4-Dinitrofluorobenzene is useful in modification of α- and ε-amino groups of enzymes:

(12)

The reaction is usually performed at pH 8, where amino groups, in uncharged form, react rapidly, as do the imidazole, tyrosyl hydroxyl, and sulfhydryl groups. When the reaction is carried out at pH 10.7 with α-chymotrypsin, at least half the reaction occurs with a specific imidazole group in the active site of chymotrypsin [18], as shown in Fig. 6.

2. Modification of Zymogen

When chymotrypsinogen A is treated with acetic anhydride, all amino and tyrosyl hydroxyl groups of the protein can be modified. On treatment of the acetylated protein with catalytic amounts of trypsin the peptide bond between arginyl residue 15 and isoleucyl residue 16 is hydrolyzed, just as it is with unmodified chymotrypsinogen. The completely acetylated chymotrypsin, except for the newly formed

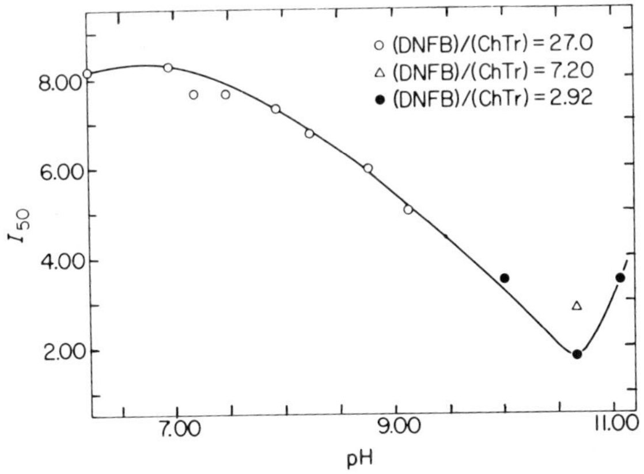

Figure 6 Effect of pH on the specificity of inactivation of α-chymotrypsin with 2,4-dinitrofluorobenzene (DNFB). At pH 7.0, addition of 8 mol of DNFB per mole of enzyme is required for 50% inactivation (I_{50}), while at pH 10.7 addition of 2 mol of DNFB per mole of enzyme produces 50% inactivation. (From Ref. 18, p. 759, by courtesy of the American Society of Biological Chemists.)

α-amino group of the isoleucyl residue, is fully active. If the α-amino group of the isoleucyl residue 16 is then acetylated, the enzyme is completely inactive. This experiment shows clearly that none of the ϵ-amino and tyrosyl hydroxyl groups are essential for activity of the enzyme but that the α-amino group of the N-terminal isoleucyl residue 16 is essential [19].

3. Modification of Enzyme Followed by Removal of Selected Groups

When carboxypeptidase A is treated with acetic anhydride or with acetylimidazole there is extensive modification of ϵ-amino and tyrosyl hydroxyl groups. The enzyme is completely inactive as a peptidase. If the modified protein is treated with hydroxylamine, the acetyl groups are removed from the tyrosyl hydroxyl groups but not from the ϵ-amino groups. Removal of two acetyl groups from tyrosyl hydroxyl groups restores full activity of the enzyme [20]. Therefore, one or both of the tyrosyl groups are essential for activity of the enzyme.

4. Correlation of Rate of Loss of Certain Groups with Rate of Loss of Activity

Aliquots are removed periodically from a reaction mixture of enzyme and inhibitor and assayed for amount of activity left, for total number of groups modified and for specific groups modified. For curve a in Fig. 7, all activity is lost when only one

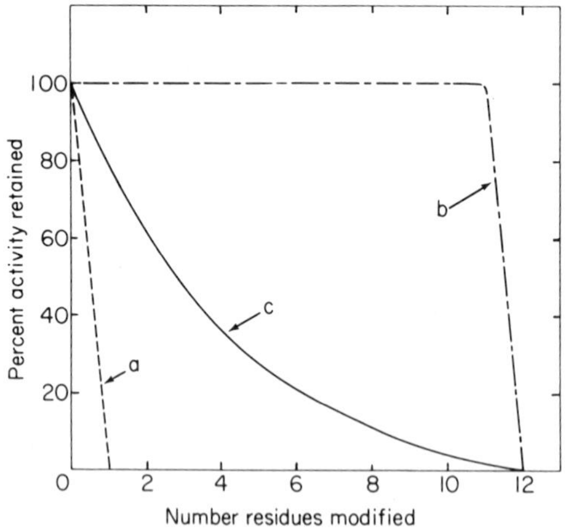

Figure 7 Schematic representation of percent activity retained as a function of number of residues modified on enzyme. Curve a, all activity is lost when one residue is modified; curve b, no activity is lost when 11 residues are modified but modification of one additional residue causes complete activity loss; curve c, activity is lost slowly as the number of residues modified is increased.

group is modified, while in curve b all activity is retained until the very last group is modified. Both of these would be ideal cases for determining which amino acid residue is involved in activity, but they are rarely achieved in practice. The most common situation found with nonselective reagents is that shown in curve c.

Koshland has shown that a determination of rate of modification of each type of group can be useful in deciding the specific group responsible for activity loss [21].

5. Modification of Enzyme in Presence of Saturating Amount of Competitive Inhibitor

When an enzyme is fully saturated with a substrate analog, the groups in the active site should be fully protected from modification. This is not always true since in the case of some enzymes, for example, ficin, phosphoglucomutase, and creatine kinase, the presence of substrate analogs increases reactivity of groups in the active site. The experiment is performed by fully saturating the enzyme with the substrate analog followed by treatment with chemical modifying agent until all available groups are modified. The substrate analog is removed by dialysis or passage through a Sephadex gel column. The enzyme, with retention of most of the original activity, is then reacted with more of the modifying agent, preferably radioactively labeled material, in the absence of substrate analog. Modification should lead to complete loss in activity. It then becomes a rather easy experiment to determine how many and what kinds of groups are modified by the radioactively labeled compound, and their location in the enzyme.

6. Differential Rates of Reaction of Groups

Frequently, the groups involved in the active site of an enzyme are superreactive because of the environment in which they are located. Therefore, by use of low concentrations of modifying agent, it is often possible to modify these groups before appreciable reaction occurs with groups outside the active site. The essential sulfhydryl group of papain is 30,000 times more reactive toward chloroacetate than is the sulfhydryl group of cysteine under the same conditions at pH 6 [8].

7. Site-Specific Reagents

The type of reaction discussed here has been referred to frequently as affinity labeling. The rationale of the method is to design a compound which the enzyme recognizes as a substrate so that it will be bound at the active site. Once bound, it then reacts with a specific group in the active site to form an irreversible derivative. A number of such compounds are available and have been described by Feeney [15]. We shall discuss only one of these compounds to illustrate the utility of the method.

α-*N*-*p*-Toluenesulfonyl-L-phenylalanine chloromethyl ketone (TPCK) has the necessary specificity, as provided by the phenylalanyl residue, to bind in the active site of chymotrypsin [22]. Once bound, it reacts with the imidazole group of His-57 in the active site [Eq. (13)]. The reaction is irreversible and leads to

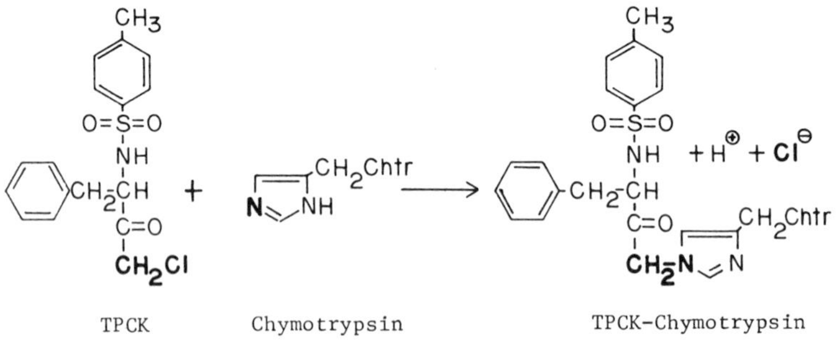

(13)

complete inactivation of enzyme. The C-N bond formed is stable to hydrolysis, so the derivative can be isolated. The inhibitor is identical in every respect to a good substrate for the enzyme, α-*N*-toluenesulfonyl-L-phenylalanine ethyl ester, except for the chloromethyl group. The peptide bond of a substrate is attacked in a nucleophilic reaction by the hydroxyl group of Ser-195, while TPCK is attacked in a nucleophilic reaction by the N of the imidazole side chain of His-57.

8. Enzyme Modification by Recombinant DNA Technology

In vivo biosynthesis of proteins is described in Chapter 2. The primary sequence of amino acids, joined by peptide bonds, of a protein is determined by the sequence of nucleotides of DNA of the gene. Therefore, by replacing one or more of the purine or pyrimidine bases of a codon for an amino acid, that amino acid can be replaced by any of the other 19 amino acids found in proteins. By use of the required enzymes, this has proved to be relatively easy. Therefore, pinpoint modification at any amino acid residue in any enzyme is possible, permitting a direct answer to questions concerning the essentiality of amino acid residues in the active sites of enzymes. The reader is referred to the general reference by Oxender and Fox [23] for detailed information.

E. Macromolecules

Inhibition of enzymes is not limited to small molecules, although interpretation of the nature of inhibition is often easier in this case. In inhibition by large polymeric molecules it is often difficult to determine whether inhibition is due to interaction with a specific group or just a masking of the active site by the large inhibitor.

1. Macroions

Nucleic acids are very large polyanions that readily interact with positively charged proteins to form complexes. Protamines are low-molecular-weight basic proteins which, because of their high arginine and lysine content, are polycations. These basic proteins react readily with acidic proteins to form complexes that usually precipitate and have little or no enzymatic activity.

2. Antienzymes

All enzymes are proteins, and antibodies can be produced against these enzymes. The pure enzyme is injected into a suitable animal (e.g., a rabbit). After a few weeks, the blood of the animal will contain antibodies (specific γ-globulins) against the specific enzyme used in eliciting their production. When antibody is combined with the enzyme used to elicit it, a complex is formed which generally precipitates. The enzyme is usually inhibited by such a reaction.

3. Naturally Occurring Protein Inhibitors

Naturally occurring proteins that form complexes with specific enzymes are widely distributed in nature. The best studied of these proteins are the protease and α-amylase inhibitors of the pancreas, egg white, and legumes. These proteins usually form very strong complexes with specific enzymes with binding constants on the order of 10^7 to 10^{11} M^{-1}. The specific mechanism(s) by which these proteins inhibit specific enzymes is not fully understood. It is known that trypsin recognizes a specific arginine or lysine residue in the trypsin inhibitors and forms a strong adsorptive complex that is inactive. Therefore an enzyme·substrate-like complex is formed, but catalysis of a peptide bond does not occur.

F. Substrate Analogs

A substrate analog is defined broadly as any compound which is structurally related to the substrate so that it binds into the active site of an enzyme. Generally speaking, a substrate analog can participate in a reaction to a greater or lesser extent than the normal substrate, or it may interfere completely with the catalytic process, usually by preventing binding of substrate.

The concept that there must be near-perfect stereospecific fit of substrate to the active site of an enzyme in order to have catalysis is generally accepted. Such a rigid restriction on fit then implies that an isomer of a substrate may bind into the active site, but its susceptible bond may not be positioned properly so that catalysis can occur (Fig. 8). On the other hand, many structural isomers of substrates do not bind to the active site, because they do not fit the shape of the active site and/or form adsorptive bonds.

1. Types of Substrate Analogs

Substrate analogs may be divided into enantiomeric analogs, anomeric analogs, positional isomers, geometrical isomers, products of the reaction, and other compounds that closely resemble the substrate (Table 5). Many enzyme inhibitors have been studied that fall into the classification of substrate analogs. At least one

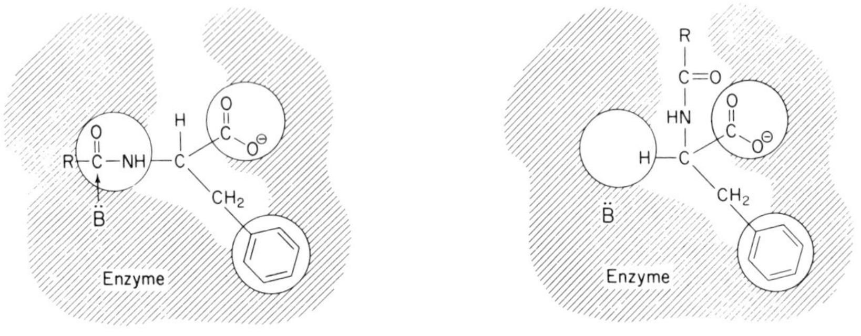

Figure 8 Schematic representation of binding of two structural isomers in the active site of an enzyme. The areas occupied by the carboxylate and phenyl groups are most important in binding of substrate, the area occupied by CONH acts only secondarily in binding but primarily in transformation as catalyzed by nucleophilic attack of B. The L-substrate shown on the left is bound so that the peptide bond is positioned correctly with respect to B; the D-substrate is bound but the peptide bond is not positioned properly. Therefore, the D-isomer will be a competitive inhibitor of the L-isomer but will not be a substrate.

inhibitor is known for practically every enzyme. In most cases the presence of a substrate analog leads to competitive inhibition which can be reversed by increasing substrate concentration to saturating levels.

Substrate analogs are ideal to use in mapping the topography of active sites of enzymes. Substituents on a molecule may be changed one at a time, and the effect of such a change on the inhibitory ability of the molecule measured. The same approach can be used with a substrate. However, interpretation is more difficult since the substituent change can affect binding or transformation to product or both.

2. Specific Examples of Enzyme Inhibition by Substrate Analogs

We shall select two examples from the wealth of data available on this subject to show the information that may be obtained about the binding sites of enzymes by this technique.

a. Inhibitors of Succinic Dehydrogenase

Table 6 shows data for the effect of several compounds on activity of succinic dehydrogenase. The substrate for this enzyme is succinate. From the data it may be concluded that those compounds that have two carboxylate groups situated about the same distance apart as in succinate will be the best inhibitors. Malondiamide, succindiamide, malondialdehyde, and diethyloxaloacetate in which the carboxylate groups are blocked are not inhibitory. Essentiality of two charged carboxylate groups is shown in that propionate is only weakly inhibitory.

Table 5 Various Types of Substrate Analogs

Type (enzyme involved)	Substrate	Inhibitor
1. Enantiomeric analogs (ficin, papain)	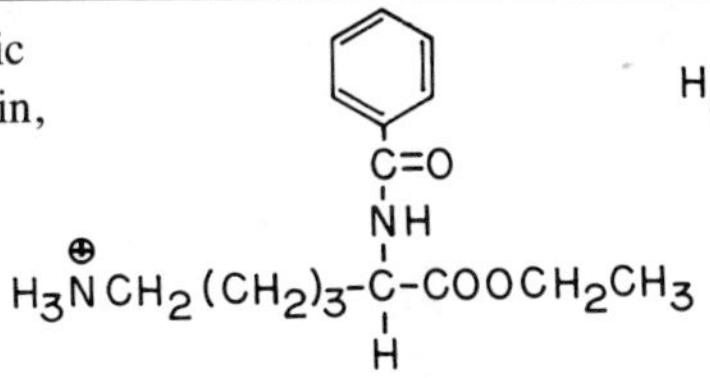 L-BLEE	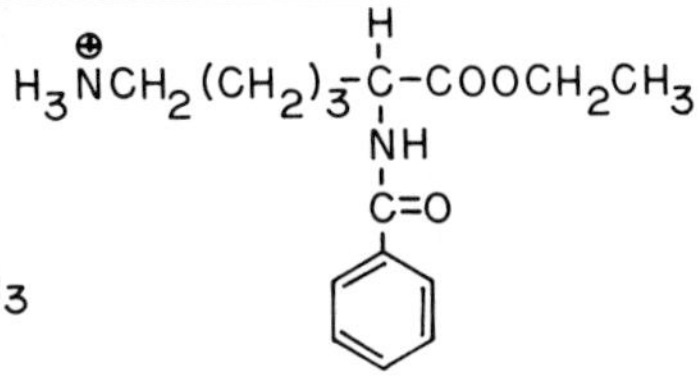 D-BLEE
2. Anomeric analogs (α-glucosidase)	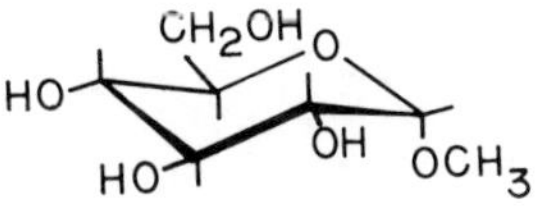 α-Methyl-D-glucose	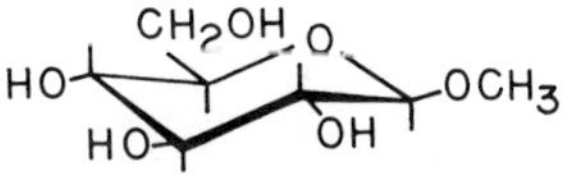 β-Methyl-D-glucose
3. Positional isomers (p-hydroxyphenyl pyruvate hydroxylase)	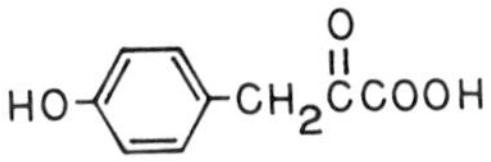 *p*-Hydroxyphenylpyruvic acid	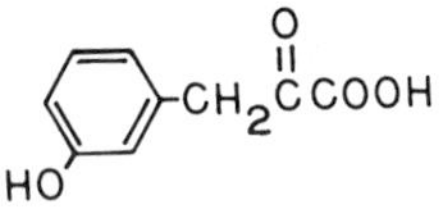 *m*-Hydroxyphenylpyruvic acid
4. Geometrical isomers (fumarase)	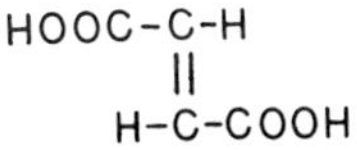 Fumaric acid	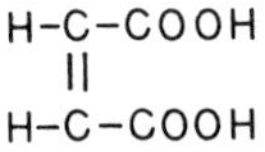 Maleic acid
5. Product of reaction (carboxy-peptidase A)	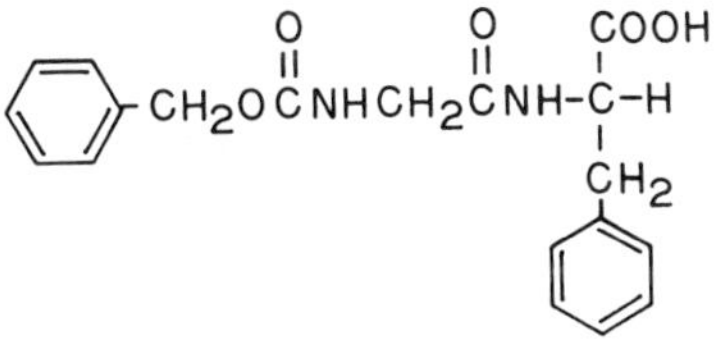 Benzyloxycarbonylglycyl-L-phenylalanine	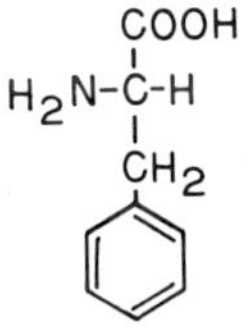 L-Phenylalanine
6. Compounds resembling substrate (enolase)	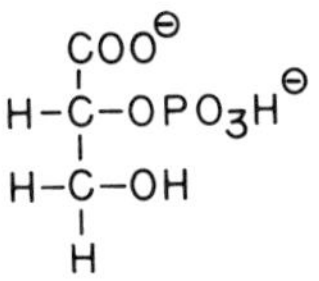 D-2-Phosphoglycerate	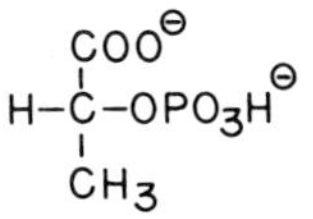 D-Phospholactate

Table 6 Inhibition of Pig Heart Succinic Dehydrogenase

Inhibitor	Structure	K_i (m*M*)
Oxaloacetate	$^{\ominus}OOC-C(=O)-CH_2-COO^{\ominus}$	0.0016
Ethyloxaloacetate	$CH_3-CH_2-O-C(=O)-C(=O)-CH_2-COO^{\ominus}$	0.04
Malonate	$^{\ominus}OOC-CH_2-COO^{\ominus}$	0.05
Fumarate	$^{\ominus}OOC-CH=CH-COO^{\ominus}$	3.5
cis-Cyclohexane-1,2-dicarboxylate	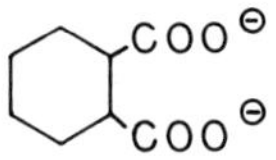	132
Diethyloxaloacetate	$CH_3-CH_2-O-C(=O)-C(=O)-CH_2-COOCH_2CH_3$	No inhibition
Propionate[a]	$CH_3-CH_2-COO^{\ominus}$	Very weak inhibition
Malondialdehyde[b]	$H-C(=O)-CH_2-C(=O)-H$	No inhibition
Malondiamide[c]	$H_2N-C(=O)-CH_2-C(=O)-NH_2$	No inhibition
Succindiamide[d]	$H_2N-C(=O)-CH_2-CH_2-C(=O)-NH_2$	No inhibition

Source: Ref. 24, except as noted.
[a]Ref. 25
[b]Ref. 26.
[c]Ref. 27.
[d]Ref. 28.

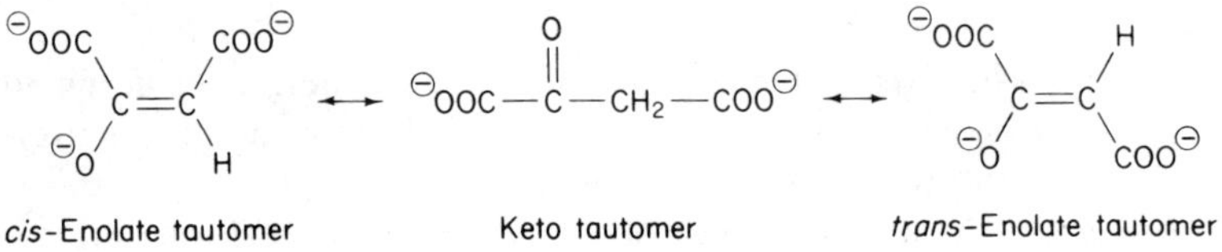

Figure 9 Tautomeric forms of oxaloacetate. The *trans*-enolate tautomer is the inhibitory species.

The necessity of having the two carboxylate groups close together is shown by the fact that glutarate and adipate are not inhibitory. Fumarate is not a particularly good inhibitor because the charge separation is too great and because of its *trans*-configuration. The charge separation in *cis*-cyclohexane dicarboxylate is 3.05 Å whereas it is 3.01 Å in succinate; therefore, *cis*-cyclohexane dicarboxylate should be a good inhibitor. It is not a good inhibitor because of the bulky nature of the cyclohexyl group. Oxaloacetate is a much better inhibitor than malonate due to the tautomeric nature of this compound (Fig. 9). On the basis of these data Hellerman et al. [24] have concluded that the binding site of succinic dehydrogenase contains two cationic groups 3 to 4 Å apart (Fig. 10).

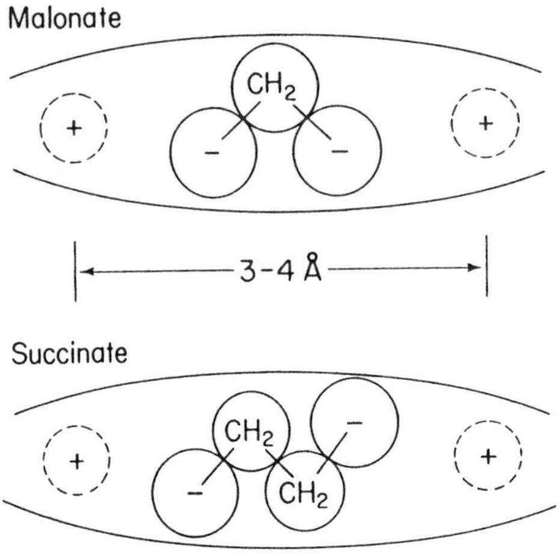

Figure 10 Schematic representation of binding of malonate and succinate in the active site of succinic dehydrogenase. The positive charges enclosed in dashed circles are on the enzyme surface and interact with negative charges on the substrate in the binding step. The oval lines depict steric barriers imposed by the peptide chain surrounding the active site. (From Ref. 29, p. 45, by courtesy of Academic Press.)

b. Inhibitors of Carboxypeptidase A

Carboxypeptidase A has (a) a preference for substrates containing aromatic amino acids, (b) an absolute requirement for a free carboxyl group on the carboxyl terminal end of the substrate, and (c) prefers that the amino group be blocked. Various competitive inhibitors of this enzyme are available (Table 7). Benzyloxycarbonylglycyl-L-phenylalanine (Fig. 11) is a good substrate for this enzyme.

The three areas in the binding locus for substrate are pictured in Fig. 12. Those compounds that fit into all three areas are the best competitive inhibitors. The very best inhibitor is 3-indoleacetate (Table 7); β-phenylpropionate, valerate, and L- and D-phenylalanine are also good inhibitors. Apparently, the third area, labeled

Table 7 Inhibitors of Carboxypeptidase A

Compound	Structure	K_i (mM)
Propionate	$CH_3CH_2COO^{\ominus}$	100
Valerate	$CH_3CH_2CH_2CH_2COO^{\ominus}$	2.7
Benzoate	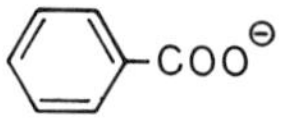	143
Phenylacetate	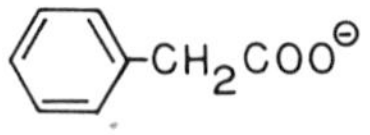	4.55
β-Phenylpropionate	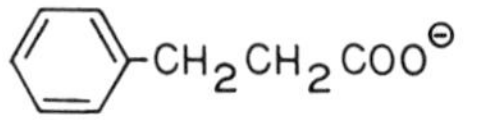	1.2
γ-Phenylbutyrate	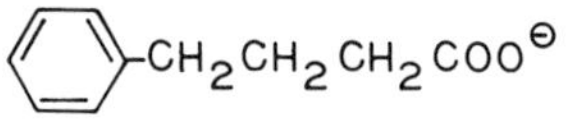	20
3-Indoleacetate	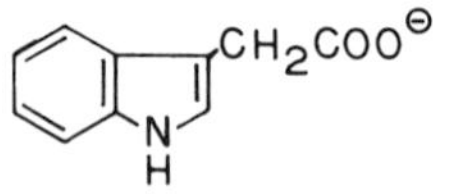	0.78
L-Phenylalanine[a]	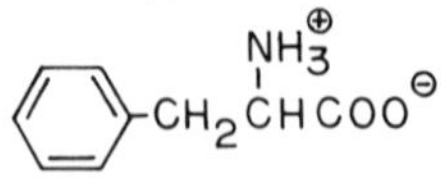	2.0
D-Phenylalanine[b]	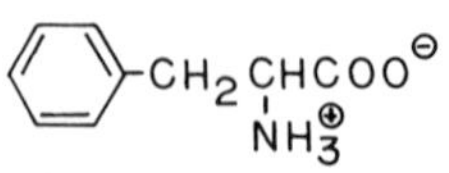	2.0

Source: Ref. 30, except as noted.
[a]Ref. 31.
[b]Ref. 32.

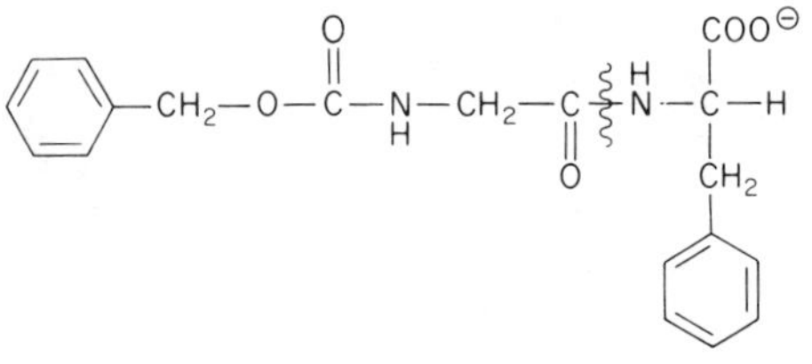

Figure 11 Benzyloxycarbonylglycyl-L-phenylalanine, a substrate of carboxypeptidase A. Substrate is hydrolyzed as indicated by the wavy line.

A on the top diagram in Fig. 12, is not very important in binding. Therefore, both L- and D-phenylalanine bind equally well. Phenylacetate is not as good an inhibitor since the distance between carboxylate and phenyl groups is too short to permit the inhibitor to bind into areas B and C simultaneously. Addition of another methylene group, as in β-phenylpropionate, permits a good fit. Addition of yet another methylene group, as in γ-phenylbutyrate, makes the distance between carboxyl and phenyl groups too large. Benzoate and propionate are poor inhibitors because their structures do not permit them to bind in areas B and C simultaneously.

G. Cofactor Analogs

Work on cofactor analogs was stimulated by discovery that the reason for in vivo inhibition by sulfonamides is their resemblance to *p*-aminobenzoate, a necessary growth factor for certain bacteria (Fig. 13). *p*-Aminobenzoic acid is a component of the vitamin folic acid.

A few cofactor analogs are listed in Table 8. The restricted listing is by no means indicative of the great number of these compounds available. Not all of these compounds are inhibitory to an enzyme in vitro. This will depend on the binding constant of the enzyme·cofactor complex. If the binding constant is small, the analog can compete successfully with cofactor for binding to apoenzyme. On the other hand, if the cofactor is tightly bound to apoenzyme, as is the case with most of the prosthetic groups, there will be no observed inhibition in vitro. However, when the analog is fed to an animal where it may be synthesized into the holoenzyme, it will cause classic symptoms of vitamin deficiency. Few if any changes can be made in the structure of a vitamin without the analog becoming inhibitory. If changes are too extensive, the derivative may not resemble the vitamin closely enough to act as an inhibitor.

V. ACTIVATION OF ENZYMES BY COMPOUNDS USUALLY CONSIDERED TO BE INHIBITORS

Occasionally, when compounds normally considered to be inhibitors, particularly of the irreversible type, are added to a crude enzyme preparation an increase

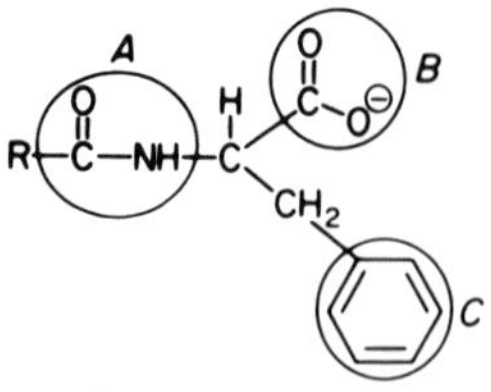

Acyl-L-phenyl-
alanine (substrate)

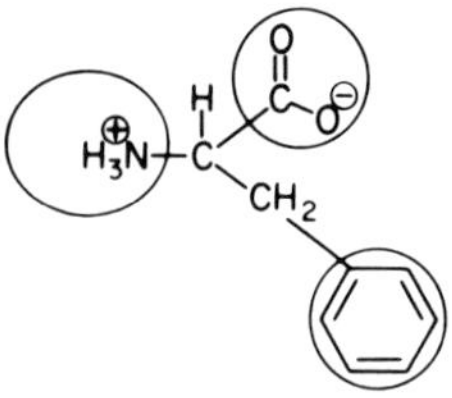

L-Phenylalanine

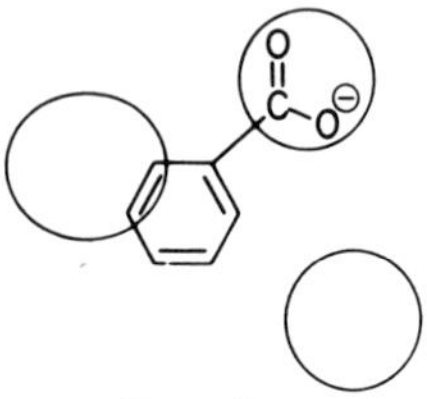

Benzoate

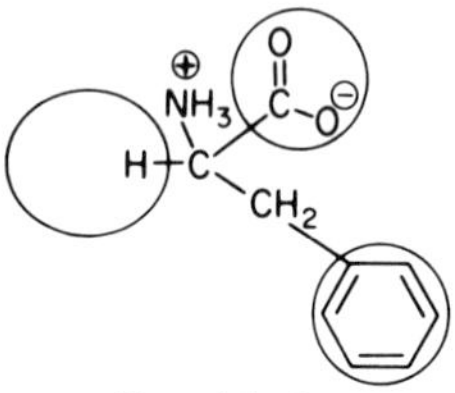

D-Phenylalanine

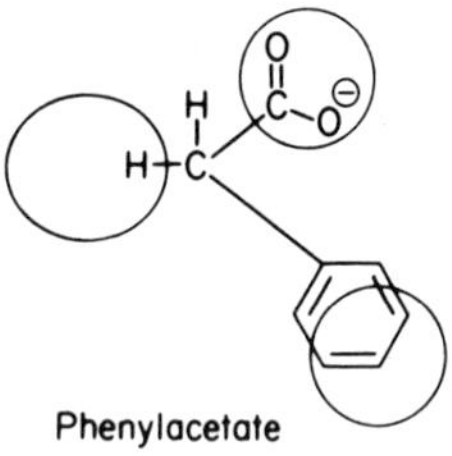

Phenylacetate

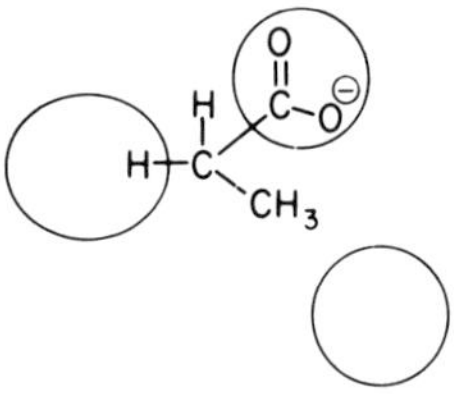

Propionate

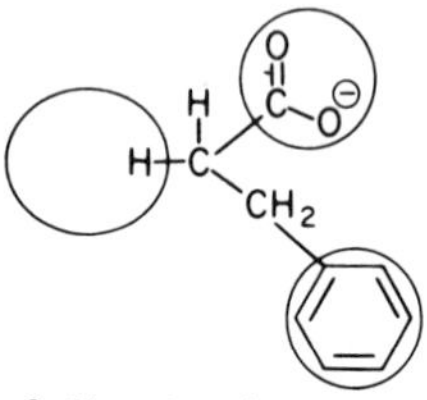

β-Phenylpropionate
(hydrocinnamate)

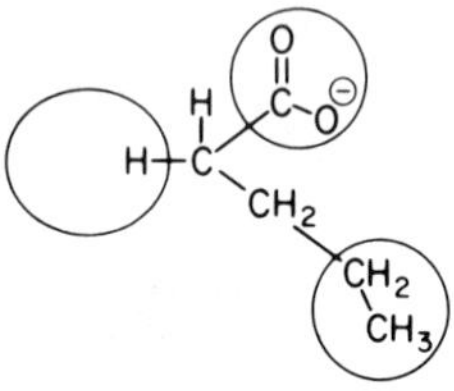

Valerate

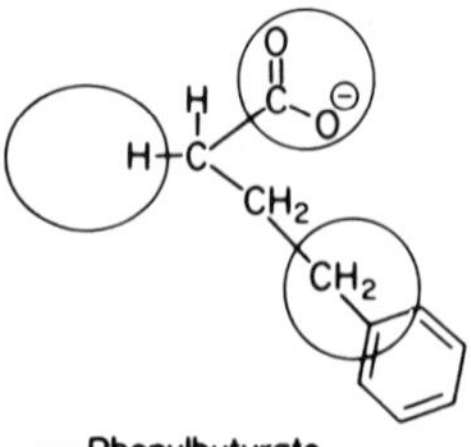

γ-Phenylbutyrate

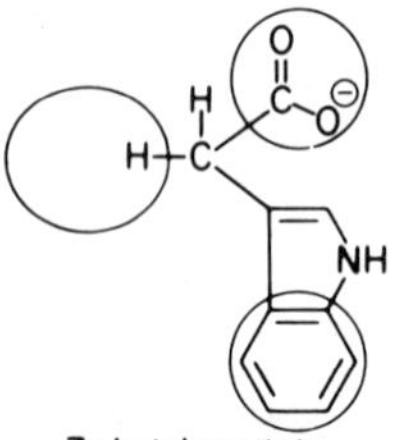

3-Indoleacetate

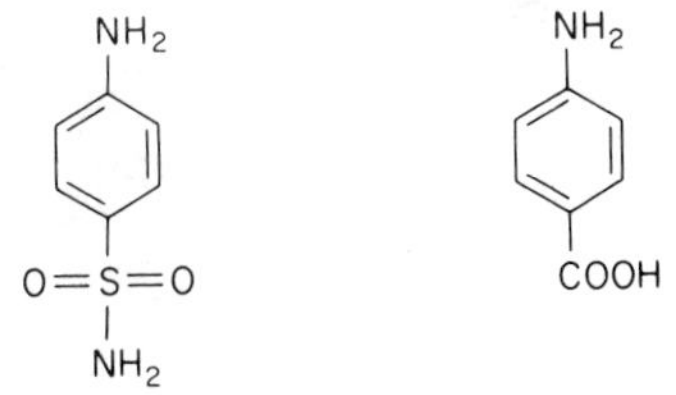

Sulfonilamide p-Aminobenzoic acid

Figure 13 Structures of sulfonilamide and *p*-aminobenzoic acid.

in activity is observed. There are a number of reasonable explanations for this: (a) The added compound combines with a naturally occurring inhibitor already present in the crude preparation, thus relieving inhibition of enzyme; (b) the compound inhibits a second enzyme which is perturbing the measured activity (e.g., in a sequential reaction where E_1 converts A to product B and E_2 converts product B to C, inhibition of E_2 will increase the observed rate of formation of B); and (c) the compound may cause an aggregate of proteins to dissociate into subunits that have less, the same, or *more* activity. Much additional research is needed in this area.

VI. INHIBITION AND STIMULATION OF ENZYME ACTIVITY BY GASES

A number of enzymes require O_2 or CO_2 as one of the substrates. To the food scientist, polyphenol oxidase and lipoxygenase are very important enzymes because of their deteriorative properties. Replacement of the necessary gaseous substrate with an inert (to the enzyme) gas will inhibit the enzyme not by a mechanism involving reaction of the inert gas with the enzyme, but rather just by elimination of substrate. Polyphenol oxidase, for example, is not catalytically active in N_2.

This principle may be put to good use in the controlled ripening of fruit. When fruit is stored in a CO_2 atmosphere, or even one in which the O_2 concentration is reduced, there is a delay in ripening. This is because O_2 is needed as the terminal acceptor in the electron transport system that supplies the majority of ATP for

---

Figure 12 Possible orientations of substrate and inhibitors at the active site of carboxypeptidase A. Binding in B and C involves electrostatic and hydrophobic bonds, respectively. Area A includes hydrolysis of a susceptible peptide bond. Molecular dimensions are only approximate. (From Ref. 33, p. 366, by courtesy of Academic Press.)

Table 8 Cofactor Analogs

Cofactor or vitamin	Analogs
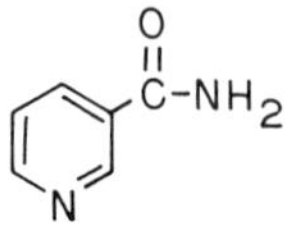 Nicotinamide	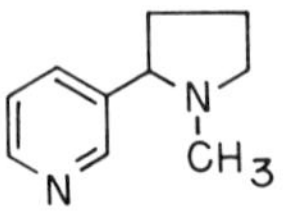 Nicotine
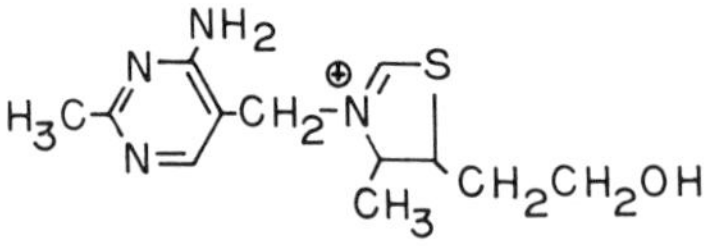 Thiamine	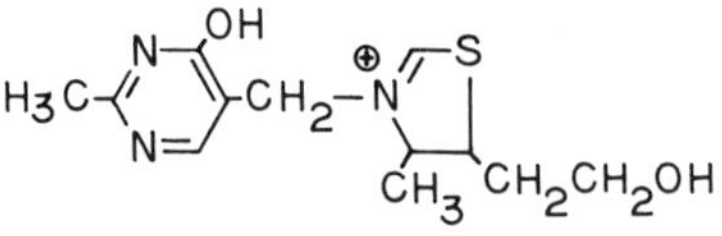 Oxythiamine
	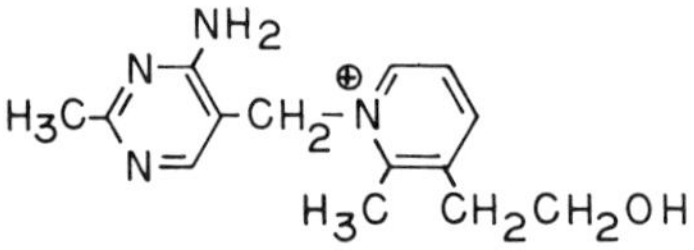 Pyrithiamine
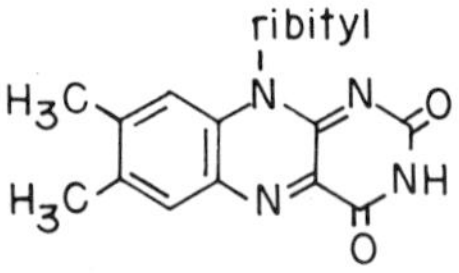 Riboflavin	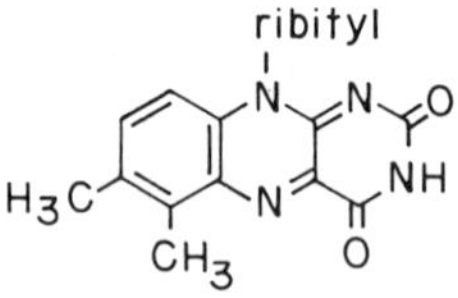 Isoriboflavin
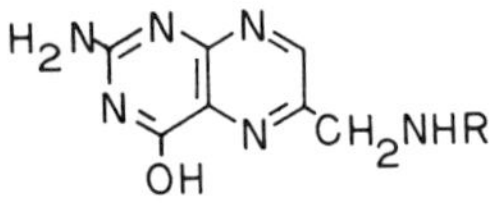 Folic acid (where R = 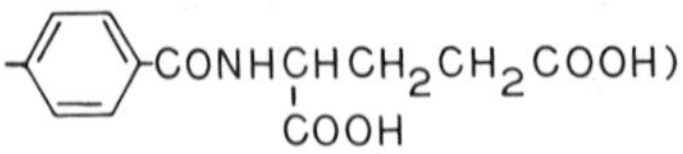)	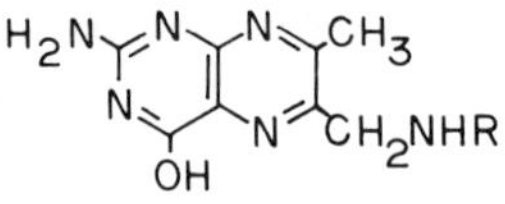 7-Methylfolic acid

Table 8 Continued

Cofactor or vitamin	Analogs	
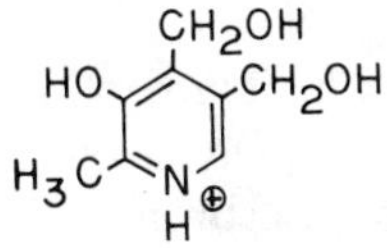 Pyridoxine (pyridoxol)	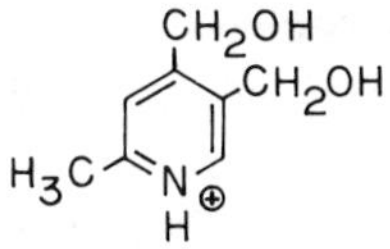 3-Deoxypyridoxol	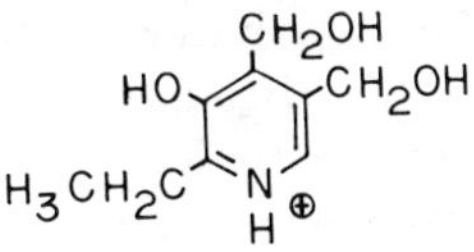 ω-Methylpyridoxol
$RNHCH_2CH_2COOH$ Pantothenic acid (where R = $HOCH_2C(CH_3)_2CHOHCO-$)	$RNHCH_2CH_2CONHCH_2CH_2SH$ Pantetheine (active)	
	$RNHCH_2CH_2SH$ Pantoylaminoethanethiol	
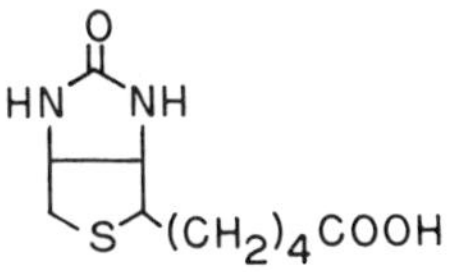 Biotin	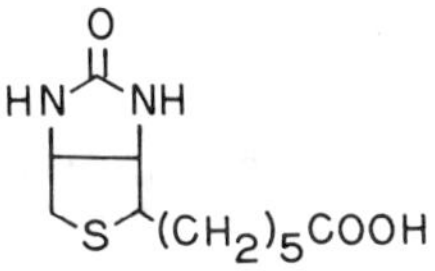 Homobiotin	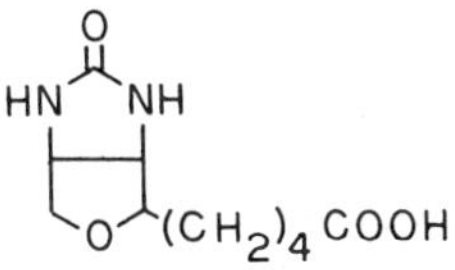 Oxybiotin

cellular activity. On the other hand, placing a fruit in an atmosphere containing a small amount of ethylene, or placing ethylene oxide on a fruit, markedly stimulates the sequence of enzymatic changes leading to ripening.

Although these processes have been known for some time, it is only recently that the mechanisms of action of adding or deleting a gas in controlling ripening have been understood even partially. This is an area where the food scientists and enzymologists might well make some very important breakthroughs in the controlled storage of plant materials.

REVIEW QUESTIONS

1. Discuss the advantages of using a competitive inhibitor rather than substrate for elucidating the nature of the binding locus of an enzyme.

2. What is unique about the seryl residue (Ser-195) in α-chymotrypsin that reacts with diisopropylfluorophosphoridate and with substrates as compared with the other 26 seryl residues in the enzyme?

3. β-Phenylpropionate and indole are good competitive inhibitors of α-chymo-

trypsin. What resemblance do these compounds have to substrates for chymotrypsin? Speculate as to the important group(s) of each for binding to enzyme.

4. What conclusions or comments can you make about each of the following? (Treat each as a separate observation.)

(a) Treatment with 0.001 *M* iodoacetamide gave complete inhibition of an enzyme; subsequent treatment with cysteine did not restore activity. On the other hand, after inhibition (85%) of enzyme by oxidized glutathione, treatment with cysteine restored all the activity.

(b) Treatment with 0.01 *M* carbon monoxide gave 80% inhibition of an enzyme with a concurrent change in absorbance at 400 nm.

(c) Treatment with 0.001 *M* diisopropylfluorophosphoridate for 10 min gave complete inhibition of an enzyme.

(d) Dialysis against distilled water for 24 hours gave no decrease in activity of an enzyme, while dialysis against 1.0×10^{-4} *M* EDTA for 24 hours gave a 95% decrease in activity.

(e) Addition of 0.1 *M* lead acetate gave complete inhibition of an enzyme.

(f) Addition of 0.01 *M* mercaptoethanol to an enzyme gave complete loss of activity.

(g) After dialysis for 24 hours, an enzyme had only 5% of its original activity.

(h) An enzyme caused maximum reduction in starch viscosity in the presence of mercaptoethanol but caused no reduction in viscosity in the presence of iodoacetate.

(i) The requirement for Mg^{2+} was found to be dependent on substrate concentration but to be independent of enzyme concentration.

(j) An enzyme was completely inhibited by 1×10^{-4} *M* HCN.

(k) Addition of 1×10^{-4} *M* *p*-chloromercuribenzoate to an enzyme did not cause any loss of activity, but an increase in absorbance at 253 nm over that of a control solution was observed.

(l) Treatment of an enzyme with 1×10^{-4} *M* *p*-chloromercuribenzoate caused complete loss of activity; however, in another experiment under the same conditions 1×10^{-4} *M* methylmercuric hydroxide did not cause any loss of activity.

5. Discuss the relative merits and relative disadvantages of iodoacetamide *vs.* *p*-chloromercuribenzoate as enzyme inhibitors.

6. What is the possible mechanism of inhibition of an enzyme by the following compounds?

(a) HCN
(b) *N*-Ethylmaleimide
(c) Sarin
(d) Silver nitrate
(e) Cysteine

7. What can you say in general about the relationship between a site-specific reagent and a substrate? About a substrate analog and a substrate? About a substrate and a metal chelating agent?

8. On treatment of a 1.0×10^{-5} M solution of an enzyme with 5,5′-dithiobis(2-nitrobenzoic acid) there was an increase in absorbance of 0.675 at 412 nm. Given that $\epsilon_m = 1.36 \times 10^4\ M^{-1}\ cm^{-1}$ and $l = 1$ cm, calculate the number of reactive sulfhydryl groups in the enzyme. On treatment of another aliquot of the same enzyme solution with 5,5′-dithiobis(2-nitrobenzoic acid) in the presence of 8 M urea, the change in absorbance was 0.995. Calculate again the number of reactive sulfhydryl groups. Explain the difference found between determinations performed in the absence and presence of urea, respectively.

REFERENCES

General References

Baker, B. R., *Design of Active-Site-Directed Irreversible Enzyme Inhibitors*, Wiley, New York, 1967.

Cohen, L. A., Group-specific reagents in protein chemistry, *Ann. Rev. Biochem.*, **37**, 695 (1968).

Feeney, R. E., Tailoring proteins for food and medical uses: state of the art and interrelationships, in *Protein Tailoring for Food and Medical Uses* (R. E. Feeney and J. R. Whitaker, eds.), Marcel Dekker, New York, 1986.

Hirs, C. H. W., and S. N. Timasheff, eds., Enzyme structure, Part B, *Methods Enzymol.*, **25**, 387 (1972).

Hirs, C. H. W., and S. N. Timasheff, eds., Enzyme structure, Part E, *Methods Enzymol.*, **47**, 407 (1972).

Hochster, R. M., and J. H. Quastel, Eds., *Metabolic Inhibitors, A Comprehensive Treatise*, 2 Vols., Academic Press, New York, 1963.

Means, G. E., and R. E. Feeney, *Chemical Modification of Proteins*, Holden Day, San Francisco, 1971.

Singer, S. J., Covalent labeling of active sites, *Advan. Protein Chem.*, **22**, 1 (1967).

Vallee, B. L., and J. F. Riordan, Chemical approaches to the properties of active sites of enzymes, *Ann. Rev. Biochem.*, **38**, 733 (1969).

Webb, J. L., *Enzyme and Metabolic Inhibitors*, Vols. I–III, Academic Press, New York, 1967.

Cited References

1. R. C. Krueger, *Arch. Biochem. Biophys.*, **57**, 52 (1955).
2. J. R. Whitaker, and L.-S. Lee, *Arch. Biochem. Biophys.*, **148**, 208 (1972).
3. J. L. Webb, *Enzyme and Metabolic Inhibitors*, Vol. II, Academic Press, New York, 1966, p. 33.
4. M. Dixon and E. C. Webb, *Enzymes*, Longmans, Green, London, 1958, pp. 374–375.
5. E. H. Fischer and E. A. Stein, α-Amylases, in *The Enzymes* (P. D. Boyer, H. Lardy, and K. Myrbäck, eds.), Vol. 4, Academic Press, New York, 1960, p. 324.
6. L. G. Sillén and A. E. Martell, *Stability Constants of Metal–Ion Complexes*, Chemical Society, London, 1964.
7. B. L. Vallee and W. E. C. Wacker, Metalloproteins, in *The Proteins: Their Structure and Composition* (H. Neurath, ed.), Vol. V, Academic Press, New York, 1970.

8. L. A. Ae. Sluyterman, *Biochim. Biophys. Acta.* **151**, 178 (1968).
9. J. L. Webb, *Enzyme and Metabolic Inhibitors*, Vol. II, Academic Press, New York, 1966, p. 830.
10. P. D. Boyer, *J. Am. Chem. Soc.*, **76**, 4331 (1954).
11. G. L. Ellman, *Arch. Biochem. Biophys.*, **82**, 70 (1959).
12. C. H. W. Hirs and S. N. Timasheff, eds., *Methods Enzymol.*, **25**, 387 (1972).
13. C. H. W. Hirs and S. N. Timasheff, eds., *Methods Enzymol.*, **47**, 407 (1977).
14. B. R. Baker, *Design of Active-Site-Directed Irreversible Enzyme Inhibitors*, Wiley, New York, 1967.
15. R. E. Feeney, in *Protein Tailoring for Food and Medical Uses* (R. E. Feeney and J. R. Whitaker, eds.), Marcel Dekker, New York, 1986, p. 1.
16. G. E. Means and R. E. Feeney, *Chemical Modification of Proteins*, Holden Day, San Francisco, 1971.
17. C. H. W. Hirs, ed., Enzyme structure, in *Methods in Enzymology* (S. P. Colowick and N. O. Kaplan, series eds.), Vol. XI, Academic Press, New York, 1967.
18. J. R. Whitaker and B. J. Jandorf, *J. Biol. Chem.*, **223**, 751 (1956).
19. B. Labouesse, H. L. Oppenheimer, and G. P. Hess, *Biochem. Biophys. Res. Commun.*, **14**, 318 (1964).
20. R. T. Simpson, J. F. Riordan, and B. L. Vallee, *Biochemistry*, **2**, 616 (1963).
21. D. E. Koshland, Jr., D. H. Strumeyer, and W. J. Ray, Jr., in *Enzyme Models and Enzyme Structure*, Brookhaven Symposia in Biology, **15**, 101 (1962).
22. G. Schoellman and E. N. Shaw, *Biochem. Biophys. Res. Commun.*, **7**, 36 (1962); *Biochemistry*, **2**, 252 (1963).
23. D. L. Oxender and C. F. Fox, eds., *Protein Engineering*, Alan R. Liss, New York, 1987.
24. L. Hellerman, O. K. Reiss, S. S. Parmar, J. Wein, and N. L. Lasser, *J. Biol. Chem.*, **235**, 2468 (1960).
25. J. L. Webb, *Enzyme and Metabolic Inhibitors*, Vol. II, Academic Press, New York, 1966, p. 15.
26. D. E. Holtkamp and R. M. Hill, *Arch. Biochem. Biophys.*, **34**, 216 (1951).
27. G. Farwaz and E. N. Farwaz, *Proc. Soc. Exptl. Biol. Med.*, **87**, 30 (1954).
28. E. V. Dietrich, A. L. Fisher, J. P. Long, and R. M. Featherstone, *Arch. Biochem. Biophys.*, **41**, 118 (1952).
29. J. L. Webb, *Enzyme and Metabolic Inhibitors*, Vol. II, Academic Press, New York, 1966, p. 45.
30. E. L. Smith, R. Lumry, and W. J. Polglase, *J. Phys. Colloid Chem.*, **55**, 125 (1951).
31. J. R. Whitaker, *Biochem. Biophys. Res. Commun.*, **22**, 1 (1966).
32. E. Elkins-Kaufman and H. Neurath, *J. Biol. Chem.*, **178**, 645 (1949).
33. J. L. Webb, *Enzyme and Metabolic Inhibitors*, Vol. II, Academic Press, New York, 1966, p. 366.

10

Effect of pH on Rates of Enzyme-Catalyzed Reactions

An enzyme has maximum activity only within a narrow pH range. The pH optimum of an enzyme is dependent on a number of experimental parameters, including time of reaction, temperature, nature and concentration of substrate, nature and concentration of buffer, ionic strength of medium, and purity of enzyme preparation, among others. Some of these effects are demonstrated by the data of Fig. 1. For the calf intestinal mucosa alkaline phosphatase-catalyzed hydrolysis of phenyl phosphate there is an increase in the pH optimum from around pH 8.3 at 2.5×10^{-5} M substrate to pH 10 at 2.5×10^{-2} M [Fig. 1(a)]. The pH optimum is plotted against $-\log(A_0)$ in Fig. 1(d). The pH optimum of alkaline phosphatase–catalyzed hydrolysis of phenyl phosphate is 7.9 when $MnCl_2$ is the activating cation and 9.3 when $MgCl_2$ is the activating cation [Fig. 1(b)]. (Alkaline phosphatase is a Zn^{2+}-enzyme but other divalent cations can replace Zn^{2+} with retention of some activity.) The pH optimum of alkaline phosphatase is also influenced by the nature of the substrate [Fig. 1(c)]. The pH optima are 9.4, 9.5, and 9.8 for hydrolysis of β-glycerophosphate, phosphocreatine, and phenyl phosphate, respectively.

These rather diverse data on the pH optimum of alkaline phosphatase all have logical explanations. For the data of Fig. 1(a) and (d), the differences in pH optima are due to different saturating levels of substrate and to substrate inhibition. The K_m for phenyl phosphate is 8.6×10^{-4} M, so that at 2.5×10^{-5} M substrate (2.5% saturation of enzyme) the observed effect of pH is on the free enzyme. As substrate concentration is increased, the effect of pH on both free enzyme and enzyme·substrate complex is observed until at high substrate concentrations (2.5

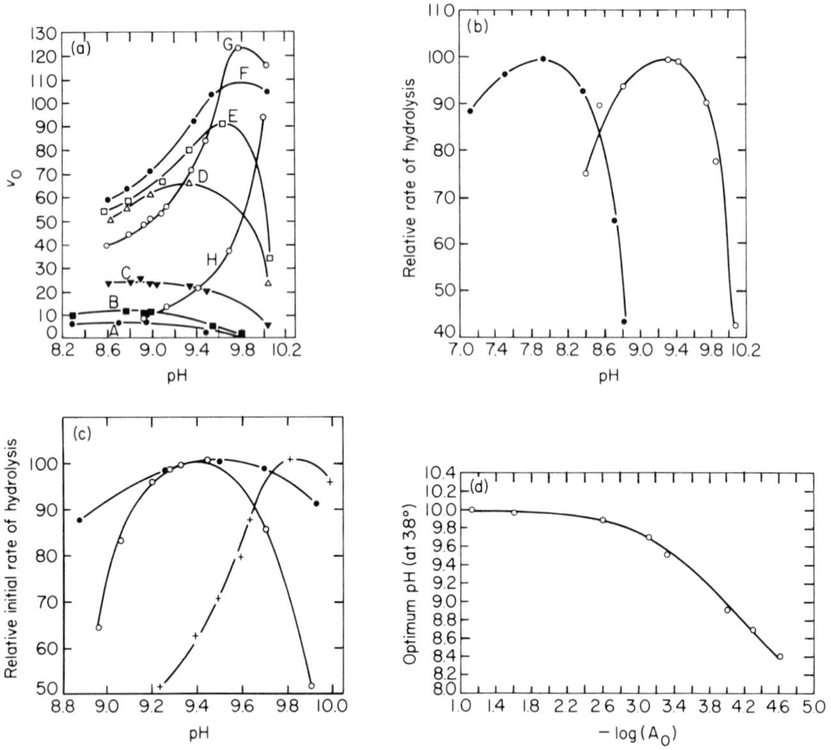

Figure 1 Effect of experimental conditions on the pH optimum of calf intestinal mucosa alkaline phosphatase. (a) Effect of initial substrate concentration on pH optimum. The curves are for the following substrate (phenyl phosphate) concentrations (*M*): A, 2.5×10^{-5}; B, 5×10^{-5}; C, 1×10^{-4}; D, 5×10^{-4}; E, 7.5×10^{-4}; F, 2.5×10^{-3}; G, 2.5×10^{-2}; H, 7.5×10^{-2}. (From Ref. 1, p. 675, by courtesy of the Biochemical Society.) (b) Effect of nature of activating cation on pH optimum. The cations were: ○, 5×10^{-3} *M* $MgCl_2$ and ●, 1×10^{-3} *M* $MnCl_2$ with 2.5×10^{-4} *M* phenyl phosphate as substrate. (From Ref. 1, p. 677, by courtesy of the Biochemical Society.) (c) Effect of nature of substrate on pH optimum. The substrates were: ●, 0.015 *M* phosphocreatine; ○, 0.02 *M* β-glycerophosphate; +, 0.025 *M* phenyl phosphate. (From Ref. 2, p. 235, by courtesy of the Biochemical Society.) (d) pH optima from data of part (a) plotted against $-\log(A_0)$. The lowest substrate concentration is at the right of the figure. (From Ref. 1, p. 675, by courtesy of the Biochemical Society.) The temperature was 38.0°C in all cases.

to 7.5×10^{-2} *M*) the effect of pH is largely on enzyme·substrate complex. Also, the observed velocity is influenced by substrate inhibition at the higher levels of substrate. Different rate-determining steps are involved in the Mg^{2+}-enzyme-versus Mn^{2+}-enzyme-catalyzed reactions [Fig. 1(b)]. With respect to the substrates shown in data of Fig. 1(c), the enzyme is 97, 54, and 30% saturated with phenyl phosphate (0.025 *M*; $K_m = 8.6 \times 10^{-4}$ *M*), β-glycerol phosphate (0.02 *M*, $K_m = 1.7 \times 10^{-2}$ *M*), and phosphocreatine (0.015 *M*, $K_m = 3.5 \times 10^{-2}$ *M*), respectively. Therefore, the effect of pH on different ratios of free enzyme to enzyme·substrate complex as well as the effect of different structures of substrates is observed in the three cases. What appears to be a hopelessly complex effect of pH on enzyme activity can be explained by proper consideration of conditions used in the experiments. It is the purpose of this chapter to offer a logical treatment of the effect of pH on enzyme-catalyzed reactions.

The pH optimum of an enzyme should be thought of only as an operational parameter unless very strict conditions are used in its measurement. The effect of pH on the observed activities of pepsin, trypsin, peroxidase, and alkaline phosphatase is shown in Fig. 2. Table 1 gives pH optima for a number of enzymes of interest to the food scientist.

Factors responsible for the effect of pH on enzyme-catalyzed reactions can be

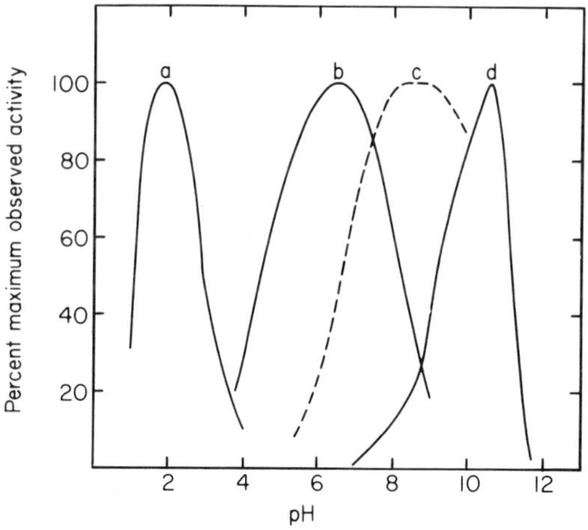

Figure 2 pH effect on activity of several enzymes. a, Pepsin acting on acetyl-L-phenylalanyl-L-diiodotyrosine at 36.7°C and 5×10^{-4} *M* pepsin, reaction time 15 min [3]. b, *Ficus glabrata* peroxidase acting on 0.03 *M* guaiacol and 0.005 *M* H_2O_2 at 30.0°C [4]. c, Trypsin acting on casein [5]. d, Hydrolysis of 5×10^{-4} *M* *p*-nitrophenyl phosphate by crude milk alkaline phosphatase at 35°C in 0.1 *M* sodium glycinate buffer, reaction time 15 min [6].

Table 1 pH–Activity Optimum of Several Enzymes[a]

Enzyme	pH optimum
Acid phosphatase (prostate gland)	5
Alkaline phosphatase (milk)	10
α-Amylase (human salivary)	7
β-Amylase (sweet potato)	5
Carboxypeptidase A (bovine)	7.5
Catalase (bovine liver)	3–10
Cathepsins (liver)	3.5–5
Cellulase (snail)	5
α-Chymotrypsin (bovine)	8
Dextransucrase (*Leuconostoc mesenteroides*)	6.5
Ficin (fig)	6.5
Glucose oxidase (*P. notatum*)	5.6
Lactate dehydrogenase (liver)	8 (forward reaction)
Lipase (pancreatic)	7
Lipoxygenase-1 (soybean)	9
Lipoxygenase-2 (soybean)	7
Pectin esterase (higher plants)	7
Pepsin (bovine)	2
Peroxidase (fig)	6
Polygalacturonase (tomato)	4
Polyphenol oxidase (peach)	6
Rennin (calf)	3.5
Ribonuclease (pancreatic)	7.7
Trypsin (bovine)	8

[a]The pH optimum will vary with source and experimental conditions (see the text). These pH values should be taken as approximate values.

divided into two categories: (a) those that influence the stability of the enzyme, and (b) those that influence binding and catalysis.

I. EFFECT OF pH ON STABILITY OF ENZYMES

All enzymes are proteins. Therefore, factors that influence the stability of an enzyme are those that affect the secondary, tertiary, and/or quaternary structures of proteins. For example, most enzymes undergo irreversible denaturation in very acid and very alkaline solutions. The pH at which this occurs varies with enzyme. Pepsin, the proteolytic enzyme of the stomach, is very rapidly inactivated at pH 7, whereas it is quite stable at pH 2. Milk alkaline phosphatase is quite stable within the region pH 7 to 9, but above pH 9.5 it is rapidly inactivated.

A. Factors That Modify the pH Stability of Enzymes

The pH stability of an enzyme is modified by a number of factors. Consider, for example, the effect of pH on the stability of trypsin (Fig. 3). The shape of the pH stability curve is a function of the incubation conditions. After 24 hours of incubation at 30°C, the pH of maximum stability is at pH 2.5, while the amount of activity left at pH 1 and 8 is quite low. When the enzyme is held at various pH values for 15 min at 0°C it retains full activity up to about pH 10. Above pH 10 it becomes less stable, with a minimum at pH 12, an increase in stability around pH 13, followed by marked activity loss above pH 13.

The observed stability behavior of trypsin is the result of two processes [5]. Instability below pH 2.5 (24 hours at 30°C) is the result of irreversible protein denaturation. Within the region of pH 2 to 8.5, primary loss in activity is due to autodigestion (i.e., trypsin hydrolyzes itself). This is a result of native enzyme molecules being in equilibrium with reversibly denatured enzyme molecules which the native enzyme can use as substrate. As the pH is increased toward pH 13, most of the enzyme is in the reversibly denatured form (15 min at 0°C), so there is little or no native enzyme to perform the autodigestion. Therefore, the enzyme is more stable at pH 13 than at pH 12. Above pH 13 the rate of irreversible denaturation becomes quite rapid, so that within 15 min at 0°C most activity is lost. The stability of trypsin has been treated in detail here because the data show the influence of protein denaturation, proteolysis, temperature, and time of incubation as well as pH on stability of an enzyme.

The pH stability of an enzyme is often influenced by the presence or absence

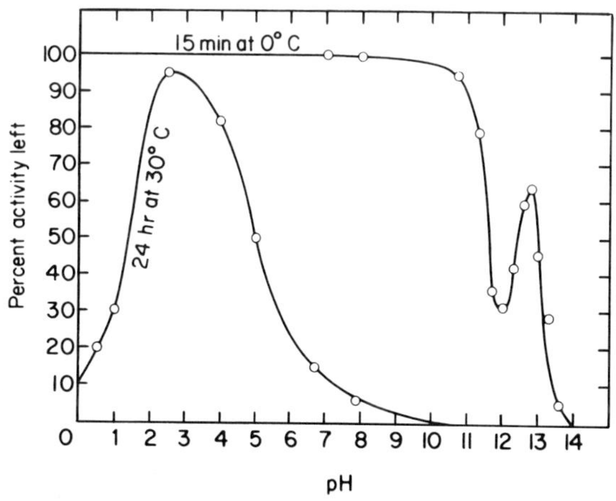

Figure 3 pH stability of trypsin under two sets of conditions. (From Ref. 5, p. 142, by courtesy of Columbia University Press.)

of a cofactor or other small molecules. As an example, consider the case of hog pancreas α-amylase (Fig. 4). α-Amylase contains one Ca(II) per molecule. This Ca(II) is not involved in either binding or transformation of substrate, but it is involved in stabilization of protein structure. Stability of the Ca(II)-free enzyme is much reduced over that of the Ca(II) enzyme, particularly in the pH range 7 to 11.

Incubation temperature is an extremely critical parameter in determining the stability of an enzyme. As indicated in Chapter 11, the activation energy, E_a, for protein denaturation is in the range 50 to 150 kcal/mol. This means that the rate of protein denaturation is increased 16 to 5000 times for each 10°C change in temperature. For this reason it is incorrect to determine the stability of an enzyme at one temperature and then to assume that it will show the same stability characteristics at a higher temperature. A determination of the effect of pH on stability of an enzyme at 0°C cannot be extrapolated easily to 25 to 35°C.

The effect of pH on enzyme stability is modified by other factors, such as buffer type and concentration, presence or absence of substrate, ionic strength and dielectric constant of the medium (modulated by the addition of organic solvents such as alcohol), and the effect of pH on the stability of cofactors and activators.

B. Experimental Determination of pH Stability of Enzymes

Before an interpretation of the effect of pH on observed activity can be made, it is necessary to determine the pH stability of an enzyme. This experiment should be performed carefully and as closely as possible under the same conditions as used in the experiment in which activity is determined. Into each of several tubes is added a fixed quantity of buffer of different pH values so as to cover the pH

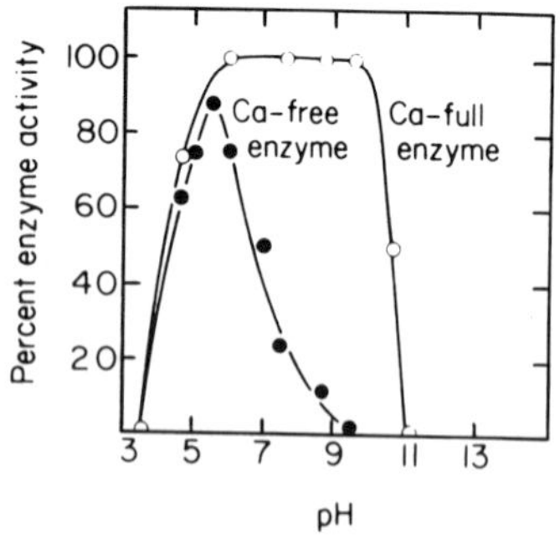

Figure 4 Stability of Ca(II)-containing and Ca(II)-free hog pancreas α-amylase. The enzyme was incubated for 20 hours at 25°C at various pH values, the pH was adjusted to a common value, and the activity was measured in the presence of Ca^{2+}. (From Ref. 6, p. 324, by courtesy of Academic Press.)

range over which activity will be determined. The tubes are equilibrated in a water bath for at least 10 min, then a constant-size aliquot of enzyme is added to each tube. At various intervals of time, aliquots removed from each tube are added to a solution of substrate in a buffer of fixed pH and of sufficient concentration so that the final pH of all the aliquots will be the same during the activity assay. The pH at which the assay for amount of activity left is performed does not have to be the pH optimum of the enzyme; however, it should be in a pH region where the enzyme is reasonably stable during the assay and has activity.

When pH affects only denaturation of an enzyme, plots of log (percent activity left) versus time will often be straight lines (Fig. 5). The slope of the line is the first-order rate constant for loss of activity. Not all rates of inactivation of enzymes follow first-order rate kinetics because more than denaturation may be involved in loss of activity, or isozymes with different stabilities are present. When the rate constants for inactivation determined at each pH are plotted versus pH, Fig. 6 is obtained. The enzyme is stable over the pH range 3 to 7 but becomes more unstable as the pH is increased above pH 7.

The investigator may not want to spend the amount of time required to get stability data as complete as those shown in Figs. 5 and 6. Aliquots of the enzyme for assay of activity left may be taken after a single interval of incubation. The time

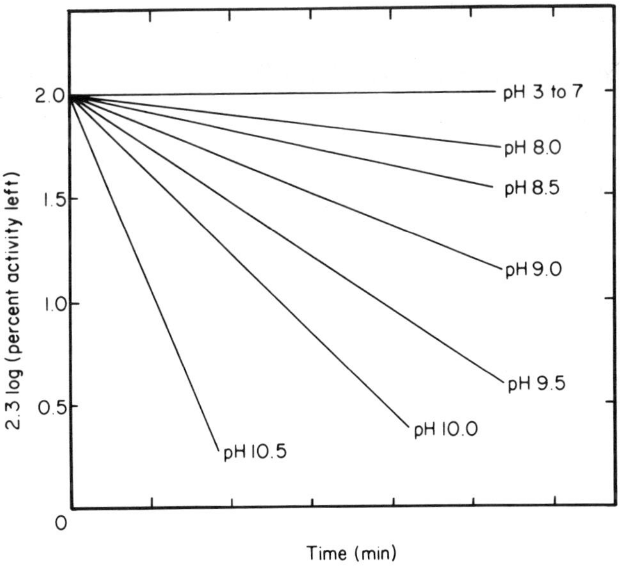

Figure 5 Schematic representation of rates of inactivation of an enzyme incubated at different pH values.

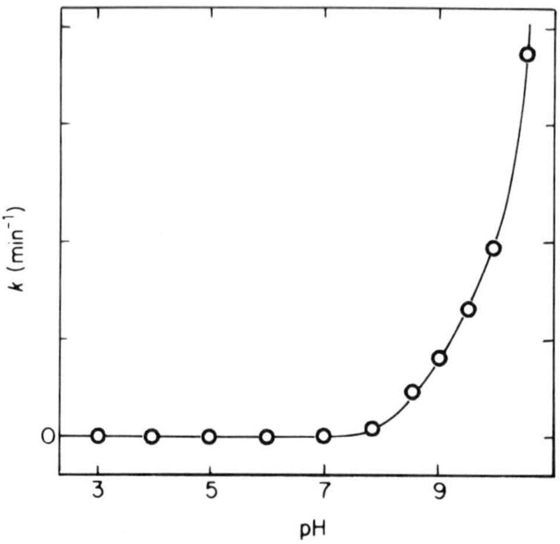

Figure 6 Schematic representation of effect of pH on enzyme stability. The rate constants for inactivation obtained from the data of Fig. 5 are plotted versus pH.

interval t chosen for stability determinations should be at least as long as the time over which the activity assay is run. The data are plotted as percent activity left at time t versus pH.

II. EFFECT OF pH ON CATALYTIC ACTIVITY OF ENZYMES

The effect of pH on catalytic activity should be clearly distinguished from the effect of pH on stability of enzymes. We shall define the effect of pH on the catalytic activity of an enzyme to mean the effect of pH on the ionization of prototropic groups involved in the active site of an enzyme.

A. Nature and Function of Prototropic Groups in Active Site

The prototropic groups (groups capable of ionizing) in the active site of an enzyme may be involved (a) in maintaining proper conformation of the active site, (b) in binding of substrate to enzyme, or (c) in transformation of substrate to products. These prototropic groups are located on the side chains of the acidic and basic amino acid residues. The nature of these groups, and their approximate pK_a and ΔH_{ion} values are given in Table 2. A range has been given for the pK_a values. The ionization of a prototropic group is sensitive to its environment, as illustrated for the pK value of acetic acid in water (pK_a = 4.71) and in 80% acetone

Table 2 Prototropic Groups That May Be Involved in Enzyme Catalysis

Group	Ionization		pK_a	ΔH_{ion} (kcal/mol)
Carboxyl	$-COOH \rightleftharpoons -COO^{\ominus} + H^{\oplus}$	α[a], β,γ,	3.0–3.2 3.0–4.7	±1.5
Imidazolium	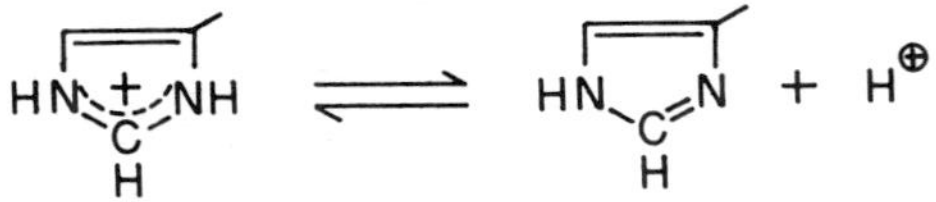		5.6–7.0	6.9–7.5
Sulfhydryl	$-SH \rightleftharpoons -S^{\ominus} + H^{\oplus}$		8.0–8.5	6.5–7.0
Ammonium	$-NH_3^{\oplus} \rightleftharpoons -NH_2 + H^{\oplus}$	α[a], ε,	7.6–8.4 9.4–10.6	10–13
Phenolic hydroxyl	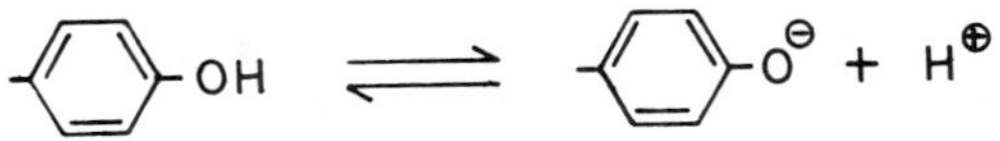		9.8–10.4	6
Guanidinium	$-NHC(=NH_2^{\oplus})NH_2 \rightleftharpoons -NHC(=NH)NH_2 + H^{\oplus}$		11.6–12.6	12–13

[a]Located at the end of polypeptide chain.

($pK_a \sim 7$) at 25°C. In a protein the pK_a value of a group may be shifted because it is in a hydrophobic environment (contributed by the side chains of the neutral and aromatic amino acids) or in an environment produced by a concentration of positive or negative charges. For this reason, conclusive assignment cannot be made of types of groups involved solely on the basis of pK_a values determined by the methods presented below. However, reasonable "guesses" about the groups involved can be made, which can be substantiated by other types of data.

B. General Discussion of Effect of pH on Ionization of Prototropic Groups in Active Sites of Enzymes

Let us consider the effect of pH on the ionization of prototropic groups involved in catalysis by an enzyme based on the model presented in Eq. (1). In this general

$$
\begin{array}{ccccccc}
E^{n+1} + A & \underset{k'_{-1}}{\overset{k'_1}{\rightleftharpoons}} & E^{n+1}\cdot A & \xrightarrow{k'_2} & E^{n+1}\cdot P & \underset{k'_{-3}}{\overset{k'_3}{\rightleftharpoons}} & E^{n+1} + P \\
\updownarrow K_1^{E} & & \updownarrow K_1^{EA} & & \updownarrow K_1^{EP} & & \updownarrow K_1^{E} \\
E^{n} + A & \underset{k_{-1}}{\overset{k_1}{\rightleftharpoons}} & E^{n}\cdot A & \xrightarrow{k_2} & E^{n}\cdot P & \underset{k_{-3}}{\overset{k_3}{\rightleftharpoons}} & E^{n} + P \\
\updownarrow K_2^{E} & & \updownarrow K_2^{EA} & & \updownarrow K_2^{EP} & & \updownarrow K_2^{E} \\
E^{n-1} + A & \underset{k''_{-1}}{\overset{k''_1}{\rightleftharpoons}} & E^{n-1}\cdot A & \xrightarrow{k''_2} & E^{n-1}\cdot P & \underset{k''_{-3}}{\overset{k''_3}{\rightleftharpoons}} & E^{n-1} + P
\end{array}
\qquad (1)
$$

scheme we have defined six dissociation constants (K_1^E, K_2^E, K_1^{EA}, K_2^{EA}, K_1^{EP}, and K_2^{EP}) and 15 rate constants (k_1, k_{-1}, k'_1, k'_{-1}, . . ., k''_3, k''_{-3}). The dissociation constants are associated with changes in states of ionization of the various species of enzyme. Because of added complexity with the many and varied combinations possible, we have not shown any ionization of substrate or product. We shall return to this point later. Note that the step in the conversion of substrate to product (k_2, k'_2, or k''_2). has been written as being irreversible. This may be so in specific examples involving many of the hydrolytic enzymes, or it may be so operationally in those cases where initial rates are used. In the discussion that follows it will be assumed that the rate-determining step is that for conversion of enzyme·substrate complex to enzyme·product complex (k_2, k'_2, or k''_2). This assumption is not valid for all cases. Release of product is rate determining for several dehydrogenases, for example.

If initial rate data are obtained, the scheme in Eq. (1) can be simplified since, kinetically, none of the processes that go on beyond the rate-determining step can be observed. Let us discuss in turn experimental methods by which all six dissociation constants can be determined.

1. pH and Ionization of Enzyme·Product Complex

The effect of pH on ionization of groups on the enzyme·product complex (K_1^{EP} and K_2^{EP}) may be obtained by adding product to the reaction initially and observing the effect of this addition on the observed initial rate. In general, the product is expected to show competitive inhibition with respect to substrate. Therefore, an analysis of the data by the methods described in Chapter 8 permits a determination of K_i (= K_p) as well as the effect of pH on K_i when the experiments are run at several different pH values.

2. pH and Ionization of Enzyme·Substrate Complex

When the reaction described by Eq. (1) is performed under conditions where (A_0) $\gg K_m$, only the effect of pH on prototropic groups involved in transformation of enzyme·substrate complex to products will be observed. Experiments should be performed at different substrate concentrations and at several pH values over the pH range of interest and the data analyzed by Lineweaver–Burk plots (Fig. 7). As shown in Fig. 7, K_m is the same at pH 5 and 6 (same intercept on x-axis) but increases at pH 7 and 8 (highest at pH 8). V_{max} is the same at pH 6 and 7 (same intercept on y-axis) but decreases at pH 5 and 8 (lowest at pH 5). Much work is required to obtain such data, but it is a prerequisite if proper interpretation of

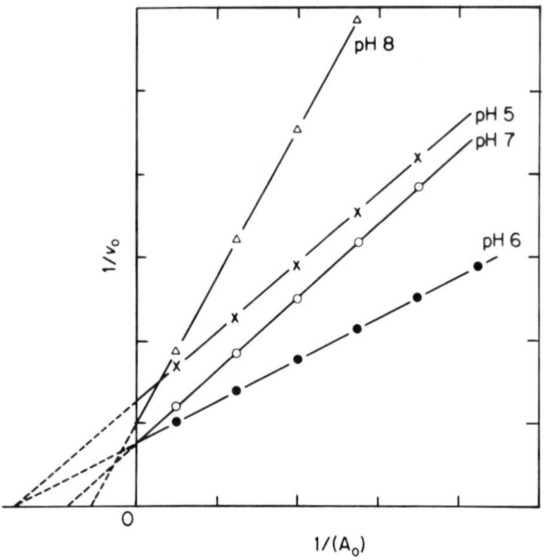

Figure 7 Initial velocity v_0 as a function of initial substrate concentration (A_0) at several pH values as plotted by the Lineweaver–Burk method.

the effect of pH on an enzyme-catalyzed reaction is to be made. It is wrong to determine the amount of substrate needed to saturate the enzyme [$(A_0) \sim 100K_m$] at one pH and then assume that this amount will be sufficient at all pH values. This would imply that K_m is pH independent, which is far from the usual case.

3. pH and Ionization of Free Enzyme

By determining initial rates of reactions under conditions where $(A_0) \ll K_m$ and constant (i.e., where the reaction is first order with respect to substrate concentration), the effect of pH on ionization of prototropic groups on the free enzyme will be determined. Under these conditions $v_0 = V_{max}(A_0)/K_m$, which is plotted against pH. Alternatively, V_{max}/K_m versus pH may be plotted to achieve the same result. Notice that the two equations differ in that one contains the fixed substrate concentration used.

4. pH and Ionization of Substrate and Product

In Eq. (1) the substrate and product were assumed to be neutral or not to ionize within the pH range of importance. This may not always be the case. The effect of pH on ionization of substrate and product can be determined by titrating the compounds with standard acid or base just as is done in determination of the pK_a value of any weak acid or weak base (Fig. 8). Because of the complexity of substrate and of pH effects on prototropic groups on the substrate it is not possible to analyze rigorously the effect of pH on the activity of an enzyme acting on a protein, nucleic acid, or pectic substance. Therefore, any rigorous analysis of the

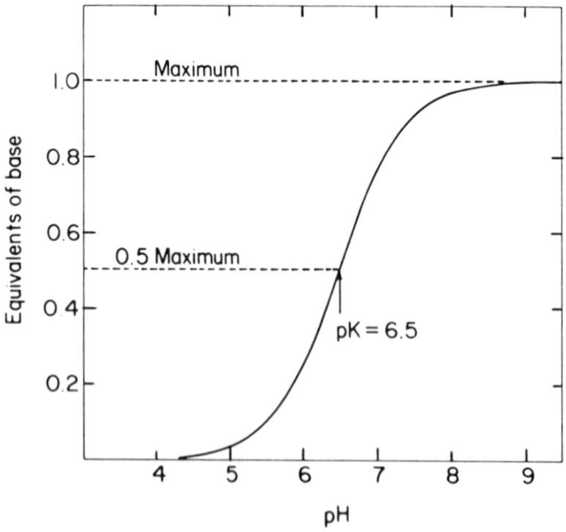

Figure 8 Titration of a compound with NaOH so as to determine the pK_a value of the prototropic group on the compound.

effect of pH on the activity of an enzyme requires the use of a low- molecular-weight substrate. Such substrates are available for most enzymes.

C. Analysis and Interpretation of Effect of pH on Catalytic Activity of Enzymes

1. Michaelis pH Functions [7]

To aid in interpretation of effect of pH on activity of enzymes we shall develop the Michaelis pH functions which permit us to write conservation equations in terms of a single prototropic form of the enzyme.

Previously, we talked about the conservation equations in terms of the various forms of enzyme (free enzyme E, enzyme·substrate complex EA, and enzyme·product complex EP) in a reaction mixture so that total enzyme concentration, (E_0), is given by

$$(E_0) = (E) + (EA) + (EP) \tag{2}$$

In an analysis of effect of pH on catalytic activity of an enzyme, we must also write conservation equations which show the various prototropic forms of each enzyme form:

$$(E) = (E^{n+1}) + (E^{n}) + (E^{n-1}) \tag{3}$$

$$(EA) = (AE^{n+1}) + (AE^{n}) + (AE^{n-1}) \tag{4}$$

$$(EP) = (PE^{n+1}) + (PE^{n}) + (PE^{n-1}) \tag{5}$$

The concentrations of different prototropic forms of an enzyme species shown in Eqs. (3), (4), or (5) are not independent of each other but are related through the equilibria shown in Eq. (6) for the free enzyme species.

$$E^{n+1} \underset{+H^+}{\overset{K_1, -H^+}{\rightleftharpoons}} E^{n} \underset{+H^+}{\overset{K_2, -H^+}{\rightleftharpoons}} E^{n-1} \tag{6}$$

The interrelationships among the different prototropic forms are best expressed quantitatively by the Michaelis pH functions. The Michaelis pH functions were originally developed for polybasic acids. However, they are universally applicable to all ionizing substances, including enzymes.

In the discussion that follows, the Michaelis pH functions will be developed using the free enzyme. However, the final derived functions are independent of the nature of the substance undergoing ionization. The derivations are also based on treatment of the region of the enzyme containing the prototropic groups involved as being symmetrical and on the assumption there are no interactions among the prototropic groups. Neither of these assumptions is probably correct. In an asymmetrical molecule (or region) there must be four forms of a species containing two prototropic groups. Consider as an example the dicarboxylic acid malic acid:

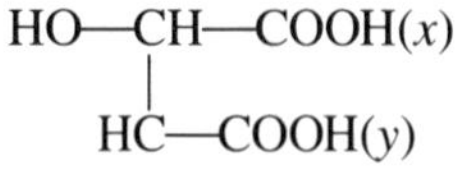

Because of asymmetry of the molecule the two carboxyl groups are not identical. Therefore, in writing Michaelis pH functions for this molecule the total malic acid concentration = $(MH_2) + (MH_y^-) + (MH_x^-) + M^{2-})$. The reader is referred to Ref. 8 for a more complete discussion of this situation. Our less rigorous treatment seems justified for the majority of cases for two reasons: (a) the kinetically observable pK_a values are usually well separated, which minimizes interaction among the prototropic groups, and (b) experimental pH data are usually too poor to justify a more rigorous treatment.

Consider that Eq. (6) properly describes the effect of pH on the free enzyme species and that the equilibrium constants are defined as

$$K_1 = \frac{(E^n)(H^+)}{(E^{n+1})} \tag{7}$$

$$K_2 = \frac{(E^{n-1})(H^+)}{(E^n)} \tag{8}$$

(It is understood that $K_1 = K_1^E$ and $K_2 = K_2^E$ in this discussion; therefore, the superscripts are omitted for simplification.)

Equations (7) and (8) may be solved in all possible combinations for (E^{n+1}), (E^n), and (E^{n-1}).

$$(E^{n+1}) = \frac{(E^n)(H^+)}{K_1} = \frac{(E^{n-1})(H^+)^2}{K_1K_2} \tag{9}$$

$$(E^n) = \frac{(E^{n-1})(H^+)}{K_2} = \frac{K_1(E^{n+1})}{(H^+)} \tag{10}$$

$$(E^{n-1}) = \frac{(E^n)K_2}{(H^+)} = \frac{K_1K_2(E^{n+1})}{(H^+)^2} \tag{11}$$

where the third equality in Eq. (9) is obtained from the second equality by substitution of definition of (E^n) from Eq. (10). In Eq. (10) the third equality is obtained by substituting the definition of $(E^{n-1}) = K_1K_2(E^{n+1})/(H^+)^2$ from Eq. (11).

These relationships are substituted into Eq. (3) so that all the free enzyme may be written in the form E^{n+1}, E^n, or E^{n-1} as shown in Eqs. (12) to (14).

1. Distribution of total free enzyme, E, in terms of E^{n+1}:

$$\begin{aligned}(E) &= (E^{n+1}) + \frac{K_1(E^{n+1})}{(H^+)} + \frac{K_1K_2(E^{n+1})}{(H^+)^2} \\ &= (E^{n+1})\left[1 + \frac{K_1}{(H^+)} + \frac{K_1K_2}{(H^+)^2}\right] = (E^{n+1})f^{n+1}\end{aligned} \tag{12}$$

2. Distribution of total free enzyme, E, in terms of E^n:

$$\begin{aligned}(E) &= \frac{(E^n)(H^+)}{K_1} + (E^n) + \frac{(E^n)K_2}{(H^+)} \\ &= (E^n)\left[1 + \frac{(H^+)}{K_1} + \frac{K_2}{(H^+)}\right] = (E^n)f^n\end{aligned} \tag{13}$$

3. Distribution of total free enzyme, E, in terms of E^{n-1}:

$$\begin{aligned}(E) &= \frac{(E^{n-1})(H^+)^2}{K_1K_2} + \frac{(E^{n-1})(H^+)}{K_2} + (E^{n-1}) \\ &= (E^{n-1})\left[1 + \frac{(H^+)}{K_2} + \frac{(H^+)^2}{K_1K_2}\right] = (E^{n-1})f^{n-1}\end{aligned} \tag{14}$$

f^{n+1}, f^n, and f^{n-1} are known as the Michaelis pH functions and have the definitions given in Eqs. (12) to (14). Since they do not include any parameter of E they are just as appropriate for the effect of pH on A, EA, EI, EP, and so on.

2. Nature of Plots of V_{max}^{H+}, V_{max}^{H+}/K_m^{H+}, and K_m^{H+} Versus pH and Assignment of pK Values

Having concluded a general discussion of the Michaelis pH functions, we examine specifically types of plots expected from experimental data and how to assign pK values to the proper form of the enzyme or substrate.

a. V_{max}^{H+} Versus pH

When the enzyme is completely saturated with substrate at all pH values, the observed effect of pH will be on the enzyme·substrate complex for the situation described by Eq. (1). V_{max}^{H+} as well as K_m^{H+} values are determined at a number of pH values by the Lineweaver–Burk treatment of data. Theoretically, one might visualize the following four situations for the influence of pH on V_{max}^{H+}, where $V_{max}^{H+} = V_{max}/f^z{}_{EA}$. V_{max}^{H+} indicates the maximum velocity at a given pH; V_{max} in this chapter is reserved for the maximum velocity that is truly pH independent and an absolute constant.

1. Figure 9(a) describes a situation where V_{max}^{H+} is independent of pH over the entire region and therefore $V_{max}^{H+} = V_{max}$. This situation is rarely seen when data are obtained from pH 3 to 9. One exception to this is the effect of pH on catalase activity (Chapter 26, Section II.E).

2. Figure 9(b) shows the dependence of V_{max}^{H+} on a single prototropic group. The active prototropic form of EA would be E^n·A, and thus $V_{max}^{H+} = V_{max}/\{1 + [(H^+)/K^{EA}]\}$.

3. Figure 9(c) shows the dependence of V_{max}^{H+} on a single prototropic group. The active prototropic species of EA is E^{n+1}·A and $V_{max}^{H+} = V_{max}/\{1 + [K^{EA}/(H^+)]\}$.

4. Figure 9(d) shows the dependence of V_{max}^{H+} on two prototropic groups. The active prototropic species of EA is E^n·A, so that V_{max}^{H+} is given by $V_{max}/\{1 + [(H^+)/K_1^{EA}] + [K_2^{EA}/(H^+)]\}$ according to Eq. (13).

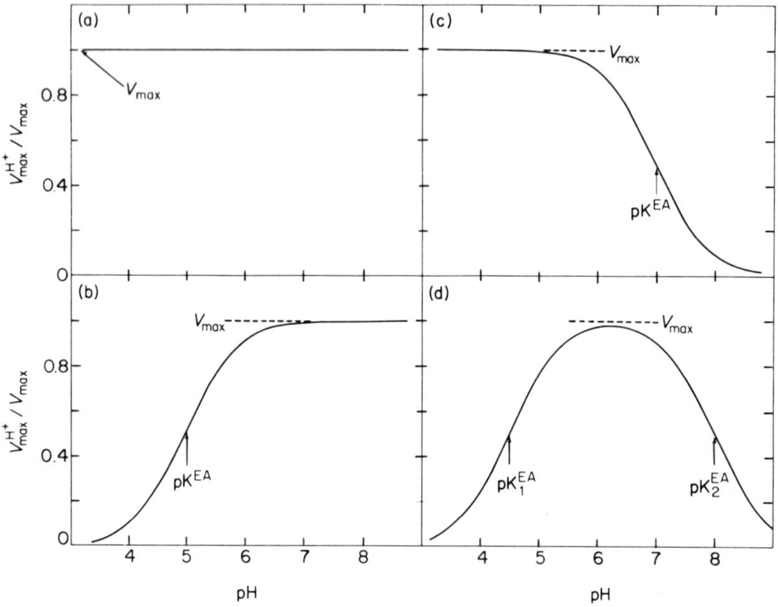

Figure 9 Possible effects of pH on V_{max}^{H+}. In (b) pK^{EA} = 5.0, in (c) pK^{EA} = 7.0, and in (d) pK_1^{EA} and pK_2^{EA} = 4.5 and 8.0, respectively.

b. V_{max}^{H+}/K_m^{H+} Versus pH

A plot of V_{max}^{H+}/K_m^{H+}versus pH permits a determination of prototropic groups on the free enzyme which are involved (a) in maintaining conformation of the active site, (b) in binding of the substrate, and/or (c) in transformation of substrate to products.

It can readily be shown that $K_m^{H+} = K_m f_E^x f_A^y / f_{EA}^z$, and since $V_{max}^{H+} = V_{max}/f_{EA}^z$, the ratio V_{max}^{H+}/K_m^{H+} is given by

$$\frac{V_{max}^{H+}}{K_m^{H+}} = \frac{V_{max}/f_{EA}^z}{K_m f_E^z f_A^y / f_{EA}^z} = \frac{V_{max}}{K_m f_E^x f_A^y} \tag{15}$$

in which the pH dependence is due to prototropic groups on the free enzyme and/or the free substrate. One may eliminate, or determine independently, the effect of pH on free substrate. Under these conditions

$$\frac{V_{max}^{H+}}{K_m^{H+}} = \frac{V_{max}}{K_m f_E^x} \tag{16}$$

Theoretically, the four possible relations between V_{max}^{H+}/K_m^{H+} and pH are shown in Fig. 10 assuming that substrate does not ionize within the region of pH 3 to 9.

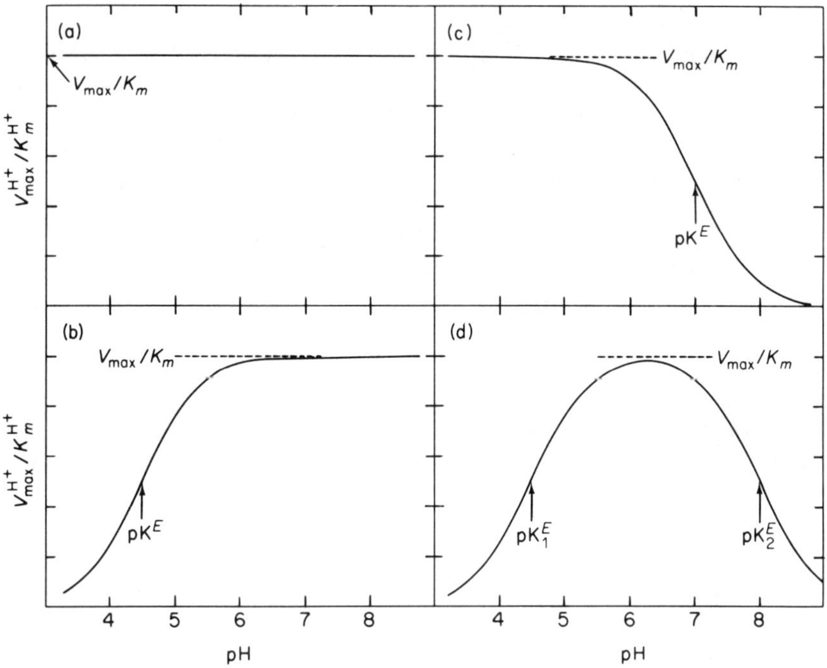

Figure 10 Possible effects of pH on $V^{H^+}_{max}/K^{H^+}_m$. In (b) $pK_E = 4.5$, in (c) $pK^E = 7$, and in (d) pK^E_1 and $pK^E_2 = 4.5$ and 8.0, respectively.

1. In Fig. 10(a) $V^{H^+}_{max}/K^{H^+}_m$ is independent of pH over the entire pH range and is equal to V_{max}/K_m. In this case there is no kinetically observable prototropic group in the active site of the free enzyme. Experimentally, this situation would be encountered only rarely.
2. In Fig. 10(b) $V^{H^+}_{max}/K^{H^+}_m$ is dependent on a single prototropic group in the active site of the free enzyme. The active form of the enzyme is E^n so that $V^{H^+}_{max}/K^{H^+}_m = V_{max}/K_m\{1 + [((H^+)/K^E]\}$.
3. In Fig. 10(c) $V^{H^+}_{max}/K^{H^+}_m$ is dependent on a single prototropic group on the free enzyme. The active form of the free enzyme is E^{n+1}, so that $V^{H^+}_{max}/K^{H^+}_m = V_{max}/K_m\{1 + [K^E/(H^+)]\}$.
4. In Fig. 10(d) $V^{H^+}_{max}/K^{H^+}_m$ is dependent on two prototropic groups on the free enzyme. The active form of the free enzyme is E^n, so that $V^{H^+}_{max}/K^{H^+}_m = V_{max}/K_m\{1 + [(H^+)/K^E_1] + [K^E_2/(H^+)]\}$. It seems to us that this will be the case most frequently encountered.

c. K_m^{H+} Versus pH

The pH dependence of K_m^{H+} is determined along with that of V_{max}^{H+} when the experimental data are treated by the Lineweaver–Burk method. In the interpretation of effect of pH on enzyme-catalyzed reactions the important variations with pH are those of V_{max}^{H+} and V_{max}^{H+}/K_m^{H+}. The pH variation of K_m^{H+} is simply the ratio of these two other variables.

3. Methods for Determination of pK Values of Prototropic Groups from Experimental Data

Several methods are available for determining the required parameters, V_{max}, K_m, pK_1^E, pK_2^E, pK_1^{EA}, and pK_2^{EA} ($pK = -\log K$) from initial rate data.

In developing the methods we shall use V_{max}^{H+} and pH as the experimental parameters. The same methods may be applied to analysis of V_{max}^{H+}/K_m^{H+} as a function of pH. In the examples given we shall use the Michaelis pH function, f_{EA}^n. It would be just as appropriate to use f_{EA}^{n+1} or f_{EA}^{n-1} for illustration of the methods. In choosing f_{EA}^n we are arbitrarily indicating that the active form of the enzyme·substrate complex is AE^n and that AE^{n+1} and AE^{n-1} are inactive. K_1^{EA} and K_2^{EA} are used to designate the first and second dissociation constants of EA, respectively. K_1^E and K_2^E are used to designate the first and second dissociation constants of E, respectively. K_1 and K_2 are used in a general sense where the specific enzyme form is not specified.

a. Method I: Determination Directly from the Michaelis pH Functions

The true maximum velocity of a reaction will be observed when all the enzyme is combined with substrate [$(E_0) = (EA)$] and all the enzyme·substrate complex is in the correct prototropic form:

$$V_{max} = k_2(E_0) = k_2(AE^n) \tag{17}$$

Here EA and AE are the same complex. The maximum velocity observed at any pH is given by

$$V_{max}^{H+} = \frac{k_2(AE^n)}{f_{EA}^n} \tag{18}$$

On substitution of $k_2(AE^n) = V_{max}$ [Eq. (17)] into Eq. (18) we obtain

$$V_{max}^{H+} = \frac{V_{max}}{f_{EA}^n} \tag{19}$$

Analysis of this equation shows that V_{max}^{H+} will range between the limits of 0 to V_{max}. Stated another way, V_{max}^{H+}/V_{max} ranges from 0 to 1.

We shall now describe several methods of analyzing experimental data based on the specific Michaelis pH function, $f^n = 1 + [(H^+)/K_1^{EA}] + [K_2^{EA}/(H^+)]$, so that

$$\frac{V_{max}}{V_{max}^{H^+}} = f_{EA}^{n} = 1 + \frac{(H^+)}{K_1^{EA}} + \frac{K_2^{EA}}{(H^+)} \tag{20}$$

1. Solution of Eq. (20) based on simultaneous equations. $V_{max}^{H^+}$ and (H^+) are known experimental quantities, while V_{max}, K_1^{EA}, and K_2^{EA} are unknowns. By use of data at three different pH values and thus three equations, V_{max}, K_1^{EA}, and K_2^{EA} may be determined. In solution of Eq. (20) by this method the best results will be obtained by using data that lie between K_1^{EA} and K_2^{EA}.

2. Plots of $1/V_{max}^{H^+}$ versus $f(H^+)$. When sufficient experimental data are available at pH values near pK_1^{EA} and/or pK_2^{EA}, V_{max}, K_1^{EA}, and K_2^{EA} may be obtained from plots as shown in Figs. 11 and 12, in which the data are plotted according to

$$\frac{1}{V_{max}^{H^+}} = \frac{1}{V_{max}} + \frac{(H^+)}{V_{max}K_1^{EA}} \tag{21}$$

$$\frac{1}{V_{max}^{H^+}} = \frac{1}{V_{max}} + \frac{K_2^{EA}}{V_{max}(H^+)} \tag{22}$$

These equations come directly from Eq. (20) by dividing both sides by V_{max}.

Equation (21) is used for data in the region of K_1^{EA} ($10^{-1}K_1^{EA}$ to $10^{1}K_1^{EA}$) and Eq. (22) is used for data in the region of K_2^{EA} ($10^{-1}K_2^{EA}$ to $10^{1}K_2^{EA}$). The true V_{max} will be obtained only if the x axis is graphed to include (H^+) and $1/(H^+)$ to zero.

Although the specific situation under consideration is that defined by Eq. (20), it is valid to ignore the contribution of the term $K_2^{EA}/(H^+)$ at low pH values and the $(H^+)/K_1^{EA}$ at high pH values, when the two K values differ by $\geqslant 1000$.

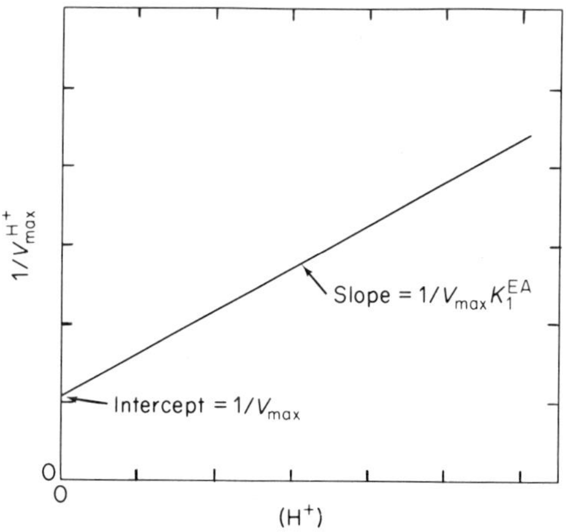

Figure 11 Determination of V_{max} and K_1^{EA} from experimental data plotted according to Eq. (21).

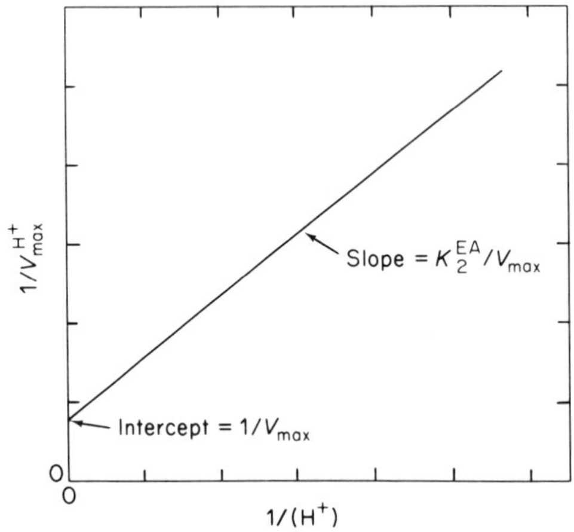

Figure 12 Determination of V_{max} and K_2^{EA} from experimental data plotted according to Eq. (22).

b. Method II: Dixon–Webb Plots [9]

The logarithm of Eq. (20) gives

$$\log V_{max}^{H^+} = \log V_{max} - \log\left[1 + \frac{(H^+)}{K_1^{EA}} + \frac{K_2^{EA}}{(H^+)}\right] \tag{23}$$

If $\log V_{max}^{H^+}$ is plotted versus pH [since $-\log(H^+) = pH$], Fig. 13 is obtained, from which pK_1^{EA} and pK_2^{EA} can be determined. Correct results will be obtained only if certain restrictions are observed in drawing tangents to the experimental data.

1. The horizontal tangent (slope = 0) must be $\log V_{max}$. In those cases where pK_1^{EA} and pK_2^{EA} are at least 3.5 pH units apart, this will offer no difficulty since a line drawn through the data points at top of the curve will be at $\log V_{max}$. Whenever pK_1^{EA} and pK_2^{EA} are less then 3.5 pH units apart, the true $\log V_{max}$ line will lie above the experimental points. This situation is treated below.

2. The slopes of the lines drawn tangent to the data at pH values below and above the "flat of the curve" must be 1 and -1, respectively. Proof of this may be shown by use of Eq. (23). $\log V_{max}$ is a constant term and contributes only in placing the curve in the vertical direction. Consider the term

$$-\log\left[1 + \frac{(H^+)}{K_1^{EA}} + \frac{K_2^{EA}}{(H^+)}\right] = -\log f_{EA}^{n}$$

At low pH only the first and second terms will affect the results, while at high pH only the first and third terms will affect the results. In the region of K_1^{EA}, $\log f_{EA}^{n} = \log(H^+) - \log K_1^{EA}$ and

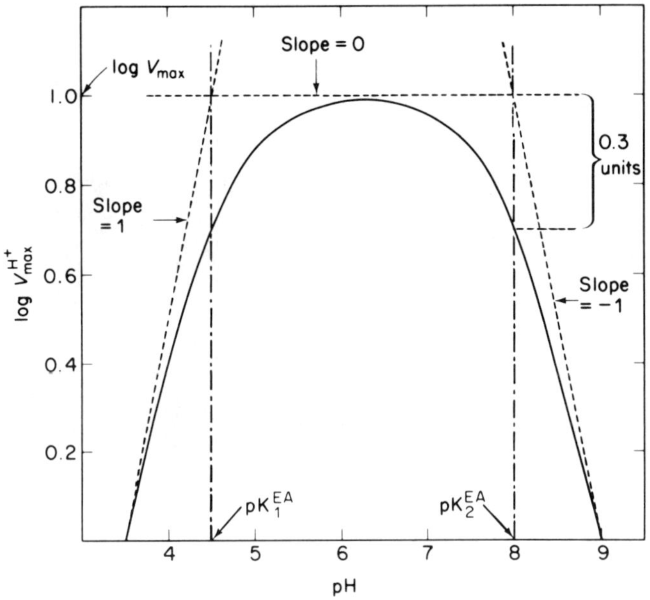

Figure 13 Determination of V_{max}, pK_1^{EA}, and pK_2^{EA} from experimental data plotted according to the Dixon–Webb method. Data are shown for $V_{max} = 10.0\ \text{min}^{-1}$, $pK_1^{EA} = 4.50$, and $pK_2^{EA} = 8.00$.

$$pf^n_{EA} = pH - pK_1^{EA} \tag{24}$$

where $pf^n_{EA} = -\log f^n_{EA}$, $pH = -\log(H^+)$, and $pK_1^{EA} = -\log K_1^{EA}$. A plot of pf^n_{EA} versus pH gives a straight line with slope of +1. In the region of K_2^{EA}, $pf^n_{EA} = pK_2^{EA} - pH$, so the slope is −1.

3. The points of intersections of the tangents should be exactly 0.3 pH unit (= log 2; 50% of maximum value) above the curvature of the line representing the experimental data. This distance will be 0.48 pH unit (= log 3) if two pK values occur together.

c. Method III: Visual Observation

When the pK values are sufficiently well separated so that a plot of $V^{H^+}_{max}$ versus pH includes V_{max}, the pK values are given at $0.5V_{max}$ (Fig. 14). In the example shown in Fig. 14, the theoretical values are 10.0 min^{-1}, 4.50, and 8.00 for V_{max}, pK_1^{EA}, and pK_2^{EA}, respectively, The values obtained visually from the graph are 9.70 min^{-1}, 4.45, and 8.05, respectively, for these parameters. The small errors are well within those expected of the usual experimental data, where $K^{H^+}_m$ and $V^{H^+}_{max}$ often have an experimental error of ±10%.

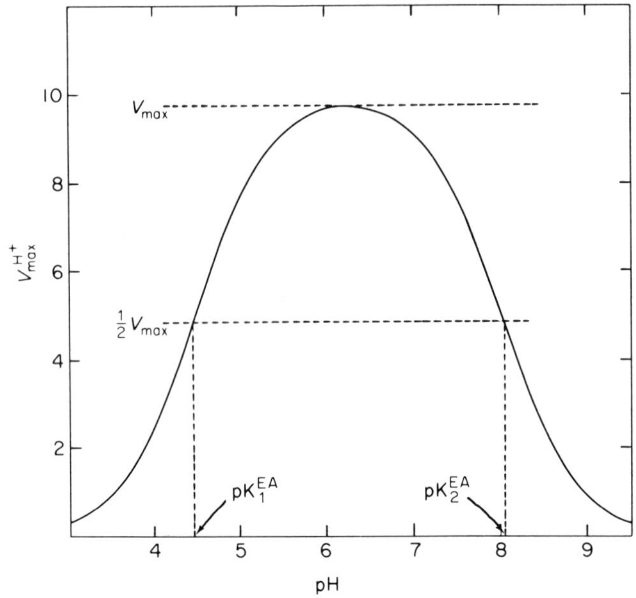

Figure 14 Determination of V_{max}, pK_1^{EA}, and pK_2^{EA} from visual inspection of data plotted as $V_{max}^{H^+}$ versus pH.

This method is less rigorous than methods I and II in that it gives most weight to data points only at V_{max} and $0.5V_{max}$. However, it does offer a rapid means of determining a rough value for the p*K* values which may then be compared to those obtained by a more rigorous treatment.

d. Cases Where pK_1 and pK_2 Are Separated by Less Than 3 pH Units

Determination of V_{max}, K_m, pK_1, and pK_2 from experimental data when pK_1 and pK_2 are close together has been discussed by Alberty and Massey [10]. This situation will exist only when the Michaelis pH function is f^n, so that two ionization constants, K_1 and K_2, are involved.

$$f^n = 1 + \frac{(H^+)}{K_1} + \frac{K_2}{(H^+)} \tag{25}$$

The (H^+) at the optimum pH, $(H^+)_{opt}$, is

$$(H^+)_{opt} = \sqrt{K_1 \cdot K_2} \tag{26}$$

which may also be written as $pH_{opt} = (pK_1 + pK_2)/2$.

Height of the peak at $(H^+)_{opt}$ is given by

$$\frac{1}{f^n} = \frac{K_1(H^+)_{opt}}{K_1(H^+)_{opt} + (H^+)^2_{opt} + K_1K_2} \tag{27}$$

which is the reciprocal of Eq. (25). Combining Eqs. (26) and (27) yields

$$\frac{1}{f^n} = \frac{1}{1 + 2\sqrt{K_2/K_1}} \tag{28}$$

At one-half of the peak where the K values will be observed,

$$\frac{1}{f^n_{1/2}} = \frac{K_2(H^+)_{1/2}}{K_1(H^+)_{1/2} + (H^+)^2_{1/2} + K_1K_2} = \frac{1}{2 + 4\sqrt{K_2/K_1}} \tag{29}$$

Equation (29) may be simplified to

$$(H^+)^2_{1/2} - [K_1 + 4(H^+)_{opt}](H^+)_{1/2} + (H^+)^2_{opt} = 0 \tag{30}$$

Equation (30) has two real roots, $(H^+)_1$ and $(H^+)_2$ (apparent K_1 and K_2), which correspond to the (H^+) at one-half of the experimentally observed optimum on each side of the curve. The sum of these two roots is given in Eq. (31), which may be used to calculate K_1.

$$(H^+)_1 + (H^+)_2 = K_1 + 4(H^+)_{opt} \tag{31}$$

Therefore, K_1 may be calculated from Eq. (31), and in turn K_2 may be calculated from Eq. (26). Unfortunately, Eq. (31) becomes less precise the closer the two apparent K's are together.

Consider the data shown by the solid line of Fig. 15, where the apparent pK_1^{EA} and apparent pK_2^{EA} are 5.5 and 7.5, respectively, corresponding to apparent K_1^{EA} and K_2^{EA} of 3.16×10^{-6} and 3.16×10^{-8} M. $(H^+)_1$ and $(H^+)_2$ are then 3.16×10^{-6} and 3.16×10^{-8} M, respectively. From Eq. (26), $(H^+)_{opt} = 3.16 \times 10^{-7}$ M,

When these values are placed in Eq. (31), K_1 is found to be 1.93×10^{-6} M. K_2, obtained from Eq. (26), is 5.18×10^{-8} M. Therefore, the true pK_1^{EA} and pK_2^{EA} are 5.71 and 7.29, respectively. V_{max}, calculated by use of Eq. (20), is 1.33 times the apparent V_{max} (Fig. 15).

D. Interpretation of Experimental Data

We shall take two cases for further interpretation. It will be assumed that the effect of pH on stability of the enzyme has been accounted for. Dixon–Webb plots are used as the best means of visualizing the total data.

1. Case I

The data to be treated are shown in Fig. 16.

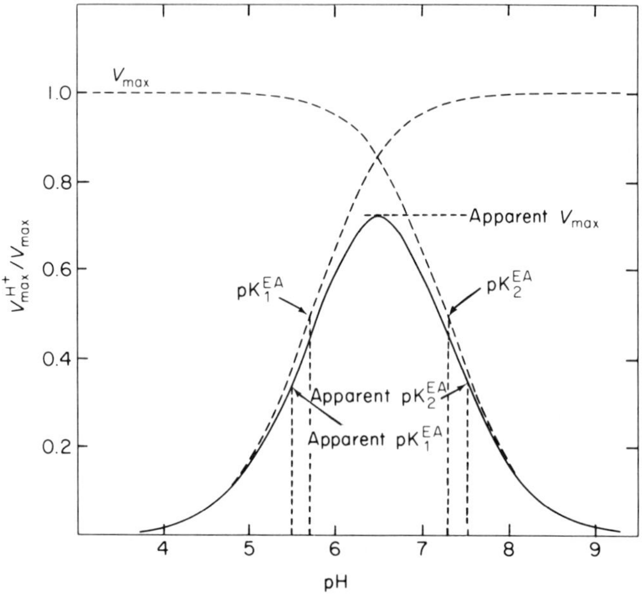

Figure 15 $V_{max}^{H^+}$ versus pH curve; pK_1^{EA} and pK_2^{EA} are close together so that the true V_{max}, pK_1^{EA}, and pK_2^{EA} cannot be obtained from visual inspection of data or by use of methods I and II. See the text for the equations and calculations involved.

In the plot of log $V_{max}^{H^+}/K_m^{H^+}$ versus pH (middle graph) there are two prototropic groups, pK_1^E and $pK_2^E = 4.5$ and 8.0, respectively, on the free enzyme which may be involved (a) in maintaining conformation of active site, (b) in binding of substrate, and/or (c) in transformation of substrate to product. The group pK_1^E is involved in transformation of substrate to product as shown by a plot of log $V_{max}^{H^+}$ versus pH where $pK_1^{EA} = pK_1^E$ (upper graph). The group of pK_2^E is involved in either binding of substrate or maintaining conformation as shown by the plot of $-\log K_m^{H^+}$ versus pH (lower graph). The enzyme species E^{n-1} is not capable of binding substrate.

The effect of pH on the reaction is summarized in Eq. (32). As shown, E^{n+1} and E^n, but not E^{n-1}, are capable of binding substrate and $E^n{\cdot}A$, but not $E^{n+1}{\cdot}A$, is capable of forming product. $E^{n-1}{\cdot}A$ does not show up in the data. This can be due either to masking of the group on formation of EA or that $E^n{\cdot}A$ and $E^{n-1}{\cdot}A$ are both formed and k_2 and k_2'' are equal, or that E^{n-1} is not formed.

2. Case II

Analysis of the data of case II (Fig. 17) shows that there are two prototropic groups on the free enzyme with pK_1^E and pK_2^E of 5.5 and 9.0, respectively (middle graph).

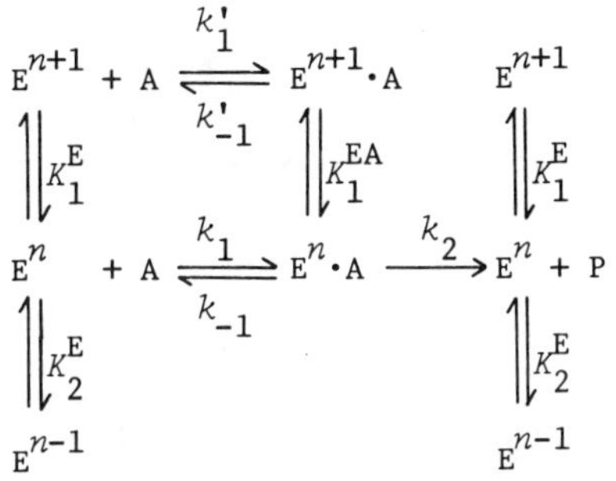

(32)

Both are involved in transformation of substrate to product as shown by log $V_{max}^{H^+}$ versus pH plot (top graph). Thus neither of these groups shows up in a plot of $-\log K_m^{H^+}$ versus pH (bottom graph).

The effect of pH on the reaction is summarized in Eq. (33), which indicates that all prototropic forms of the free enzyme bind substrate equally well but only the prototropic form, $E^n \cdot A$ is effective in transformation of substrate to product.

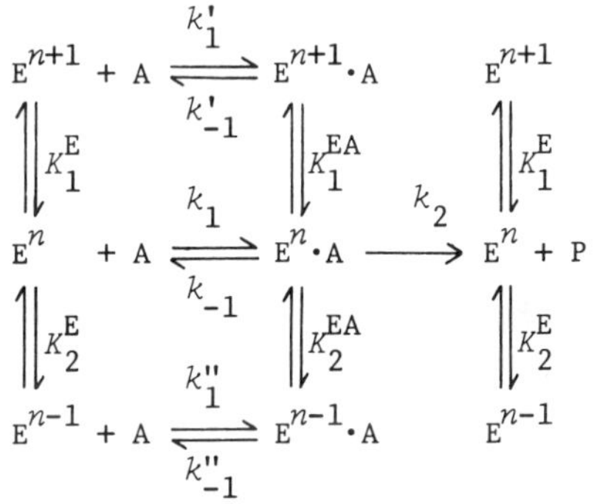

(33)

E. Limitations of Mathematical Treatment Presented on Effect of pH on Catalytic Activity of Enzymes

The effect of pH on enzyme activity is a powerful tool in the elucidation of the mechanism of action of an enzyme based on the ionizable groups involved. However, its limitations must not be overlooked. A change in pH may affect many factors, which will result in a change in the observed activity of an enzyme. Effect of pH on many of these factors can be determined independently. In other cases, the factors can be eliminated from consideration by proper experimental design.

We have used a simple scheme to depict an enzyme-catalyzed reaction in development of the theory. In many cases the system is much more complex, so that $K_m^{H^+}$ and $V_{max}^{H^+}$ may be composite terms involving several more rate constants than in the case taken. Then the effect of pH on each of the rate constants must be

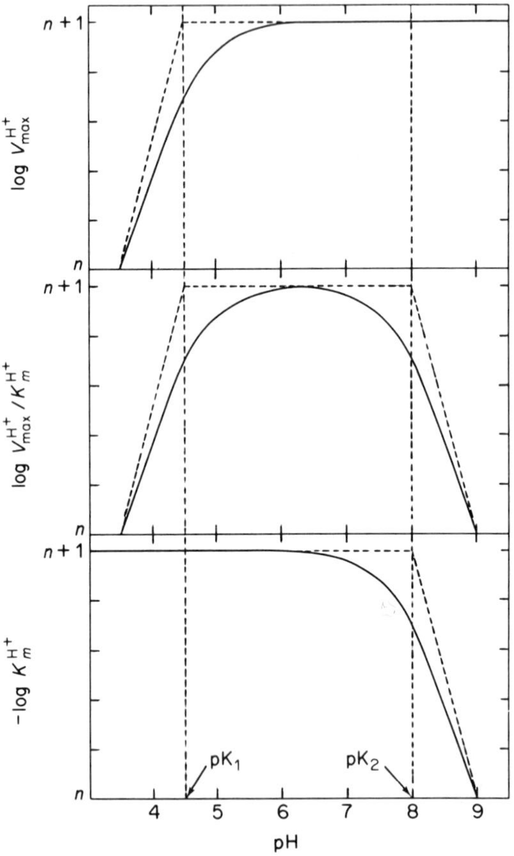

Figure 16 Experimental data showing effect of pH on $V_{max}^{H^+}$, $V_{max}^{H^+}/K_m^{H^+}$, and $K_m^{H^+}$ plotted by method of Dixon and Webb. See the text for interpretation of case I.

determined; alternatively, conditions may be found that will permit one step in a reaction to be totally rate determining. In that case the situation will reduce to that described in this chapter. In some cases ionization (or protonation) of a group on an enzyme leads to only partial activity loss. As a result, a plot of $V_{max}^{H^+}$ and $V_{max}^{H^+}/K_m^{H^+}$ versus pH shows a decrease followed by a leveling off at a new level. This situation is the result of both prototropic forms having activity, one being more efficient than the other.

The prototropic groups likely to be involved in the activity of an enzyme are listed in Table 2. Each has a characteristic pK value. Therefore, as a first approximation, one may assess a kinetically determined pK value in terms of

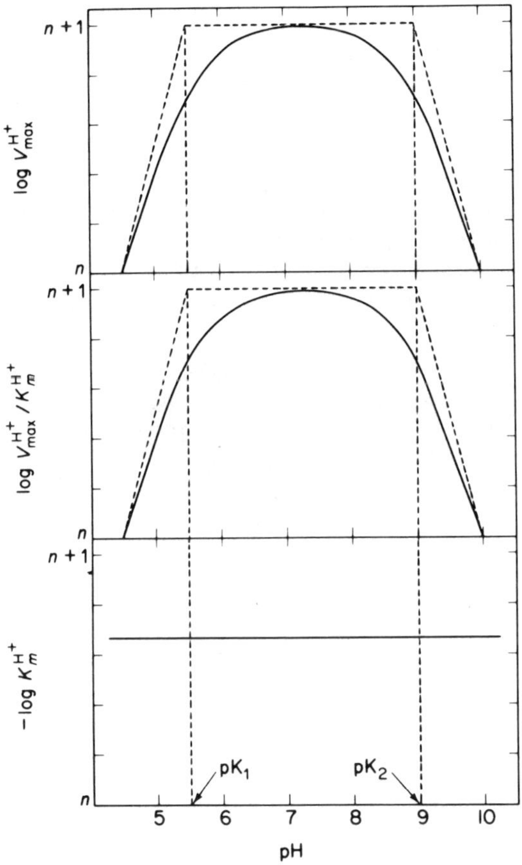

Figure 17 Experimental data showing effect of pH on $V_{max}^{H^+}$, $V_{max}^{H^+}/K_m^{H^+}$, and $K_m^{H^+}$ plotted by method of Dixon and Webb. See the text for an interpretation of case II.

one of these groups. It would be hazardous, however, to insist that such a group is present without additional information which could be obtained from ΔH_{ion}, from chemical modification, from x-ray crystallography, or by solvent–buffer perturbation of the pK.

In some enzymes a step other than a catalytic step may be rate determining. For example, with most dehydrogenases, the rate-determining step is rate of dissociation of the coenzyme product from the enzyme. Under these conditions the observed pK of a prototropic group involved in catalysis ($V_{max}^{H^+}$) will be displaced in a direction and amount depending on its active form and the ratio of rate of catalysis to the rate-determining step. For example, the pK of the group will be displaced to a lower value if it becomes inactive on protonation or to a higher pH if it becomes inactive on ionization. If the rate of a catalytic step involving the prototropic group is 100 times faster than rate of release of product, for example, the displacement of the observed pK will be 2 pH units. The V_{max}^{H+}/K_m^{H+} profile usually reflects true pK values since these are for the free enzyme.

The shift in apparent pK upon addition of an organic solvent, such as dioxane, in the presence of carboxylic acid buffer or imidazole buffer can be quite useful in determining the nature of a prototropic group involved. For example, in the presence of dioxane, the pK of a carboxyl group on the enzyme will not change much in the presence of a carboxylic acid buffer, but will change drastically in the presence of an imidazole buffer. The converse is true for the pK of an imidazole group. pK values of groups can shift substantially depending on whether they are in a hydrophobic or hydrophilic environment of the enzyme.

Despite the reservations that must be kept in mind in interpreting the effect of pH on the enzyme system, there are good reasons why one should attempt to describe the effect of pH on an enzyme in as rigorous a fashion as possible. Knowing the limitations of such data, the experiments should be carefully designed from the beginning. As pointed out in Fig. 1, the observed effect of pH on activity can appear to be complex and uninterpretable. However, by proper analysis of the experimental conditions, all the data fit a logical pattern to explain what is going on in the system.

REVIEW QUESTIONS

1. By plotting percent maximum observed activity (assume that 100% = maximum activity for case a) versus pH, determine the activity–pH profiles from pH 2 to 11 (at intervals of 1 pH) under the following conditions. Treat each as a separate case.

(a) pK_1^E = 3.50, where the active form of enzyme is E^n.

(b) pK_1^{EA} and pK_2^{EA} = 3.50 and 8.20, respectively, where the active form is $E^n{\cdot}A$.

(c) pK_1^E and pK_2^E = 6.50 and 7.50, respectively, where the active form of enzyme is E^n.

2. Explain in detail how you would determine whether a decrease in observed activity above and below the pH optimum of an enzyme is due to the effect of pH on stability of enzyme or on ionization of essential prototropic groups in the active site of the enzyme.

3. In an enzyme-catalyzed reaction described by the equation

$$E + A \underset{k_{-1}}{\overset{k_1}{\rightleftharpoons}} E{\cdot}A \overset{k_2}{\rightarrow} \underset{\displaystyle P_1}{\underset{+}{E'A}} \overset{k_3}{\rightarrow} E + P_2$$

k_2 was the rate-determining step in the hydrolysis of substrate 1 and k_3 was the rate-determining step in the hydrolysis of substrate 2. E·A contained two essential prototropic groups with pK_1^{EA} and pK_2^{EA} of 5 and 8, respectively, while E′A contained one essential prototropic group of pK 5. Assuming that the active form is E^n·A, calculate the V_{max}^{H+}–pH profiles for the two substrates. What can you conclude about the pH optima for the two substrates?

4. How is it that the pH optimum of pepsin can be 2 whereas that of alkaline phosphatase is 10?

5. Using the appropriate Michaelis pH function, calculate the relative percentage of the isoelectric species, $H_3^+N{-}CH_2{-}COO^-$, present at pH values from 2 to 12 (at intervals of 1 pH). Assume that $pK_1 = 3.12$ and $pK_2 = 8.07$.

6. Explain how, experimentally, you would distinguish among the following:
(a) Effect of pH on stability of enzyme.
(b) Effect of pH on essential ionizable groups on the free enzyme.
(c) Effect of pH on essential ionizable groups on the enzyme·substrate complex.
(d) Effect of pH on ionizable groups on the substrate.

7. A competitive inhibitor was found to react only with E^{n+1} form of the enzyme. Given that pK_1^E is 6.5, calculate the effect of pH on percent inhibition. Assume that $(I_0) \gg K_i$ at all values of pH.

8. Discuss the meaning of the term *optimum pH* of an enzyme-catalyzed reaction. Include in the discussion why *it is or is not* a unique characteristic of an enzyme and some of the parameters that affect the pH optimum.

9. The initial velocity of an enzyme-catalyzed reaction was determined at a number of pH values under conditions where $(A_0) \gg K_m$. The pH optimum of the reaction was 6.5 with the activity dependent on apparent pK values of 6.0 and 7.0, respectively.
(a) What is the relative value of V_{max}^{H+} to V_{max} at pH 6.5?
(b) What are the correct values for pK_1 and pK_2?

10. pK_1^E and pK_2^E are 4 and 8, respectively, while pK_1^{EA} is 4. K_m^{H+} is equal to K_m at all pH values below 6 but increases above pH 6 with dependence on an apparent pK of 8. Calculate the observed velocity versus pH curves when
(a) $(A_0) \ll K_m$
(b) $(A_0) = K_m$
(c) $(A_0) \gg K_m$
Plot all the data on the same graph. What conclusions can you draw from the data?

REFERENCES

General References

Alberty, R. A., and V. Massey, *Biochim. Biophys. Acta*, **13**, 347 (1954).
Dixon, M., and E. C. Webb, *Enzymes*, 3rd Ed., Academic Press, San Diego, 1980.
Peller, L., and R. A. Alberty, *J. Am. Chem. Soc.*, **81**, 5907 (1959).

Cited References

1. R. K. Morton, *Biochem. J.*, **65**, 674 (1957).
2. R. K. Morton, *Biochem. J.*, **61**, 232 (1955).
3. L. E. Baker, *J. Biol. Chem.*, **193**, 809 (1951).
4. S. Kon and J. R. Whitaker, *J. Food Sci.*, **30**, 977 (1965).
5. J. H. Northrop, M. Kunitz, and R. M. Herriott, *Crystalline Enzymes*, Columbia University Press, New York, 1948, p. 12.
6. E. H. Fischer and E. A. Stein, α-Amylases, in *The Enzymes* (P. D. Boyer, H. Lardy, and K. Myrbäck, eds.), Vol. IV, Academic Press, New York, 1960, p. 313.
7. L. Michaelis, *Die Wasserstaffionenkonzentration*, Springer-Verlag, Berlin, 1922, p. 48.
8. M. Dixon and E. C. Webb, *Enzymes*, Longmans, Green, London, 1958, p. 128.
9. M. Dixon and E. C. Webb, *Enzymes*, Longmans, Green, London, 1958, p. 126f.
10. R. A. Alberty and V. Massey, *Biochim. Biophys. Acta*, **13**, 347 (1954).

11

Effect of Temperature on Rates of Enzyme-Catalyzed Reactions

Temperature modification as a means of controlling rates of enzyme-catalyzed reactions is very important. Storage at low temperatures decreases the effect of enzymes on softening, off-flavor development, maturation, and so on. On the other hand, high temperatures destroy all enzyme activity, as in processing. Control of temperature is a prerequisite for reproducible analyses of enzyme activity.

I. PARAMETERS AFFECTED BY TEMPERATURE CHANGES

Changes in temperature affect enzymatic reactions in a number of ways. These include the effect on (a) stability of enzymes, (b) changes in solubility of gases, (c) pH of buffer, (d) affinity of enzyme for activators and inhibitors, (e) competing reactions, (f) ionization of prototropic groups of the system, (g) enzyme–substrate affinity, (h) velocity of conversion of substrate to product, and (i) degree of association of multipolypeptide enzymes.

II. STABILITY OF ENZYMES

An enzyme is a protein and a large, complicated molecule. Integrity of the three-dimensional structure of the active site is essential for maintenance of activity. Any factor that influences the integrity of the secondary, tertiary, and quaternary structures of an enzyme will affect its activity.

A. Factors That Affect Stability

Stability of an enzyme is a function not only of temperature but also of pH, ionic strength and nature of buffer, presence or absence of substrate, concentration of enzyme as well as other proteins in the system, time of incubation, and the presence or absence of activators and inhibitors. Data on the temperature stability of an enzyme have meaning only when all of these factors are controlled and stated explicitly.

In general, enzymes are more stable the lower the temperature. There are a few enzymes that are more unstable at 0 to 10°C than at 20 to 30°C because of the types of bonds holding the subunits together. There are also great differences among enzymes in their susceptibility to heat. For example, beef liver catalase is unstable at 35°C, whereas ribonuclease can withstand exposure to 100°C for a few minutes. Adenylate kinase can be boiled for a long time without loss of activity. Milk alkaline phosphatase and plant peroxidase are both relatively heat stable near pH 7 and are used in the dairy and processing industries for measuring adequacy of pasteurization and blanching, respectively. Because of their high-temperature stability, other enzymes and microorganisms are destroyed by heat by the time these enzymes are inactivated.

Some correlation may be drawn between the size and complexity of an enzyme and its susceptibility to heat. In general, those enzymes that have molecular weights ranging from 12,000 to 50,000, are composed of single polypeptide chains, and have disulfide bonds are more resistant to heat treatment. The larger an enzyme and the more complex its structure, the more susceptible it is to high temperatures.

In general, an enzyme is more stable to temperature in an intact tissue or in a homogenate, where its structure is protected by the presence of other colloidal material (proteins, carbohydrates, pectins, etc.), than it is in a purified form. Most of this increased stability in crude preparations over purified preparations is an apparent one and is simply a reflection of the relative concentration of total protein present in each type of preparation. Added gelatin or bovine serum albumin protect dilute enzyme preparations against denaturation.

Conditions for maximum stability of crude enzyme preparations may be quite different from that of the purified enzyme. For example, crude trypsin preparations have maximum stability at pH 5 and become more stable the higher the concentration of protein. However, they lose all activity irreversibly above 70°C. On the other hand, solutions of crystallized trypsin are most stable at pH 2 to 3 and can be heated to boiling briefly at this pH without permanent loss in activity.

B. Determination of Effect of Temperature on Stability

In the absence of perturbing factors such as proteolysis, the rate of denaturation of an enzyme generally follows a first-order process. This makes calculation of time necessary to destroy all enzymatic activity quite convenient since it is only necessary to know the half-life or the rate constant of the process (see Chapter 5). For example, if 50% of an enzyme activity is destroyed in 10 min ($t_{1/2}$ = 10 min) the activity remaining after 20, 40, 60, 80, and 100 min will be 25, 6.2, 1.6, 0.4, and 0.10%, respectively. This is readily calculated as $100\%/2^n$, where n is the number of half-lives.

The *temperature optimum* of an enzyme is an operational parameter rather than a true characteristic (Figs. 1 and 2). At the lower temperatures the rate of product formation is constant, but at higher temperatures there is a decrease in rate the longer the reaction time (Fig. 1). This is a result of a decrease in the total amount of enzyme present because it is being denatured as a function of time and temperature.

A different effect of temperature on the observed velocity, dP/dt, of the reaction is obtained depending on whether initial rates or rates taken at time t are used (Fig. 2). The decrease in rate at higher temperatures is a result of temperature on stability of the enzyme, while the increase in rate at the lower temperatures is due to the effect of temperature on increasing the rate of conversion of substrate to product. The temperature where optimum activity is seen is above 60°C when initial rates are used but is at 56°C and 50°C when activities at times t_1 and t_2 are used (Fig. 2).

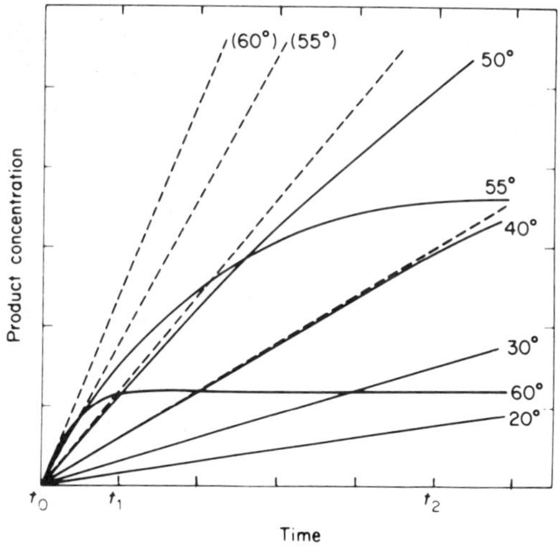

Figure 1 Effect of temperature on rate of product formation. The solid lines are for experimental data; the dashed lines are based on initial rates (i.e., lines drawn tangent to experimental data at time = 0). The data shown were calculated for the following conditions: $(A_0) \gg K_m$; E_a for transformation of reactant to product, 12,000 cal/mol; E_a for denaturation of enzyme, 60,000 cal/mol; first-order rate constant, k, for denaturation of enzyme, 9.0×10^{-4} min^{-1} at 40°C.

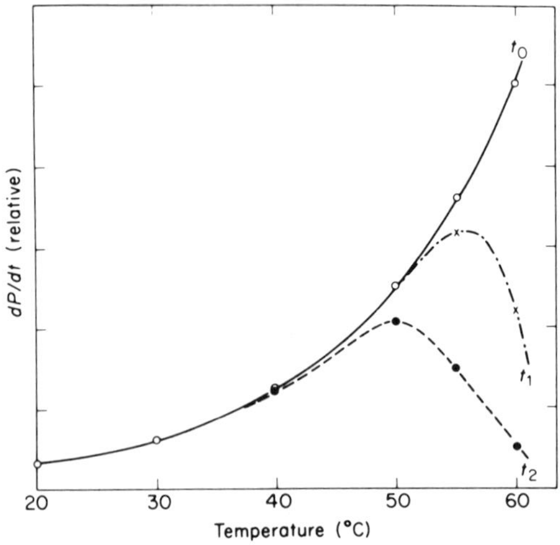

Figure 2 Rate of formation of product as function of temperature. Data taken from Fig. 1 at t_0, t_1, and t_2 are plotted, where t is time.

In general, activation energies for transformation of reactants to products (catalysis) in enzyme-catalyzed reactions are within the range 6000 to 15,000 cal/mol while activation energies for denaturation of enzymes are within the range 50,000 to 150,000 cal/mol. What this means from a practical point is that at lower temperatures enzymes will be relatively stable. But at higher temperatures denaturation will become very rapid because relatively larger numbers of molecules have sufficient energy to achieve the denatured state. This is illustrated in Table 1, where E_a, the activation energy, is considered to be 6000 and 60,000 cal/mol for conversion of reactants to products and denaturation, respectively. At 60°C the rate of conversion of reactant to product is only 11.4 times faster than at −10°C, while the rate of denaturation is 3.16×10^{10} faster at 60°C than at −10°C.

The effect of temperature on stability of an enzyme can be determined readily. Solutions of the enzyme are incubated at various temperatures in the absence of substrate while all other conditions are kept constant. Aliquots are removed from the incubation mixture at intervals and placed into an ice bath (for storage until activity analyses can be performed) or directly into reaction mixtures containing substrate. The amount of activity left is determined at a constant pH and temperature where the enzyme is stable. Typical plots of data are shown in Figs. 3 and 4. The enzyme is completely stable from 20 to 35°C, but above 35°C there is loss of activity and the rate of activity loss is greater the higher the temperature.

Table 1 Relative Effect of Temperature on Rate of Transformation of Reactants and on Rate of Denaturation[a]

Temperature (°C)	Relative rates at E_a of: 6000 cal/mol	60,000 cal/mol
−10	1.0	1.0
0	1.55	7.94×10^1
20	3.24	1.26×10^5
40	6.31	1.0×10^8
60	11.4	3.16×10^{10}

[a]Calculated by use of Eq. (21) of this Chapter. The absolute rates of the two reactions would be very different at 0°C, with rate of denaturaton being very much slower than rate of transformation of reactants to product.

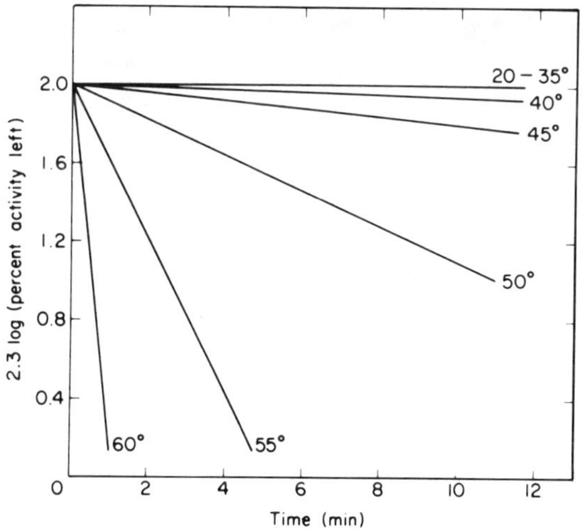

Figure 3 Rate of denaturation of an enzyme at various temperatures. The data are calculated for E_a of 60,000 cal/mol, where the first-order rate constant, k, of denaturation is 0.005, 0.020, 0.090, 0.395, and 1.80 min^{-1} at 40, 45, 50, 55, and 60°C, respectively.

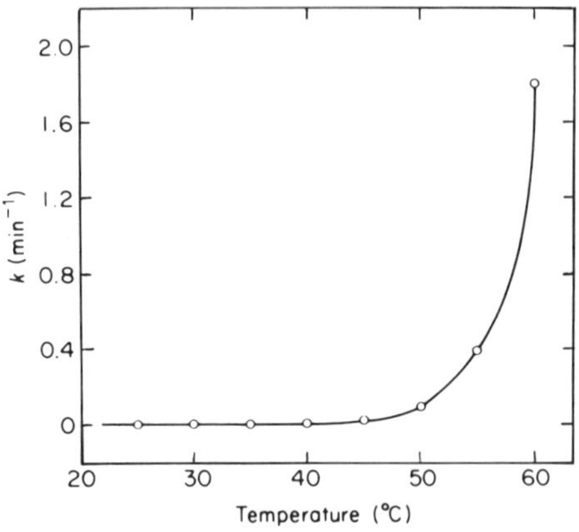

Figure 4 Effect of temperature on stability of an enzyme. The first-order rate constants, *k*, from Fig. 3 are plotted as a function of temperature.

C. Reversible Inactivation of Enzymes

The method described will determine only irreversible loss of activity (irreversible denaturation). In a number of cases (e.g., trypsin and peroxidase) it has been shown that the enzyme can also exist in a reversibly denatured form. This would occur most frequently near the temperature optimum of the enzyme. When the temperature is lowered to determine the amount of activity left, the reversibly denatured enzyme will return to the active form. The method also does not take into account any influence the substrate may have on enzyme stability.

In the case of trypsin there is a marked difference between the solubility of active native enzyme and reversibly denatured enzyme in 0.5 *M* NaCl at pH 2.0 [1]. The active enzyme is completely soluble while the inactive enzyme is completely insoluble. There will also be some irreversibly denatured enzyme present which is also insoluble. The reversibly denatured enzyme may regain activity and solubility in 0.5 *M* NaCl, pH 2.0 on storage at 20°C.

The idea of reversibly inactivated enzymes is not a strange one for the food scientist. Numerous observations have been made on the partial recovery of peroxidase and alkaline phosphatase activities in fruits and vegetables and milk, respectively, on standing after heat treatment. There are several ways in which activity may be reversibly lost during heat treatment and subsequently recovered in part on prolonged storage at lower temperatures. These ways include changes in the secondary and tertiary structures of enzymes, dissociation of subunits in multichain enzymes, and dissociation of essential cofactors during heat treatment.

The process of reversible inactivation may be visualized with an enzyme occurring in several reversible conformational states in solution (Fig. 5) with one of these conformational states active as an enzyme. As the temperature is

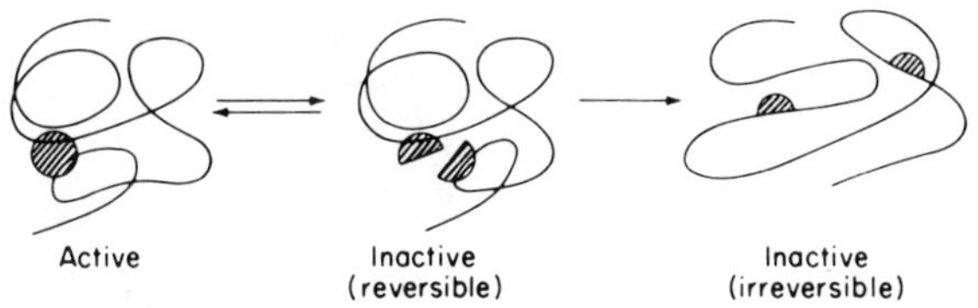

Temperature	Active	Inactive (reversible)	Inactive (irreversible)
1	90 %	10 %	0 %
2	20 %	70 %	10 %
3	0 %	30 %	70 %
4	0 %	0 %	100 %

Figure 5 Schematic representation of reversible and irreversible thermal inactivation of an enzyme. Higher temperatures ($4 > 3 > 2 > 1$) favor reversible and irreversible forms of molecules and the percentage of each type is shown at various temperatures. The shaded circle includes the groups essential for activity. In the reversibly inactive form the active-site groups are not too far separated. In the irreversibly inactive form the active-site groups are widely separated and cannot re-form the active site on cooling. Aggregation of the irreversibly denatured enzyme may also occur.

increased, more of the enzyme is converted to conformational states which are inactive, both reversibly and irreversibly. The amount of enzyme in the irreversibly denatured form will be a function of both temperature and time of application of elevated temperatures. At a sufficiently high temperature all the enzyme will be converted to irreversibly denatured form and there will be no recovery of activity when the solution is stored at a lower temperature. It follows that recovery of activity in a solution where activity is zero immediately after heat treatment will be observed only under those conditions where application of the inactivating reagent (pH, temperature, high salt concentration, etc.) is marginal in destruction of enzyme activity (temperature 3, for example, Fig. 3).

The problem of influence of a reversibly denatured form of enzyme on interpretation of effect of temperature on transformation of substrate to product can be avoided by studying the effect of temperature on this transformation at temperatures below the temperature optimum of the enzyme. The rationale for this is based on the quite different temperature dependences of the two processes. At a temperature below the temperature optimum, little if any denaturation of the enzyme will be observed, particularly if initial rates are used.

III. CHANGES IN SOLUBILITY OF GASES

Enzymes such as polyphenol oxidase, lipoxygenase, and glucose oxidase require O_2 as one of the substrates. A change in temperature of the reaction will cause a change in solubility of O_2 (Table 2).

Table 2 Effect of Temperature on Solubility of CO_2 and O_2 in Water

Temperature (°C)	CO_2 ($M \times 10^2$)	O_2 ($M \times 10^3$)
0	7.60	2.17
10	5.26	1.68
20	3.84	1.35
30	2.86	1.12
40	2.21	0.96
50	1.73	0.83
60	1.31	0.71

Source: Ref. 2.

The effect of temperature on solubility of O_2 (Table 2) can be related to some reported values of K_m for O_2. For wheat lipoxygenase [3] K_m for O_2 is 3×10^{-4} M at 7.2×10^{-3} M sodium linoleate, pH 9.0, and 20°C. The K_m for O_2 is low because the other substrate, sodium linoleate, is in less than saturating amounts. The rates of O_2 uptake in ascorbate oxidation by ascorbic acid oxidase are 8.8 and 16.1 μL/min at 21 (normal composition of air) and 100% oxygen concentration, respectively [4]. K_m for glucose is 0.0042 and 0.013 M at 20 and 100% oxygen concentration, respectively [5], in the glucose oxidase-catalyzed reaction.

IV. EFFECT OF TEMPERATURE ON EQUILIBRIA

A. Theoretical Considerations

In an enzyme·substrate system there are a number of equilibria involved. These might be indicated schematically as

$$E + A \rightleftharpoons EA \tag{1}$$

$$E + P \rightleftharpoons EP \tag{2}$$

$$E + CoF \rightleftharpoons E\cdot CoF \tag{3}$$

$$E + I \rightleftharpoons EI \tag{4}$$

$$H_2E^{n+1} \underset{H^+}{\overset{K_1^E,\, -H^+}{\rightleftharpoons}} HE^n \underset{H^+}{\overset{K_2^E,\, -H^+}{\rightleftharpoons}} E^{n-1} \tag{5}$$

$$A\cdot H_2E^{n+1} \underset{H^+}{\overset{K_1^{EA},\, -H^+}{\rightleftharpoons}} A\cdot HE^n \underset{H^+}{\overset{K_2^{EA},\, -H^+}{\rightleftharpoons}} A\cdot E^{n-1} \tag{6}$$

$$HB^{n+1} \underset{H^+}{\overset{-H^+}{\rightleftharpoons}} B^n \tag{7}$$

where E, A, CoF, P, I, and B are enzyme, substrate, cofactor, product, inhibitor, and buffer, respectively. All of these equilibria are temperature sensitive. In addition, there is an effect on ionization of substrate and product if they have prototropic groups.

The effect of temperature on an equilibrium between two states is quantitatively expressed by the Van't Hoff relationship,

$$\frac{d\ln K}{dT} = \frac{\Delta H}{RT^2} \tag{8}$$

where K is the equilibrium constant, ΔH is change in enthalpy, and R is the universal gas constant (1.98 cal/mol degree). Integration of Eq. (8) between the limits of K_1 and K_2 at T_1 and T_2 yields

$$\log\frac{K_2}{K_1} = \frac{\Delta H}{2.3R}\frac{T_2 - T_1}{T_2T_1} = \mathrm{p}K_1 - \mathrm{p}K_2 \tag{9}$$

Experimentally, ΔH is obtained by determining the effect of temperature on K. An example will illustrate this. The pK_a value of Tris [tris(hydroxymethyl)aminomethane] was determined at several temperatures in the usual manner by titrating Tris, at several temperatures, with a standard solution of HCl and by analyzing the data by one of the methods described in Chapter 10 so as to determine the pK_a values at the different temperatures (Fig. 6).

When the data obtained from Fig. 6 are plotted as $-\log K$ versus $1/T$, Fig. 7 is obtained. ΔH_{ion} calculated from the slope of the line is 11,200 cal/mol from the data presented for ionization of amino group of Tris. Phosphate and carboxylate groups have ΔH values near zero.

The method is equally valid for determination of ΔH of any of the other equilibria described by Eqs. (1) to (7).

The change in Gibbs free energy of the system, ΔG, is calculated by use of

$$\Delta G = RT\ln K_{\text{eq}} \tag{10}$$

where K_{eq} is $(\text{product})_{\text{eq}}/(\text{reactant})_{\text{eq}} = k_1/k_{-1}$:

$$\text{reactant} \underset{k_{-1}}{\overset{k_1}{\rightleftharpoons}} \text{product} \tag{11}$$

The change in entropy of the system, ΔS, is calculated from the relationship

$$\Delta G = \Delta H - T\Delta S \tag{12}$$

B. Methods of Determination of Effect of Temperature on Equilibria

Since a change in temperature affects so many equilibria in an enzyme-catalyzed reaction, is it ever possible to measure the true effect of temperature on transfor-

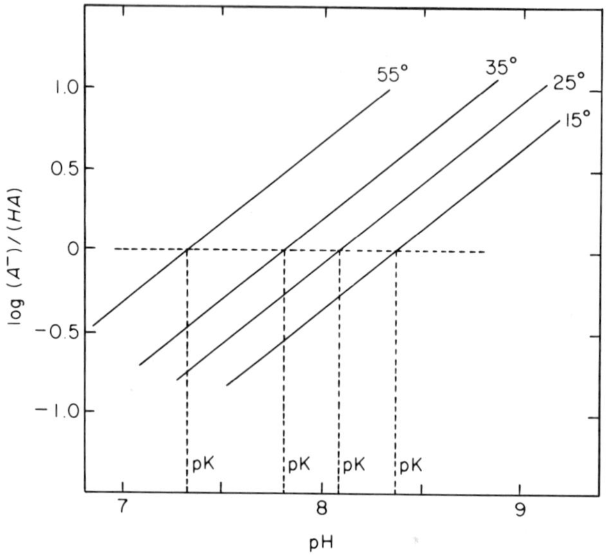

Figure 6 Titration curves for Tris at several different temperatures. The data are plotted according to the Henderson–Hasselbalch equation $\log(A^-)/(HA) = pH - pK$, where (A^-) and (HA) are concentrations of R—NH_2 and R—NH_3^+ species, respectively. The pK values are given at pH value where $\log(A^-)/(HA) = 0$ (i.e., where the lines cross the zero horizontal line).

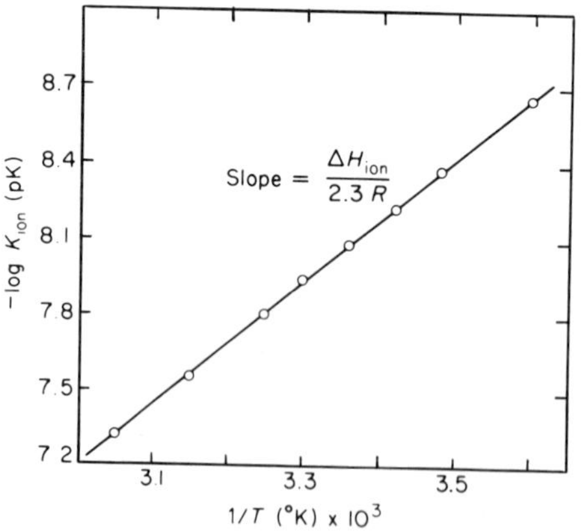

Figure 7 Effect of temperature on ionization constant, K_{ion}, of Tris. pK values determined as in Fig. 6 are plotted as a function of $1/T$(K) to permit calculation of ΔH_{ion}.

mation of reactant to product? It is possible. Let us discuss methods of how to minimize or account for the influence of these perturbations on the rate of transformation of reactant to product.

Effect of changes in temperature on pH of a buffer is easily taken care of experimentally. The pH of the reaction mixture should be measured and adjusted to the desired pH at the temperature at which the experiment is to be performed. With amino or imidazole buffers, for example, the pK_a, and thus the pH, will change 0.24 or 0.17 unit, respectively, per 10°C change in temperature (between 27 and 37°C). The pH values of phosphate and carboxylate buffers are not appreciably affected by changes in temperature.

Effect of temperature on reaction of enzyme with inhibitor [Eq. (4)] can be eliminated by removal of all inhibitors from the reaction. For purified system this is no problem, but in crude systems there may be a number of inhibitory substances present.

Contribution of the reactions described by Eqs. (1), (3), and (5) to the observed effect of temperature on rate of reaction may be eliminated by saturation of the enzyme with substrate [$(A_0) \gg K_m$] and cofactor [$(CoF_0) \gg K_d$] at all temperatures or by determination of V_{max} in the usual manner. It is necessary to determine the concentrations needed for saturation at all temperatures under conditions of the experiment since the degree of saturation will change with temperature. Influence of temperature changes on the equilibrium involving enzyme and product [Eq. (2)] can be eliminated by use of initial rates.

Analysis of data with respect to the equilibria described by Eqs. (5) and (6) can give valuable information on the nature of the prototropic groups involved in the active site of the free enzyme or enzyme·substrate complex. The data are also useful in suggesting how the contribution of this ionization to the observed effect of temperature on transformation of substrate to product can be minimized.

Data for hydrolysis of α-*N*-benzoyl-L-argininamide by papain is used to illustrate the effect of temperature on ionization of prototropic groups in the active site of an enzyme (Fig. 8). As shown in Chapter 10, a plot of $V_{max}^{H^+}/K_m^{H^+}$ versus pH permits a determination of the essential prototropic groups in the active site of the free enzyme. Data of Fig. 8 were obtained by determining $V_{max}^{H^+}/K_m^{H^+}$ at a number of pH values over the range 3 to 9 and at three different temperatures: 5, 38, and 66°C. At any pH, $V_{max}^{H^+}/K_m^{H^+}$ is higher the higher the temperature. In order to compare the data, relative k_1 values were calculated (where $k_1 = V_{max}^{H^+}/K_m^{H^+}$) by letting the highest value at any temperature be 100 and calculating the others relative to this.

The curves are bell shaped at all temperatures, indicating the involvement of two prototropic groups in the active site of the free enzyme. The pK value of the prototropic group on the acid side does not change with temperature, whereas the one on the basic side does. Both the pK value of 4 and the ΔH_{ion} value of zero support the interpretation of this group as a carboxyl group. The other prototropic

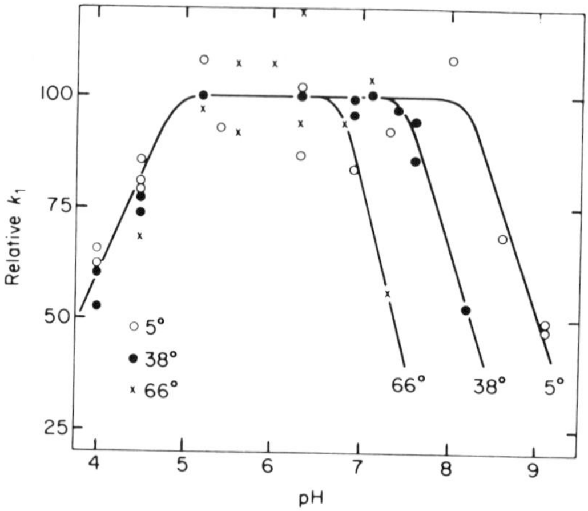

Figure 8 Effect of temperature (and pH) on relative k_1 for papain-catalyzed hydrolysis of α-*N*-benzoyl-L-argininamide. The k_1 values were calculated from $V_{max}/K_m^{H^+}$, where the maximum value at each temperature was set to 100. (From Ref. 6, p. 20, by courtesy of the American Society of Biological Chemists.)

group has pK values of 9.0, 8.2, and 7.4 at 5, 38, and 66°C, respectively. ΔH_{ion} calculated by use of Eq. (9) is 7900 cal/mol. Both the pK and ΔH_{ion} support the postulation that this prototropic group is a sulfhydryl group. ΔH_{ion} values for several prototropic groups are given in Table 2, Chapter 10.

In order to determine the effect of temperature on the rate of conversion of $EA \xrightarrow{k_2} E + P$, the experiment should be run with $(A_0) \geqslant 100K_m$ and at the pH optimum of the enzyme. Between pH 5.5 and 6.5 (pH optimum) there is no effect of temperature on ionization of prototropic groups. The only effect of temperature in this region is on the rate of transformation of substrate to products. Therefore, in determination of effect of temperature on conversion of enzyme·substrate complex to products, it is essential to work at the pH optimum to eliminate as much as possible corrections for effect of temperature on ionization of prototropic groups.

V. EFFECT OF TEMPERATURE ON TRANSFORMATION OF ENZYME-SUBSTRATE COMPLEX TO PRODUCTS

When the effects of temperature on stability and equilibria have been accounted for as described above and in the presence of saturating amounts of substrate (V_{max} determined), the effect of temperature on the rate-determining step in conversion of enzyme·substrate complex to products can be determined. This rate-determining step may involve covalent bond formation, covalent bond breakage, dissociation of product from enzyme, or enzyme conformational changes. The rate-determining step may be different with different substrates, at different pH values, and may change with temperature. If more than one step in a reaction contributes to the observed rate of a reaction, the observed effect of temperature will be a composite of effect of temperature on each of those steps.

A. Quantitation of Effect of Temperature on Rates of Reactions

A number of methods are available for quantitatively expressing the effect of temperature on rate of transformation of substrate to product. Each of these methods will be examined briefly for what it has to offer.

1. Q_{10}

A term often used in biology is that of Q_{10}. Q_{10} is defined as the increase in rate of a reaction for a 10°C increase in temperature:

$$Q_{10} = \frac{\text{rate}_{T+10°}}{\text{rate}_{T}} \tag{13}$$

It is determined by observing the rates of a reaction at two temperatures 10°C apart. The Q_{10} values for most chemical and enzymatic reactions fall within the range 1.5 to 3. Although the difference between a Q_{10} of 2 and 3 does not look large, consider the relative rates (based on 1 at 0°C) of 243 versus 32 at 50°C for a Q_{10} of 3 and 2, respectively (Table 3). Such comparisons point out that the relationship between Q_{10} and rate is logarithmic rather than linear:

Table 3 Relative Rates at Different Q_{10} Values[a]

Temperature (°C)	Relative rates at Q_{10} of: 2	2.5	3	4
0	1	1	1	1
10	2	2.5	3	4
20	4	6.25	9	16
30	8	15.6	27	64
50	32	97.7	243	1,020
70	128	610	2,190	16,400
100	1,024	9,540	59,000	1,050,000

[a]Calculated by use of Eq. (14). The absolute rates at different Q_{10} values are not to be interpreted as being equal at 0°C.

$$\Delta\text{rate} = Q_{10}^n, \qquad \text{where } n = \Delta T/10°\text{C} \tag{14}$$

When $Q_{10} = 2$, a 10°C rise in temperature will double the rate of a reaction. It takes a doubling of reactant concentration to double the rate of a first-order process. Therefore, the effect of temperature on rates of reactions are extremely important.

2. Arrhenius Equation

The first quantitative, experimental formulation of the dependence of reaction rates on temperature was made by Hood. His formulations were later extended and made into a general equation by Arrhenius. The dependence of the specific reaction rate constant, k, on temperature is given by

$$k = Ae^{-E_a/RT} \tag{15}$$

which may be rewritten as [taking logarithms of both sides of Eq. (15)]

$$\log k = \log A - \frac{E_a}{2.3RT} \tag{16}$$

where k is a specific reaction rate constant at some temperature, T(K), and A and E_a are empirical constants. E_a is usually referred to as the Arrhenius *activation energy* and A is called the *frequency factor* or Arrhenius factor.

Equation (15) may be integrated between the limits of k_1 and k_2 at T_1 and T_2 to give

$$\log \frac{k_2}{k_1} = \frac{E_a}{2.3R} \frac{T_2 - T_1}{T_2 T_1} \tag{17}$$

The similarity of this equation to the Van't Hoff equation [Eq. (9)] should be noted. Any parameter that is proportional to k may be used in Eq. (17) *as long as the order of the reaction is the same at both temperatures*.

E_a is best evaluated by plotting log k versus $1/T$ (Fig. 9). This plot is valid for the effect of temperature on any rate process. An approximate relationship between Q_{10} and E_a can be calculated when $T_2 - T_1 = 10°\text{C}$:

$$Q_{10} = e^{10E_a/RT^2} \tag{18}$$

However, since Q_{10} is dependent on temperature whereas E_a is essentially independent of temperature, a constant relationship is not found over a wide range of temperature.

What is E_a? E_a may be defined as the minimum quantity of energy that a reactant molecule must have in order to be converted to a product molecule.

3. Collision Theory of Chemical Reactions

According to the collision theory of chemical reactions, the relationship between T and the specific reaction rate constant, k, is

$$k = PZe^{-E_a/RT} \tag{19}$$

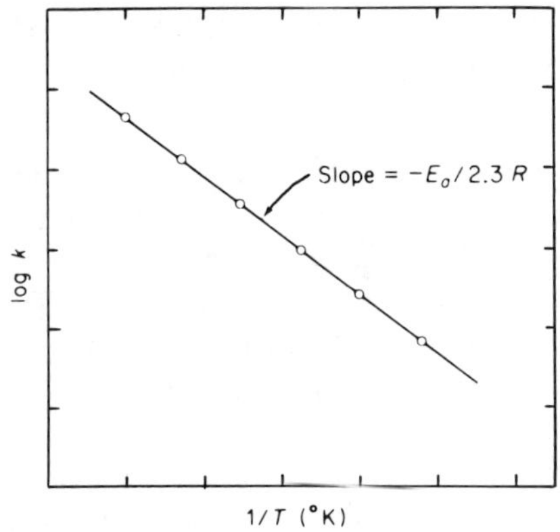

Figure 9 Effect of temperature on rate constant of a reaction. Plotted as log k versus $1/T$ (K) to permit determination of E_a as shown.

where P is the probability of a reaction occurring when molecules collide, Z is the number of collisions per second per unit volume (collision rate) and $e^{-E_a/RT}$ is the fraction of the total number of molecules with an energy of E_a or greater. P is a collection of factors, including the probability of proper orientation of the molecules when they collide.

Equation (19) permits us to take a closer look at what temperature really affects most in changing the rate of a reaction. Given that E_a is 12,000 cal/mol, what is the relative rate of a reaction at 27 and 37°C? Assume that P is temperature independent, which is nearly but not entirely true for an enzyme-catalyzed reaction. The collision rate, Z, is a kinetic energy factor and as such is changed by $\Delta T/273$. A 10°C change in temperature will change Z by 10/273 or 3.7%. This is a negligible change and certainly not capable of explaining a Q_{10} of 2 to 3.

What about the factor $e^{-E_a/RT}$, which is the fraction n' of the total number of molecules n with energy E_a or greater?

$$\frac{n'}{n} = e^{-E_a/RT} \tag{20}$$

At 27°C,

$$\log\frac{n'}{n} = -\frac{E_a}{2.3RT} = -\frac{12{,}000 \text{ cal/mol}}{2.303 \times 1.98 \text{ cal/mol/deg} \times 300°} = -8.8$$

and $n'/n = 1.6 \times 10^{-9}$. At 37°C $n'/n = 3.2 \times 10^{-9}$ and the ratio of the two $[(n'/n)_{37°C}/(n'/n)_{27°C}]$ is 2.

Therefore, the rate of the reaction will be doubled by a 10°C increase in temperature. All except a very small amount of the effect of change in temperature on rate is on the fraction of the total number of molecules that have energy E_a or greater.

4. Theory of Absolute Reaction Rates

The Arrhenius empirical equation and the collision theory equation are not completely adequate to explain all the experimental observations concerning the effect of temperature on the rate of a reaction, particularly in solution. Consider the course of events in conversion of a reactant to product depicted in Fig. 10. The energy content of A and P can be obtained by calculation from their heats of formation or by measuring their heat of combustion in a calorimeter. The difference in energy between A and P may also be determined by measuring K_{eq}, ΔH, ΔG, and ΔS by methods described earlier (Section IV.A). This difference in energy content of A and P is a constant dependent only on the chemical nature of A and P and is independent of the mechanism (pathway) by which A is converted to P.

According to the theory of absolute reaction rates (transition-state theory), the energy content of A must be increased until it has sufficient energy to be at the top of the peak as $A^{\ddagger}$ (Fig. 10). For A to be converted to P, it must attain the amount of energy to become $A^{\ddagger}$. Molecules of A with less than this amount of energy will "slide" back down the peak on the wrong side. This, then, implies an equilibrium between molecules "climbing" to the top of the peak and those falling back down:

$$A \underset{k_{-1}}{\overset{k_1}{\rightleftharpoons}} A^{\ddagger} \rightarrow P \tag{21}$$

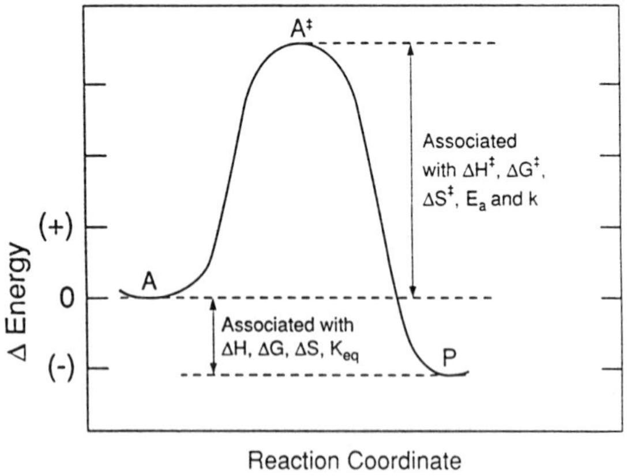

Figure 10 Change in energy in the pathway of reaction during conversion of reactant A to product P, according to theory of absolute reaction rates. $A^{\ddagger}$ indicates the activated molecules, which are intermediates in conversion of A to P.

According to the theory, all molecules, once they are committed to form product (i.e., are in form $A^{\ddagger}$), do so at a rate independent of temperature and the nature of the molecule. Therefore, a rate for the reaction may be written

$$k = MK^{\ddagger} = M\frac{(A^{\ddagger})_{eq}}{(A)_{eq}} = M\frac{k_1}{k_{-1}} \tag{22}$$

where M is a proportionality constant and $K^{\ddagger}$ is an equilibrium constant [Eq. (11)].

In analogy with the laws of thermodynamics for unactivated molecules, we may define an expression for the transition state where

$$K^{\ddagger} = \frac{(A^{\ddagger})_{eq}}{(A)_{eq}} = \frac{k_1}{k_{-1}} = e^{-\Delta G^{\ddagger}/RT} \tag{23}$$

According to the theory of absolute reaction rates M has the definition of $RT/Nh = k_BT/h$, where R is the universal gas constant per mole (8.31×10^7 erg deg^{-1} mol^{-1} for this particular equation RT/Nh), N is Avogadro's number (6.023×10^{23} molecules mol^{-1}), h is the Planck constant (6.624×10^{-27} erg sec), and k_B ($= R/N$) is the Boltzmann constant (1.380×10^{-16} erg deg^{-1}). Therefore, Eq. (22) may be written as

$$k = \frac{k_BT}{h} e^{-\Delta G^{\ddagger}/RT} \tag{24}$$

which can be rearranged to the form of $\Delta G^{\ddagger} = -RT \ln (kh/k_BT)$.

In an analogous fashion one can, according to the laws of thermodynamics, define a relations among $\Delta G^{\ddagger}$, $\Delta H^{\ddagger}$, and $\Delta S^{\ddagger}$:

$$\Delta G^{\ddagger} = \Delta H^{\ddagger} - T\Delta S^{\ddagger} \tag{25}$$

Combining Eqs. (24) and (25), dividing through by T, and taking logarithms of both sides gives Eq. (26), the equation of a straight line.

$$2.3 \log \frac{k}{T} = 2.3 \log \frac{k_B}{h} + \frac{\Delta S^{\ddagger}}{R} - \frac{\Delta H^{\ddagger}}{RT} \tag{26}$$

To determine $\Delta H^{\ddagger}$, the specific reaction rate constant is determined at a number of temperatures and plotted according to Eq. (26) (Fig. 11). $\Delta H^{\ddagger}$ is independent of the rate parameter plotted as long as the parameter is proportional to the specific reaction rate constant at all temperatures. Further calculations to obtain $\Delta S^{\ddagger}$ and $\Delta G^{\ddagger}$, however, require use of the specific reaction rate constant. A plot of $\log k$ versus $1/T$ gives E_a, not $\Delta H^{\ddagger}$. The two are related in that $E_a = \Delta H^{\ddagger} + RT$.

$\Delta G^{\ddagger}$ is calculated from the specific reaction rate constant k at a given temperature according to Eq. (24). It cannot be obtained from a plot of $\log k$ versus $1/T$, which gives E_a. Having $\Delta H^{\ddagger}$ and $\Delta G^{\ddagger}$, $\Delta S^{\ddagger}$ can be calculated from Eq. (25).

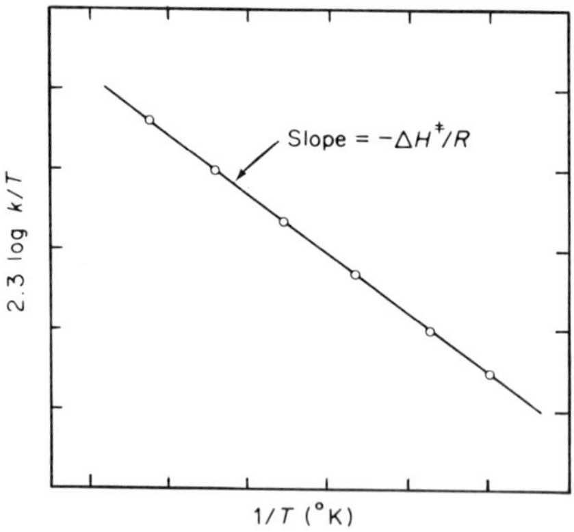

Figure 11 Effect of temperature on rate constant of a reaction plotted according to Eq. (26) so as to determine $\Delta H^{\ddagger}$.

B. Application of Theory to an Enzyme-Catalyzed Reaction

The simplest equation (assuming reversibility) to describe an enzyme-catalyzed reaction is

$$E + A \rightleftharpoons EA \rightleftharpoons EP \rightleftharpoons E + P \tag{27}$$

However, this equation deals only with unactivated molecules and intermediates. Incorporation of activated-state intermediates in the equation gives

$$E + A \rightleftharpoons (E \cdot\cdot A)^{\ddagger} \rightleftharpoons EA \rightleftharpoons (EA \cdot\cdot EP)^{\ddagger} \rightleftharpoons EP \rightleftharpoons (E \cdot\cdot P)^{\ddagger} \rightleftharpoons E + P \tag{28}$$

Plotted in terms of an energy versus reaction coordinate, this give the diagram shown in Fig. 12. Following the pathway of the reaction in the *forward direction* there are seven distinct groups of parameters to be dealt with. Each of these will be considered separately according to the numbering used in Fig. 12.

1. $E + A \overset{K_{eq}}{\rightleftharpoons} E + P$ (or $A \rightleftharpoons P$)

This represents the overall energy difference between unactivated reactant and product. The difference in energy is constant regardless of mechanism of reaction. The overall reaction is described by the thermodynamic parameters, K_{eq}, ΔH, ΔS, and ΔG and the relationships discussed in Eqs. (9), (10), and (12). ΔH may also be obtained from the heats of formation of A and P.

2. $E + A \overset{K_{eq}}{\rightleftharpoons} EA$

This equation shows unactivated molecules in formation of enzyme·substrate complex from enzyme plus substrate. The parameters involved are K_s, ΔH, ΔG,

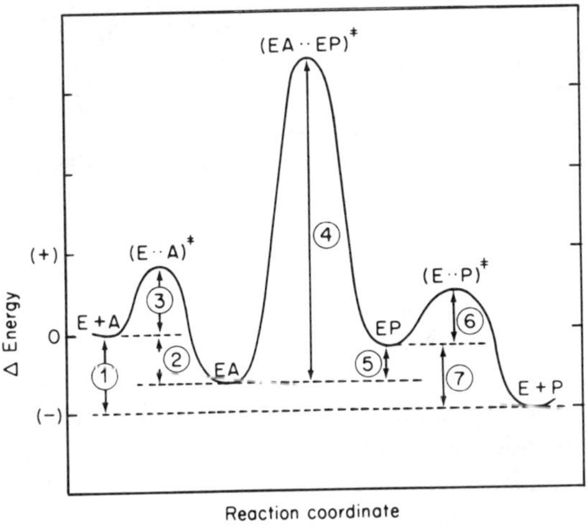

Figure 12 Change in energy in the pathway of reaction during conversion of A to P in an enzyme-catalyzed reaction. The numbers refer to groups of thermodynamic and transition-state constants in the forward direction.

and ΔS, which may be obtained by use of Eqs. (9), (10), and (12). If the experimental conditions can be arranged so that $K_m = K_s$, the effect of temperature of K_s may be determined as shown in Fig. 7 so as to obtain ΔH.

K_s is usually in the range 1×10^{-4} to 1×10^{-1} M. K_s = (E)(A)/(EA), whereas the K_{eq} constant in Eq. (11) would be defined as (EA)/(E)(A) [(products)/(reactants)] for the forward reaction. Therefore, in calculating ΔG one uses $1/K_s = K_{eq}$; therefore, ΔG would be in the range 1.4 to 5.5 kcal/mol at 25°C.

3. $E + A \xrightarrow{k} (E \cdot\cdot A)^{\ddagger}$

This equation shows formation of the activated-state intermediate $(E \cdot\cdot A)^{\ddagger}$. The description E··A is used to indicate the approach of E to A during formation of the adsorptive complex, EA. In most enzyme-catalyzed reactions the rate of formation of the adsorptive complex is rapid, 10^6 to 10^9 M^{-1} sec^{-1} and is diffusion controlled. The appropriate parameters involved here are k_1 (specific reaction rate constant for formation of complex), $\Delta H^{\ddagger}$, $\Delta G^{\ddagger}$, and $\Delta S^{\ddagger}$. $\Delta H^{\ddagger}$, $\Delta G^{\ddagger}$, and $\Delta S^{\ddagger}$ are obtained by use of Eqs. (24) to (26). The difficulty here is in measuring the effect of temperature on k_1 since the rate is so fast. Stopped-flow or other rapid reaction techniques must be used.

4. $EA \xrightarrow{k} (EA \cdot\cdot EP)^{\ddagger}$

This step includes the bond-making and/or bond-breaking step in the enzyme-catalyzed reaction. When it is the rate-determining step in conversion of reactant to product, it is the step that is measured whenever the activation energy of an enzyme-catalyzed reaction is determined properly. From the effect of temperature on the specific reaction rate constant for this process, $\Delta H^{\ddagger}$, $\Delta G^{\ddagger}$, and $\Delta S^{\ddagger}$ are calculated from Eqs. (24) to (26). $\Delta H^{\ddagger}$ for this step is usually in the range 6 to 15 kcal/mol. If more than a single step is rate determining, an "average" effect of temperature is observed experimentally.

5. $EA \overset{K_{eq}}{\rightleftharpoons} EP$

This reaction involves unactivated complexes and represents the energy difference between EA and EP (Fig. 12). Neither of these adsorptive complexes is directly observable in an experiment, so that determination of K_{eq} and the effect of temperature on K_{eq} so as to determine ΔH, ΔG, and ΔS is difficult experimentally. Fortunately, these values can be calculated when all the other steps are fully described. In those enzyme systems where the reaction is freely reversible and the bond-making and/or bond-breaking step is rate determining, the necessary data can be obtained directly.

6. $EP \overset{k}{\rightarrow} (E \cdot\cdot P)^{\ddagger}$

This step represents the transition state in dissociation of enzyme and product. Generally, the specific reaction rate constants for dissociation of enzyme·substrate complexes are in the range 10^3 to 10^5 sec^{-1}. As a first approximation, the rate constant for dissociation of enzyme and product is expected to be on the same order of magnitude, if not faster. The effect of temperature on the rate constant of this step is needed so that $\Delta G^{\ddagger}$, $\Delta H^{\ddagger}$, and $\Delta S^{\ddagger}$ can be calculated. Values for this step are shown below for α-chymotrypsin.

7. $EP \overset{K_{eq}}{\rightleftharpoons} E + P$

This step represents the equilibrium between unactivated enzyme and product and unactivated enzyme·product adsorptive complex. It is a reasonably easy step to characterize thermodynamically since K_i [= (E)(P)/(EP)] can be determined by using the product as an inhibitor in initial rate studies. The data are then plotted as $1/v_0$ versus $1/(A_0)$ so as to obtain K_i at different temperatures. From these data, ΔH, ΔG, and ΔS can be calculated.

Figure 12 shows only the steps in the forward direction. A complete description of the system also requires a determination of the parameters for the seven steps in the reverse direction; see Fig. 13, where the steps are marked from P to A. K_{eq} for the forward reaction is equal to $1/K_{eq}$ for the back reaction and the absolute magnitudes of ΔH, ΔG, and ΔS are the same. They differ only in sign. Similarly, steps 1, 2, 5, and 7 for the forward reaction are related to steps 14, 13, 10, and 9, respectively, for the back reaction. These relationships are summarized in Table 4.

No similar direct relationship exists for steps 3, 4, and 6 for the forward reaction and 8, 11, and 12 for the back reaction involving activated-state species. The energy

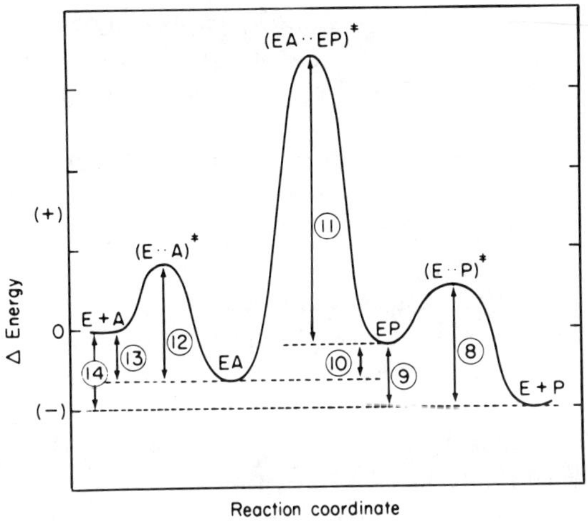

Figure 13 Change in energy and pathway of reaction during conversion of P to A in an enzyme-catalyzed reaction. The diagram is to be read from right to left. The numbers refer to groups of thermodynamic and transition state constants in the reverse direction from that in Fig. 12.

Table 4 Some Relationships Among Thermodynamic and Transition-State Parameters in Reactions Shown in Figs. 12 and 13[a]

Step	Relationship
1, 2, 5, and 7	$\Delta H_1 = \Delta H_2 + \Delta H_5 + \Delta H_7^{b}$
14, 13, 10, and 9	$\Delta H_{14} = \Delta H_{13} + \Delta H_{10} + \Delta H_9^{c}$
1 and 14	$\Delta H_1 = -\Delta H_{14}$
2 and 13	$\Delta H_2 = -\Delta H_{13}$
5 and 10	$\Delta H_5 = -\Delta H_{10}$
7 and 9	$\Delta H_7 = -\Delta H_9$
12, 2, and 3	$\Delta H^{\ddagger}_{12} = \Delta H_2 + \Delta H^{\ddagger}_3$
4, 10, and 11	$\Delta H^{\ddagger}_4 = \Delta H_{10} + \Delta H^{\ddagger}_{11}$
8, 6, and 7	$\Delta H^{\ddagger}_8 = \Delta H_7 + \Delta H^{\ddagger}_6$

[a]The steps are numbered as in Figs. 12 and 13. ΔH values are used to illustrate the relationships. Similar relationships could be written for ΔS and ΔG values.
[b]Numerical values of ΔH_1, ΔH_2, and ΔH_7 are negative, while ΔH_5 is positive.
[c]Numerical values of ΔH_{14}, ΔH_{13}, and ΔH_9 are positive while ΔH_{10} is negative.

requirement for formation of (E··P)‡ in the back reaction is considerably greater than that for formation of (E··A)‡ in the forward reaction (Figs. 12 and 13). However, there are relationships between the forward and back reaction parameters (Table 4).

It takes 14 sets of thermodynamic and transition-state constants to describe fully the effect of temperature on the reaction shown by Eq. (28). It is not surprising that very few complete studies on the effect of temperature on rates of enzyme-catalyzed reactions have been published. Two such reports are those for fumarase [7] and for α-chymotrypsin [8]. Changes in free energy ($\Delta G^‡$ or ΔG for the appropriate steps) for hydrolysis of *N*-acetyl-L-tryptophan methyl ester and *N*-acetyl-L-tryptophanamide at 25°C by α-chymotrypsin are shown in Figs. 14 and 15.

It is readily seen from Figs. 14 and 15 why the rate-determining step in the forward direction is that controlled by k_2 for amide hydrolysis while it is k_3 for ester hydrolysis. By use of Eq. (24) we can calculate that the rate of formation of acylenzyme will be 2.96×10^4 times more rapid for the ester as for the amide at 25°C. Similarly, we can calculate the relative rates of acylation (k_2) to deacylation (k_3) for the two substrates where we find relative rates of 1.48×10^3 for the ester and 0.05 for the amide.

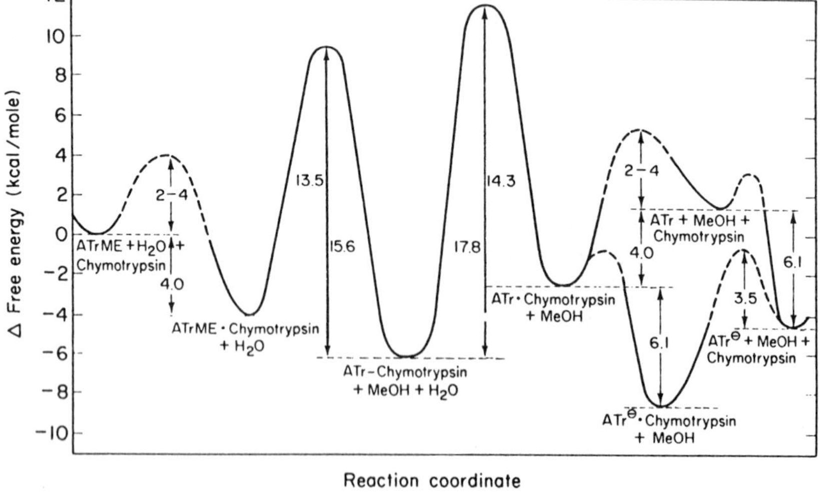

Figure 14 Standard free-energy change (ΔG^o or $\Delta G^‡$) versus reaction coordinate for the α-chymotrypsin-catalyzed hydrolysis of *N*-acetyl-L-tryptophan methyl ester at 25°C. (From Ref. 8, p. 3719, by courtesy of the American Chemical Society.)

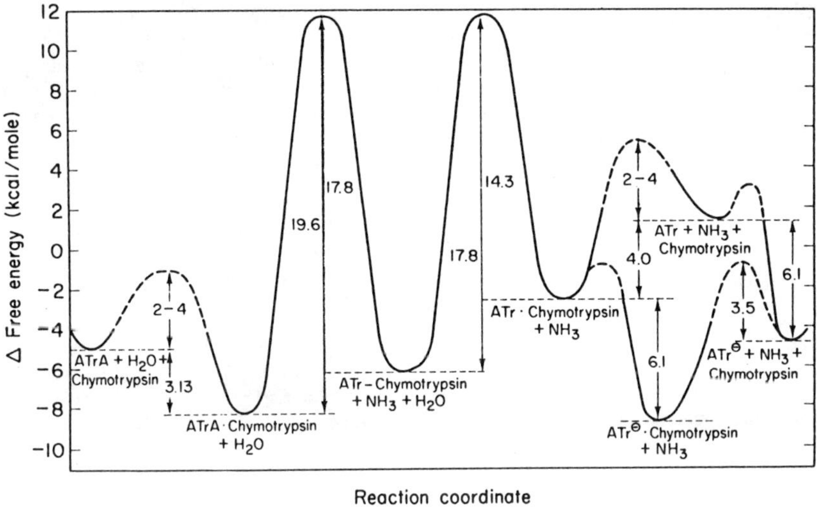

Figure 15 Standard free-energy change ($\Delta G°$ or $\Delta G^{\ddagger}$) versus reaction coordinate for the α-chymotrypsin-catalyzed hydrolysis of *N*-acetyl-L-tryptophan amide at 25°C. (From Ref. 8, p. 3720, by courtesy of the American Chemical Society.)

C. Importance of Lowering Activation Energy

Enzymes decrease the activation energy required for conversion of reactants to products. By whatever mechanism the enzyme uses to accomplish this feat, it is this lowering of E_a that is responsible for the large rate enhancement observed. The great difference in rate that is brought about by a small change in activation energy is demonstrated by the data in Table 5. For example, in the decomposition of H_2O_2, lowering E_a from 18.0 to 6.4 kcal/mol increases the rate by a factor of 3.47×10^8.

VI. QUANTITATION OF THERMAL DENATURATION OF ENZYMES

The effect of temperature on stability of enzymes was covered in Section II in a descriptive fashion. It is essential to quantitate this process based on the background of Section V. The effect of temperature on rate of an enzyme-catalyzed reaction over a wide temperature range is shown in Fig. 16. The data points follow two straight lines, one with a positive slope, the other with a negative slope, connected with a curvilinear portion. The point at which $dy/dx = 0$ is the temperature optimum of the enzyme.

The positive slope of the data gives a measure of the effect of temperature on

Table 5 Effect of Catalyst on E_a and on Relative Rates of Some Reactions

Substrate	Catalyst	E_a (kcal/mol)	n'/n (25°C)	Relative rates[a] (25°C)
H_2O_2	None	18.0	5.62×10^{-14}	1.00
	I^-	13.5	1.16×10^{-10}	2.07×10^{3}
	Catalase	6.4	1.95×10^{-5}	3.47×10^{8}
Sucrose	H^+	25.6	1.44×10^{-19}	1.00
	Invertase	11.0	8.04×10^{-9}	5.58×10^{10}
Carbonic acid	None	20.5	8.32×10^{-16}	1.00
	Carbonic anhydrase	11.7	2.46×10^{-9}	2.96×10^{6}
Urea	H^+	24.5	9.33×10^{-19}	1.00
	Urease	8.7	3.96×10^{-7}	4.25×10^{11}

[a]Relative rates were calculated by use of Eq. (21) and are approximate only, since differences in $\Delta S^{\dagger}$ are ignored.

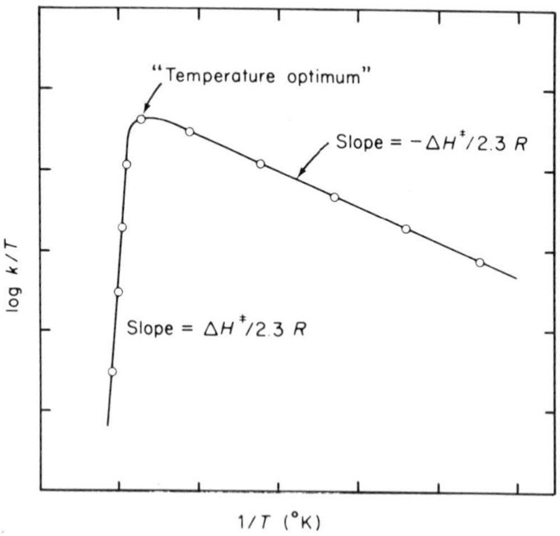

Figure 16 Effect of temperature on rate of an enzyme-catalyzed reaction. The data presented cover a wide range of temperatures so as to demonstrate effect of temperature on denaturation of enzyme as well as the effect on conversion of substrate to product. Data plotted according to Eq. (26).

Table 6 Transition State Denaturation Constants for Various Proteins

Substance	$\Delta H^{\ddagger}$ (cal/mol)	$\Delta S^{\ddagger}$ (eu)[a]	Number of bonds broken[b]	$\Delta G^{\ddagger}$ (25°C) (cal/mol)
Lipase, pancreatic	45,400	68.2	9	25,100
Amylase, malt	41,600	52.3	8	26,000
Pepsin	55,600	113.3	11	21,800
Egg albumin	132,000	315.7	26	37,900
Hemoglobin	75,600	152.7	15	30,100
Peroxidase, milk	185,300	466.0	37	46,400
Rennin	89,300	208.0	18	27,300
Trypsin	40,200	44.7	8	26,900
Invertase, yeast				
pH 5.7	52,400	84.7	10	27,200
pH 5.2	86,400	185.0	17	31,300
pH 4.0	110,400	262.5	22	32,200
pH 3.0	74,400	152.4	15	29,000

Source: Modified from Ref. 9.

[a]$\Delta S^{\ddagger}$ in cal/mol degree.

[b]Number of noncovalent bonds broken on denaturation = $\Delta H^{\ddagger}/5000$, where the average $\Delta H^{\ddagger}$ per bond is assumed to be 5000 cal/mol.

the rate of enzyme denaturation. The slope of the line is very steep, indicating that effect of temperature on rate of denaturation is quite pronounced. $\Delta H^{\ddagger}$ values range from 40,200 cal/mol for trypsin to 185,300 cal/mol for peroxidase (Table 6).

The entropy values, $\Delta S^{\ddagger}$, are also included in Table 6. $\Delta G^{\ddagger}$ can be calculated from $\Delta H^{\ddagger}$ and $\Delta S^{\ddagger}$. The $\Delta G^{\ddagger}$ values for the enzymes are much more similar than are $\Delta H^{\ddagger}$ values. They are much smaller and can account for the rates of denaturation observed experimentally.

A number of noncovalent bonds are broken when a protein is denatured. These bonds involve electrostatic interactions, hydrophobic interactions, hydrogen bonds, and van der Waals forces. The exact number of bonds broken during denaturation cannot be calculated. However, if an average value of 5000 cal/mol is used for each bond, quite a few bonds are broken in the denaturation process (Table 6).

$\Delta H^{\ddagger}$ is a function of pH (see the data for invertase in Table 6). This is because of the importance of electrostatic bonds in maintaining the structure of a protein.

VII. CONCLUSIONS

The effect of temperature on rates of enzyme-catalyzed reactions has been treated in a rigorous fashion, which must be done if such data are to be used to draw conclusions about the mechanism of a reaction or about specific steps within the catalytic process. This treatment should not frighten one away from using temperature as an important parameter in describing the behavior of an enzyme system in vivo, in crude homogenates, or in model systems. The researcher should be aware of limitations involved in interpreting the data and realize that data are valid only for the special conditions under which they were obtained. It is a mistake to extrapolate such data to draw conclusions about what may be happening in steps of the reaction not adequately delineated by the experimental design. For these reasons, the research worker is encouraged to state explicitly all experimental conditions used so that other workers may draw the proper conclusions about the application of the data to their system.

REVIEW QUESTIONS

1. An enzyme is known to catalyze the hydrolysis of a substrate by the following reaction scheme:

$$E + A \underset{k_{-1}}{\overset{k_1}{\rightleftharpoons}} EA \xrightarrow{k_2} \underset{\displaystyle P_1}{\underset{+}{E'A}} \xrightarrow{k_3} E + P_2$$

E_a for the k_2 and k_3 steps were determined to be 3000 and 12,000 cal/mol, respectively.

(a) Assuming the enzyme to be stable, $\Delta S^‡$ to be independent of change in temperature, and k_2 to be 100 sec^{-1} at 10°C, calculate the relative rates of all possible combinations of rate constants, k_2/k_3, k_2'/k_2, k_3'/k_3, and so on, where k_2 and k_3 are at 10°C and k_2' and k_3' are at 50°C.

(b) Draw a graph of log k_{cat} (= log $[k_2k_3/(k_2 + k_3)]$) versus $1/T$. State all assumptions made in drawing the graph and how you would interpret the graph.

2. Explain the following observations or calculate the results requested.

(a) All other conditions being constant, an enzyme-catalyzed reaction was performed at two temperatures which differed by 20°C. The initial velocities of the reactions were the same at both temperatures.

(b) $\Delta H^‡$ for an enzyme-catalyzed reaction was independent of temperature.

(c) At 25°C, ΔG for the conversion of A to P was −12,000 cal/mol for both an enzyme-catalyzed reaction and for the same reaction performed with HCl as the catalyst. Calculate the relative rates of these two reactions at 35°C.

3. ΔH_{ion} of a group on an enzyme was 7500 cal/mol. If the pK_a value of this group were 7.00 at 20°C, what would be the pK_a value of the group at 30°C?

4. E_a for an enzyme-catalyzed reaction was 7000 cal/mol and for denaturation of enzyme was 50,000 cal/mol. The enzyme had an observed temperature optimum of 40°C. Calculate the two temperatures at which the enzyme would have 30% of the maximum observed initial activity.

5. An enzyme-catalyzed reaction was 2.5 times as fast at 40°C as at 20°C.

(a) Calculate the activation energy for this reaction assuming linearity of $\log k$ versus $1/T$.

(b) What is the Q_{10} of the reaction?

(c) What can you say about ΔG of the reaction?

(d) Suppose that the activation energy had been measured between 25 and 35°C. What would be the value of E_a in relation to the answer given in part (a)?

(e) Calculate $\Delta H^{\ddagger}$.

6. An enzyme-catalyzed reaction followed the reaction scheme shown in Question 1. The E_a values for k_2 and k_3 were measured to be 5000 and 15,000 cal/mol, respectively. Assume that the relative rate of k_2/k_3 at 20°C = 5.

(a) At what temperature would the rates be equal?

(b) At what temperature would $k_3/k_2 = 5$ assuming that the enzyme is stable at that temperature?

(c) Draw a $\log k$ versus $1/T$ diagram for k_2 and k_3 over the range of temperatures from 20°C to the temperature calculated in part (b).

(d) On the same graph as (c) plot $\log k_{cat}$ (= $\log [k_2k_3/(k_2 + k_3)]$) versus $1/T$. What conclusions do you draw from these data?

7. An enzyme reaction A had an activation energy of 6000 cal/mol. There was also present in the enzyme preparation a proteolytic enzyme that hydrolyzed the other enzyme. E_a for proteolysis was 12,000 cal/mol. In an effort to minimize the proteolytic reaction with respect to reaction A, the investigator decided to carry out the experiments at 27°C rather than at 37°C. Was this an appropriate decision? Justify your answer by the appropriate calculations.

8. Two reactions had identical rates at 30°C, but the activation energies were 5000 and 13,000 cal/mol, respectively. Calculate the relative rates of the two reactions at 0°C.

9. In an experiment designed to determine the effect of temperature on the ionization of a group involved in the activity of an enzyme, a Tris chloride buffer was used. Assume that ΔH_{ion} of the group on the enzyme as well as ΔH_{ion} of Tris is 11,000 cal/mol.

(a) Calculate the pK_a value of the group on the enzyme at 35 and 45°C if pK_a is 8.0 at 20°C. For the moment, ignore the effect of the presence of Tris.

(b) Calculate the pK_a value of Tris at 35 and 45°C if pK_a is 8.0 at 20°C.

(c) Calculate the experimentally observed effect of temperature on ionization of the group on the enzyme when the reaction is performed in Tris buffer. (Two answers are needed.)

10. What effect would the following changes in conditions have on the alkaline phosphatase–catalyzed reaction: disodium *p*-nitrophenyl phosphate + $H_2O \rightarrow$ *p*-nitrophenol + Na_2HPO_4? Why? (Assume that the enzyme is stable under the conditions employed.)

(a) The effect on the activation energy E_a of changing the temperature from 25°C to 35°C (Q_{10} = 1.5 for this reaction).

(b) The effect on ΔG of using 0.1 M HCl rather than 1×10^{-6} M alkaline phosphatase to hydrolyze the substrate.

(c) The effect on K_{eq} of increasing the enzyme concentration from 1×10^{-6} to 1×10^{-5} M.

(d) The effect on E_a of using 0.1 M HCl rather than 1×10^{-6} M alkaline phosphatase to hydrolyze the substrate.

(e) The effect on E_a of adding 1×10^{-5} M urease to the reaction.

11. Compound A can be converted to products either enzymatically or nonenzymatically. For the nonenzymatic reaction, E_a and ΔG were 30,000 and −6000 cal/mol, respectively. If the rate of the enzymatic reaction at 40°C is four times that at 20°C, what is:

(a) The energy of activation E_a of the enzymatic process?

(b) The free energy ΔG of the overall reaction catalyzed by the enzyme?

REFERENCES

General References

Alberty, R. A., and R. J. Silbey, *Physical Chemistry*, Wiley, New York, 1992.

Benson, S. W., *The Foundations of Chemical Kinetics*, McGraw-Hill, New York, 1960, p. 66.

Bull, H. B., *An Introduction to Physical Biochemistry*, F.A. Davis, Philadelphia, 1964, pp. 31, 362.

Dixon, M., and E. C. Webb, *Enzymes*, 3rd Ed., Academic Press, San Diego, 1980.

Frost, A. A., and R. G. Pearson, *Kinetics and Mechanism*, 2nd Ed., Wiley, New York, 1961.

Gutfreund, H., *Enzymes: Physical Principles*, Wiley-Interscience, London, 1972.

Klotz, I. M., *Energy Changes in Biochemical Reactions*, Academic Press, New York, 1967.

Cited References

1. J. H. Northrop, M. Kunitz, and R. M. Herriott, *Crystalline Enzymes*, 2nd Ed., Columbia University Press, New York, 1948, p. 138.
2. N. A. Lange, ed., *Handbook of Chemistry*, Handbook Publishers, Sandusky, Ohio, 1946, pp. 1238–1240.
3. A. L. Tappel, P. D. Boyer, and W. O. Lundberg, *J. Biol. Chem.*, **199**, 267 (1952).
4. K. V. Thimann, C. S. Yocum, and D. P. Hackett, *Arch. Biochem. Biophys.*, **53**, 239 (1954).
5. H. Laser, *Proc. Roy. Soc. B*, **140**, 230 (1952).
6. A. Stockell and E. L. Smith, *J. Biol. Chem.*, **227**, 1 (1957).
7. V. Massey, *Biochem. J.*, **53**, 72 (1953).
8. M. L. Bender, F. J. Kézdy, and C. R. Gunter, *J. Am. Chem. Soc.*, **86**, 3714 (1964).
9. A. E. Stearn, *Advan. Enzymol.*, **9**, 25 (1949).

12

Enzyme Cofactors

I. GENERAL NATURE OF COFACTORS

Many enzymes, on hydrolysis, give only amino acids; therefore, their catalytic properties must reside in a unique arrangement of amino acid residues. Other enzymes contain, in addition to the polypeptide chain(s), small molecules that are essential for activity of the enzyme. These small molecules, referred to as cofactors, vary from the complexity of the B_{12} coenzymes to the simplicity of inorganic ions. We shall discuss these cofactors under the general groupings of (a) coenzymes, (b) prosthetic groups, and (c) inorganic ions.

A. Distinguishing Features of Coenzymes and Prosthetic Groups

In general, coenzymes are attached less firmly to the protein portion of the enzyme than are prosthetic groups. However, there is considerable overlap between the two types of cofactors in the degree of binding to protein, so that classification based on this distinction alone frequently is equivocal. The best method of assigning the organic cofactors to one of the two groups is on the basis of the mechanism involved in their repetitive turnover in the functioning system.

For repetitive turnover of coenzymes, two substrates (in addition to coenzyme) and two enzymes are needed. If only one enzyme or substrate is present, the coenzyme will be used up in one passage through the system. Consider as an example the enzyme alcohol dehydrogenase (ADH) involved in the oxidation of ethanol to acetaldehyde.

$$CH_3CH_2OH + NAD^+ \overset{ADH}{\rightleftharpoons} CH_3CHO + NADH + H^+ \qquad (1)$$

NAD^+ (nicotinamide adenine dinucleotide) is an essential coenzyme for the oxidation of ethanol to acetaldehyde. In the process, NAD^+ is reduced to NADH, serving as the second substrate. NADH cannot function to convert more ethanol to acetaldehyde and thus is used up. For NAD^+ to function in a repetitive fashion in vivo it must dissociate from alcohol dehydrogenase and reassociate with a second enzyme, which then uses the reduced coenzyme to reduce a second substrate molecule. Equations (2) and (3) show the general reactions involved. The reactions

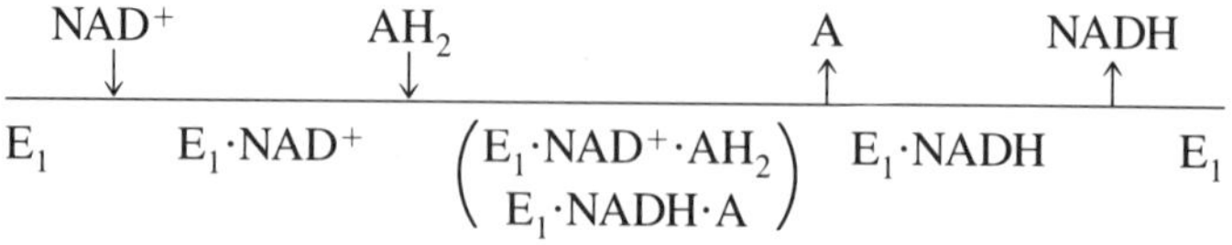

(2)

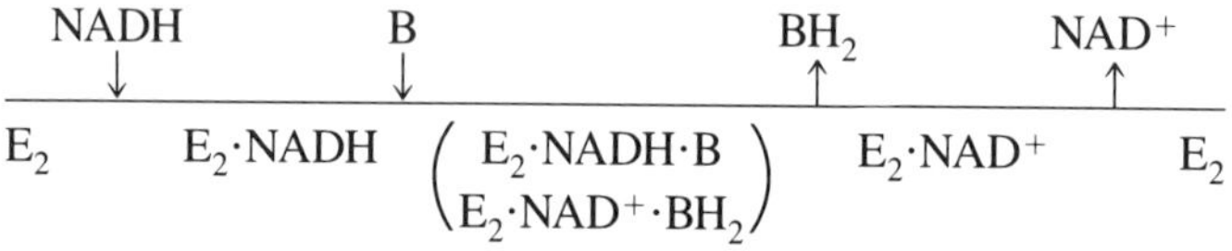

(3)

are shown as ordered sequential reactions in this particular case. This is not to imply that all reactions involving coenzymes will be ordered. The point to be made is that two different enzymes, E_1 and E_2, and two different substrates, AH_2 and B, are involved in the complete cycle $NAD^+ \rightarrow NADH \rightarrow NAD^+$.

With prosthetic groups, the repetitive cycle is performed while the cofactor is attached tightly to one enzyme. Two substrates are involved. This is shown in a general fashion in Eq. (4), where a specific cofactor has been used in the general scheme. This in no way should detract from the general features being discussed.

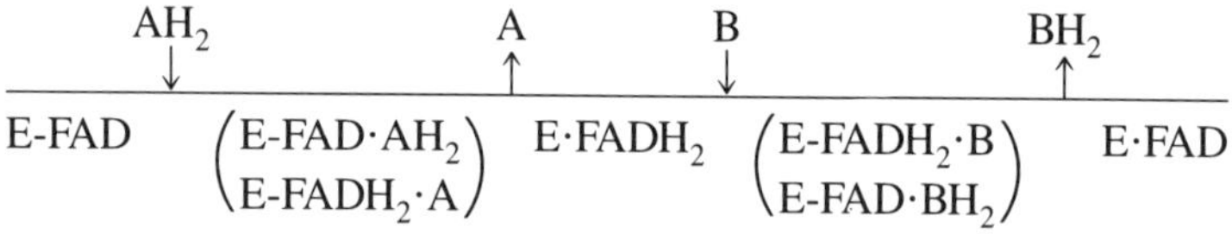

(4)

B. Cofactors and Nutrition

Vitamins and essential trace metals are integral parts of cofactors. Nutritional requirements for vitamins and essential trace metals were recognized long before it was generally appreciated that these perform their roles in vivo as essential components of enzyme systems. As early as 1882 it was recognized that beriberi could be cured by a balanced diet. In 1901, Eijkman demonstrated the presence

Table 1 Importance of Phosphate, Ribose, and Purine and Pyrimidine Bases in Cofactors

Cofactor	Vitamin	Phosphate	Ribose	Base
NAD^+	Niacin	+	+	Adenine
$NADP^+$	Niacin	+	+	Adenine
ATP	—	+	+	Adenine
UTP	—	+	+	Uridine
CTP	—	+	+	Cytidine
CoA	Pantothenic acid	+	+	Adenine
Acetyl phosphate	—	+	—	—
Carbamyl phosphate	—	+	—	—
S-Adenosyl methionine	—	—	+	Adenine
Adenosine-3′-phosphate-5′-phosphosulfate	—	+	+	Adenine
FMN	Riboflavin	+	+	—
FAD	Riboflavin	+	+	Adenine
Pyridoxal phosphate	Pyridoxine	+	—	—
Thiamine pyrophosphate	Thiamine	+	—	—

of an anti-beriberi factor in the bran of rice. It was not until 1926 that this factor, thiamine, was isolated by Jansen and Donath. Proof of the structure of this factor culminated in its synthesis in 1936. Recognition that thiamine performs its role as a part of a cofactor came in 1937 through work of Lohmann and Schuster. The first recognition of the function of a vitamin in an enzyme system was in 1932 and involved the role of riboflavin in "the old yellow enzyme" of Warburg.

C. Importance of Phosphate, Ribose, and Nucleoside Bases in Cofactors

The central role play by phosphate, ribose, and purine and pyrimidine bases in a number of the cofactors is summarized in Table 1.

II. COENZYMES IN ENZYME-CATALYZED REACTIONS

A. Carriers of Hydrogens or Electrons

The enzymes associated with this group of coenzymes are known as *dehydrogenases* and belong to the oxidoreductase group of enzymes.

1. Pyridine Coenzymes

The pyridine coenzymes are essential cofactors for a number of enzymes involved in the oxidation and reduction of substrates by the transfer of hydrogen. The structures of NAD^+ and $NADP^+$ are shown in Fig. 1. Starting at the left-hand

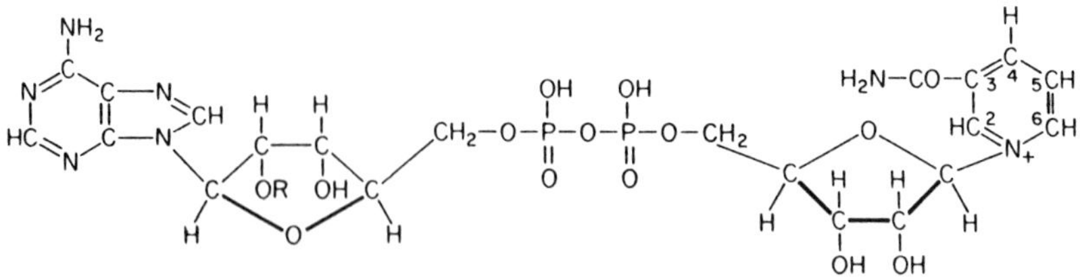

Figure 1 NAD^+ and $NADP^+$. In NAD^+ the R group on C(2) of the left-hand ribose is H, in $NADP^+$ it is a phosphate moiety.

side of the formula, NAD^+ is composed of adenine, ribose, pyrophosphate, ribose, and nicotinamide. $NADP^+$ differs from NAD^+ by having a third phosphate group on the C(2) hydroxyl group of the ribose attached to adenine.

NAD^+ and $NADP^+$ are synthesized in the mammal from nicotinamide (niacin) or from nicotinic acid. Niacin is a dietary requirement of humans. Inadequate intakes of niacin result in pellagra, characterized by a smooth red tongue, diarrhea, dermatitis, and dementia associated with neurological lesions and is most prevalent in humans who have a diet composed almost entirely of poor-quality protein such as corn. Pellagra is usually the result of a combination of factors, which include (a) low tryptophan content of protein, (b) presence of antiniacin factors or toxic materials, and (c) deficiencies in thiamine, riboflavin, and pyridoxine.

Most dehydrogenases are specific for either NAD^+ or $NADP^+$. However, a few dehydrogenases can use either cofactor. The enzymes are specific for the other substrate involved.

The part of NAD^+ and $NADP^+$ involved in transfer of hydrogen is the nicotinamide residue. Two hydrogens are removed from, or added to, a substrate when it undergoes oxidation or reduction by the pyridine coenzyme-containing dehydrogenases. One hydrogen is removed stereospecifically from the substrate and is transferred to the C(4) position of the nicotinamide residue to give either the A or B form of the coenzyme, as shown in Eq. (5) for alcohol dehydrogenase.

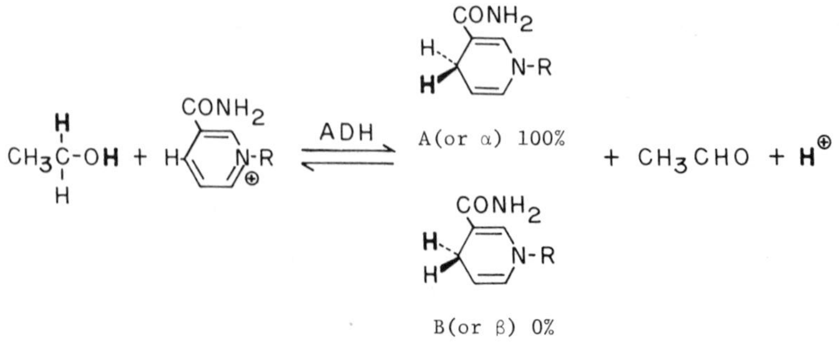

(5)

One group of dehydrogenases, the A-dehydrogenases, removes the hydrogen from one side of the planar pyridine ring; the other group, the B-dehydrogenases, removes the hydrogen from the opposite side of the ring. Both ethanol and the C(4) position of nicotinamide are symmetrical, yet the same hydrogen atom of ethanol is always transferred to give either the A or B stereoisomer of the coenzyme. This means that the enzyme bestows on this system the observed stereospecificity (see Chapter 4).

Oxidation and reduction of the pyridine coenzymes are easily followed spectrophotometrically. On reduction of NAD^+ (or $NADP^+$) there is a decrease in absorbance at 260 nm and the appearance of a new absorbance peak at 340 nm (Fig. 2). This new absorbance peak is due to the quinoid structure of the reduced nicotinamide ring [Eq. (5)].

2. Ascorbic Acid

Although ascorbic acid is present in all plant materials and is a dietary requirement of humans and guinea pigs, little is known about its function at the molecular level. In humans it plays an important role in formation of the intercellular proteins reticulin and collagen in bone metabolism and in wound healing. It appears to be important in maintenance of the integrity of membranes such as those of the capillaries and red blood cells. It is also involved in metabolism of the aromatic amino acids phenylalanine and tyrosine.

Ascorbic acid appears to serve as a coenzyme, along with vitamin B_{12}, in the oxidation of 4-hydroxyphenylpyruvate to 2,5-dihydroxyphenylpyruvate [Eq. (6)].

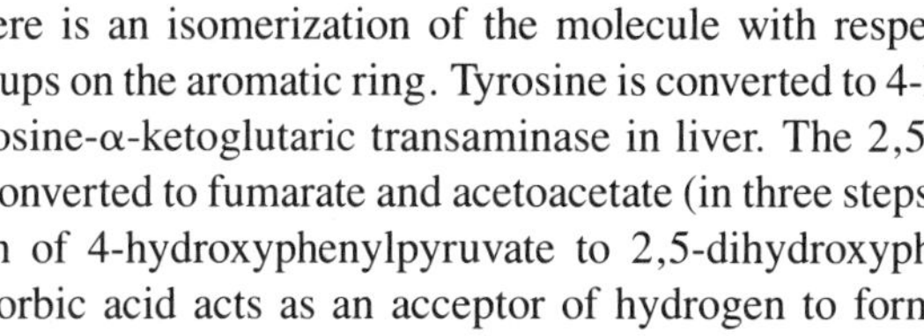

(6)

There is an isomerization of the molecule with respect to location of hydroxyl groups on the aromatic ring. Tyrosine is converted to 4-hydroxyphenylpyruvate by tyrosine-α-ketoglutaric transaminase in liver. The 2,5-dihydroxyphenylpyruvate is converted to fumarate and acetoacetate (in three steps). It appears that in conversion of 4-hydroxyphenylpyruvate to 2,5-dihydroxyphenylpyruvate, L-dehydroascorbic acid acts as an acceptor of hydrogen to form L-ascorbic acid (Fig. 3).

3. Ubiquinone

Ubiquinone (coenzyme Q) was isolated from beef mitochondria in 1957 [2] and shown to be an integral part of the electron transport system. Ubiquinone serves as an acceptor system between $FADH_2$ and cytochrome *b* in the respiratory system. The hydrogens of $FADH_2$ are transferred, via a two-step reduction mechanism, to the 1,4-benzoquinone ring of ubiquinone (Fig. 4). Ubiquinone

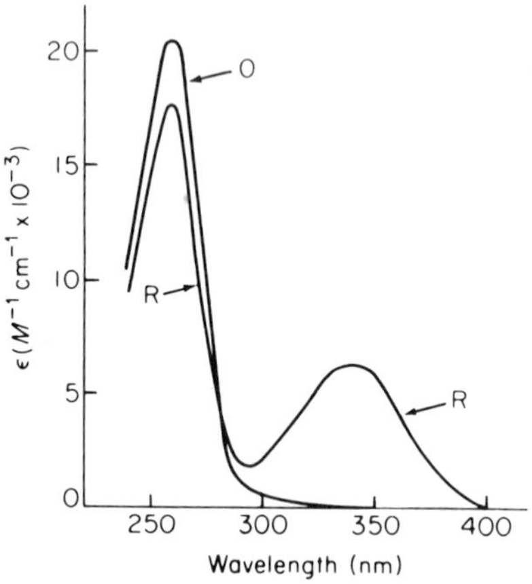

Figure 2 Spectra of oxidized (O) and reduced (R) forms of NAD^+ and $NADP^+$. (From Ref. 1, p. 363, by courtesy of Academic Press.)

isolated from liver has $n = 10$ (i.e., there are 10 isoprenoid units). Four related compounds, all with similar coenzyme activity, have been isolated from microbial sources. These have been designated as UQ_9, UQ_8, UQ_7, and UQ_6 since they contain 9, 8, 7, and 6 isoprenoid units, respectively.

B. Amino Group Carriers

α-Ketoglutarate, an intermediate in the citric acid cycle, is involved in the transfer of amino groups from amino acids to α-keto acids [Eq. (7)]. Also involved in

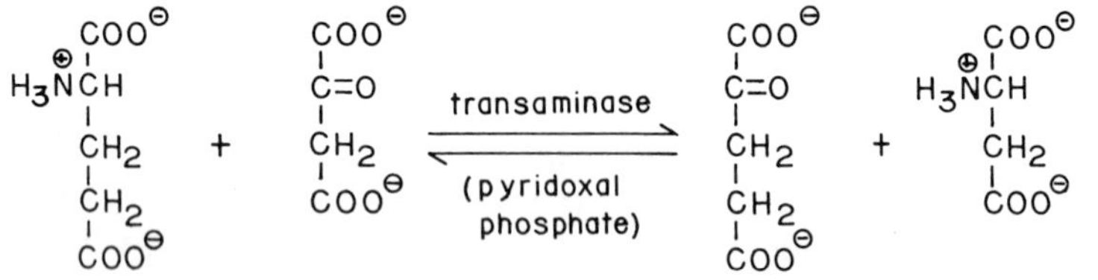

(7)

L-Glutamate Oxaloacetate α-Ketoglutarate L-Aspartate

this reaction is the prosthetic group, pyridoxal phosphate, which is firmly attached to the *transaminase* involved in the reactions. Glutamate is the only amino acid that serves as a general carrier of amino groups.

C. Phosphate Carriers

The phosphate transfer systems are concerned with transfer of energy from one process to another. These coenzymes permit the in vivo coupling of endogonic and

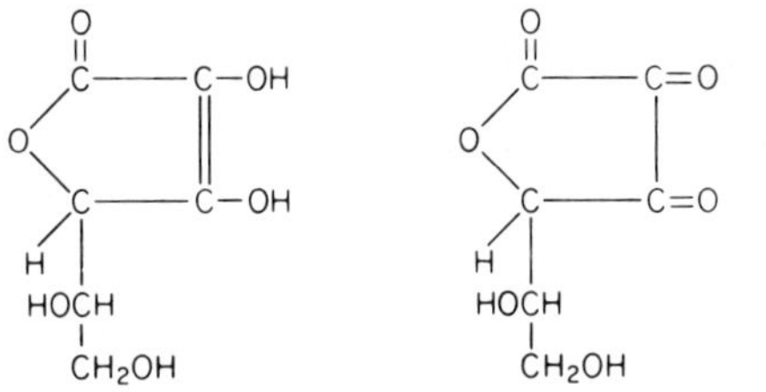

Figure 3 L-Ascorbic and L-dehydroascorbic acids.

exogonic reactions so that an organism can synthesize proteins, nucleic acids, complex carbohydrates, and lipids. Muscular contraction, photosynthesis, and bioluminescence are also dependent on one of these coenzymes, adenosine triphosphate (ATP; Fig. 5).

1. Adenosine Nucleotide Coenzymes

ATP is composed of one adenine, one ribose, and three phosphate residues (Fig. 5). ATP is involved in two general types of reactions, one in which cleavage of the ATP molecule provides the energy for an otherwise energetically unfavorable reaction (see above) and a second type in which some portion of the ATP molecule is transferred to a suitable acceptor molecule.

ATP is involved in transfer of a specific group in four ways.

1. Transfer of a phosphate group as in

$$\text{ATP} + \text{acetate} \rightleftharpoons \text{acetyl phosphate} + \text{ADP} \quad (8)$$

or

$$\text{ATP} + \text{creatine} \rightleftharpoons \text{ADP} + \text{creatine phosphate} \quad (9)$$

The enzymes involved in this type of reaction are called *phosphokinases* (or phosphotransferases) and require a divalent metal ion, usually Mg^{2+}.

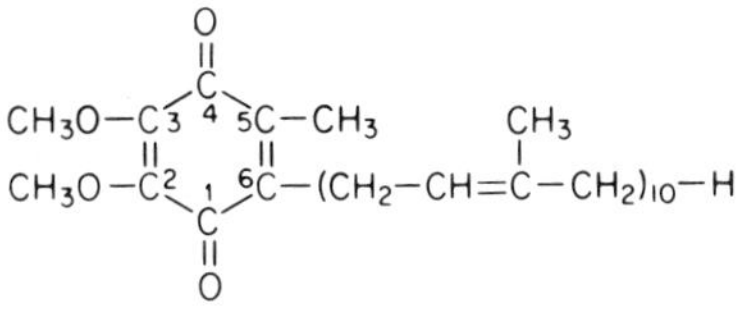

Figure 4 Ubiquinone.

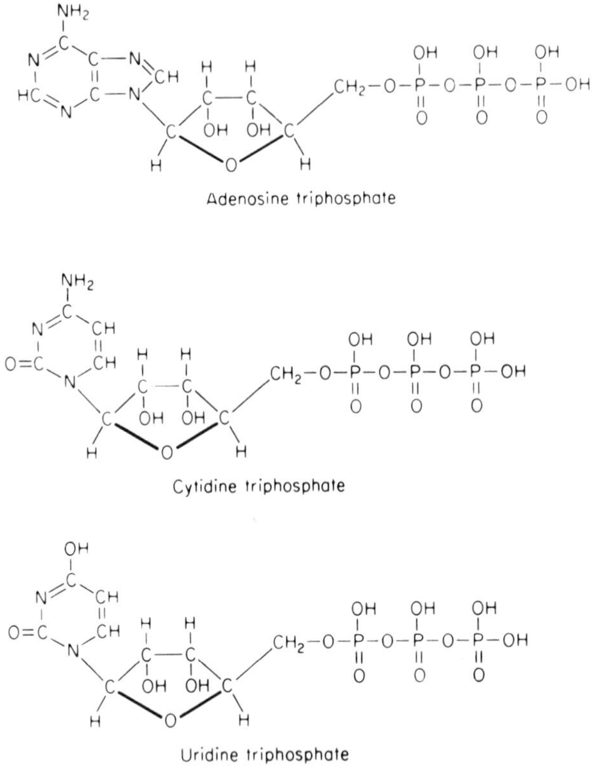

Figure 5 Triphosphates of adenosine, cytidine, and uridine.

2. Transfer of a pyrophosphate group as in

$$\text{ATP} + \text{D-ribose-5-phosphate} \rightleftharpoons \text{AMP} + \text{5-phospho-1-pyrophosphoryl ribose} \tag{10}$$

The enzymes involved are called *pyrophosphotransferases*.

3. Transfer of adenosine-5′-phosphate as in formation of dinucleotide coenzymes.

$$\text{ATP} + \text{FMN} \rightleftharpoons \text{FAD} + \text{PP}_i \tag{11}$$

The enzyme involved is a *nucleotidyl transferase*.

4. Transfer of adenosine in formation of the cofactor *S*-adenosyl methionine (see Section II.E).

$$\text{ATP} + \text{methionine} \rightleftharpoons S\text{-adenosyl methionine} + \text{trimetaphosphate} \tag{12}$$

2. Uridine Nucleotide Coenzymes

The uridine nucleotide coenzymes (Fig. 5) are involved in carbohydrate metabolism. The general structure of all the uridine nucleotide coenzymes is the same, the only variation being the specific nature of the sugar attached to the terminal phosphate. Uridine diphosphate glucose (UDPG) is shown in Fig. 6. The uridine nucleotide coenzymes are formed biosynthetically,

$$\text{sugar-1-PO}_4 + \text{UTP} \rightleftharpoons \text{UDP-sugar} + \text{PP}_i \tag{13}$$

and the enzymes involved are called *UDP-sugar pyrophosphorylases* or *uridyltransferases*. The specific sugar is named in a specific enzyme system.

The uridine nucleotide coenzymes are involved in two general reactions. The first reaction involves the transfer of glycosyl residues. Using UDPG as a specific example, the glucosyl group is transferred to fructose in the biosynthesis of sucrose in plants:

$$\text{UDPG} + \text{fructose} \rightarrow \text{sucrose} + \text{UDP} \tag{14}$$

The enzyme involved in this reaction is *sucrose UDP-transglucosylase*.

Another reaction that involves glycosyl transfer is the chain-lengthening step in the biosynthesis of glycogen in animals:

$$\text{UDPG} + (\text{glucose})_n \rightarrow (\text{glucose})_{n+1} + \text{UDP} \tag{15}$$

The enzyme involved in the biosynthesis of glycogen is *glycogen synthetase*. UDP is regenerated to UTP by transfer of a phosphate group from ATP.

$$\text{UDP} + \text{ATP} \rightarrow \text{UTP} + \text{ADP} \tag{16}$$

The second type of reaction involving the uridine nucleotide coenzyme is chemical conversion of a sugar while attached to the UDP handle. Typical reactions include the conversion of UDP-glucose to UDP-galactose [Eq. (17)] and the oxidation of UDP-D-glucose to UDP-D-glucuronic acid followed by decarboxylation to UDP-D-xylose [Eq. (18)].

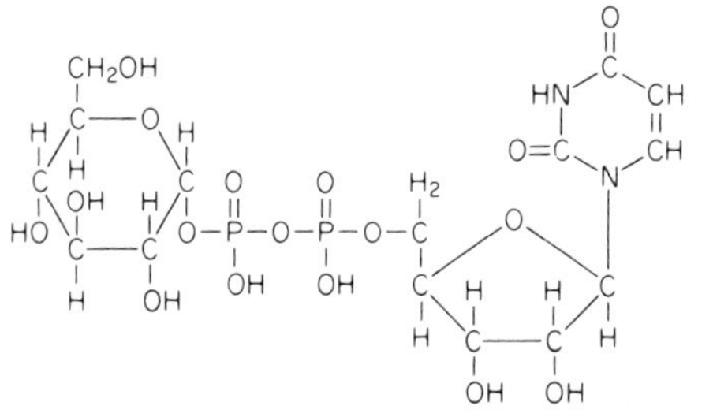

Figure 6 Uridine diphosphate glucose (UDPG).

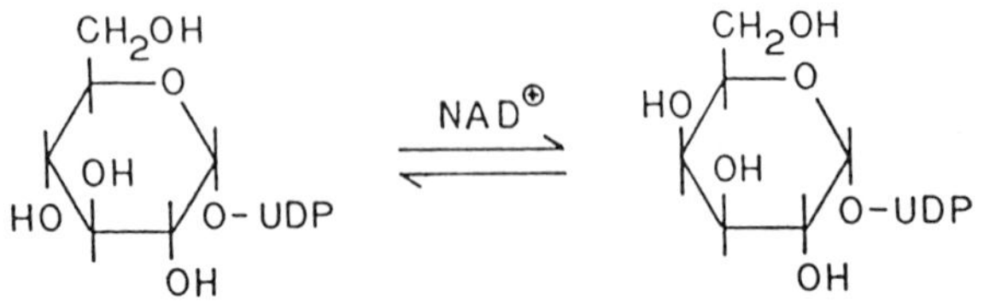

(17)

UDP-D-glucose UDP-D-galactose

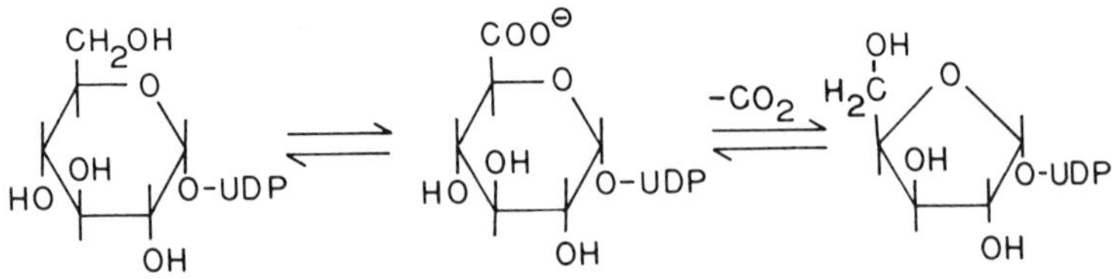

(18)

UDP-D-glucose UDP-D-glucuronate UDP-D-xylose

3. Cytidine Nucleotide Coenzymes

The cytidine nucleotide coenzymes (Fig. 5) are involved in biosynthesis of phosphoglycerides. The first step involves the biosynthesis of the cytidine nucleotide coenzyme by reaction of cytidine triphosphate (CTP) with ethanolamine phosphate or choline phosphate.

$$\text{CTP} + \underbrace{(CH_3)_3N^+CH_2CH_2OPO_3H^-}_{\text{Choline phosphate}} \rightleftharpoons \text{CDP-choline} + PP_i \qquad (19)$$

The structure of the coenzyme is shown in Fig. 7.

The second step in the reaction is transfer of the choline phosphate moiety of the coenzyme to a 1,2-diglyceride to give a lecithin [Eq. (20)].

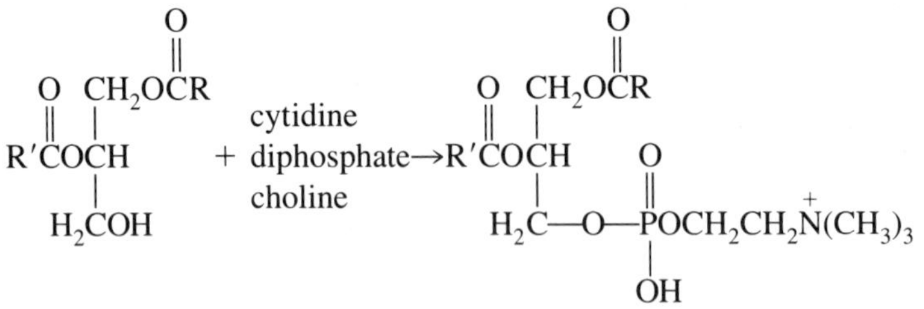

(20)

D. Acyl Group Carriers

1. Coenzyme A

The most important acyl group carrier is coenzyme A (Fig. 8). It is found in all organisms and is involved in a large number of reactions. Coenzyme A (CoASH) is composed of several residues. Starting from the adenine residue these are: adenine, ribose 3′-phosphate, pyrophosphate, pantoic acid, β-alanine, and cysteamine. The vitamin pantothenic acid (pantoyl β-alanine) is a component of this

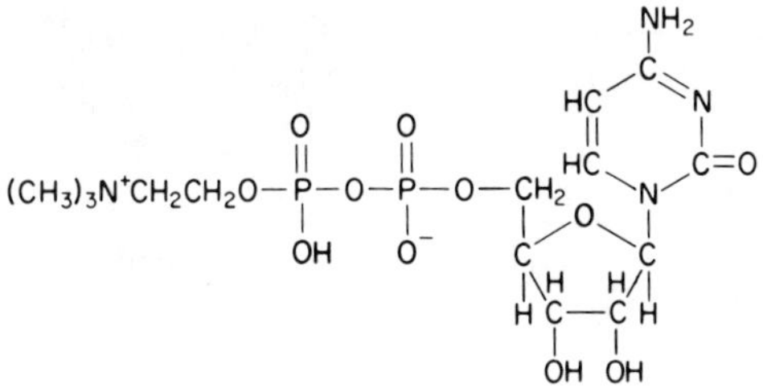

Figure 7 Cytidine diphosphate choline.

coenzyme. The vitamin is an essential dietary component needed for maintenance of normal skin, normal growth, and normal development of the central nervous system.

Coenzyme A is involved in a number of reactions, all of which involve the form

$$\mathrm{CoA{-}S{-}\overset{\overset{\displaystyle O}{\|}}{C}{-}R.}$$

1. One type of reaction involves nucleophilic attack by phosphates, amines, alcohols, water, and so on, to release the coenzyme A. A specific example is the biosynthesis of triglycerides:

$$\mathrm{2CoA{-}S{-}\overset{\overset{\displaystyle O}{\|}}{C}{-}R + HO{-}\begin{array}{l}\mathrm{H_2C{-}O{-}P{-}O_3H^-}\\ \quad | \\ \mathrm{CH}\\ \quad | \\ \mathrm{H_2C{-}OH}\end{array}}$$

$$\mathrm{\rightarrow R{-}\overset{\overset{\displaystyle O}{\|}}{C}{-}O{-}\begin{array}{l}\mathrm{H_2C{-}O{-}P{-}O_3H^-}\\ \quad | \\ \mathrm{CH}\\ \quad | \\ \mathrm{H_2C{-}O{-}\overset{\overset{\displaystyle O}{\|}}{C}{-}R}\end{array} + 2CoASH} \qquad (21)$$

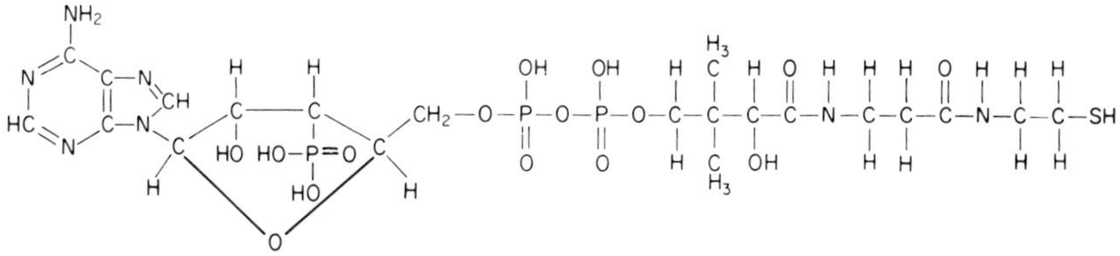

Figure 8 Coenzyme A.

After removal of the phosphate group by a phosphatase, the remaining acyl group is put on to give the finished triglyceride by reaction with a third $\mathrm{CoA{-}S{-}\overset{O}{\overset{\|}{C}}{-}R}$. The R groups are different fatty acids.

2. A second type of reaction involves condensation at the α-carbon of acetyl-*S*-CoA, as in the conversion of oxaloacetate to citrate in the citric acid cycle:

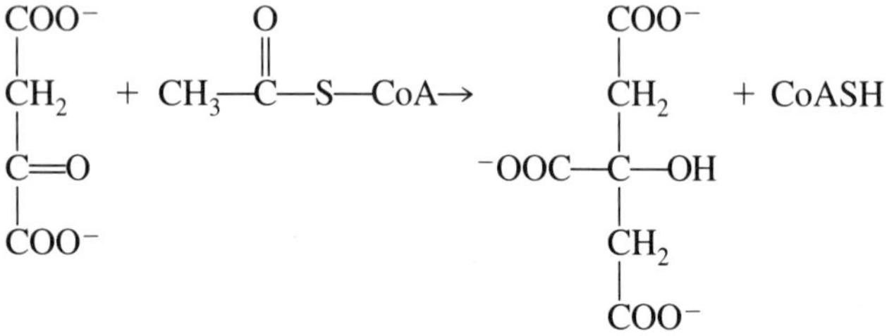 (22)

Oxaloacetate Acetyl CoA Citrate

3. A third type of reaction involves addition to unsaturated acyl derivatives of CoASH:

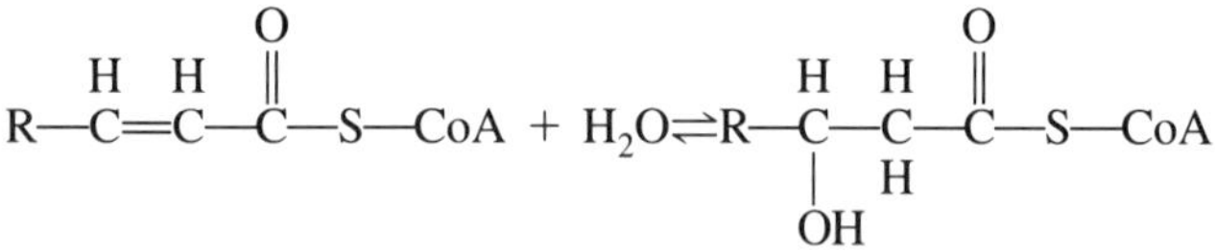 (23)

4. A fourth type of reaction involves interchange of acyl groups:

acetoacetyl-*S*-CoA + succinate ⇌ acetoacetate + succinyl-*S*-CoA (24)

5. CoASH also serves as a "handle" during the β-degradation of fatty acids.

2. Lipoic Acid

Lipoic acid (1,2-dithiolane-3-valeric acid; Fig. 9) was first isolated from yeast and then was obtained in crystalline form in 1953 from beef liver by Reed and coworkers [3]. They isolated 30 mg of crystalline lipoic acid from 10 tons of liver! It participates, along with two other cofactors, thiamine pyrophosphate (TPP) and coenzyme A, in the oxidative decarboxylation of α-keto acids, such as the

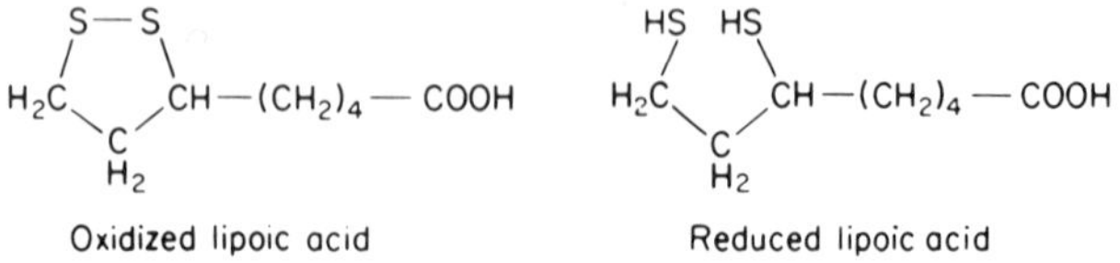

Figure 9 Oxidized and reduced lipoate.

conversion of pyruvate to acetate and carbon dioxide [Eqs. (25) to (28)]. In the first reaction the "acetyl" group from pyruvate is covalently bound to position 2 of the thiazolium ring in TPP. The acetyl group is then in turn transferred from TPP to

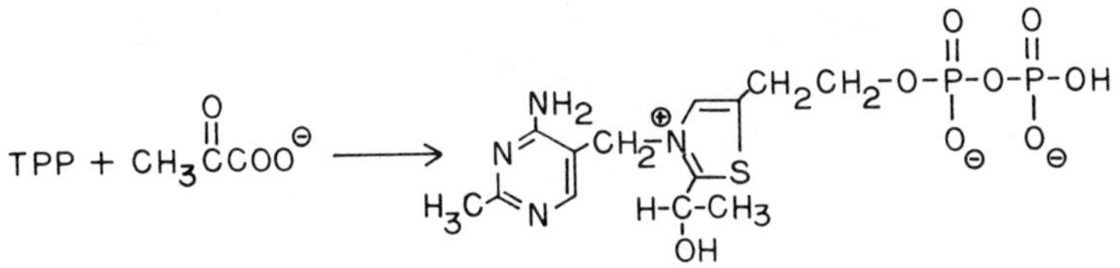

(25)

α-Hydroxyethyl thiamine pyrophosphate (HETPP)

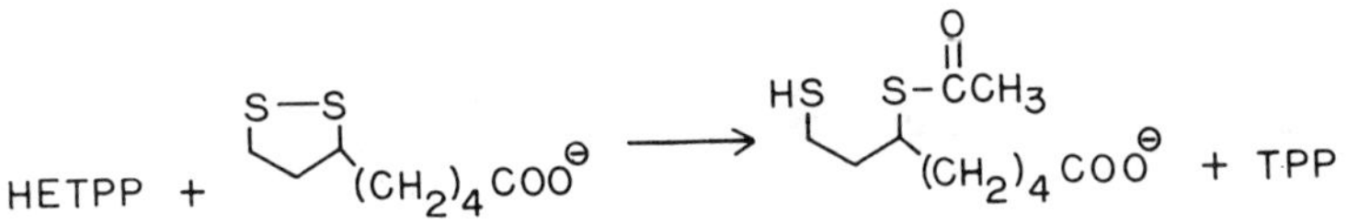

(26)

Lipoate 6-S-Acetyl lipoate

6-S-Acetyl lipoate + CoASH→

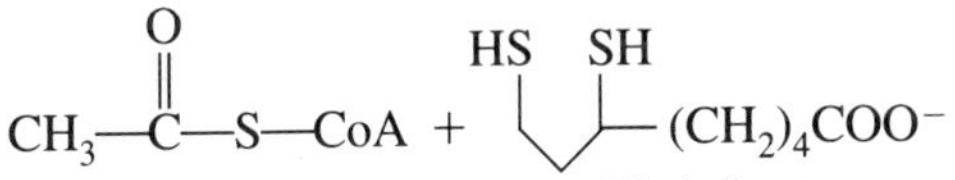

(27)

Dihydrolipoate

$$\text{Dihydrolipoate} + NAD^{+} \rightarrow \text{lipoate} + NADH + H^{+} \tag{28}$$

lipoic acid to coenzyme A, where it is used in the biosynthesis of fatty acids, oxaloacetate, and so on. A different enzyme is involved in each of the four reactions. The series of reactions take place in a complex of enzymes (see Chapter 2, Section VII.A).

3. Glutathione

Glutathione (GSH; Fig. 10) is probably found in all plant and animal tissues. It serves a general function of protecting the sulfhydryl groups of intracellular proteins from oxidation and as a coenzyme in a few reactions. Glutathione is a

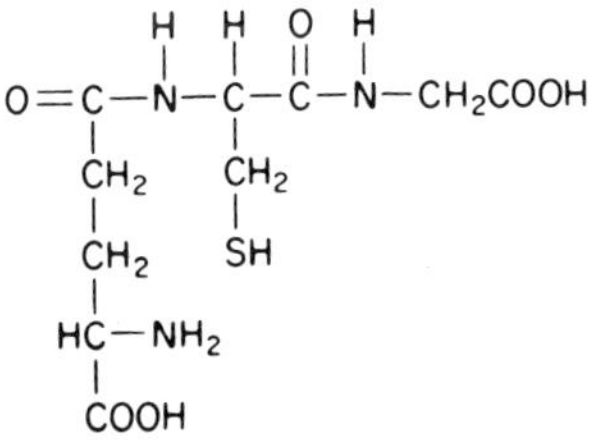

Figure 10 Glutathione (γ-L-glutamyl-L-cysteinylglycine) in the reduced form.

tripeptide, γ-L-glutamyl-L-cysteinylglycine. The functional group is the sulfhydryl group.

One of the best studied reactions involving glutathione as a coenzyme is the conversion of methylglyoxal to L-lactate:

$$CH_3{-}\overset{O}{\overset{\|}{C}}{-}\overset{O}{\overset{\|}{C}}H + GSH \xrightarrow{\text{glyoxylase I}} CH_3{-}\underset{H}{\overset{OH}{\overset{|}{C}}}{-}\overset{O}{\overset{\|}{C}}{-}S{-}G \qquad (29)$$

$$CH_3{-}\underset{H}{\overset{OH}{\overset{|}{C}}}{-}\overset{O}{\overset{\|}{C}}{-}S{-}G \xrightarrow[H_2O]{\text{glyoxylase II}} CH_3{-}\underset{H}{\overset{OH}{\overset{|}{C}}}{-}COO^- + GSH + H^+ \qquad (30)$$

In the reaction catalyzed by glyoxylase I, glutathione acts as an acyl acceptor.

4. Acyl Phosphate

In a number of reactions phosphate is the initial acceptor of acetyl groups and these acetyl groups are then transferred to CoASH by an enzyme, *phosphotransacetylase*. In humans and other ureotelic organisms, the initial step in urea formation is the formation of carbamoyl phosphate [Eq. (31)]. The carbamoyl group is then

$$CO_2 + NH_3 + 2ATP \rightleftharpoons \underbrace{NH_2{-}\overset{O}{\overset{\|}{C}}{-}OPO_3H^-}_{\text{Carbamoyl phosphate}} + 2ADP + P_i \qquad (31)$$

transferred to ornithine to give citrulline [Eq. (32)]. The enzyme involved in this

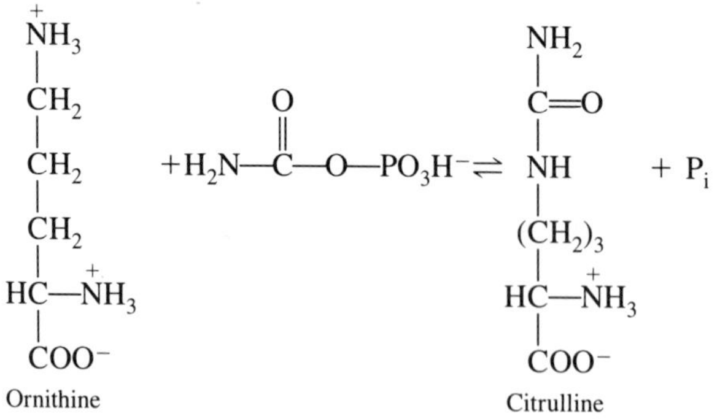

(32)

reaction is *ornithine transcarbamylase*.

E. Carriers of One-Carbon Groups

1. Tetrahydrofolic Acid

During the metabolism of certain amino acids and purines, one-carbon fragments in the form of hydroxymethyl (—CH_2OH), formyl (—CHO), formimino

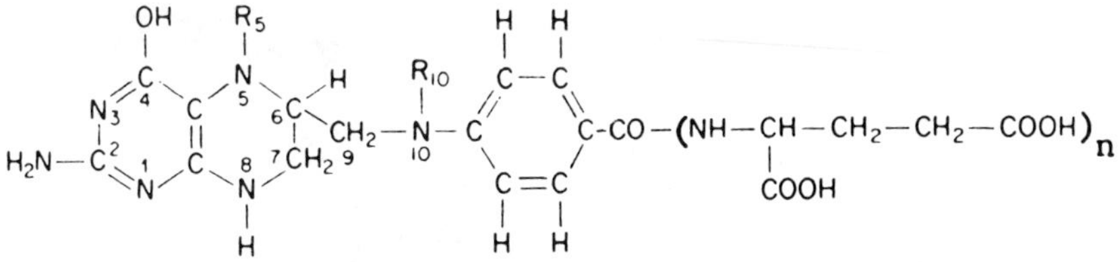

Figure 11 Tetrahydrofolic acid (R_5 and R_{10} are hydrogens in tetrahydrofolic acid). *n* varies from 1 to 7.

(—CHNH), and methyl (—CH_3) groups are transferred to acceptor molecules. Tetrahydrofolic acid is involved in the transfer of methyl, hydroxymethyl, formyl, and formimino groups, and coenzyme B_{12} in the transfer of methyl groups (see Section II.G). Tetrahydrofolic acid (Fig. 11) is composed of a reduced pteridine, *p*-aminobenzoic acid, and L-glutamic acid (one to seven residues).

Two vitamins are involved in tetrahydrofolic acid, folic acid (oxidized form of tetrahydrofolic acid) and *p*-aminobenzoic acid. Folic acid deficiencies produced with sulfonamides or other folic acid antagonists such as aminopterin (in which the fourth hydroxyl group of pterdine is replaced with an amino group) lead to macrocytic anemia in humans. Since folic acid is involved in purine biosynthesis, there is a shortage of intermediates for DNA biosynthesis in a folic acid–deficient organism.

The one-carbon fragments are carried by the coenzyme at positions 5 and/or 10 (R_5 and R_{10}; Fig. 11). Both vitamin B_{12} and ascorbic acid are involved in formation of the folic acid coenzymes.

2. S-Adenosyl Methionine

S-Adenosyl methionine (Fig. 12) is involved in transfer of the methyl group of methionine in reactions such as the formation of creatine and choline. Therefore, *S*-adenosyl methionine is the coenzyme for these *transmethylases*.

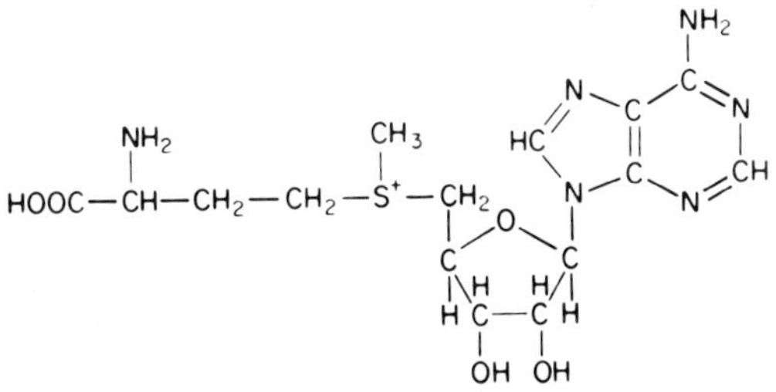

Figure 12 *S*-Adenosyl methionine ("active methionine").

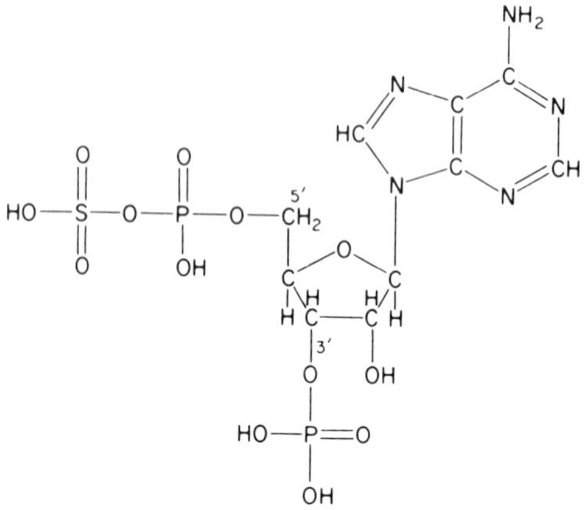

Figure 13 Adenosine-3′-phosphate-5′-phosphosulfate ("active sulfate").

F. Carriers of Sulfate Groups

Lipmann [4] showed that sulfate is carried in an active form in the coenzyme adenosine-3′-phosphate-5′-phosphosulfate (Fig. 13). This coenzyme is involved in enzymatic sulfation of phenols (detoxification) and of hexosamine derivatives, as in the formation of chondroitin sulfates and in other sulfate transfer reactions. The chondroitin sulfates, bound to proteins, are among the principal mucopolysaccharides in the ground substances of mammalian tissues and cartilages. One of the principal detoxification mechanisms in animals involves the sulfation of phenolic compounds. The enzymes that use this coenzyme are *sulfatases*.

G. Coenzyme B_{12}

In 1926, Minot and Murphy showed that there is a substance in liver which, when added to the diet, cures pernicious anemia [5]. Coenzyme B_{12} is the most complex of all the enzyme cofactors (Fig. 14). The molecule is composed of two principal parts, the porphyrin-like corrin ring containing the Co(III) and the nucleotide residues.

Coenzyme B_{12} is involved in many reactions, but only a few reactions have been studied mechanistically. Coenzyme B_{12} is the prosthetic group of methylmalonyl-*S*-CoA mutase, which converts methylmalonyl-*S*-CoA to succinyl-*S*-CoA [Eq. (33)] and in glutamate mutase, which converts glutamate to β-methyl aspartate (by isomerization) [Eq. (34)]. Both of these reactions are similar in that there is a transfer of a group from one part of the molecule to another. Both enzymes belong to the isomerase group of enzymes (see Chapter 13).

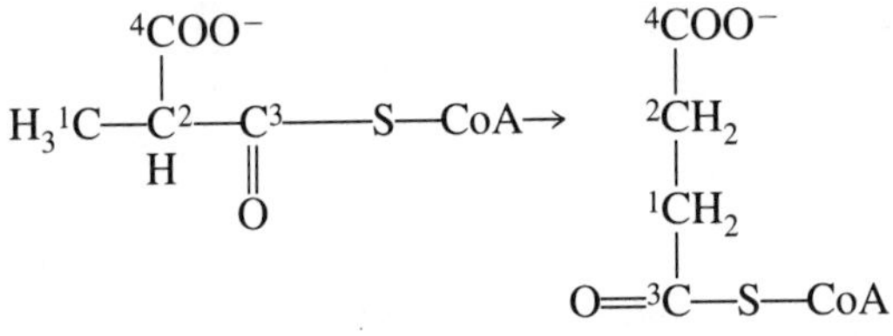

(33)

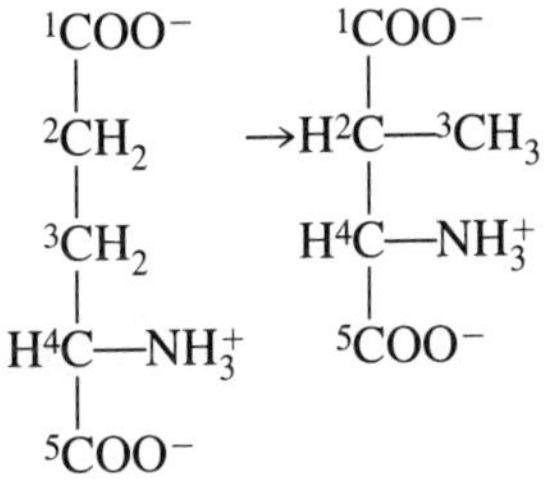

(34)

III. PROSTHETIC GROUPS IN ENZYME-CATALYZED REACTIONS

A. Flavin Cofactors

In 1932, Warburg and Christian isolated a yellow enzyme from brewer's yeast which catalyzes the oxidation of NADPH [6]. The yellow color is due to riboflavin 5′-phosphate (Fig. 15). This cofactor, flavin mononucleotide (FMN), is composed of a substituted isoalloxazine ring linked to D-ribitol-5′-phosphate.

In 1938, Warburg and Christian found a second riboflavin-containing cofactor which differed from FMN in containing an additional phosphate residue, D-ribose, and adenine [7]. This cofactor is flavin adenine dinucleotide (FAD; Fig. 16).

Riboflavin deficiency in humans is usually manifested as abnormal ectodermal tissue maintenance. This shows up as inflammation of the tongue, cracking of the corners of the mouth and lips, a seborrheic dermatitis, and occasionally, corneal vascularization.

As prosthetic groups of specific enzymes, FMN and FAD perform their function in oxidation-reduction reactions. Two hydrogens, removed from oxidizable substrates, are added at the 1 and 10 positions of the isoalloxazine ring. The large number of oxidation-reduction reactions in which these prosthetic groups participate are subdivided on the basis of the type of acceptor of these two hydrogens from the isoalloxazine ring.

In one type of reaction the hydrogens are transferred to O_2 only. The enzymes involves are the flavoprotein oxidases and include glucose oxidase (see Chapter 23) and D-amino acid oxidase.

In a second type of reaction catalyzed by FMN- and FAD-containing enzymes the hydrogens are transferred to an acceptor other than oxygen. Examples are

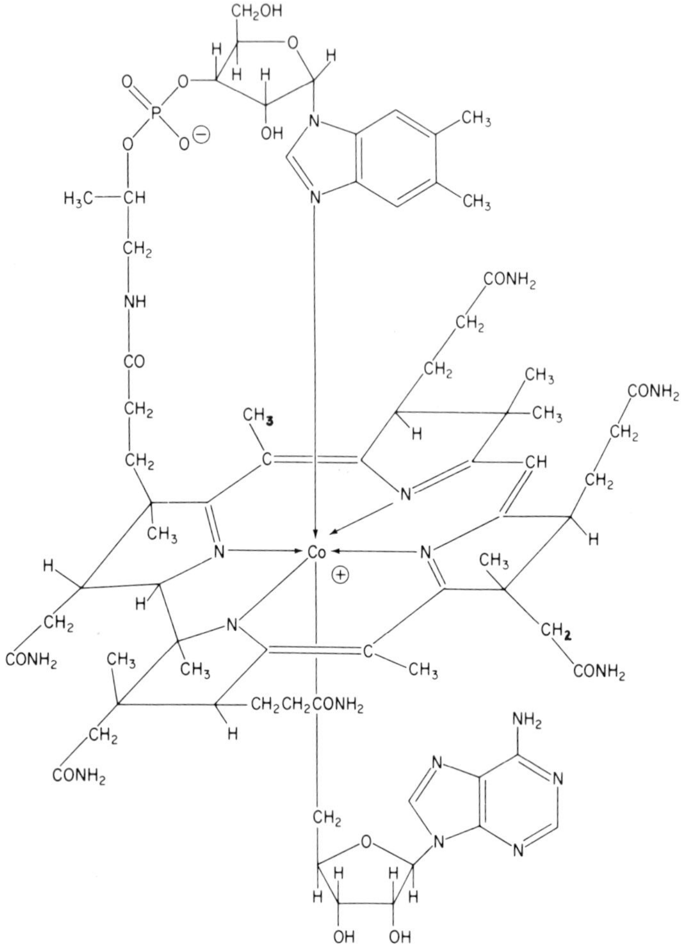

Figure 14 Coenzyme B_{12}.

succinic dehydrogenase and NADH dehydrogenase of mitochondria and NADPH dehydrogenase of microsomes. This reaction is presented generally as shown in Eqs. (35) to (37).

$$\text{NADH} + \text{E-FAD} \rightleftharpoons \text{NAD}^+ + \text{E-FADH}\cdot \tag{35}$$

$$\text{E-FADH}\cdot + \text{cyto } b(\text{Fe}^{3+}) \rightleftharpoons \text{E-FAD} + \text{cyto } b(\text{Fe}^{2+}) + \text{H}^+ \tag{36}$$

$$\text{Overall:} \quad \text{NADH} + \text{cyto } b(\text{Fe}^{3+}) \rightleftharpoons \text{NAD}^+ + \text{cyto } b(\text{Fe}^{2+}) + \text{H}^+ \tag{37}$$

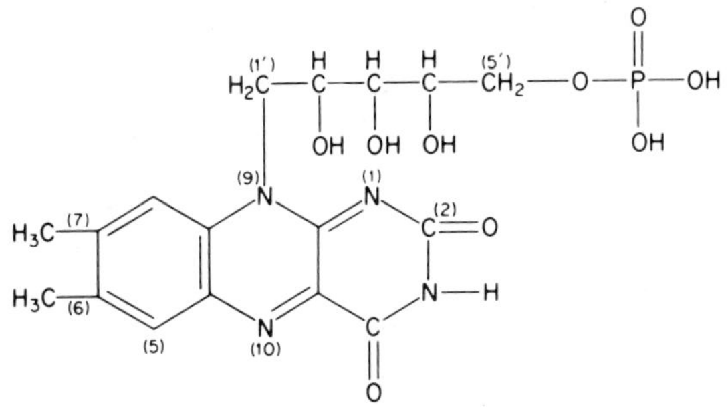

Figure 15 Riboflavin 5′-phosphate (FMN).

Many of the flavin-containing enzymes also require metals such as iron and molybdenum. For example, milk xanthine oxidase contains 2 FAD, 8 atoms of iron, and 2 atoms of molybdenum per molecule (see Chapter 25).

There is a marked decrease in the spectrum of FMN and FAD at 450 nm when the isoalloxazine ring is reduced (Chapter 23, Section II). Unlike the case with NAD^+ and NADH, this spectral change is not useful for following the activity of these enzymes in turnover experiments because of rapid reoxidation by O_2. However, the half reaction can be followed, in a single turnover reaction, at 450 nm, demonstrating that the enzyme follows a ping-pong mechanism.

B. Metal–Porphyrin Cofactors

A large number of enzymes utilize cofactors that contain a metal ion coordinated to the nitrogens of porphyrins. Chlorophyll (Fig. 17), with Mg^{2+}, is involved in the

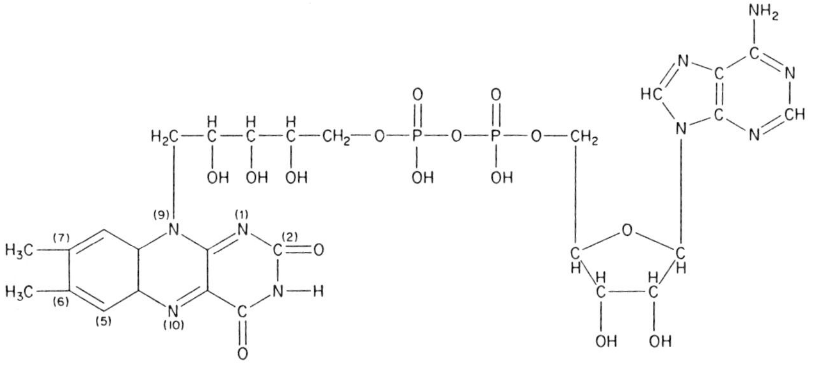

Figure 16 Flavin adenine dinucleotide (FAD).

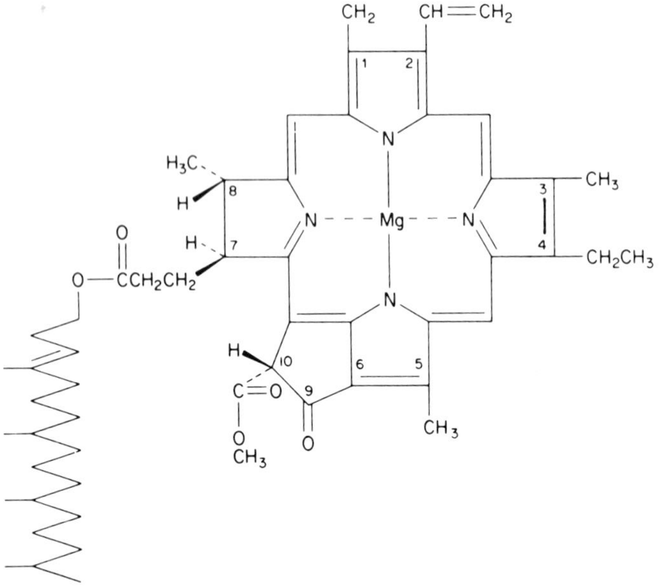

Figure 17 Chlorophyll *a*. Chlorophyll *b* differs from chlorophyll *a* in having the —CH_3 group on carbon 3 replaced with a formyl group. Removal of Mg^{2+} gives pheophytins *a* and *b*. Removal of the phytol group gives chlorophyllides *a* and *b*.

photosynthetic processes of all green plants. A large number of cytochromes are found in various organisms. All of the cytochromes have in common the tetrapyrrole porphyrin but differ one from the other in the nature of the side chains attached at positions 1 to 8 (Fig. 18).

These iron–porphyrin cofactors serve a number of roles in the living organism. They are involved (a) in O_2 and CO_2 transport (hemoglobin and myoglobin), (b) in transport of electrons in the electron transport system (cytochromes), and (c) in the oxidation of a great variety of substrates (catalases and peroxidases).

C. Pyridoxal Phosphate and Pyridoxamine Phosphate

Pyridoxal phosphate and its amine form, pyridoxamine phosphate (Fig. 19), are involved as the cofactors in a variety of enzymatic reactions; most of the reactions involve amino acids. Pyridoxal phosphate is bound to the enzyme as a Schiff base formed through reaction of the carbonyl group of the cofactor with the ϵ-amino group of a lysyl residue of the enzyme [Eq. (38)]. The first step in reaction of an

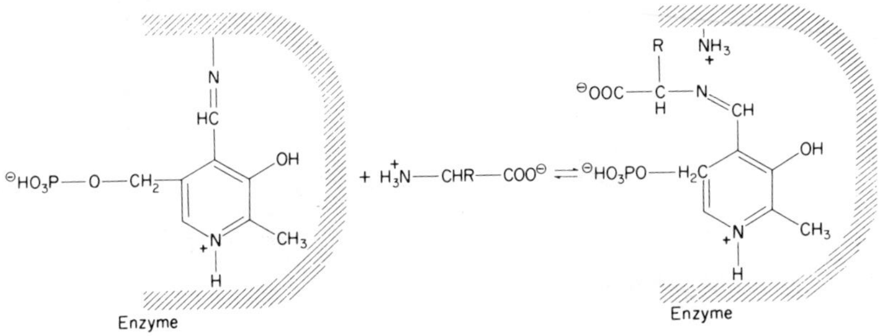

(38)

amino acid with the enzyme involves replacement of the internal Schiff base with one involving the amino acid in a Schiff base. In the process the ε-amino group of the lysyl residue of the enzyme is freed from its binding with the prosthetic group.

Once the amino acid has formed a Schiff base with the enzyme cofactor it can undergo several types of reactions. Depending on the particular type of enzyme involved, the amino acid may undergo (a) decarboxylation, (b) deamination, (c) racemization, (d) α,β-elimination, (e) β,γ-elimination, or (f) have groups added to the β carbon as in the biosynthesis of tryptophan from serine and indole. These reactions are summarized in Fig. 20.

Pyridoxal is one of the B vitamins (vitamin B_6). The most common deficiency syndrome is seborrbeic dermatitis about the eyes, in the eyebrows, and at the corners of the mouth. There is usually a pronounced increase in the excretion of xanthurenic acid in the urine since tryptophan metabolism is impaired.

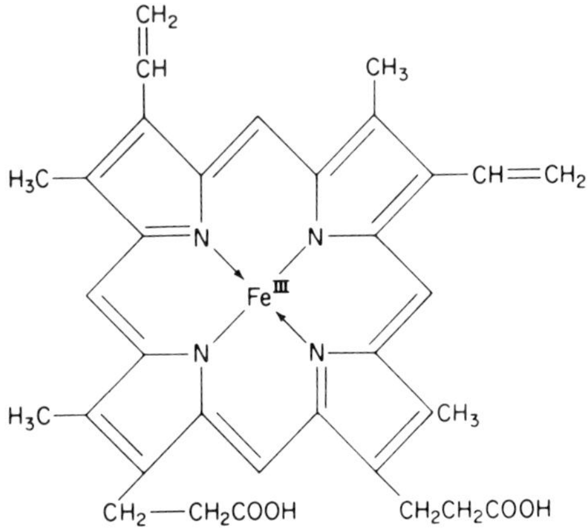

Figure 18 Ferriprotoporphyrin III (hemin).

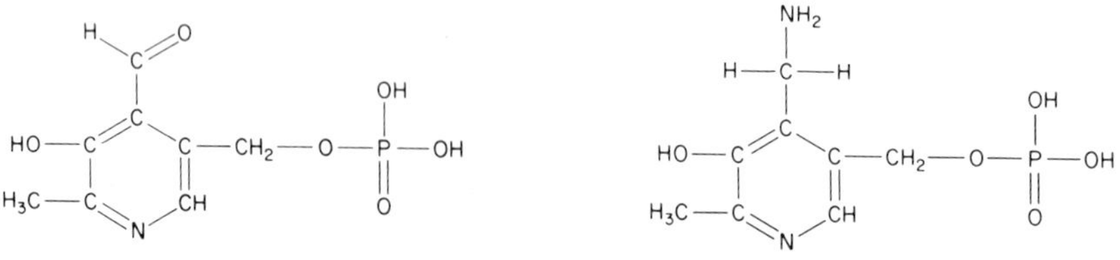

Pyridoxal phosphate

Pyridoxamine phosphate

Figure 19 Pyridoxal phosphate and pyridoxamine phosphate.

D. Thiamine Pyrophosphate

Dietary diseases due to deficiencies of thiamine were among the first noted. The cofactor thiamine pyrophosphate (Fig. 21) is a prosthetic group in enzymes that perform three types of reactions: (a) nonoxidative decarboxylation of α-keto acids, (b) oxidative decarboxylation of α-keto acids, and (c) the formation of α-ketols.

The first type of reaction described for thiamine pyrophosphate was the nonoxidative decarboxylation of pyruvate by yeast cells:

$$CH_3{-}\overset{\overset{\displaystyle O}{\|}}{C}{-}COO^- + H^+ \rightarrow CH_3{-}\overset{\overset{\displaystyle O}{\|}}{C}H + CO_2 \qquad (39)$$

The enzyme involved is *pyruvate decarboxylase*.

In the oxidative decarboxylation of α-keto acids O_2 is involved and one product is an acid:

$$2R{-}\overset{\overset{\displaystyle O}{\|}}{C}{-}COO^- + O_2 \rightarrow 2R{-}\overset{\overset{\displaystyle O}{\|}}{C}{-}O^- + 2CO_2 \qquad (40)$$

The enzyme involved is an α-keto acid oxidase.

Carbon dioxide is liberated in both the nonoxidative and oxidative reactions. However, the other product is an aldehyde and an acid, respectively, in the two types of reactions.

The α-ketol reactions involve condensation of two molecules one or both of which are α-keto acids or aldehydes:

$$CH_3{-}\overset{\overset{\displaystyle O}{\|}}{C}{-}COO^- + CH_3{-}\overset{\overset{\displaystyle O}{\|}}{C}H + H^+ \rightarrow CH_3{-}\overset{\overset{\displaystyle O}{\|}}{C}{-}\underset{\displaystyle H}{\overset{\overset{\displaystyle OH}{|}}{C}}{-}CH_3 + CO_2 \qquad (41)$$

The three types of reactions in which thiamine pyrophosphate participates are similar. In each case there is cleavage or condensation of a carbon–carbon bond

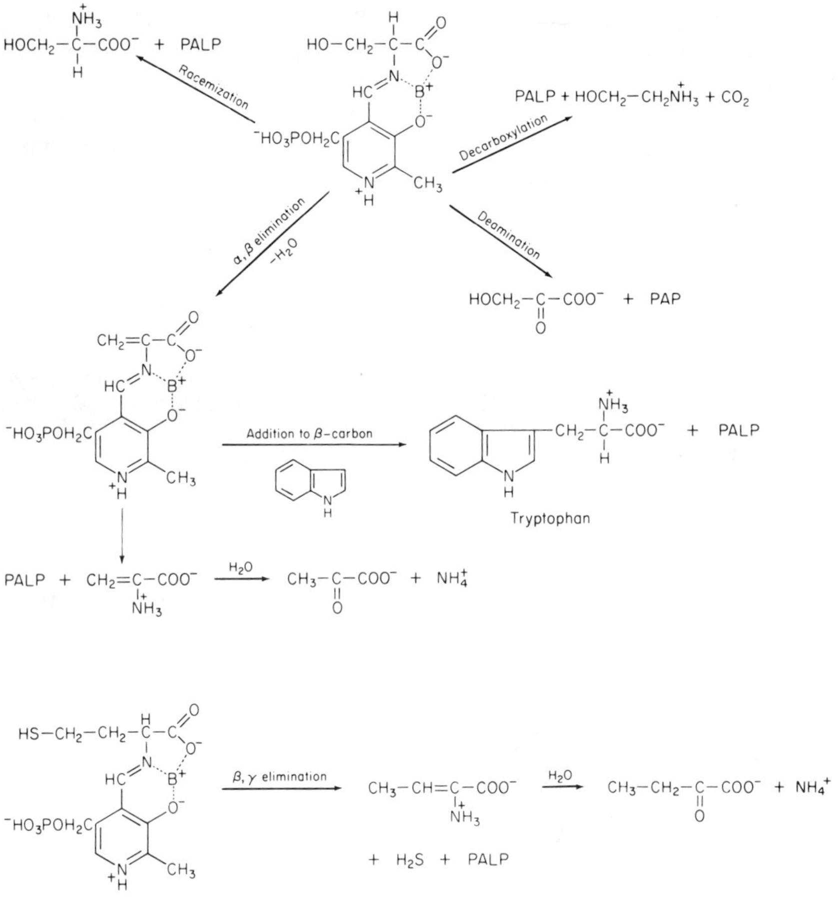

Figure 20 Schematic representation of reactions catalyzed by pyridoxal phosphate. A common Schiff base intermediate between serine and pyridoxal phosphate is shown in the upper center. This may then break down to give five types of products. The lower scheme shows β,γ elimination involving the —SH group of cysteine. The intermediate is also a Schiff base. B^+ is an acidic group on the enzyme. PALP and PAP are pyridoxal phosphate and pyridoxamine phosphate, respectively.

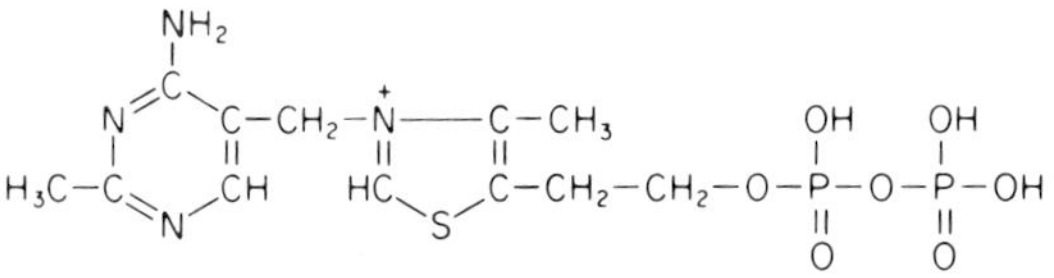

Figure 21 Thiamine pyrophosphate.

adjacent to a keto group. The initial step in all these reactions is addition of the α-keto compound to position 2 of the thiazolium ring of thiamine pyrophosphate (Fig. 22).

Thiamine deficiency results in beriberi in humans. There is a loss of appetite, weight loss, fatigue, and gastrointestinal disturbances. After prolonged thiamine deficiency there is cardiac impairment.

E. Biotin

Biotin is a cofactor for yeast and humans. In humans a deficiency of biotin results in skin lesions, grayish skin color, atrophy of the lingual papillae, loss of appetite, extreme lassitude, sleeplessness, and muscle pain. Biotin functions as a cofactor in carboxylation reactions [Eq. (42)].

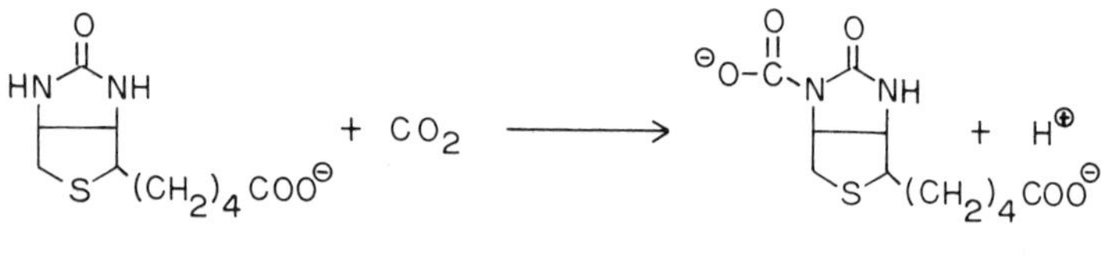

(42)

Biotin CO_2-biotin

The types of reactions in which biotin serves as a cofactor are twofold. The first type of reaction involves the transfer of CO_2 from one substrate to another, as in the conversion of pyruvate to oxaloacetate [Eq. (43)]. The enzyme involved is *methylmalonyl-CoA carboxyltransferase*.

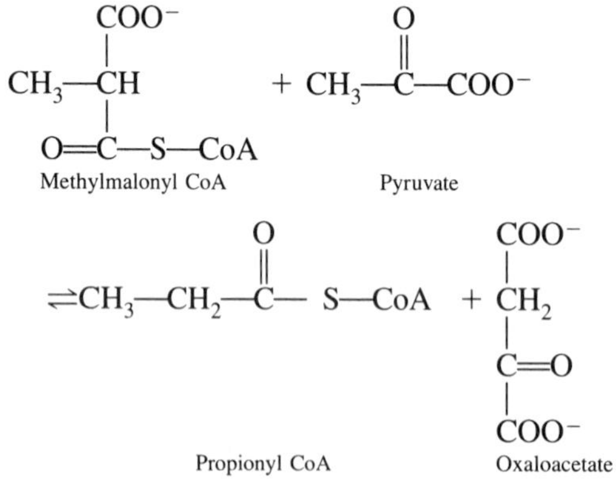

(43)

The second type of reaction includes ATP-dependent carboxylations in which ATP is cleaved to ADP and inorganic phosphate. A typical reaction is the conversion of propionyl-S-CoA to methyl malonyl-S-CoA:

$$\mathrm{CH_3{-}CH_2{-}\overset{\overset{\displaystyle O}{\|}}{C}{-}S{-}CoA} + \mathrm{CO_2} + \mathrm{ATP} + \mathrm{H_2O} \tag{44}$$

$$\rightarrow \mathrm{CH_3{-}\underset{H}{\overset{\overset{\displaystyle COO^-}{|}}{C}}{-}\underset{\underset{\displaystyle O}{\|}}{C}{-}S{-}CoA} + \mathrm{ADP} + \mathrm{P_i}$$

The enzyme involved is *propionyl-CoA carboxylase* (*ATP hydrolyzing*). In both types of reactions an intermediate CO_2-biotin compound is formed [Eq. (42)].

IV. ROLE OF SOME OTHER ORGANIC COMPOUNDS

There are a number of other compounds that may or may not function as cofactors. We shall mention briefly only a few of these.

A. Vitamin A

Vitamin A is a precursor of retinene, which is involved in the visual cycle. In this process all-*trans*-vitamin A (either A_1 or A_2) is converted to 11-*cis*-vitamin A, which is then oxidized to the aldehyde before combining with a specific protein, opsin, to produce the visual pigment of the eyes. The conversion of all-*trans*-vitamin A_1 to 11-*cis*-vitamin A_1 aldehyde is shown in Eq. (45).

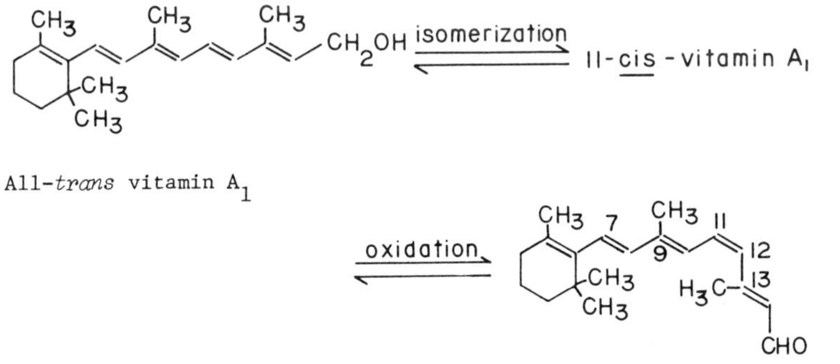

(45)

All-*trans* vitamin A_1

11-*cis* vitamin A_1 aldehyde
(11-*cis* retinene$_1$)

A conformational change produced by isomerization of 11-*cis*-retinene$_1$ to *trans*-retinene$_1$ while attached to the protein, opsin, is responsible for visual excitation (Fig. 23). A vitamin A derivative also functions as a cofactor in the ATP sulfurase reaction.

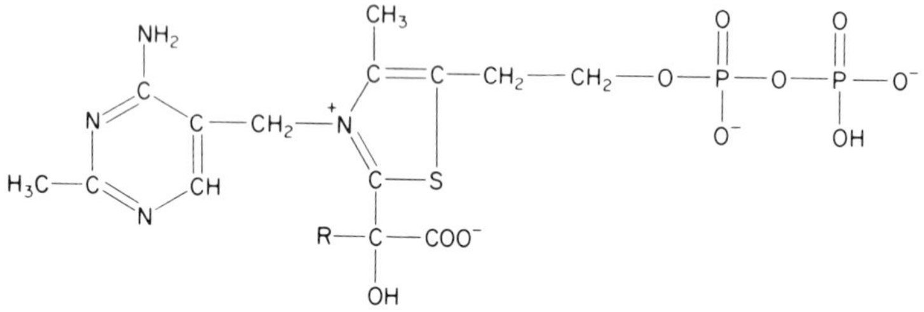

Figure 22 Thiamine pyrophosphate showing the addition of an α-keto compound to carbon 2 of the thiazolium ring.

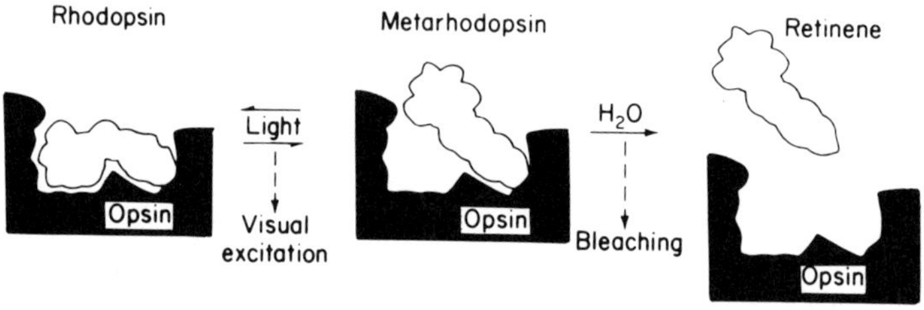

Figure 23 Schematic representation of binding of retinene$_1$ to opsin and the visual process. 11-*cis*-Retinene$_1$ fits the binding site of opsin perfectly to give rhodopsin. Under light stimulation 11-*cis*-retinene$_1$ isomerizes to all-*trans*-retinene$_1$ to give metarhodopsin and visual excitation. Interaction between *trans*-retinene$_1$ and opsin is reduced and the compound is readily hydrolyzed to *trans*-retinene$_1$ and opsin. (From Ref. 8, p. 279, by courtesy of the New York Academy of Sciences.)

B. Tetrahydrobiopterin

Tetrahydrobiopterin (Fig. 24) is a cofactor for the hydroxylation of phenylalanine:

$$\begin{aligned}&\text{tetrahydrobiopterin} + \text{phenylalanine} + O_2\\&\quad\rightarrow \text{dihydrobiopterin} + \text{tyrosine} + H_2O\end{aligned} \tag{46}$$

Biopterin is reduced to tetrahydrobiopterin by a specific NADPH-requiring enzyme.

C. Vitamin K

Vitamin K_2 (Fig. 25) is a cofactor for a number of bacterial steroidal-1-dehydrogenases and is involved in the blood-clotting process in animals. It probably functions in the production of some of the intermediates, such as

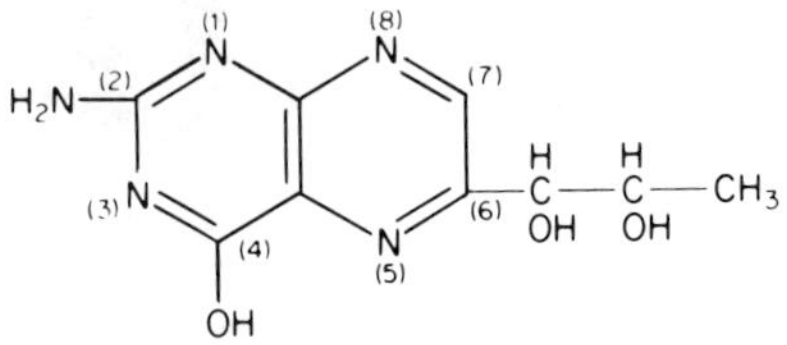

Biopterin

Figure 24 Biopterin. Tetrahydrobiopterin has hydrogens on positions 5, 6, 7, and 8.

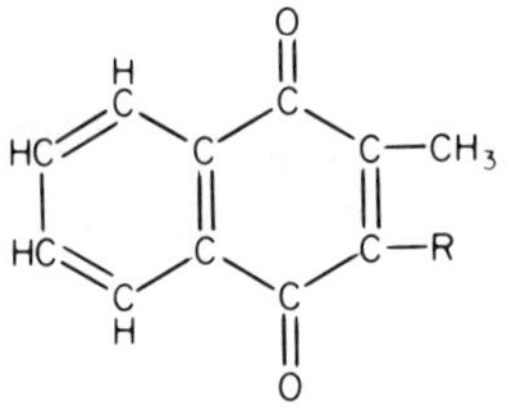

Vitamins K

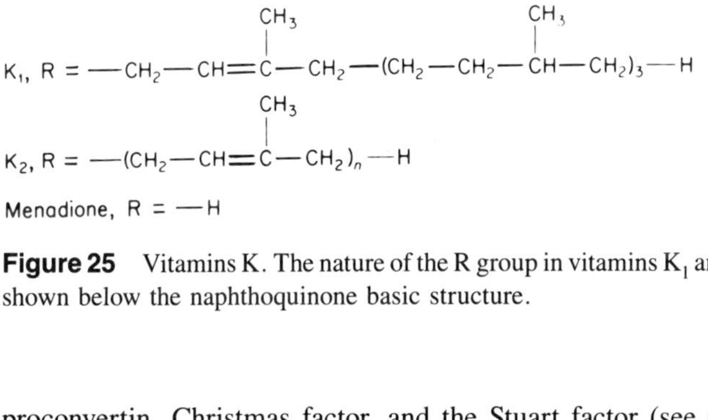

Menadione, R = —H

Figure 25 Vitamins K. The nature of the R group in vitamins K_1 and K_2 and menadione is shown below the naphthoquinone basic structure.

proconvertin, Christmas factor, and the Stuart factor (see Chapter 20, Section II.E). Whether it serves as a cofactor in one or more of these processes is not clear.

D. Vitamin D

Vitamin D (Fig. 26) deficiency leads to rickets in children; the disease can be prevented by giving them adequate amounts of vitamin D in the diet. The behavior of vitamin D is complex in the body, as it is (a) involved in the action of the hormones of the parathyroid glands; (b) apparently involved in the control of the various forms of calcium and phosphorus in the blood, tissue, urine, and feces; and (c) involved in bone growth and repair. Vitamin D *has not* been shown to function as a cofactor in any enzyme system.

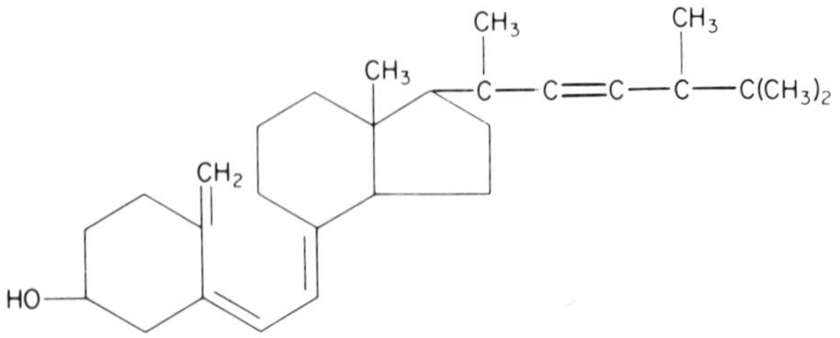

Figure 26 Vitamin D_2 (calciferol, activated ergosterol).

E. Vitamin E

The function of vitamin E (Fig. 27) is as an antioxidant to prevent peroxidation in vivo, particularly of the membranes of cellular organelles, *not* as a cofactor of any enzyme system. Deficiency of vitamin E leads to fragility of membrane structures and to nutritional muscular dystrophy.

V. INORGANIC IONS IN ENZYME-CATALYZED REACTIONS

Inorganic ions play several very key roles in the body. Calcium, magnesium, phosphorus, sodium, and potassium make up in bulk the majority of the inorganic ions. Most calcium, magnesium, and phosphorus is found in the bones, but these elements also play very important roles in several other reactions. Sodium plays an important role in the regulation of neutrality in body fluids, where the majority of sodium is found. Sodium and potassium both play important roles in muscle contraction and in nerve excitability. Unlike sodium, potassium is found predominately inside cells.

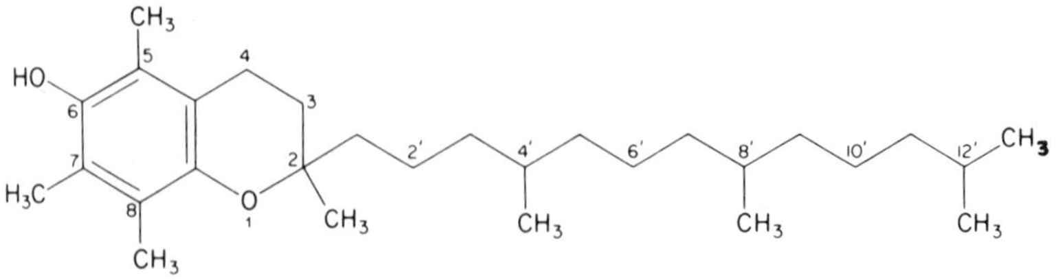

Figure 27 α-Tocopherol (vitamin E).

A. Role of Cations; Specific Effects

Many trace elements are needed by an organism and most of these play a role in enzyme-catalyzed reactions. These compounds include Zn^{2+}, Fe^{2+}, Fe^{3+}, Cu^{2+}, Mo^{5+}, Mg^{2+}, Mn^{2+}, Ni^{2+}, and CO^{2+}. More than 150 enzymes contain Zn^{2+}. In general, the cations perform a specific role in these reactions; the anions play a more general role in modulation of the activity of enzymes.

The role of cations in enzyme-catalyzed systems may be divided into the following four categories: (a) a cation may function as an essential component of the active site of an enzyme; (b) a cation may be a part of an essential cofactor, such as in chlorophyll, the heme cofactors, and coenzyme B_{12}; (c) a cation may be necessary for the maintenance of conformation of the active site of an enzyme; (d) a cation may form part of the substrate for the enzyme.

1. Component of Active Site

As part of the active site of an enzyme, a cation may be involved in binding of substrate, in transformation of substrate to products, or may function in both capacities. Zn^{2+} is essential for the activity of carboxypeptidase A, where it is involved in transformation of substrate to product. Zn^{2+} is a cofactor for alkaline phosphatase as well as helping to bind the two subunits together.

Milk xanthine oxidase contains 8 iron and 2 molybdenum atoms per molecule (MW = 275,000). Electron spin resonance and spectroscopic studies have shown that these atoms are involved in the transfer of electrons from the substrate to O_2 (see Chapter 25).

2. Maintenance of Conformation of Active Site

Yeast alcohol dehydrogenase (MW = 150,000) is composed of four identical polypeptide chains and contains four Zn^{2+} atoms per molecule, each associated with one of the four active sites of the molecule. Zn^{2+} plays two roles in the enzyme. It helps hold the quaternary structure of the molecule as well as participating in the removal of a hydrogen in transformation of substrate to product (Fig. 28). Horse liver alcohol dehydrogenase contains two classes of Zn^{2+}. Two Zn^{2+} atoms are required for conversion of substrate to product and correspond to the two coenzyme-binding sites. The other two Zn^{2+} atoms are required to maintain the quaternary structure of the enzyme. α-Amylase contains one Ca^{2+} atom per molecule, which serves to maintain the conformational integrity of the active site of the enzyme.

3. Part of the Substrate

In a number of cases, a metal ion is required for formation of one of the substrates. For example, the kinases all require Mg^{2+}, which reacts with ATP to form the $MgATP^{2-}$ complex, the true second substrate for the enzyme.

$$Mg^{2+} + ATP^{4-} \rightarrow MgATP^{2-} \xrightarrow{E} E\cdot MgATP^{2-} \xrightarrow{\text{substrate}} \text{products} \quad (47)$$

The pH optimum of polymethylgalacturonic acid lyase of *Aspergillus fonsecaeus* is 8.5 and 5.5 in the presence and absence of Ca^{2+}, respectively [10]. When all the carboxyl groups on pectin are esterified, the pH optimum in the absence of Ca^{2+} is also pH 8.5. These data support the view that the role of Ca^{2+} is to interact with the carboxyl groups of pectin so as to eliminate the negative charge of the carboxylate anion at neutral and alkaline pH and to facilitate binding of substrate to enzyme.

4. Some Specific Metalloproteins and Nature of Binding to Apoenzyme

A small, selected list of enzymes that require metal ions are given in Table 2. Except in the case of catalase and peroxidase, the metal ion is bound to the protein through direct interaction with the polypeptide chain. The most likely, although not always proven methods of attachment would be through interaction with carboxyl, imidazole, sulfhydryl, and amino groups and the carbonyl oxygen and amide nitrogen of the peptide bonds. For example, in carboxypeptidase A the Zn^{2+} is coordinated with the imidazole nitrogens of His-69 and His-196 and the carboxyl group of Glu-72.

B. Role of Anions; General Effect of Cations

The role of anions in activation of enzymes appears to be more general than that for cations. In some cases they stabilize enzyme structure. In other cases they appear to mask the influence of a prototropic group on the enzyme or substrate in binding or transformation of substrate to product.

α-Amylases from mammals and some bacteria require Cl^- for activity. However, the enzyme is not specific for Cl^-, and a number of other monovalent (but not divalent) anions also activate the enzyme. The effect of monovalent ions on the

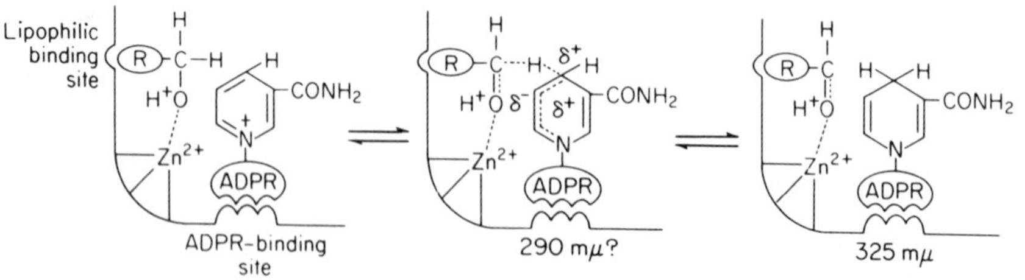

Figure 28 Schematic representation of the catalytic role of Zn^{2+} in the oxidation of ethanol to acetaldehyde by yeast alcohol dehydrogenase. (From Ref. 9, p. 552, by courtesy of Springer-Verlag.)

Table 2 Representative Metal-Requiring Enzymes

Metal	Enzyme	Metal	Enzyme
Zn	Alkaline phosphatase	Mg	Pyruvate decarboxylase
	Carboxypeptidase A		Phosphoenolpyruvate carboxylase
	Carboxypeptidase B		Gluconokinase
	Alcohol dehydrogenase		Pyridoxal kinase
	Carbonic anhydrase	Co	Acetylornithinase
	Leucine aminopeptidase		Glycylglycine dipeptidase
Mn	Arginase	Fe	Catalase
	Mevalonate kinase		Peroxidase
Mo	Xanthine oxidase		Xanthine oxidase
	Nitrate reductase	Ca	Lecithinases A and C
Cu	Polyphenol oxidase		α-Amylase
	Ascorbic acid oxidase		

Source: Ref. 11.

activity of α-amylase has been studied as a function of pH (Fig. 29). There is a difference among the anions in their ability to activate α-amylase, with Cl^- being the most effective. All of the anions cause the pH optimum of the enzyme to shift from a pH near 6 in the absence of monovalent anions to near pH 7 in their presence.

The activity of fumarase is increased two- to threefold by di- and trivalent anions and inhibited by monovalent anions. Like α-amylase, there is a shift in the pH optimum of the enzyme from near 7 in the absence of anions to near 8 in the presence of anions.

The effects of ions on the pH optimum of an enzyme are not restricted to anions. β-Galactosidase (lactase) is activated by sodium ions. The pH optimum of β-galactosidase is 7.7 in the absence of NaCl and 6.6 in the presence of 0.05 *N* NaCl (Fig. 30). The increase in activity in the presence of 0.05 *N* NaCl is dependent on the pH at which the comparison is made. At pH 8 and above, there is little or no increase in activity. However, at pH 6.5 and below there is a tremendous effect on activity. When the reaction is studied at pH 5.0 or below, the enzyme has little or no activity in the absence of NaCl.

The effect of ions on α-amylase, fumarase, and β-galactosidase activities appears to be due to masking of undesirable prototropic groups in the active sites of the enzymes. There is no effect on binding of substrate to the enzyme. All of the effect is due to the influence of these ions on the rate of breakdown of the enzyme·substrate complex. For example, in the case of β-galactosidase, Na^+ is thought to decrease the electrostatic interaction of a nonessential, negatively

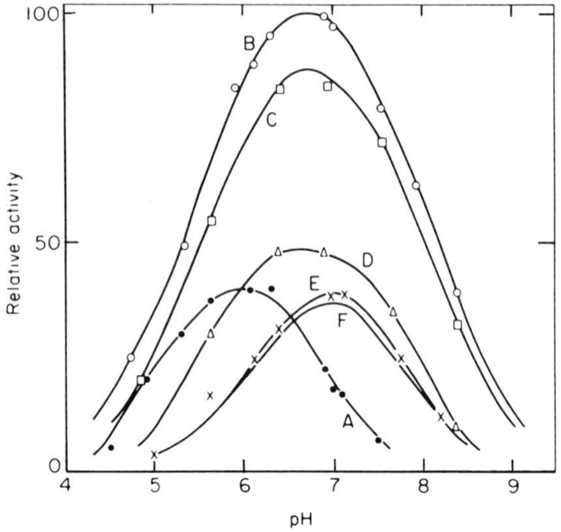

Figure 29 Activation of human salivary α-amylase by anions at various pH values. All reactions contained 0.03 *M* sodium phosphate buffer. The curves are for: A, in absence of monovalent ions; B, in 0.005 or 0.04 *M* chloride; C, in 0.04 *M* bromide; D, in 0.04 *M* iodide; E, in 0.1 *M* nitrate; F, in 0.1 *M* chlorate. (From Ref. 12, p. 35, by courtesy of Walter de Gruyter and Co.)

charged group in or near the active site with one or more of the prototropic groups involved in transformation of substrate (Fig. 31).

VI. APOENZYME-COFACTOR ASSOCIATIONS

There are marked differences among cofactors in the stability of the apoenzyme·cofactor complex. In this section we explore briefly the nature of the interaction between the two. In some cases the exact nature of the interaction is not known and may become known only by extensive x-ray crystallographic work.

One is impressed with the large number of cofactors that contain adenine (or other nucleotide base), ribose, and phosphate (Table 1). In many cases these groups are not involved directly in functioning of the cofactor. These groups play some role in binding of cofactor to protein.

Lipoic acid and biotin are covalently bound to protein through amide linkages formed by interaction of the carboxyl group of cofactor with an ϵ-amino group of a lysyl residue on the enzyme. In some cases, if not all, pyridoxal phosphate is bound to protein through formation of a Schiff base which involves the carbonyl group of the cofactor and an ϵ-amino group of a lysyl residue. On treatment with

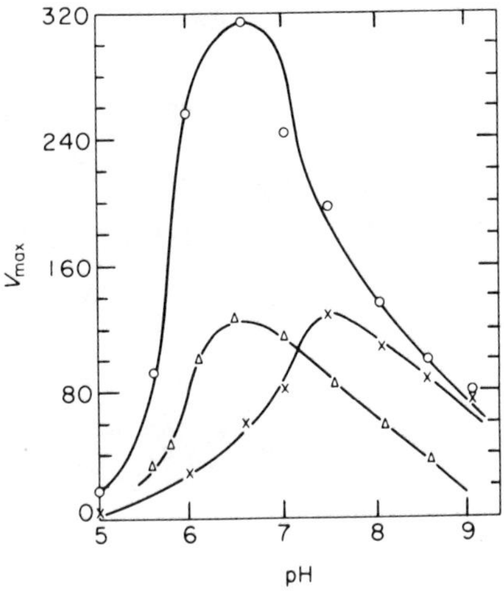

Figure 30 Effect of Na^+ and K^+ on the pH optimum of β-galactosidase-catalyzed hydrolysis of *o*-nitrophenyl-β-galactoside at 20°C. Rate of hydrolysis is expressed as μmoles of substrate hydrolyzed per minute per milligram of enzyme. ○, with 0.05 *M* NaCl; △, with 0.05 *M* KCl; ×, with no NaCl or KCl. (From Ref. 13, p. 381, by courtesy of Springer-Verlag.)

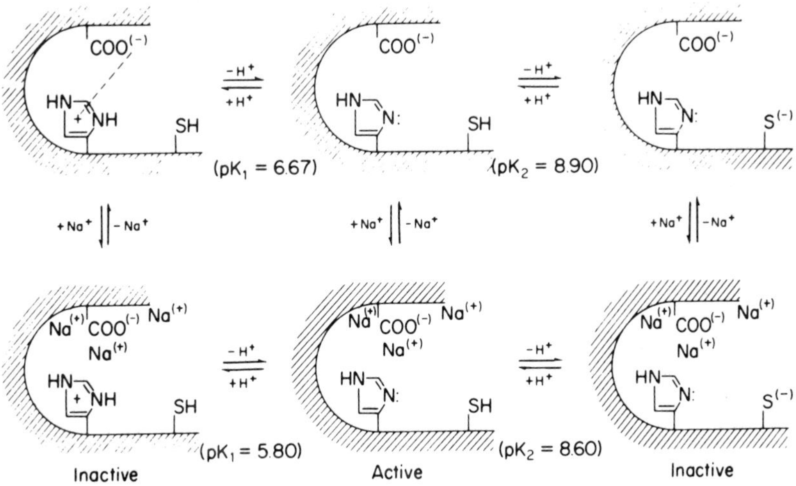

Figure 31 Schematic representation of proposed effect of Na^+ on ionization of groups in the catalytic locus of β-galactosidase. The presence of Na^+ reduces the electrostatic interaction between the COO^- and the imidazolium group, which is reflected as a decrease in pK_a from 6.67 to 5.80 and permits the imidazole group (center) to participate with the —SH group in conversion of substrate to product. (From Ref. 14, p. 429, by courtesy of Academic Press.)

$NaBH_4$, the Schiff base is converted to a stable compound that can be isolated after degradation of the protein [Eq. (48)].

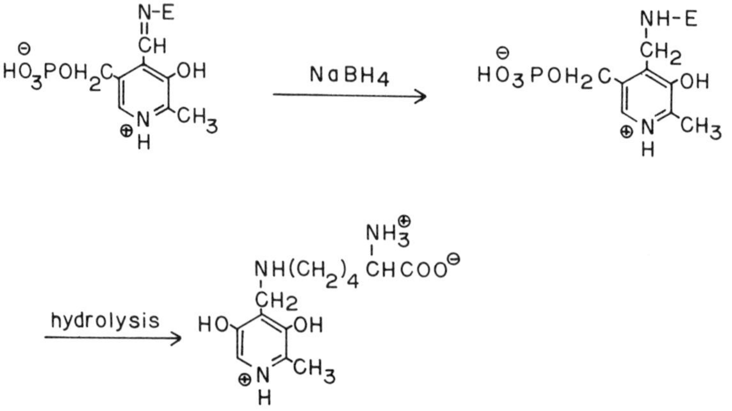

(48)

Binding of heme in catalase and peroxidase involves coordination of the iron atom with one (or two) imidazole groups of the protein, electrostatic interaction of the two carboxylate groups of the propionic acid side chains of the pyrroles with positively charged groups on the protein and hydrophobic interaction of some of the other side chains with aromatic amino acid residues on the protein. In cytochrome *c*, the heme is attached through two thiolether linkages to the protein.

Ca^{2+} is held very tightly ($K_b = 10^{12}$ to $10^{15}\ M^{-1}$) in porcine pancreatic α-amylase in the pH range 6 to 9, and prolonged dialysis against EDTA is not successful in removing it. Above and below this pH range EDTA can remove the Ca^{2+}, suggesting the involvement of one or more prototropic groups in its binding. The Zn^{2+} in carboxypeptidase A is bound through its interaction with the carboxyl group of Glu-72 and the imidazole groups of His-69 and His-196.

VII. ROLE OF PROTEIN IN ENZYMES THAT CONTAIN COFACTORS

In discussions of the functions of cofactors we have ignored the role of the protein part (apoenzyme) of the enzyme. A few of the cofactors have small but measurable activity in the absence of protein. For example, hemin decomposes H_2O_2 at about 6×10^{-7} the rate of peroxidase. Pyridoxal phosphate, at elevated temperatures and in the presence of metal ions, slowly decarboxylates an α-amino acid. However, the rates are very much slower than in the enzymatic reaction. The cofactors alone show little or no specificity in the reactions they perform.

The protein part of the enzyme confers on the system the stereospecific properties characteristic of an enzyme, due largely to the strict requirements imposed for correct binding of substrate; second, it greatly increases the rate of the reaction, due in part to the stereospecific binding of substrate at the active site. Factors that contribute to the enhanced enzymatic rates are treated in detail in Chapter 4 and elsewhere in the book.

REVIEW QUESTIONS

1. Define the following: (a) holoenzyme; (b) apoenzyme; (c) prosthetic group; (d) coenzyme; (e) activator.
2. A protein preparation with catalytic activity contains 0.1% iron. Suggest several experiments that might prove whether the catalytic activity is due to an iron-requiring enzyme or to the iron salts alone.
3. How would you determine experimentally the dissociation constant of a cofactor·apoenzyme complex?
4. What cofactor(s) would you expect to be involved in each of the following enzyme-catalyzed reactions? Classify the cofactor as to coenzyme or prosthetic group. Show, by use of chemical structures, how the cofactor participates in the reaction.
   (a) The oxidation of succinic acid to fumaric acid and the concurrent reduction of O_2.
   (b) The conversion of α-keto-β-hydroxypropionic acid to serine.
   (c) The conversion of stearic acid to acetic acid.
   (d) The hydrolysis of casein by chymotrypsin.
   (e) The hydrolysis of starch by α-amylase.
   (f) The decomposition of hydrogen peroxide by catalase.
   (g) The conversion of ethanol to acetaldehyde.
   (h) The synthesis of sucrose from glucose and fructose.
5. Write complete chemical equations showing at least one reaction catalyzed by an enzyme containing each of the following cofactors. What would be the general name of the enzyme?
   (a) Nicotinamide adenine dinucleotide (NAD^+)
   (b) α-Ketoglutaric acid
   (c) Pyridoxal phosphate
   (d) Thiamine pyrophosphate
   (e) Lipoic acid
   (f) Coenzyme A
6. Coenzyme B_{12} is involved in a number of reactions involving isomerization of the substrate. Two of these are given in Section II.G. Speculate as to the mechanism(s) involved in these isomerizations.
7. Discuss the general effect of pH on a cofactor in relation to its function in an enzyme system.

8. Would you expect an apoenzyme to be more or less resistant to denaturation in the presence of its specific cofactor? Explain.

9. Of the several possible functions of metal ions in enzyme systems, which is the most probable function of copper in polyphenol oxidase and ascorbic acid oxidase? Write chemical equations illustrating this function.

10. For each of the cofactors described in this chapter, suggest one or more compounds that might react with the cofactor to inhibit enzyme activity.

REFERENCES

General References

Boguth, W., Aspects of the actions of vitamin E, in *Vitamins and Hormones* (R. S. Harris, I. G. Wool, J. A. Loraine, and P. L. Munson, eds.), Vol. 27, Academic Press, New York, 1969, p. 1.

Chaykin, S., Nicotinamide coenzymes, *Ann. Rev. Biochem.*, **36**, 149 (1967).

Daves, G. D., Jr., P. Firis, R. K. Olsen, and K. Folkers, The chemistry of ubiquinone, in *Vitamins and Hormones* (R. S. Harris, I. G. Wool, and J. A. Loraine, eds.), Vol. 24, Academic Press, New York, 1966, p. 427.

DeLuca, H. F., Mechanism of action and metabolic fate of vitamin D, in *Vitamins and Hormones* (R. S. Harris, I. G. Wool, and J. A. Loraine, eds.), Vol. 25, Academic Press, New York, 1967, p. 316.

Fasella, P., Pyridoxal phosphate, *Ann. Rev. Biochem.*, **36**, 185, (1967).

Griminger, P., Biological activity of the various K forms, in *Vitamins and Hormones* (R. S. Harris, I. G. Wool, and J. A. Loraine, eds.), Vol. 24, Academic Press, New York, 1966, p. 605.

Hogenkamp, H. P. C., Enzymatic reactions involving corrinoids, *Ann. Rev. Biochem.*, **37**, 225 (1968).

Kaufman, S., Pteridine cofactors, *Ann. Rev. Biochem.*, **36**, 171 (1967).

Krampitz, L. O., Catalytic functions of thiamine diphosphate, *Ann. Rev. Biochem.*, **38**, 213 (1969).

Olson, J. A., Some aspects of vitamin A metabolism, in *Vitamins and Hormones* (R. S. Harris, I. G. Wool, and J. A. Loraine, eds.), Vol. 26, Academic Press, New York, 1968, p. 1.

Page, M. I., and A. Williams, eds., *Enzyme Mechanisms*, Royal Society of Chemistry, London, 1987.

Reed, L. J., Biochemistry of lipoic acid, in *Vitamins and Hormones* (R. S. Harris and I. G. Wool, eds.), Vol. 20, Academic Press, New York, 1962, p. 1.

Vallee, B. L., and W. E. C. Wacker, Metalloproteins, in *The Proteins, Composition, Structure and Function* (H. Neurath, ed.), Vol. V, Academic Press, New York, 1970.

Walsh, W., *Enzymatic Reaction Mechanisms*, W.H. Freeman, San Francisco, 1979.

Wong, D. W. S., *Mechanism and Theory in Food Chemistry*, Chapter 10, Van Nostrand Reinhold, New York, 1989.

Cited References

1. M. Dixon and E. C. Webb, *Enzymes*, 2nd Ed., Academic Press, New York, 1964.
2. F. L. Crane, Y. Hatefi, R. L. Lester, and C. Widmer, *Biochim. Biophys. Acta*, **25**, 220 (1957).
3. L. J. Reed, I. C. Gunsalus, G. H. F. Schnakenberg, Q. F. Soper, H. E. Boaz, S. F. Kern, and T. V. Parke, *J. Am. Chem. Soc.*, **75**, 1267 (1953).
4. F. Lipmann, *Science*, **128**, 575 (1958).
5. G. R. Minot and W. P. Murphy, *J. Am. Med. Assoc.*, **87**, 470 (1926).
6. O. Warburg and W. Christian, *Biochem. Z.*, **254**, 438 (1932).
7. O. Warburg and W. Christian, *Biochem. Z.*, **298**, 368 (1938).
8. A. Kropf and R. Hubbard, *Ann. N. Y. Acad. Sci.*, **74**, 266 (1958).
9. H. Theorell and T. Yonetani, *Biochem. Z.*, **338**, 537 (1963).
10. R. D. Edstrom and H. J. Phaff, *J. Biol. Chem.*, **239**, 2403 (1964).
11. B. L. Vallee and J. E. Coleman, Metal coordination and enzyme action, in *Comprehensive Biochemistry* (M. Florkin and E. C. Stotz, eds.), Vol. 12, Elsevier, Amsterdam, 1964, p. 165.
12. K. Myrbäck, *Z. Physiol. Chem.*, **159**, 1 (1926).
13. K. Wallenfels, O. P. Malhotra, and D. Dabich, *Biochem. Z.*, **333**, 377 (1960).
14. K. Wallenfels and O. P. Malhotra, β-Galactosidase, in *The Enzymes* (P. D. Boyer, H. Lardy, and K. Myrbäck, eds.), Vol. IV, Academic Press, New York, 1960, p. 409.

13
Classification and Nomenclature of Enzymes

I. BASES FOR NAMING OF ENZYMES AND PROBLEMS WITH NOMENCLATURE

A. Miscellaneous

In 1833, Payen and Persoz found that an alcohol precipitate of malt extract contained a thermolabile substance that converted starch to sugar. They named this substance *diastase* because of its ability to separate soluble dextrin from the insoluble envelopes of starch grains. This set a general unorganized pattern for naming enzymes. The name given to an enzyme often had nothing to do with the type of reaction catalyzed. For example, the names *diaphorase*, *Zwischenferment*, *catalase*, *pepsin*, and *trypsin* tell nothing about the chemical reaction. In other cases, the name is completely misleading.

B. Discoverer; Color

An enzyme could be named after its discoverer or the name could be based on its color when highly purified. The first riboflavin-containing enzyme discovered by Warburg and Christian in 1932 became known as "the yellow enzyme," "the yellow enzyme of Warburg and Christian," "the yellow enzyme of Warburg," and "the old yellow enzyme of Warburg," which was undoubtedly most confusing to chemists and other scientists not versed in enzymology.

C. Genus or Species from Which Isolated

A number of enzymes were named after the species or genus of the organism in which they are found. For example, the group of proteolytic enzymes from the genus *Ficus* have all been called ficin, even though there are multiple molecular forms of these enzymes within any one plant. One of the more intensively studied plant proteolytic enzymes, papain, is named after the species of plant from which it is obtained, *Carica papaya*. The names of a number of the plant enzymes end in "-ain."

D. Methods of Isolation

In the early 1960s when work started on the enzymes that catalyze reaction between sRNA and the carboxyl group of amino acids, the group of enzymes involved were called the "pH 5 enzymes" because they had their maximum activity at this pH. Just as misleading, they soon became known as the "activating enzymes." Since all enzymes catalyze the transformation of substrate to product by activation, such a designation is meaningless. However, to date, the enzyme that converts H_2O_2 to H_2O and O_2 is known as "catalase."

E. Substrate Utilized

One attempt at a systematic nomenclature in the early days of enzymology was the suggestion that "-ase" be added to the stem of the substrate upon which the enzyme acts. The name clearly indicated what the substrate was. This system worked well for the very specific enzymes and as long as there were very few substrates for an enzyme. It is clear that maltase is the enzyme that acts on maltose, urease is the enzyme that acts on urea, and amylase is the enzyme that acts on amylose. But there are at least three difficulties with this nomenclature.

First, as more and more substrates became available for an enzyme it became more difficult to know which substrate to use for the name. For example, an enzyme of plants which oxidizes phenolic compounds has been called cresolase, catecholase, phenolase, polyphenol oxidase, tyrosinase, and so on, depending on the nature of the substrate used.

Second, the nomenclature does not communicate the nature of the chemical reaction catalyzed by the enzyme. Without prior knowledge, one cannot tell that lactase, maltase, and amylase are all enzymes that hydrolyze glycosidic bonds in the appropriate substrate. It is equally difficult to tell that the enzymes pepsin, trypsin, chymotrypsin, and carboxypeptidase A all hydrolyze peptide bonds. In fact, the first three names do not even indicate that they are enzymes.

The last problem with the nomenclature is demonstrated with the group of enzymes above. All four enzymes—pepsin, trypsin, chymotrypsin, and carboxypeptidase A—act on proteins. Therefore, they are all called proteinases. As a generic name this is all right, but as a specific name it does not permit one to distinguish among these very different enzymes. As more was learned of the mechanism of action of these four enzymes it became clear that pepsin, trypsin, and chymotrypsin have specificity toward selected peptide bonds in the interior of the protein, while carboxypeptidase A hydrolyzes only the peptide bond at the carboxyl terminal end of the protein. Therefore, pepsin, trypsin, and chymotrypsin are endopeptidases or endoproteinases; carboxypeptidase A is an exopeptidase or exoproteinase.

F. Same Specificity, Different Source and Physical and Chemical Properties

As more and more enzymes were described, other problems developed in their systematic nomenclature. It became clear that enzymes from different organisms that catalyzed the same enzymatic reaction often differ markedly in their chemical and physical properties. For example, yeast alcohol dehydrogenase is a protein of 150,000 molecular weight, contains four subunits, and has four Zn^{2+} atoms per molecule. Horse liver alcohol dehydrogenase also catalyzes the conversion of ethanol to acetaldehyde in the presence of NAD^+; however, it has a molecular weight of 70,000, contains two subunits and four atoms of Zn^{2+}, and has quite a different amino acid composition from that of yeast alcohol dehydrogenase.

There are two classes of fructose-bisphosphate aldolases. The class I aldolases ("lysine aldolases") are found in animals, plants, protozoa, green algae, *Euglena*, and *Chlamydomonas*, involve a Schiff base intermediate (lysine residue) and carboxyl and tyrosyl groups for activity, and have molecular weights of 140,000 to 160,000. The class II aldolases ("metalloaldolases") are found in bacteria, yeast, fungi, blue-green algae, *Euglena*, and *Chlamydomonas*, require a divalent metal ion (probably Zn^{2+}), contain an essential sulfhydryl group, and have molecular weights of approximately 70,000. These two classes of fructose-bisphosphate aldolases catalyze the same overall reaction, converting D-fructose 1,6-bisphosphate to glycerone phosphate and D-glyceraldehyde 3-phosphate.

α-Lytic protease from the microorganism *Sorangium* sp. has kinetic, physical, and chemical properties that are nearly identical to those of the mammalian proteolytic enzyme α-chymotrypsin. Bovine trypsin and human trypsin have the same substrate specificities and the same general physical and chemical properties. However, bovine trypsin is strongly inhibited by ovomucoid trypsin inhibitor, whereas human trypsin is only weakly inhibited.

For these reasons the name of an enzyme is not complete unless its source is indicated. The source should specify the organism, the organ, and the organelle (if known). Enzymes that catalyze the same chemical reaction often differ in physical and chemical properties from one type organelle to another.

G. Multiple Molecular Forms

Another major problem concerns the multiple molecular forms of enzymes found in tissues. There is need for a meaningful way of differentiating among the various forms so that other workers in the field will clearly know which multiple molecular form is being described. The nomenclature will also need to distinguish among the genetically engineered forms (based on recombinant DNA technology).

II. COMMISSION ON ENZYMES

All these problems, with the exception of that of multiple molecular forms, had become abundantly clear by 1955. At that time the General Assembly of the International Union of Biochemistry decided during its Third International Congress of Biochemistry in Brussels that something had to be done to bring order out of chaos. An international Commission on Enzymes of the International Union of Biochemistry was established with 10 members chosen from among the biochemists of the United Kingdom, the United States, the USSR, Denmark, France, Germany, and Austria. The charge to the Enzyme Commission was "to consider the classification and nomenclature of enzymes and coenzymes, their units of activity and standard methods of assay, together with the symbols used in the description of enzyme kinetics."

The members of the Enzyme Commission did not take their work lightly; they seemed to realize that this might be the last opportunity to bring order to a field that was growing by leaps and bounds. In 1961 the Enzyme Commission submitted its report to the Council of the International Union of Biochemistry during the Fifth International Congress of Biochemistry in Moscow [1]. The report has been updated four times since 1961 [2,3] and is continuously under revision as new enzymes are described; however, it is hoped the basic features of the system will be permanent. The *Report of the Enzyme Commission* (1961) contained 712 enzymes; *Enzyme Nomenclature* (1964), 875; *Enzyme Nomenclature* (1972), 1770; *Enzyme Nomenclature* (1978), 2122; *Enzyme Nomenclature* (1978) plus *Supplements* 1–4, 2252; *Enzyme Nomenclature* (1984), 2477 enzymes.

The Enzyme Commission considered several alternative systems of classifying and naming enzymes: (a) the chemical nature of the enzyme (flavoprotein, pyridoxal phosphate enzyme, hemoprotein, etc.), (b) the chemical nature of the substrate (proteins, carbohydrates, nucleotides, etc.), and (c) the type of reaction catalyzed. The problem of classifying enzymes on the basis of cofactors involved is that not all enzymes have recognizable cofactors and even those with the same cofactor do not always catalyze the same type of reaction. The second system, based on the chemical nature of substrate, would be inadequate, as it could not distinguish among the many diverse enzymes acting on any one polymeric substrate. A classification based on the type of chemical reaction catalyzed seems to be a logical classification since this is really the basis for distinguishing one enzyme from another. Therefore, the basis for the classification adopted was to *divide the enzymes into groups on the basis of the type of reaction catalyzed, and this, together with the name(s) of the substrate(s), provided a basis for naming individual enzymes.*

III. RECOMMENDED CLASSIFICATION OF ENZYMES

A. Rules for Naming

We shall consider first some of the rules established by the Enzyme Commission for naming enzymes. All the rules cannot be reproduced here; the original reports should be consulted for more details [1–3].

1. Use of "-ase"

It is recommended that the names, especially those ending in "-ase," be for single enzymes only. These names are not to be applied to complex biological systems containing more than one enzyme. For example, the single enzyme that converts succinate to fumarate is "succinate dehydrogenase" (trivial name), while the system of enzymes that is responsible for the oxidation of succinate and the transfer of electrons all the way to O_2 is called the "succinate oxidase system." All of the systematic names of enzymes end in "-ase." However, the suffix "-ase" will not be attached to the name of the substrate directly but rather to words that describe the specific type of reaction.

2. Overall Reaction as Basis of Nomenclature

The overall reaction, as expressed by the formal equation, is the basis for the nomenclature. Intermediate steps in the reaction reflecting the intimate mechanism are not taken into account. An enzyme cannot be named properly under this system until it is known what reaction the enzyme catalyzes. Special care must be taken in naming enzymes that catalyze reversible reactions. For example, alcohol dehydrogenase catalyzes both the reaction of ethanol with NAD^+ and the reaction of acetaldehyde with NADH. The rule to be followed is that the direction chosen for naming should be the same for all the enzymes in a given class, even if this direction has not been demonstrated. Therefore, the names of the dehydrogenases are based on the *substrate in the reduced form* and the *cofactor in the oxidized form*. Thus alcohol dehydrogenase is alcohol:NAD oxidoreductase. The Enzyme Commission recommended the use of NAD, NADH, NADP, and NADPH rather than DPN^+, DPNH, TPN, and TPNH. This recommendation was adopted belatedly by biochemists in the United States.

In a transfer type of reaction a system based either on the nature of the group transferred or upon the compound involved in breakage of the bond could be used:

$$X{-}Y + Z \rightleftharpoons X + Y{-}Z \quad (1)$$

Because of the large number of enzymes involved in these types of reactions, the Enzyme Commission chose to divide the enzymes into two groups, the hydrolases and the transferases. The *hydrolase* group is defined by the requirement that water (Z) be the compound involved in breaking the X—Y bond. Naming of the second group of enzymes, the *transferases*, is based on the type of group (Y) transferred.

3. Three-System Designation

Each well-defined enzyme is given three designations: a systematic name, a trivial name, and a number. In general, a systematic name of an enzyme is composed of two parts. The first part consists of the name of the substrate, or in the case of a two-substrate reaction, of the two substrate names separated by a colon. The second part of the name, ending in "-ase," indicates the nature of the chemical reaction, such as "oxidoreductase," "racemase," "isomerase," "epimerase," "lyase," "ligase," "hydrolase," "transferase" and "mutase." When the overall reaction includes two different chemical reactions (e.g., an oxidative demethylation), the second type of reaction is indicated by adding a suitable participle in parentheses: for example, "sarcosine:oxygen oxidoreductase (demethylating)."

A trivial name is one that is generally recognized and is in common usage for an enzyme, such as urease, α-amylase, cellulase, papain, and pepsin. It is a short name, whereas the systematic name is often quite long. In most cases, the trivial name has remained as the one in common use.

Suggested use of the systematic and trivial names was given. When an enzyme is the main subject of a paper or abstract, its systematic name, number, and source should be given when first mentioned in the text. An enzyme that oxidizes ethanol in the presence of NAD^+ would be "alcohol:NAD^+ oxidoreductase (EC 1.1.1.1) from yeast." Thereafter, the trivial name, alcohol dehydrogenase, is used. If an enzyme is not the main subject of a paper (e.g., it is mentioned in connection with reagents used), the trivial name followed by the code number in parentheses should be used at first mention. This is shown as "yeast alcohol dehydrogenase (EC 1.1.1.1)."

The code numbers come directly from the classification scheme and each number contains four elements, separated by periods. The numbers have been arranged so as to be permanent. New enzymes, when properly described, will be placed at the end of the list under the appropriate headings.

B. Basis for Numbering System and Nomenclature

The scheme for the numbering of enzymes is based on the following principles [3].

1. First Number

The first number (1 through 6) shows to which of the six main groups of enzymes the particular enzyme belongs. The six groups are made on the basis of the general chemical reaction catalyzed as:

1. Oxidoreductases
2. Transferases
3. Hydrolases
4. Lyases
5. Isomerases
6. Ligases

a. Oxidoreductases

The oxidoreductases are enzymes that oxidize or reduce substrates by transfer of hydrogen or electrons. The systematic name is formed on the pattern "donor: acceptor oxidoreductase."

Examples are (trivial name in parentheses followed by code number):

1. alcohol + NAD^+ = aldehyde + NADH
   alcohol:NAD^+ oxidoreductase (alcohol dehydrogenase, EC 1.1.1.1)
2. β-D-glucose + O_2 = D-glucono-1,5-lactone + H_2O_2
   β-D-glucose:oxygen 1-oxidoreductase (glucose oxidase, EC 1.1.3.4)
3. $H_2O_2 + H_2O_2 = O_2 + 2H_2O$
   hydrogen peroxide:hydrogen-peroxide oxidoreductase (catalase, EC 1.11.1.6)

b. Transferases

The transferases are enzymes that remove groups (not including H) from substrates and transfer them to acceptor molecules (not including water). The systematic name is formed on the pattern "donor:acceptor group-transferred-transferase."

Examples are (trivial name in parentheses followed by code number):

1. acetyl-CoA + choline = CoA + *O*-acetylcholine
   acetyl-CoA:choline *O*-acetyltransferase (choline acetyltransferase, EC 2.3.1.6)
2. L-aspartate + 2-oxoglutarate = oxaloacetate + L-glutamate
   L-aspartate:2-oxoglutarate aminotransferase (aspartate aminotransferase, EC 2.6.1.1; formerly known as glutamic-oxaloacetic transaminase).
3. ATP + D-glucose = ADP + D-glucose 6-phosphate
   ATP:D-glucose 6-phosphotransferase (glucokinase, EC 2.7.1.2)

Note that the position to which the group is transferred is specified when more than one possibility exists.

c. Hydrolases

The hydrolases are enzymes in which water participates in the breakage of covalent bonds in compounds, with concurrent addition of the elements of water to the participles of those bonds. The systematic name is formed on the pattern "substrate hydrolase." Where the enzyme is specific for the removal of a specific group, the group is named as a prefix: for example, "adenosine aminohydrolase."

Examples are (trivial name in parentheses followed by the code number):

1. Triacylglycerol + H_2O = diacylglycerol + a fatty acid anion
   triacylglycerol acylhydrolase (triacylglycerol lipase, EC 3.1.1.3)
2. D-glucose 1-phosphate + H_2O = D-glucose + orthophosphate
   D-glucose-1-phosphate phosphohydrolase (glucose-1-phosphatase, EC 3.1.3.10)
3. Preferential cleavage: Tyr-, Trp-, Phe-, Leu-
   Chymotrypsin A and B (chymotrypsin, EC 3.4.21.1)
   Preferential cleavage means cleavage at the carbonyl end of the amino acid whose symbol is given.
4. urea + H_2O = CO_2 + $2NH_3$
   urea amidohydrolase (urease, EC 3.5.1.5)

d. Lyases

The lyases are enzymes that remove groups from their substrates (not by hydrolysis) to leave a double bond, or which conversely add groups to double bonds. The name is formed on the order "substrate prefix-lyase." Prefixes such as "hydro-" and "ammonia-" are used to indicate the type of reaction: for example, "L-malate hydro-lyase" (EC 4.2.1.2). Decarboxylases are named as carboxy-lyases. A hyphen is always written before "lyase" to avoid confusion with hydrolases, carboxylases, and so on.

Examples are (trivial name in parentheses followed by code number):

1. oxalate = formate + CO_2
   oxalate carboxy-lyase (oxalate decarboxylase, EC 4.1.1.2)
2. (*S*)-malate = fumarate + H_2O
   (*S*)-malate hydro-lyase (fumarate hydratase, EC 4.2.1.2; formerly known as fumarase)
3. L-histidine = urocanate + NH_3
   L-histidine ammonia-lyase (histidine ammonia-lyase, EC 4.3.1.3; formerly known as histidine α-deaminase and histidase)

e. Isomerases

The isomerases are enzymes that bring about an isomerization of substrate. The name is formed on the order "substrate prefix-isomerase." The prefix indicates the type of isomerization involved: for example, "maleate *cis-trans*-isomerase" (EC 5.2.1.1), "phenylpyruvate keto-enol-isomerase" (EC 5.3.2.1). Enzymes that catalyze an aldose-ketose interconversion are known as "ketol-isomerases": for example, "L-arabinose ketol-isomerase" (EC 5.3.1.4). When the isomerization consists of an intramolecular transfer of a group, such as 2-phospho-D-glycerate = 3-phospho-D-glycerate, the enzyme is named a "mutase": for example, "D-phosphoglycerate 2,3-phosphomutase" (EC 5.4.2.1). The isomerases that catalyze inversions of asymmetric groups are termed "racemases" or "epimerases," depending on whether the substrate contains one or more than one center of asymmetry, respectively. A numerical prefix is attached to the word "epimerase" to show the position of inversion.

Examples are (trivial name in parentheses followed by code number):

1. L-alanine = D-alanine
   alanine racemase (alanine racemase, EC 5.1.1.1)
2. D-ribulose 5-phosphate = D-xylulose 5-phosphate
   D-ribulose-5-phosphate 3-epimerase (ribulose-phosphate 3-epimerase, EC 5.1.3.1)
3. D-mannose = D-fructose
   D-mannose ketol-isomerase (mannose isomerase, EC 5.3.1.7)

f. Ligases

The ligases are enzymes that catalyze the linking together of two molecules, coupled with the breaking of a pyrophosphate bond as in ATP. This group of enzymes has previously been referred to as the "synthetases." The systematic name is formed on the pattern "X:Y ligase (Z)," where X and Y are the two molecules to be joined together. The compound in parentheses (Z) is the *product* formed from the triphosphate during the reaction.

Examples are (trivial name in parentheses followed by code number):

1. ATP + L-tyrosine + *t*RNA^{Tyr} = AMP + pyrophosphate + L-tyrosyl-*t*RNA^{Tyr}
   L-tyrosine:*t*RNA^{Tyr} ligase (AMP-forming) (tyrosyl-*t*RNA ligase, EC 6.1.1.1)
2. ATP + L-aspartate + NH_3 = AMP + pyrophosphate + L-asparagine
   L-aspartate:ammonia ligase (AMP-forming) (asparate-ammonia ligase, EC 6.3.1.1)
3. ATP + pyruvate + HCO_3^- = ADP + orthophosphate + oxaloacetate
   pyruvate:carbon-dioxide ligase (ADP-forming) (pyruvate carboxylase, EC 6.4.1.1)

2. Second Number

The second number in the code number indicates the subclass. For the *oxidoreductases* it indicates the type of group on the hydrogen *donor* which undergoes oxidation. For example, 1 denotes an alcohol group (first two elements are 1.1), 2 an aldehyde or keto-group (1.2), and so on. For the *transferases* the second number indicates the nature of the group transferred: one-carbon group (2.1), aldehydic or ketonic residues (2.2), acyl groups (2.3), and so on. For the *hydrolases* the second number indicates the type of bond hydrolyzed: ester (3.1), glycosyl (3.2), peptide (3.4), and so on. For the *lyases* the second number indicates the type of linkage which is broken between the group removed and the remainder of the molecule: carbon–carbon (4.1), carbon–oxygen (4.2), carbon–nitrogen (4.3), and so on. For the *isomerases* the second number indicates the type of isomerization involved: racemization and epimerization (5.1), *cis-trans* (5.2), intramolecular oxidoreduction (5.3), and so on. For the *ligases* the second number indicates the type of bond formed: carbon–oxygen (6.1), carbon–sulfur (6.2), carbon–nitrogen (6.3), and so on. The nature of the second number is illustrated in Table 1.

3. Third Number

The third number indicates the sub-subclass. For the *oxidoreductases* the third number indicates for each type of donor the type of acceptor involved. For example, 1 denotes NAD(P) (the first three numbers are 1.1.1), 2 a cytochrome (1.1.2), 3 a molecular oxygen (1.1.3), and so on. For the *transferases* the third number subdivides the type of group transferred: whether the one-carbon group transferred (2.1) is a methyl (2.1.1), a hydroxymethyl (2.1.2) or formyl (2.1.3) group, and so on. In the case of the phosphotransferases, the third number indicates the type of acceptor. For the *hydrolases* the third number indicates more precisely the type of bond hydrolyzed. For example, with an ester bond (3.1) it indicates whether the ester is carboxylic ester (3.1.1), thiolester (3.1.2), monophosphate ester (3.1.3), and so on. For the *lyases* the third number defines more precisely the nature of the group eliminated. For example, with the carbon–carbon lyases (4.1) it indicates whether the bond breakage involves release of CO_2 (4.1.1), aldehyde (4.1.2), ketoacid (4.1.3), and so on. For the *isomerases* the third number defines more precisely the nature of the transformation. For example, in the racemases and epimerases (5.1), it indicates whether the substrates involved are amino acids and derivatives (5.1.1), hydroxy acids and derivatives (5.1.2), or carbohydrates and derivatives (5.1.3). For the *ligases*, the third number defines the nature of the substance formed. For example, those ligases that form carbon–nitrogen bonds (6.3) may produce amide bonds (6.3.1), peptide bonds (6.3.2), and so on. The nature of the third number is illustrated in detail in Table 1.

4. Fourth Number

The fourth element in the code number is the serial number of the enzyme in its sub-subclass. For example, with the *oxidoreductases* (1), which act on the —CHOH group of donors (1.1) with NAD^+ or $NADP^+$ as acceptor (1.1.1), the fourth element denotes the specific type of alcohol and coenzyme (NAD^+ or $NADP^+$) involved. This is shown in the three examples given.

1. alcohol + NAD^+ = aldehyde or ketone + NADH
   alcohol:NAD^+ oxidoreductase (alcohol dehydrogenase, EC 1.1.1.1)
2. alcohol + $NADP^+$ = aldehyde + NADPH
   alcohol:$NADP^+$ oxidoreductase (alcohol dehydrogenase ($NADP^+$), EC 1.1.1.2)
3. L-homoserine + $NAD(P)^+$ = L-aspartate 4-semialdehyde + NAD(P)H
   L-homoserine:$NAD(P)^+$ oxidoreductase (homoserine dehydrogenase, EC 1.1.1.3)

Table 1 Key to Numbering and Classification of Enzymes

1. Oxidoreductases
   - 1.1 Acting on the CH—OH group of donors
     - 1.1.1 With NAD^+ or $NADP^+$ as acceptor
     - 1.1.2 With a cytochrome as acceptor
     - 1.1.3 With oxygen as acceptor
     - 1.1.5 With a quinone or related compound as acceptor
     - 1.1.99 With other acceptors
   - 1.2 Acting on the aldehyde or oxo group of donors
     - 1.2.1 With NAD^+ or $NADP^+$ as acceptor
     - 1.2.2 With a cytochrome as acceptor
     - 1.2.3 With oxygen as acceptor
     - 1.2.4 With a disulfide compound as acceptor
     - 1.2.7 With an iron-sulfur protein as acceptor
     - 1.2.99 With other acceptors
   - 1.3 Acting on the CH—CH group of donor
     - 1.3.1 With NAD^+ or $NADP^+$ as acceptor
     - 1.3.2 With a cytochrome as acceptor
     - 1.3.3 With oxygen as acceptor
     - 1.3.5 With a quinone or related compound as acceptor
     - 1.3.7 With an iron-sulfur protein as acceptor
     - 1.3.99 With other acceptors
   - 1.4 Acting on the CH—NH_2 group of donors
     - 1.4.1 With NAD^+ or $NADP^+$ as acceptor
     - 1.4.2 With a cytochrome as acceptor
     - 1.4.3 With oxygen as acceptor
     - 1.4.4 With a disulfide compound as acceptor
     - 1.4.7 With an iron-sulfur protein as acceptor
     - 1.4.99 With other acceptors
   - 1.5 Acting on the CH—NH group of donors
     - 1.5.1 With NAD^+ or $NADP^+$ as acceptor
     - 1.5.3 With oxygen as acceptor
     - 1.5.99 With other acceptors
   - 1.6 Acting on NADH or NADPH
     - 1.6.1 With NAD^+ or $NADP^+$ as acceptor
     - 1.6.2 With a cytochrome as acceptor
     - 1.6.4 With a disulfide compound as acceptor
     - 1.6.5 With a quinone or related compound as acceptor
     - 1.6.6 With a nitrogenous group as acceptor
     - 1.6.8 With a flavin as acceptor
     - 1.6.99 With other acceptors
   - 1.7 Acting on other nitrogenous compounds as donors
     - 1.7.2 With a cytochrome as acceptor
     - 1.7.3 With oxygen as acceptor
     - 1.7.7 With an iron-sulfur protein as acceptor
     - 1.7.99 With other acceptors

Table 1 Continued

1.8	Acting on a sulfur group of donors	
	1.8.1	With NAD^+ or $NADP^+$ as acceptor
	1.8.2	With a cytochrome as acceptor
	1.8.3	With oxygen as acceptor
	1.8.4	With a disulfide compound as acceptor
	1.8.5	With a quinone or related compound as acceptor
	1.8.7	With an iron-sulfur protein as acceptor
	1.8.99	With other acceptors
1.9	Acting on a heme group of donors	
	1.9.3	With oxygen as acceptor
	1.9.6	With a nitrogenous group as acceptor
	1.9.99	With other acceptors
1.10	Acting on diphenols and related substances as donors	
	1.10.1	With NAD^+ or $NADP^+$ as acceptor
	1.10.2	With a cytochrome as acceptor
	1.10.3	With oxygen as acceptor
	1.10.99	With other acceptors
1.11	Acting on hydrogen peroxide as acceptor	
1.12	Acting on hydrogen as donor	
	1.12.1	With NAD^+ or $NADP^+$ as acceptor
	1.12.2	With a cytochrome as acceptor
1.13	Acting on single donors with incorporation of molecular oxygen (oxygenases)	
	1.13.11	With incorporation of two atoms of oxygen
	1.13.12	With incorporation of one atom of oxygen (internal monooxygenases or internal mixed function oxidases)
	1.13.99	Miscellaneous (requires further characterization)
1.14	Acting on paired donors with incorporation of molecular oxygen	
	1.14.11	With 2-oxoglutarate as one donor, and incorporation of one atom each of oxygen into both donors
	1.14.12	With NADH or NADPH as one donor, and incorporation of two atoms of oxygen into one donor
	1.14.13	With NADH or NADPH as one donor, and incorporation of one atom of oxygen
	1.14.14	With reduced flavin or flavoprotein as one donor, and incorporation of one atom of oxygen
	1.14.15	With a reduced iron-sulfur protein as one donor, and incorporation of one atom of oxygen
	1.14.16	With reduced pteridine as one donor, and incorporation of one atom of oxygen
	1.14.17	With ascorbate as one donor, and incorporation of one atom of oxygen
	1.14.18	With another compound as one donor, and incorporation of one atom of oxygen
	1.14.99	Miscellaneous (requires further characterization)

Table 1 Continued

	1.15		Acting on superoxide radicals as acceptor
	1.16		Oxidizing metal ions
		1.16.1	With NAD^+ or $NADP^+$ as acceptor
		1.16.3	With oxygen as acceptor
	1.17		Acting on CH_2 groups
		1.17.1	With NAD^+ or $NADP^+$ as acceptor
		1.17.3	With oxygen as acceptor
		1.17.4	With a disulfide compound as acceptor
		1.17.99	With other acceptors
	1.18		Acting on reduced ferredoxin as donor
		1.18.1	With NAD^+ or $NADP^+$ as acceptor
		1.18.6	With dinitrogen as acceptor
		1.18.99	With H^+ as acceptor
	1.19		Acting on reduced flavodoxin as donor
		1.19.6	With dinitrogen as acceptor
	1.97		Other oxidoreductases
2.			Transferases
	2.1		Transferring one-carbon groups
		2.1.1	Methyltransferases
		2.1.2	Hydroxymethyl-, formyl-, and related transferases
		2.1.3	Carboxyl- and carbamoyltransferases
		2.1.4	Amidinotransferases
	2.2		Transferring aldehyde or ketone residues
	2.3		Acyltransferases
		2.3.1	Acyltransferases
		2.3.2	Aminoacyltransferases
	2.4		Glycosyltransferases
		2.4.1	Hexosyltransferases
		2.4.2	Pentosyltransferases
		2.4.99	Transferring other glycosyl groups
	2.5		Transferring alkyl or aryl groups, other than methyl groups
	2.6		Transferring nitrogenous groups
		2.6.1	Aminotransferases
		2.6.3	Oximinotransferases
		2.6.99	Transferring other nitrogenous groups
	2.7		Transferring phosphorus-containing groups
		2.7.1	Phosphotransferases with an alcohol group as acceptor
		2.7.2	Phosphotransferases with a carboxyl group as acceptor
		2.7.3	Phosphotransferases with a nitrogenous group as acceptor
		2.7.4	Phosphotransferases with a phosphate group as acceptor
		2.7.6	Diphosphotransferases
		2.7.7	Nucleotidyltransferases
		2.7.8	Transferases for other substituted phosphate groups
		2.7.9	Phosphotransferases with paired acceptors

Table 1 Continued

	2.8	Transferring sulfur-containing groups	
		2.8.1	Sulfurtransferases
		2.8.2	Sulfotransferases
		2.8.3	CoA-transferases
3.	Hydrolases		
	3.1	Acting on ester bonds	
		3.1.1	Carboxylic ester hydrolases
		3.1.2	Thiolester hydrolases
		3.1.3	Phosphoric monoester hydrolases
		3.1.4	Phosphoric diester hydrolases
		3.1.5	Triphosphoric monoester hydrolases
		3.1.6	Sulfuric ester hydrolases
		3.1.7	Diphosphoric monoester hydrolases
		3.1.11	Exodeoxyribonucleases producing 5′-phosphomonoesters
		3.1.13	Exoribonucleases producing 5′-phosphomonoesters
		3.1.14	Exoribonucleases producing other than 5′-phosphomonoesters
		3.1.15	Exonucleases active with either ribo- or deoxyribonucleic acids and producing 5′-phosphomonoesters
		3.1.16	Exonucleases active with either ribo- or deoxyribonucleic acids and producing other than 5′-phosphomonoesters
		3.1.21	Endodeoxyribonucleases producing 5′-phosphomonoesters
		3.1.22	Endodeoxyribonucleases producing other than 5′-phosphomonoesters
		3.1.25	Site-specific endodeoxyribonucleases: specific for altered bases
		3.1.26	Endoribonucleases producing 5′-phosphomonoesters
		3.1.27	Endoribonucleases producing other than 5′-phosphomonoesters
		3.1.30	Endonucleases active with either ribo- or deoxyribonucleic acids and producing 5′-phosphomonoesters
		3.1.31	Endonucleases active with either ribo- or deoxyribonucleic acids and producing other than 5′-phosphomonoesters
	3.2	Glycosidases	
		3.2.1	Hydrolyzing O-glycosyl compounds
		3.2.2	Hydrolyzing N-glycosyl compounds
		3.2.3	Hydrolyzing S-glycosyl compounds
	3.3	Acting on ether bonds	
		3.3.1	Thioether hydrolases
		3.3.2	Ether hydrolases
	3.4	Acting on peptide bonds (peptide hydrolases)	
		3.4.11	α-Aminoacylpeptide hydrolases
		3.4.13	Dipeptide hydrolases
		3.4.14	Dipeptidylpeptide hydrolases
		3.4.15	Peptidyldipeptide hydrolases
		3.4.16	Serine carboxypeptidases
		3.4.17	Metallo-carboxypeptidases

Table 1 Continued

		3.4.18	Cysteine carboxypeptidases
		3.4.19	Omega peptidases
		3.4.21	Serine proteinases
		3.4.22	Cysteine proteinases
		3.4.23	Aspartic proteinases
		3.4.24	Metalloproteinases
		3.4.99	Proteinases of unknown catalytic mechanism
	3.5	Acting on carbon–nitrogen bonds, other than peptide bonds	
		3.5.1.	In linear amides
		3.5.2	In cyclic amides
		3.5.3	In linear amidines
		3.5.4	In cyclic amidines
		3.5.5	In nitriles
		3.5.99	In other compounds
	3.6	Acting on acid anhydrides	
		3.6.1	In phosphorus-containing anhydrides
		3.6.2	In sulfonyl-containing anhydrides
	3.7	Acting on carbon–carbon bonds	
		3.7.1	In ketonic substances
	3.8	Acting on halide bonds	
		3.8.1	In C-halide compounds
		3.8.2	In P-halide compounds
	3.9	Acting on phosphorus–nitrogen bonds	
	3.10	Acting on sulfur–nitrogen bonds	
	3.11	Acting on carbon–phosphorus bonds	
4.	Lyases		
	4.1	Carbon–carbon lyases	
		4.1.1	Carboxy-lyases
		4.1.2	Aldehyde-lyases
		4.1.3	Oxo-acid-lyases
		4.1.99	Other carbon–carbon lyases
	4.2	Carbon–oxygen lyases	
		4.2.1	Hydro-lyases
		4.2.2	Acting on polysaccharides
		4.2.99	Other carbon–oxygen lyases
	4.3	Carbon–nitrogen lyases	
		4.3.1	Ammonia-lyases
		4.3.2	Amidine-lyases
	4.4	Carbon–sulfur lyases	
	4.5	Carbon–halide lyases	
	4.6	Phosphorus–oxygen lyases	
	4.99	Other lyases	
5.	Isomerases		
	5.1	Racemases and epimerases	

Table 1 Continued

		5.1.1	Acting on amino acids and derivatives
		5.1.2	Acting on hydroxy acids and derivatives
		5.1.3	Acting on carbohydrates and derivatives
		5.1.99	Acting on other compounds
	5.2		*Cis-trans*-isomerases
	5.3		Intramolecular oxidoreductases
		5.3.1	Interconverting aldoses and ketoses
		5.3.2	Interconverting keto and enol groups
		5.3.3	Transposing C═C bonds
		5.3.4	Transposing S—S bonds
		5.3.99	Other intramolecular oxidoreductases
	5.4		Intramolecular transferases
		5.4.1	Transferring acyl groups
		5.4.2	Phosphotransferases
		5.4.3	Transferring amino groups
		5.4.99	Transferring other groups
	5.5		Intramolecular lyases
	5.99		Other isomerases
6.			Ligases
	6.1		Forming carbon–oxygen bonds
		6.1.1	Ligases forming aminoacyl-tRNA and related compounds
	6.2		Forming carbon–sulfur bonds
		6.2.1	Acid–thiol ligases
	6.3		Forming carbon-nitrogen bonds
		6.3.1	Acid–ammonia (or amine) ligases (amide synthases)
		6.3.2	Acid–amino-acid ligases (peptide synthases)
		6.3.3	Cyclo-ligases
		6.3.4	Other carbon–nitrogen ligases
		6.3.5	Carbon–nitrogen ligases with glutamine as amido-N donor
	6.4		Forming carbon–carbon bonds
	6.5		Forming phosphoric ester bonds

Source: Ref. 3, pp. V–XI, by courtesy of Academic Press.

It should be noted that all enzymes, from whatever source, that catalyze the same overall reaction will be assigned the same code number, systematic name, and trivial name. This is why it is very important to specify the exact source of the enzyme. It should also be recalled that details of the mechanism of the reaction are not considered in the naming. For example, there are two classes of fructose-bisphosphate aldolases, class I ("lysine") aldolases and class II ("metallo") aldolases. Both classes of aldolases would be named "D-fructose-1,6-bisphosphate D-glyceraldehyde-3-phosphate-lyase" and have the code number 4.1.2.13. To distinguish between the two classes it is necessary to specify "class I or class II." The reader should consult Ref. 3 for a list of specific enzymes.

C. Multiple Molecular Forms

The nomenclature of the multiple forms of an enzyme is a difficult problem. A subcommittee of the International Union of Biochemistry on Isoenzymes recommended that (a) when multiple forms of an enzyme are identified by electrophoretic separation, they should be given consecutive numbers, the form having the highest mobility toward the anode being number 1, and that (b) multiple enzyme forms in a single species should be known as isoenzymes [4]. Isoenzymes and isozymes are used synonymously to designate multiple molecular forms of enzymes.

Excellent reviews are available on the subject of multiple molecular forms of enzymes [5,6]. One of the best is that of Markert and Whitt [5], which considers the molecular basis of some 53 of the more than 100 enzymes shown by 1968 to occur in multiple molecular forms.

Mutated forms of enzymes, via recombinant DNA technology, are of more recent origin. In the current literature it is handled by the designation "chymotrypsin Ser 195 Thr," indicating that amino acid residue Ser 195 in wild-type chymotrypsin has been mutated to Thr 195.

REVIEW QUESTIONS

1. For each of the reactions (a) write a balanced equation using structural formulas, (b) indicate the type of chemical reaction involved in terms of one of the six possibilities (see the text), (c) give a systematic name, and (d) check your answers with *Enzyme Nomenclature* (1984).

(a) a β-D-glucoside + H_2O = an alcohol + D-glucose
(b) Xanthine + H_2O + O_2 = urate + H_2O_2
(c) L-Asparagine + H_2O = L-aspartate + NH_3
(d) UDP-glucose + D-fructose = UDP + sucrose
(e) ATP + pyruvate + HCO_3^- = ADP + orthophosphate + oxaloacetate
(f) Uridine + orthophosphate = uracil + D-ribose 1-phosphate
(g) Glyoxylate + CoA + $NADP^+$ = oxalyl-CoA + NADPH
(h) L-Cysteinylglycine + H_2O = L-cysteine + glycine
(i) L-Aspartate + 2-oxoglutarate = oxaloacetate + L-glutamate
(j) Acetyl-CoA + H_2O = CoA + acetate
(k) Acetylcholine + H_2O = choline + acetate
(l) Chlorophyll + H_2O = phytol + chlorophyllide
(m) Methylmalonyl-CoA + pyruvate = propionyl-CoA + oxaloacetate
(n) Ribitol + NAD^+ = D-ribulose + NADH

2. For each of the following enzymes attempt a complete writing of the reaction catalyzed.

(a) Acetate:CoA ligase (AMP-forming)
(b) α-*N*-Acetyl-L-ornithine amidohydrolase
(c) Alanine racemase
(d) Glutarate-semialdehyde:NAD$^+$ oxidoreductase
(e) D-Glucuronate ketol-isomerase
(f) Glycine:2-oxoglutarate aminotransferase
(g) Oxalate:oxygen oxidoreductase
(h) NAD$^+$ glycohydrolase
(i) L-Phenylalanine ammonia-lyase
(j) Oxalyl-CoA carboxy-lyase
(k) Phosphoserine phosphohydrolase
(l) Phenylpyruvate keto-enol-isomerase
(m) Vitamin-A-acetate hydrolase
(n) UTP:ammonia ligase (ADP-forming)

3. For each of the following enzymes for which the trivial or obsolete name (underlined) is given, write the reaction catalyzed and give the systematic name.

(a) <u>Lecithinase</u> <u>A</u>
(b) <u>Laccase</u>
(c) <u>Maltase</u>
(d) Lysozyme
(e) Alanine dehydrogenase
(f) Aspartate kinase
(g) <u>Aspartase</u>
(h) Alkaline phosphatase
(i) <u>Condensing</u> <u>enzyme</u>
(j) Mannose phosphate isomerase
(k) <u>Lipoxidase</u>
(l) <u>Disproportionating</u> <u>enzyme</u>
(m) Glucoamylase
(n) Choline acetyltransferase

4. (a) What is the basis for the primary classification of enzymes by the Enzyme Commission of the International Union of Biochemistry?

(b) List the divisions under this primary classification and define the general function of each group of enzymes illustrated with a general equation.

(c) Using one of these groups, indicate the meaning of each number in the four-number systematic numbering system.

5. Give the *group* and *specific* name of the enzyme that catalyzes each of the following reactions. Write a complete balanced equation using structural formulas. The enzyme converts

(a) ethyl phosphate to ethyl alcohol + phosphate

(b) pectin to pectate
(c) dioleostearin to glycerol, oleic acid, and stearic acid
(d) glycylproline to glycine and proline
(e) phenol to hydroquinone in the presence of hydrogen peroxide
(f) phosphoarginine to arginine and phosphate
(g) phenylsulfate to phenol and sulfate
(h) D-glucose to D-δ-gluconolactone

REFERENCES

Cited References

1. *Report of the Commission on Enzymes of the International Union of Biochemistry, 1961.* Pergamon Press, Elmsford, N.Y., 1961.
2. Enzyme nomenclature, in *Comprehensive Biochemistry*, 2nd Ed. (M. Florkin and E. H. Stotz, eds.), Vol. 13, Elsevier, Amsterdam, 1965.
3. *Enzyme Nomenclature*, recommendations (1984) of the Nomenclature Committee of the International Union of Biochemistry, published for the International Union of Biochemistry, Academic Press, San Diego, Calif., 1984.
4. E. C. Webb, *Enzymol. Biol. Clin.*, **5**, 124 (1965).
5. C. L. Markert and G. S. Whitt, *Experientia*, **24**, 977 (1968).
6. E. S. Vesell, consulting ed., *Multiple Molecular Forms of Enzymes*, *Ann. N.Y. Acad. Sci.*, **151**, 1 (1968).

14
Introduction to the Hydrolases

In the following chapters we describe the properties of several enzymes that have practical or potential importance to the food sciences. The enzymes selected for discussion have been chosen on the basis of the following criteria: (a) that the enzyme have practical or potential importance to the food sciences, (b) that the properties of the enzyme have been elucidated in sufficient detail in terms of the general features of enzymes described to this point in the book, and (c) that the enzyme illustrate one or several of the general features of enzymes. Obviously, only a few of the 2477 enzymes listed in *Enzyme Nomenclature* (1984) will be discussed [1]. Several enzymes from the hydrolases and oxidoreductases fit these requirements.

The hydrolases are a large group of enzymes (641 in number, Ref. 1) that have in common the involvement of water as a second substrate. The reaction may be described as transfer of the components of the substrate to water:

$$X—Y + HOH \rightarrow HX + YOH \qquad (1)$$

The use of ^{18}O-labeled water has played a major role in elucidation of the mechanism of action of several of these enzymes. In the reaction shown in Eq. (1), the OH component of water has joined the Y component of the substrate. A priori the OH might join with X, and therefore a means of distinguishing between the two possibilities is needed. As an illustration of the utility of using ^{18}O-labeled water, consider the hydrolysis of sucrose [Eq. (2)].

Hydrolysis of the glucoside can involve breakage of the C(1)—O or O—C(2) bond. If the C(1)—O bond is broken, glucose will contain the OH residue of water,

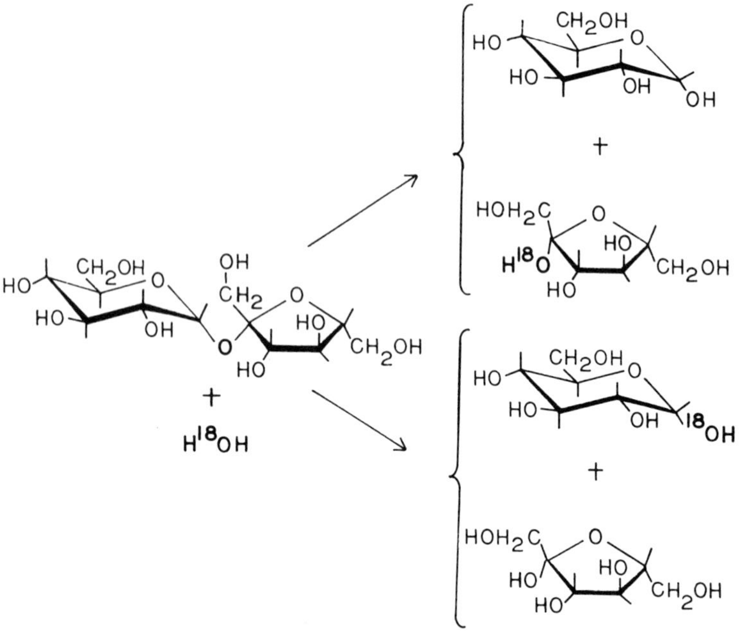

(2)

Table 1 Subdivision of the Hydrolases on Basis of Type of Bond Hydrolyzed

EC number[a]	Subdivision name	Number of enzymes
3.1	Acting on ester bonds	195
3.2	Acting on glycosyl bonds	118
3.3	Acting on ether bonds	8
3.4	Acting on peptide bonds	156
3.5	Acting on C—N bonds other than peptide bonds	109
3.6	Acting on acid anhydride bonds	40
3.7	Acting on C—C bonds	6
3.8	Acting on halide bonds	5
3.9	Acting on P—N bonds	1
3.10	Acting on S—N bonds	2
3.11	Acting on C—P bonds	1

Source: Ref. 1.
[a]Commission on Enzymes number.

while in breakage of the O—C(2) bond fructose will contain the OH residue. By use of ^{18}O-labeled water, isolation of glucose and of fructose and their analysis by mass spectrometry, the position of splitting of the bond can be established. The hydrolases are subdivided on the basis of the type of bond hydrolyzed (Table 1).

REFERENCES

Cited Reference

1. *Enzyme Nomenclature*, recommendations (1984) of the Nomenclature Committee of the International Union of Biochemistry, published for the International Union of Biochemistry, Academic Press, San Diego, Calif., 1984.

15
The Glycoside Hydrolases

I. SPECIFICITY

Enzymes are highly selective for compounds that bind into the active site and are converted to product. Therefore, specificity depends on how tightly the enzyme binds a compound and how rapidly it converts the compound to product. Quantitatively, specificity is measured by V_{max}/K_m, where V_{max} is best expressed as rate per mole of enzyme. The higher the value V_{max}/K_m, called the *specificity coefficient*, the more specific the enzyme is for a substrate.

Specificity of an enzyme is defined, therefore, by the nature of the compound converted to product. We shall use the glycosidases to illustrate this *general property of all enzymes*. Features of the substrate that are important in the action of a glycosidase may be defined in the following ways.

A. Nature of Monosaccharide Residue (Glycone) Which Donates the Reducing Group Involved in the Glycosidic Bond

In maltose the residue that donates the reducing group is glucose, while in lactose it is galactose (Fig. 1). Two separate enzymes are involved in hydrolyzing the two substrates. The enzyme (α-D-glucoside glucohydrolase, EC 3.2.1.20; trivial name "maltase"), which hydrolyzes maltose to glucose is a *glucosidase* and has specificity for the glucose residue which furnishes C(1) in the glucosidic bond. The second enzyme (β-D-galactoside galactohydrolase, EC 3.2.1.23; trivial name "lactase"), which hydrolyzes lactose to glucose and galactose, is a *galactosidase*

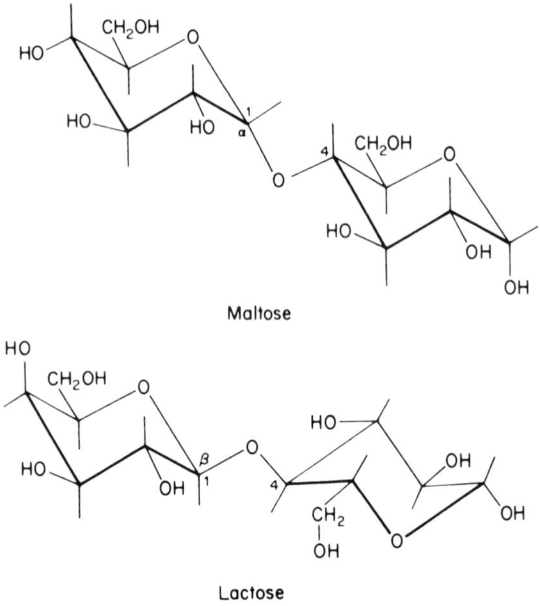

Figure 1 Structures of maltose [*O*-α-D-glucopyranosyl-(1→4)-*O*-α-D-glucopyranoside] and lactose [*O*-β-D-galactopyranosyl-(1→4)-*O*-α-D-glucopyranoside].

since it recognizes the galactose residue of the substrate. In both cases the glycosidic bond is split at the C(1)-0 position as shown by use of $H_2{}^{18}O$. The situation with sucrose is a unique one since the reducing groups of both glucose and fructose are involved in formation of the glycosidic bond [Chapter 14, Eq. (2)]. Two different enzymes can hydrolyze the glycosidic bond of sucrose; one enzyme recognizes the glucose moiety and the other the fructose moiety.

The nature of the residue (the aglycone) attached to the glycone is usually not of primary importance in meeting the specificity requirements of the enzyme. The aglycone may be a monosaccharide, a methyl group, or a nitrophenyl group, among others. The size of the aglycone permitted seems to vary among different enzymes which otherwise seem to have identical specificity. For example, *Saccharomyces cerevisiae* α-glucosidase hydrolyzes maltose and maltotriose but not higher polymers; crude extracts of *S. warum* α-glucosidase hydrolyze maltose but not maltotriose; and amyloglucosidase from various molds and bacteria and equine serum maltase hydrolyze polysaccharides in addition to maltose.

B. Configuration, α or β, About the Carbon Atom of the Glycone Involved in the Glycosidic Bond

In maltose (Fig. 1) the configuration is α with respect to the C(1)-0 bond, whereas in lactose it is β. Therefore, the enzyme that hydrolyzes maltose is an α-glucosidase, and the enzyme that hydrolyzes lactose is a β-galactosidase.

Maltose and cellobiose are both disaccharides composed of two molecules of glucose (Figs. 1 and 2). The glycone portions of the two molecules are identical except for configuration about the C(1) position. Notice the quite marked difference in orientation of the C(1)-O bond with respect to the ring O, and the side-chain substituents on the glycone and aglycone rings. These differences are sufficient that cellobiose is not hydrolyzed by α-glucosidase, nor does it act as a competitive inhibitor of α-glucosidase. The same is true of β-glucosidase (which hydrolyzes cellobiose) and maltose.

C. Configuration, D or L, of the Glycone

The hexoses have five asymmetric centers, at carbon atoms 1, 2, 3, 4, and 5. The α and β forms of a monosaccharide are the result of asymmetry around C(1). The D- or L-configuration is established by comparing the configuration of the *highest*-numbered asymmetric carbon with glyceraldehyde. In the Haworth projection formula the CH_2OH group on C(6) is written up for D-glucose and down for L-glucose (Fig. 3). The naturally occurring polymers of the hexoses all contain D-sugars. Thus, the enzymes involved in the hydrolysis of these polymers require that the sugar be of the D-configuration.

Some confusion results in understanding the ability of the β-galactosidases to hydrolyze substrates containing α-L-arabinose. In arabinose the highest-numbered asymmetric carbon is C(4) and the configuration about this carbon is L. The two

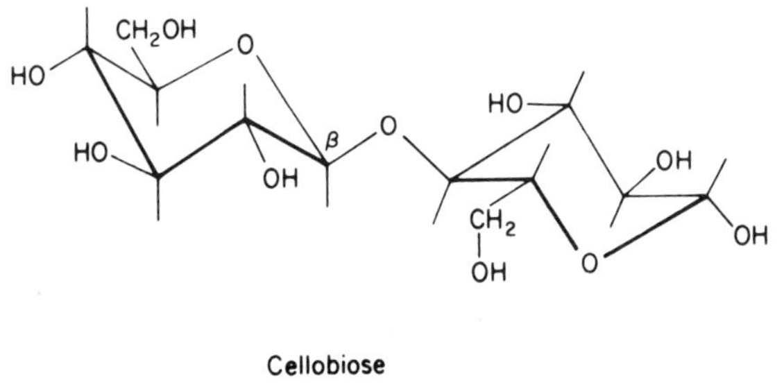

Figure 2 Structure of cellobiose [*O*-β-D-glucopyranosyl-(1→4)-*O*-α-D-glucopyranoside].

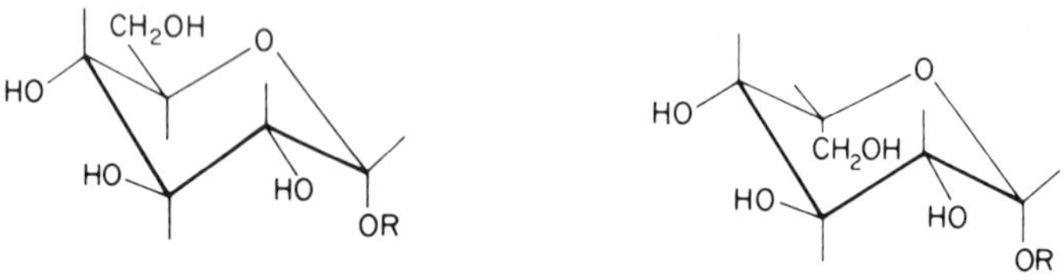

An α-D-glucoside An α-L-glucoside

Figure 3 D- and L-forms of an α-glucoside.

substrates are thus identical except for the missing —CH_2OH group in α-L-arabinosides (Fig. 4).

D. Ring Size of the Glycone

In a monosaccharide where the hydrogen of the OH on C(1) is not substituted, the molecule exists in three forms: α, β, and the open-chain forms. Ring formation can occur between the C(1) hydroxyl and the C(4) or C(5) hydroxyl to form a five- or six-membered ring, respectively. The terms *pyranose* and *furanose* are used to designate the six- and five-membered heterocyclic rings of sugars, respectively, because of their resemblance to pyran and furan (e.g., the naming of maltose and lactose in Fig. 1). D-Glucose, L-arabinose, D-xylose, D-galactose, and D-mannose occur naturally as pyranoses, while D-ribose (in combined form) and D-fructose occur as furanoses.

Since the glycone portions of naturally occurring substrates for α-glucosidase, β-glucosidase, and β-galactosidase all contain pyranose rings, it is not surprising that the enzymes do not have activity on synthetically prepared substrates in which the glycone is in the furanose form. Conversely, β-fructofuranosidase has no activity on the isomer of sucrose in which the fructose is in the pyranose form.

A β-D-galactoside An α-L-arabinoside

Figure 4 Structures of a β-D-galactoside and an α-L-arabinoside. The carbon atoms for assigning D- and L-configuration are 5 and 4 in the galactoside and arabinoside, respectively.

E. Position of Attachment of Glycone to Aglycone

All of the glycoside hydrolases that have been characterized have specificity for compounds in which the aglycone is attached to C(1) of the glycone. Potentially, the glycoside could involve the C(1), C(2), C(3), C(4), C(5), or C(6) position of a hexose aglycone in formation of the glycosidic bond. In Section I.A we implied that the nature of the aglycone does not affect the ability of an enzyme to hydrolyze the glycosidic bond provided the specificity features of the glycone portion are preserved. This statement needs modification to the extent that when the aglycone is a pentose or hexose, it makes a difference as to the position of attachment of the glycone to the aglycone. Apparently, this is due to (a) steric hindrances presented by the substrate which prevent its proper attachment to the enzyme, and/or (b) specificity toward the aglycone portion by some enzymes.

The α- and β-amylases and α-glucosidase will not hydrolyze α-1,6-linkages that occur in amylopectin. There are a number of α-1,6-glucosidases that hydrolyze these bonds. α-Dextrin endo-1,6-glucosidase (α-dextrin 6-glucanohydrolase, EC 3.2.1.41; trivial names, limit dextrinase, debranching enzyme, amylopectin 6-glucanohydrolase, pullulanase) appears to hydrolyze this linkage readily in polysaccharides but not in the lower polymers and in the two dimers, isomaltose and panose. On the other hand, there is an intestinal enzyme, oligo-1,6-glucosidase (dextrin 6-α-D-glucanohydrolase, EC 3.2.1.10) which can hydrolyze isomaltose, panose, and isomaltotriose but not higher polymers and which is specific for α-1,6-glucosides.

An enzyme isolated from insects hydrolyzes the α-1,1-glucosidic bond of trehalose (Fig. 5), which is the major sugar in the hemolymph of insects. Methyl and phenyl α-glucosides are not hydrolyzed by α,α-trehalase (α,α-trehalose 1-glucohydrolase, EC 3.2.1.28), indicating that the enzyme has some specificity requirements that must be met by the aglycone.

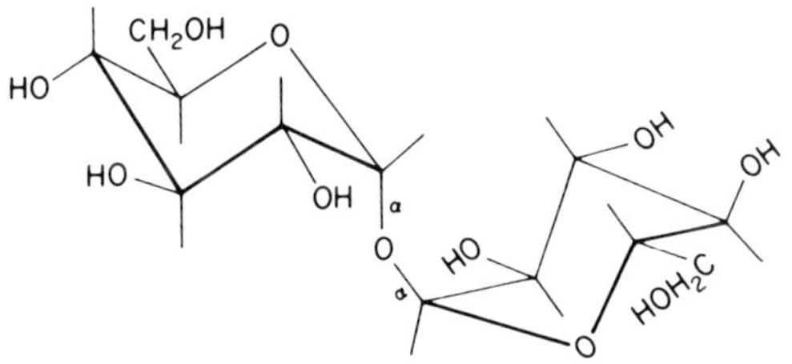

Figure 5 Structure of trehalose [*O*-α-D-glucopyranosyl-(1→1)-*O*-α-D-glucopyranoside].

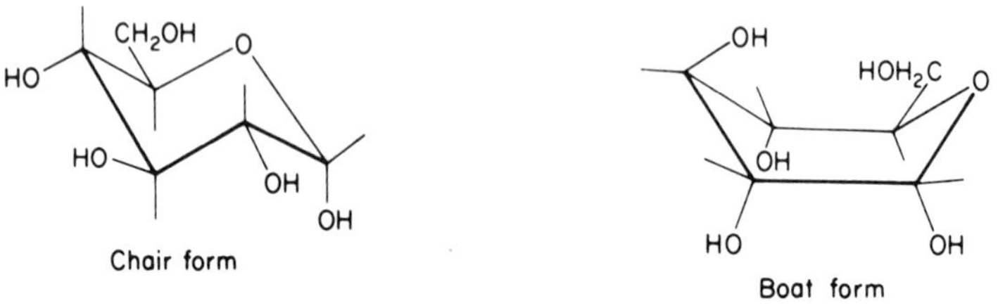

Figure 6 Chair and boat forms of α-D-glucose. There are two possible chair forms and six possible boat forms. Naturally occurring glycoside polymers probably all contain the chair configuration.

F. Conformation of the Glycone

Potentially, the pyranose configuration of the pentose or hexose moiety can exist in either the chair or the boat form (Fig. 6). In glycoside derivatives of the monosaccharides, the conformation is the chair structure. While there has been little definitive work on this point, it would appear that the enzymes would be specific for the chair conformation since that is the form that occurs in nature.

G. Size of the Molecule

The α- and β-amylases and amylopectin-1,6-glucosidase have activity on the large starch and glycogen molecules, but activity on maltose and maltotriose is very low or nonexistent. Cellulase (β-1,4-) (1,3; 1,4)-β-D-glucan 4-glucanohydrolase, EC 3.2.1.4) has activity on cellulose but not on cellobiose. The nature of these polymeric molecules is shown in Fig. 7. Starch and glycogen, and cellulose are not hydrolyzed by maltase and cellobiase, respectively. Maltose and cellobiose are not hydrolyzed by α- and β-amylases and by cellulase, respectively, but these compounds do act as competitive inhibitors. The conclusions to be drawn from these data are that starch, glycogen, and cellulose are too large to permit binding to the active sites of maltase and cellobiase, respectively, and therefore, no activity is measured. Maltose and cellobiose can bind to the active sites of α- and β-amylases and cellulase, respectively, but the binding does not position the 1,4-glucosidic bond properly with respect to the catalytic groups so that it can be hydrolyzed.

Figure 7 Structures of amylose, glycogen, and cellulose. Amylose is a coiled polymer with a repeating unit of approximately six glucose units. At the top of the structure the bottom residue is several angstrom units below the top one. The segment of the molecule is viewed from the top and slightly to the bottom edge. Cellulose is a linear chain that presents an opportunity for interchain hydrogen-bond formation and the inertness of cellulose.

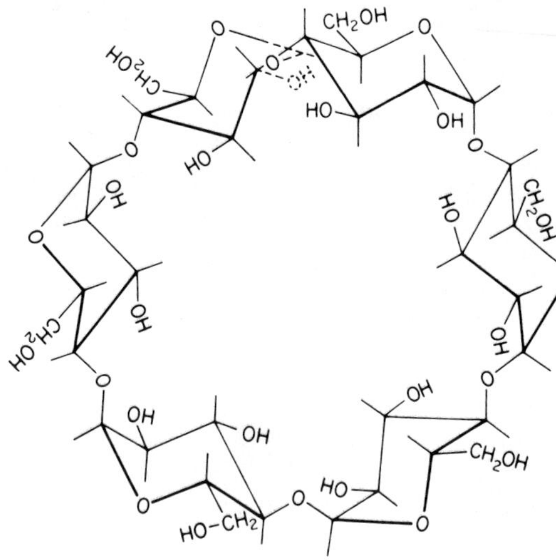

Amylose

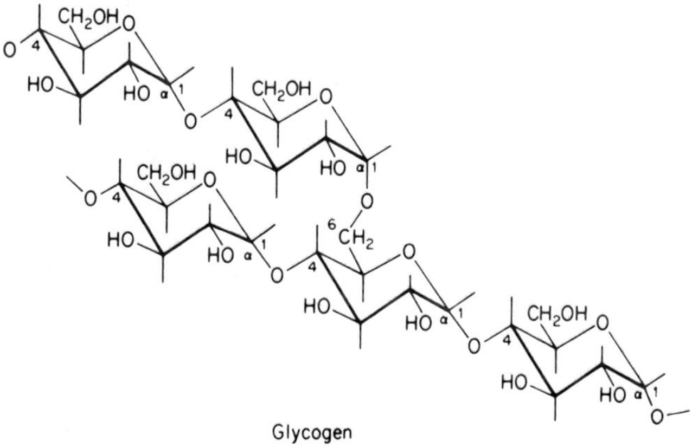

Glycogen

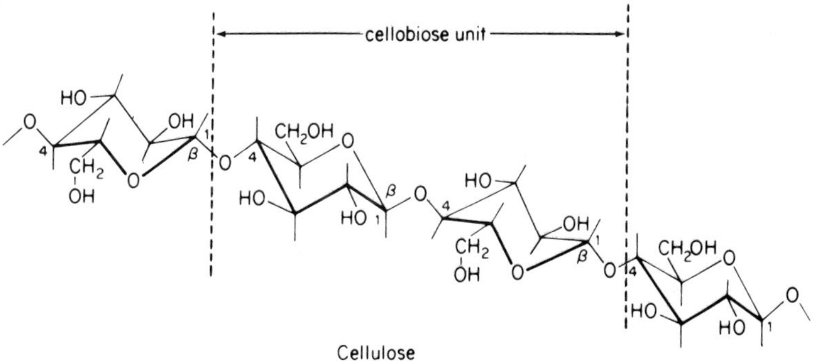

Cellulose

Table 1 Effect of Substrate Size on Rates of Hydrolysis by Three Glycosidases

Enzyme	Substrate	Relative rate[a]
Cellulase[b]	Cellobiose	1.2
	Cellotriose	46
	Cellotetraose	79
	Cellopentaose	100
	Cellohexaose	96
Polygalacturonase[c]	Digalacturonic acid	0
	Trigalacturonic acid	1.7
	Tetragalacturonic acid	2.2
	Pentagalacturonic acid	12.8
	Hexagalacturonic acid	15.8
	Polygalacturonic acid (D.P. = 12)[d]	52.0
	Polygalacturonic acid (large polymer)	100
α-Amylase[d,e]	Maltohexaose (α-1,4)	1.2
	Maltodecaose (α-1,4)	2.2
	Amylose (large polymer)	100

[a]Rate relative to substrate hydrolyzed at fastest rate.
[b]Ref. 1; data are for the enzyme from *Myrothecium verrucaria*.
[c]Ref. 2; data are for the enzyme from *Erwinia carotovora*.
[d]D.P. = degree of polymerization; D.P. = 12 means there are 12 monomer units.
[e]Ref. 3; data are for the enzyme from Bacillus subtilis.

The effect of substrate size on activity is shown in Table 1 for three glycosidases. For cellulase, there is an influence of size up to cellopentaose, but higher polymers are hydrolyzed at the same rate. Polygalacturonase and α-amylase do not hydrolyze dodecagalacturonic acid [degree of polymerization (D.P.) = 12] and maltodecaose (D.P. = 10), respectively, at the rate of the large polymeric substrates.

H. Nature of Bridging Atom in the Glycosidic Bond

Replacement of the glycosidic oxygen with sulfur prevents a compound from serving as a substrate for the *O*-glycosidases. In a number of cases, the thioglycoside is a competitive inhibitor of the enzyme, indicating that it can be bound in the active site. However, the size or electronic configuration of the sulfur atom prevents the *O*-glycosidases from hydrolyzing the bond.

Thioglucosidases (thioglucoside glucohydrolase, EC 3.2.3.1; trivial names myrosinase, sinigrinase, sinigrase) occur in several plants. These enzymes are responsible for production of such compounds as allyl isothiocyanate [Eq. (1)] which are characteristic of black mustard seeds and the root of the horseradish.

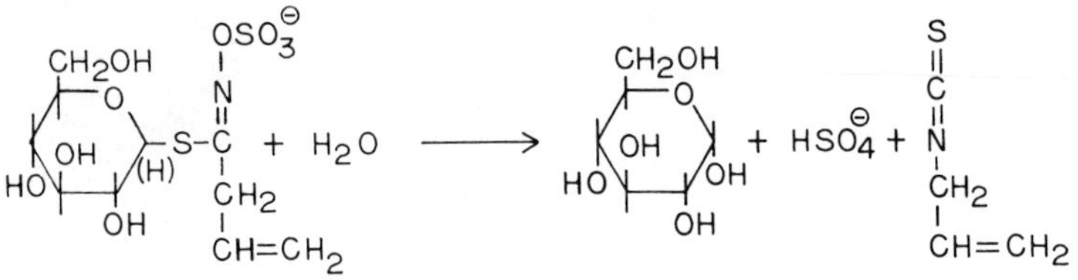 (1)

Sinigrin D-Glucose Allyl isothiocyanate

The isothiocyanate gives the potent flavor characteristics of these food products. These enzymes are very important to the food scientist but have received little study to this point.

The *N*-glycosides, in which the monosaccharide residue is D-ribose (RNA) or 2-deoxy-D-ribose (DNA) and the aglycone is attached to the glycone by a C—N bond, are very important (Fig. 8). The aglycones are usually pyrimidine or purine groups. The enzyme that hydrolyzes adenosine is a purine nucleosidase (*N*-D-ribosyl-purine ribohydrolase, EC 3.2.2.1) while the enzyme that hydrolyzes uridine is a uridine nucleosidase (uridine ribohydrolase, EC 3.2.2.3).

I. Modification of Groups on Carbons 2, 3, 4, and 5 of the Glycone

The groups on carbons 2, 3, 4, and 5 of the glycone may be systematically modified and the effect of such modification on the ability of an enzyme to hydrolyze the compound measured. Since these are asymmetric carbons, the configuration may also be changed. For example, inversion of configuration of the hydroxyl substituent at C(4) of the glycone of maltose converts the compound from a substrate for an α-glucosidase to one for an α-galactosidase (α-D-galactoside galactohydrolase, EC 3.2.1.22). The substrate would be melibiose [*O*-α-D-galactopyranosyl-(1→4)-D-glucopyranoside].

α-D-Galacturonic acid units [oxidation of C(6) of D-galactose] linked α-1,4

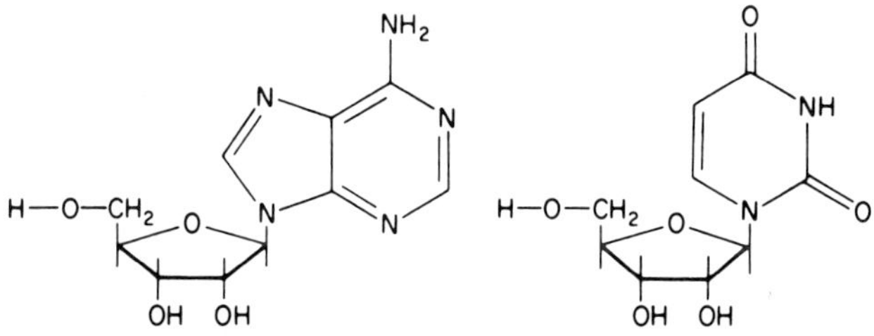

Adenosine Uridine

Figure 8 Structures of two *N*-glycosides, adenosine and uridine.

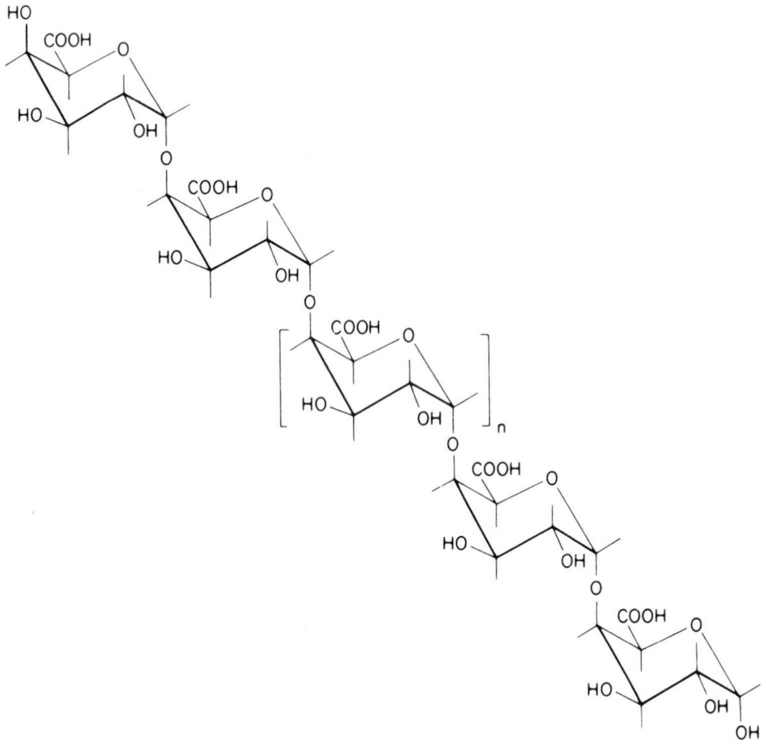

Figure 9 Structure of pectic acid. Esterification of approximately 75% of the carboxyl groups with methanol gives pectin. Conformation of the polymer is not shown.

give rise to pectic acid (Fig. 9) and when the majority of the carboxyl groups are methoxylated, to pectin. These pectic substances serve as substrates for three types of enzymes: polygalacturonase [poly(1,4-α-D-galacturonide) glycanohydrolase, EC 3.2.1.15], pectate lyase [poly(1,4-α-D-galacturonide) lyase, EC 4.2.2.2], and pectinesterase (pectin pectylhydrolase, EC 3.1.1.11).

The heterocyclic oxygen atom of the glycone may be replaced with sulfur. Although very few data have been obtained with such compounds, it appears that they are not substrates for the usual glycosidases.

II. THE AMYLASES

The amylases act on starch, glycogen, and derived polysaccharides to hydrolyze the α-1,4 and/or α-1,6 glucosidic linkages. The amylases may be divided into several types depending on (a) exo or endo mode of hydrolysis; (b) retention or inversion of configuration; (c) hydrolysis of α-1,4 versus α-1,6-glycosidic bonds; and (d) hydrolysis or transfer-type reaction. Unfortunately, it is not always possible to classify all starch-degrading enzymes specifically into one of the types.

A. Types of Amylases

Several types of amylases are shown in Table 2. Because of their special importance, the α-amylases, β-amylases, glucoamylases, and pullulanases will be discussed in detail. α-Amylase (1,4-α-D-glucan glycanohydrolase, EC 3.2.1.1) is most often an endo-splitting enzyme that hydrolyzes the α-1,4-glycosidic bonds of the substrate in a more or less random fashion. Theoretically, the splitting is entirely random; however, the action patterns of a number of α-amylases, particularly from microorganisms, show consistent and reproducible product patterns characteristic of an α-amylase from a particular source [3].

β-Amylase (1,4-α-D-glucan maltohydrolase, EC 3.2.1.2) is an exo-splitting enzyme and removes successive maltose units from the nonreducing ends of the polysaccharide chains. When the enzyme comes to an α-1,6 linkage, as is found in amylopectin and glycogen, the action on that particular chain stops.

Glucoamylase (1,4-α-D-glucan glucohydrolase, EC 3.2.1.3; recommended name, glucan 1,4 α-glucosidase) is an exo-splitting enzyme and removes successive glucose units from the nonreducing ends of the substrate chains. The action of the enzyme on a substrate chain decreases when an α-1,6 linkage is encountered as in amylopectin and glycogen, but the α-1,6 linkage is hydrolyzed. Pullulanase (α-dextrin 6-glucanohydrolase, EC 3.2.1.41) is an endo-splitting enzyme, hydrolyzing the α-1,6-glucosidic bonds of pullulan (a linear polymer of about 250 maltotriosyl units linked together by α-1,6-glucosidic bonds) and of the α- and β-amylase limit dextrins of amylopectin and glycogen.

B. Nature of Substrate and Hydrolysis Products

Starch is composed of two fractions, amylose and amylopectin. The molecular size of these fractions is variable; the number of glucose units in amylose varies from 300 to 400. Since the amylose fraction contains only α-1,4 glucosidic linkages, α- and β-amylases should hydrolyze it completely to maltose. However, some maltotriose usually remains. Glucoamylase should hydrolyze amylose completely to glucose, although some maltose and maltotriose may remain at the end because of slow hydrolysis of these compounds.

Less complete hydrolysis is obtained with amylopectin. When amylopectin (Fig. 10) is subjected to exhaustive hydrolysis by α-amylase, only small oligomers (2 to 6 glucose units) will be left. On the other hand, β-amylase hydrolyzes the chains from the nonreducing end and its action stops when all the chains are hydrolyzed to the branch points. Thus large polymers remain. Glucoamylase can slowly hydrolyze α-1,6 linkages.

Table 2 Some Starch- and Glycogen-Degrading Enzymes

Type	Configuration of glucosidic bond	Comments
Endo-splitting (configuration retained)		
α-Amylase (EC 3.2.1.1)	α-1,4	Initial major products are dextrins; final major products are maltose and maltotriose.
Isoamylase (EC 3.2.1.68)	α-1,6	Products are linear dextrins.
Isomaltase (EC 3.2.1.10)	α-1,6	Acts on products of α-amylase hydrolysis of amylopectin.
Cyclomaltodextrinase (EC 3.2.1.54)	α-1,4	Acts on cyclodextrins and linear dextrins to give maltose and maltotriose.
Pullulanase (EC 3.2.1.41)	α-1,6	Acts on pullulan to give maltotriose and on starch to give linear dextrins.
Isopullulanase (EC 3.2.1.57)	α-1,4	Acts on pullulan to give isopanose and on starch to give unknown products.
Neopullulanase	α-1,4	Acts on pullulan to give panose and on starch to give maltose.
Amylopullulanase	α-1,6	Acts on pullulan to give maltotriose.
	α-1,4	Acts on starch to give D.P. 2 to 4 products.
Amylopectin 6-glucanohydrolase (EC 3.2.1.41)	α-1,6	Acts only on amylopectin to hydrolyze α-1,6-glucosidic linkages.
Exo-splitting (nonreducing end)		
β-Amylase (EC 3.2.1.2)	α-1,4	Product if β-maltose.
α-Amylase	α-1,4	Product is α-maltose. There are specific exo α-amylases that produce maltotriose, maltotetraose, maltopentaose, and maltohexaose, with retention of configuration.
Glucoamylase (EC 3.2.1.3)	α-1,6	β-Glucose is produced.
α-Glucosidase (EC 3.2.1.20)	α-1,4	α-Glucose is produced. There are a number of α-glucosidases.
Transferase		
Cyclomaltodextrin glucanotransferase (EC 2.4.1.19)	α-1,4	α- and β-Cyclodextrins formed from starch with 6 to 12 glucose units.

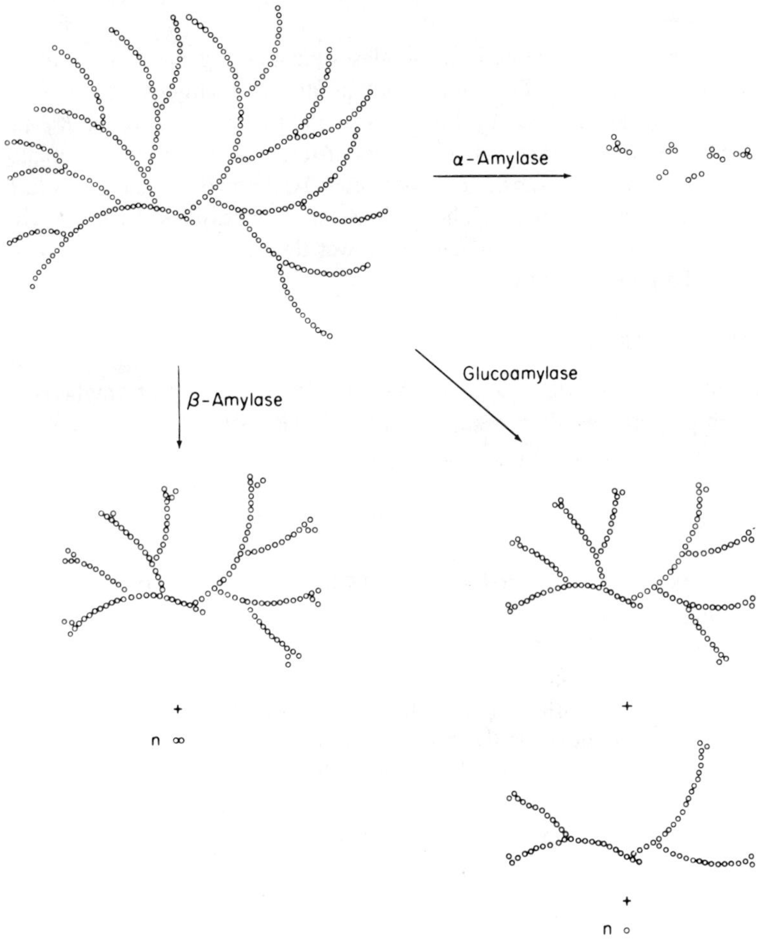

Figure 10 Schematic structure of amylopectin and representation of action of the amylases. Each circle represents a glucose unit. α-Amylase hydrolyzes amylopectin to oligosaccharides containing two to six monomers. Some α-amylases may leave larger fragments. β-Amylase hydrolyzes each chain to produce maltose until it reaches an α-1,6-linkage. Glucoamylase hydrolyzes preferentially α-1,4 linkages to give glucose but it can also hydrolyze α-1,6 linkages at a slower rate. Structure of amylopectin drawn according to Ref. 4. (From Ref. 5, p. 390, by courtesy of Wiley.)

C. Kinetics

A study of the kinetics of action of the amylases on starch, even on the amylose fraction alone, is difficult. The initial split products of amylose continue to serve as substrates down to the level of maltose and maltotriose, yet the K_m and V_{max} values for the various size compounds are different. α-*p*-Nitrophenyl maltosides have been proposed as substrates. Unfortunately, the enzymes have very little activity on the low-molecular-weight *p*-nitrophenyl maltosides; their rates of hydrolysis are several orders of magnitude lower than the rate of hydrolysis of α-1,4-glucosidic bonds in starch.

D. Activity Assays

Four general assay procedures are used for studying the action of amylases on starch. During hydrolysis there is (a) a decrease in viscosity; (b) a loss in ability to give a blue color with iodine; (c) an appearance of reducing groups; and (d) an increase in maltose, glucose, or dextrins. The three types of amylases can be distinguished from each other on the basis of two or more of these criteria (Table 3).

Because they hydrolyze the α-1,4-glucosidic linkages of the polysaccharide at random, α-amylases cause a very rapid loss in viscosity and iodine color-forming ability because both of these parameters are dependent on the integrity of the large polymer. Hydrolysis of one or a very few glucosidic bonds near the center of the substrate causes a marked change in these parameters, whereas hydrolysis of one or a very few bonds near the end of the polysaccharide chains by the exo-splitting enzymes β-amylase and glucoamylase has little effect.

The rate of decrease in viscosity or iodine color cannot be used alone to differentiate among the three enzymes because the rate of change in these parameters is dependent not only on the type of enzyme but also on the concentra-

Table 3 Relative Rates of Change of Assay Parameter for Some of the Amylases

	Relative rates		
Criterion	α-Amylase	β-Amylase	Glucoamylase
Reducing group formation	Fixed as equal	Fixed as equal	Fixed as equal
Loss in viscosity	Fast	Slow	Slow
Loss in iodine color	Fast	Slow	Slow
Maltose production	Slow	Fast	None
Glucose production	None	None	Fast

tion of that enzyme. A very high concentration of the exo-splitting enzymes will cause a rapid loss in viscosity just because there are so many bonds hydrolyzed rapidly. Conversely, a low concentration of α-amylase will cause a slow change in these parameters. Therefore, the rate of change in the number of reducing groups formed must be determined as a basis of comparison. The expected type of results for the endo- and exo-splitting enzymes are shown in Figs. 11 and 12.

Assay methods based on rates of decrease in viscosity or iodine color are very sensitive but do not measure the number of bonds hydrolyzed. Therefore, it is much better to use a method that measures reducing-group formation, or maltose or glucose production. Any number of methods, including those that use 3,5-dinitrosalicylic acid, potassium ferricyanide, and alkaline cupric solutions, may be used to determine the number of reducing groups formed. A specific measurement of maltose and glucose formation, respectively, is preferred for following the action of β-amylase and glucoamylase.

Because there are so many types of amylases (Table 2), it is essential to determine the initial and final products formed in order to determine the type of amylase present. In most cases, column separation and identification of the initial products may be sufficient. However, in the case of the exo-α-amylase that produces maltose (Table 2), the initial products are predominately maltotriose and maltotetraose, with maltose being formed more slowly from hydrolysis of the initial products.

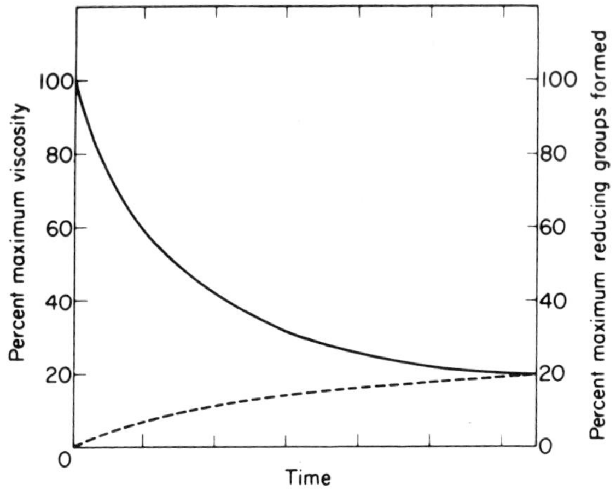

Figure 11 Schematic representation of rate of change in viscosity compared with rate of reducing-group formation for an α-amylase. The reducing groups are formed when glycosidic bonds are hydrolyzed. Solid curve, viscosity; dashed curve, reducing groups.

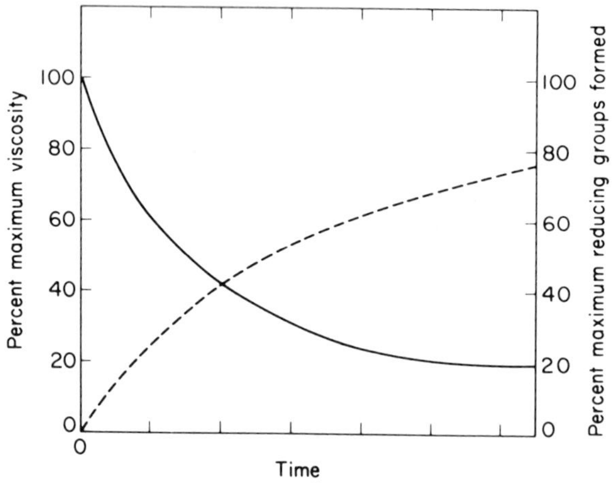

Figure 12 Schematic representation of rate of change in viscosity compared with rate of reducing-group formation for a β-amylase. The reducing groups are formed when glycosidic bonds are hydrolyzed. Solid curve, viscosity; dashed curve, reducing groups.

E. α-Amylase

We shall now look very briefly at some of the distinctive features of four types of amylases, including possible mechanisms of action. α-Amylase is found almost universally distributed throughout the animal, plant, and microbial kingdoms. In humans, salivary and pancreatic α-amylases play a major role in hydrolyzing the starches of ingested food to oligosaccharides, which in turn can be hydrolyzed by the di- and trisaccharidases found in the small intestine.

Several α-amylases have been crystallized. Their molecular weights are in the range of 50,000 and all contain 1 gram atom of Ca(II) per molecule. The activities of the mammalian and perhaps some of the bacterial α-amylases are stimulated by halogen ions. Requirement of halogen ions is not absolute since the enzyme may have some activity in the absence of such ions (Fig. 13).

Ca^{2+} binds to the protein very tightly with a K_b value of 10^{12} to 10^{15} M^{-1} and can be removed at lower pH only by use of chelating agents. On complete removal of Ca(II) the enzymes are fully active but are much less stable (Fig. 14). The role of Ca(II) in α-amylase is to maintain the secondary and tertiary stability of the molecule. There is no evidence to indicate that Ca(II) plays a direct role in binding or transforming the substrate.

The pH-rate profile for the α-amylases is a bell-shaped curve with the pH optimum dependent on the source of enzyme. For example, the mammalian α-amylases have a pH optimum of 6 to 7, depending on the absence or presence of

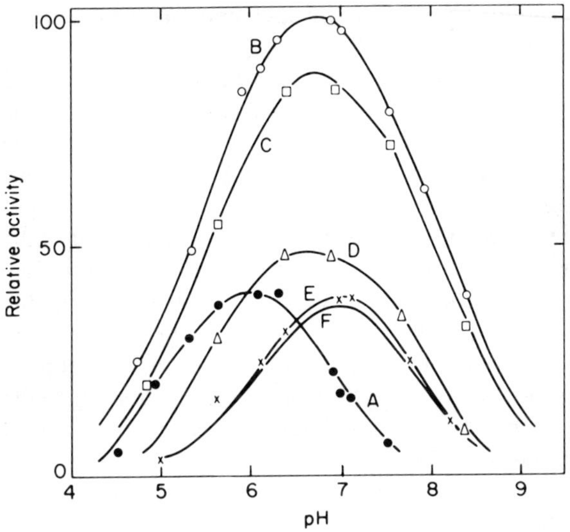

Figure 13 Effect of anions on α-amylase activity as a function of pH: (A) in absence of added anions other than 0.03 *M* phosphate present in all reactions; (B) 0.005 *M* or 0.04 *M* chloride; (C) 0.04 *M* bromide; (D) 0.04 *M* iodide; (E) 0.1 *M* nitrate; (F) 0.1 *M* chlorate. (From Ref. 6, p. 35, by courtesy of Walter de Gruyter and Co.)

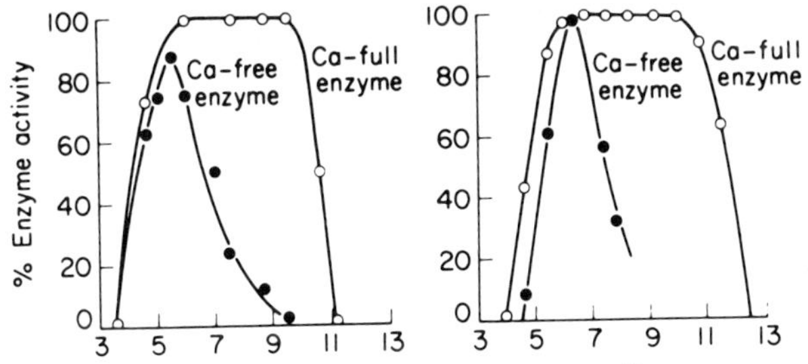

Figure 14 Effect of Ca^{2+} on stability of α-amylases. The enzymes were incubated for 20 hours at 25°C at various pH values, then tested for activity left at a common pH in the presence of Ca^{2+}. Left, hog pancreas α-amylase; right, *B. subtilis* α-amylase. (From Ref. 7, p. 324, by courtesy of Academic Press.)

halogen ions (Fig. 13). *B. subtilis* and malt α-amylases have pH optima at 5.8 to 6.0 and 4.8 to 5.4, respectively.

Koshland [8] proposed the following mechanism for the action of α-amylase (Fig. 15). As shown, the enzyme contains a carboxyl and an imidazole group in the transforming component of the active site. In some α-amylases there is some uncertainty about the nature of the nitrogen-containing group. The pK_a value of 6.5 to 8.0 could indicate either an imidazole or an amino group. The effects of photooxidation and chloride ion on the activity of porcine α-amylase [9] and the heat of ionization (4 kcal/mol) favor this being an imidazole group. The substrate forms an adsorptive complex with the enzyme, which positions the susceptible glucosidic bond in juxtaposition with the carboxyl anion and imidazolium group. In the proposed scheme the carboxyl anion serves as an attacking nucleophile on the C(1) position of the substrate, and this attack is aided by protonation of the linkage by the general acid (imidazolium ion). A glucosylenzyme intermediate involving covalent bonds forms as a result. In the deglucosylation reaction, the imidazole group (unprotonated) assists as a general base to remove a proton from water, leaving the OH^- to attack at the C(1) position of the glucosylenzyme. The configuration about C(1) is not changed by this double displacement mechanism, so the product of the reaction is of α-configuration [9–11].

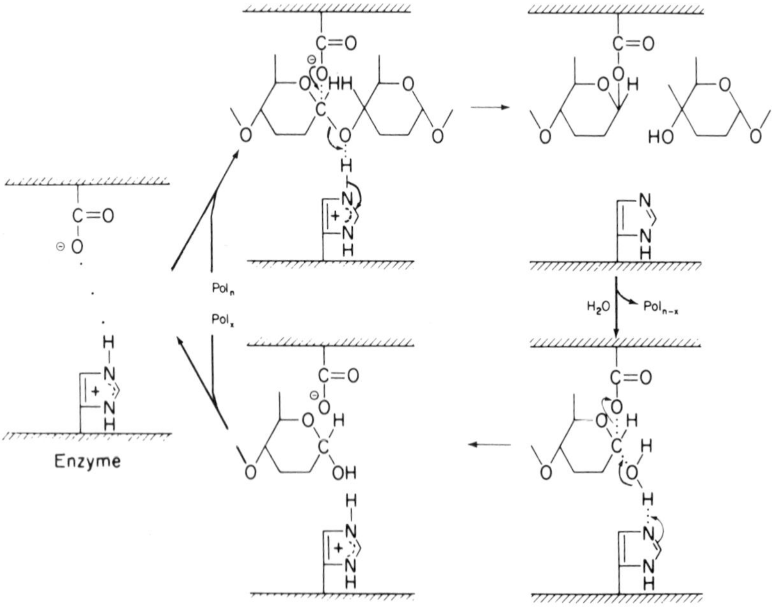

Figure 15 Proposed mechanism for α-amylase [8].

F. β-Amylase

β-Amylase hydrolyzes α-1,4-glucosidic linkages in starch and glycogen with an inversion of configuration about the C(1) position of the glycone from α to β. This has been clearly shown by several modern instrumental methods [10,11].

β-Amylase has great technological importance in the baking, brewing, and distilling industries, where starch is converted to maltose, a fermentable sugar. It is of interest that during the slow cooking of sweet potatoes, β-amylase converts a large amount of starch to maltose.

β-Amylase is found primarily in higher plants and in some microorganisms. β-Amylases from barley, wheat, rye, soybeans, and sweet potatoes have been studied extensively and have been crystallized. These plants also contain appreciable amounts of α-amylase, so that the action on starch seen during malting is a combination of the actions of these enzymes. Sweet potato β-amylase has a molecular weight of 152,000. In general, it appears that all the β-amylases have higher molecular weights than those of the α-amylases. The pH optimum is between 5 and 6.

The pH activity curves of the β-amylases are bell shaped. There appear to be two groups, with pK values of 2.5 to 3.5 and 8.0 to 8.5 involved in enzyme activity. Activities of most of the β-amylases are inhibited by sulfhydryl reagents, indicating that a sulfhydryl group is essential for activity.

In developing a mechanism for the action of β-amylase, one must explain why the enzyme splits off only maltose units and why there is inversion of configuration around C(1) of the hydrolyzed bond. Initially, Myrbäck [12] and later Thoma and Koshland [13] developed a possible arrangement of attachment of substrate to the enzyme which would account for this (Fig. 16). It is postulated that the enzyme has at least three specific groups, X, A, and B, in the active site which are involved in binding and transformation of substrate. The X group recognizes the C(4) OH group at the nonreducing end of the polysaccharide chain. When there is proper interaction between X and OH of C(4), the second glucosidic linkage of the substrate will be in proper juxtaposition with respect to the catalytic groups A and B as shown in the EA complex (reactive). The scheme also accounts for the formation of unreactive complexes where X has failed to perform its role and for inhibition by cycloamyloses.

The mechanism for inversion of configuration of the C(1) position of glucose by β-amylase may be postulated to occur as shown in Fig. 17. The proposed mechanism takes into account the four groups experimentally demonstrated to be in the active site. After formation of the enzyme·substrate complex, a nucleophilic attack of the sulfhydryl group on C(1) is facilitated by the carboxylate group acting as a general base while the imidazolium group acts as a general acid to donate a hydrogen to the glycosidic oxygen. This leads to formation of a maltosylenzyme intermediate. In hydrolysis of the intermediate, the carboxylate group acts as a

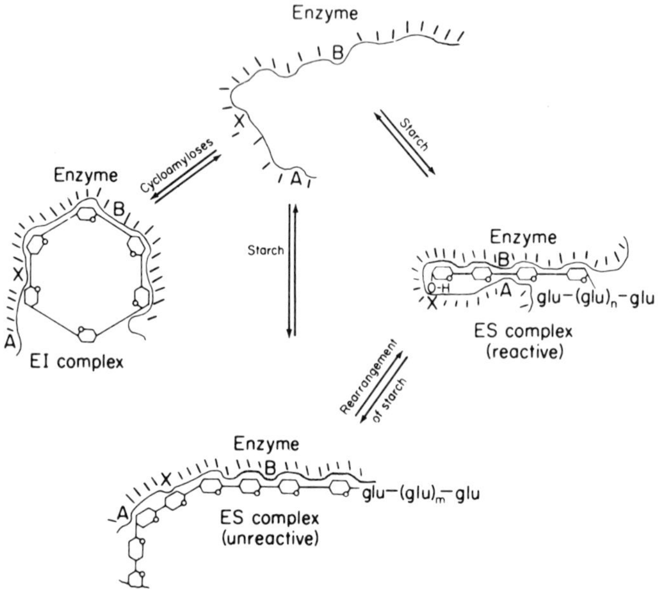

Figure 16 Schematic representation of β-amylase mechanism involving a flexible active site. Group X is responsible for interacting with the terminal C(4) hydroxyl group of glucose to align the starch chain properly with respect to catalytic groups A and B. Interior positions of the starch chain or of cycloamyloses can bind to enzyme but do not allow the folding required for reaction. (From Ref. 13, p. 3332, by courtesy of the American Chemical Society.)

general base to facilitate a backside attack of water at C(1) to release β-maltose and regenerate the enzyme. Group X serves to position the substrate properly at the active site.

G. Glucoamylase

Glucoamylase is an exo-splitting enzyme that removes glucose units consecutively from the nonreducing end of the substrate chains. The product of its action is glucose, which clearly differentiates it from α- and β-amylases. The enzyme has been reported only in bacteria and molds. The enzyme has a preference for large substrates. The K_m values for maltose, amylose, and amylopectin are 1.6×10^{-3} *M*, 4.4×10^{-5} *M*, and 4.1×10^{-7} *M*, respectively. The enzyme also hydrolyzes α-1,6 glucosidic linkages but more slowly than α-1,4 linkages are hydrolyzed. Methyl α-D-glucoside is hydrolyzed at only 7.2×10^{-4} the rate of maltose.

Glucoamylase hydrolyzes α-1,4 glucosidic linkages but the product is

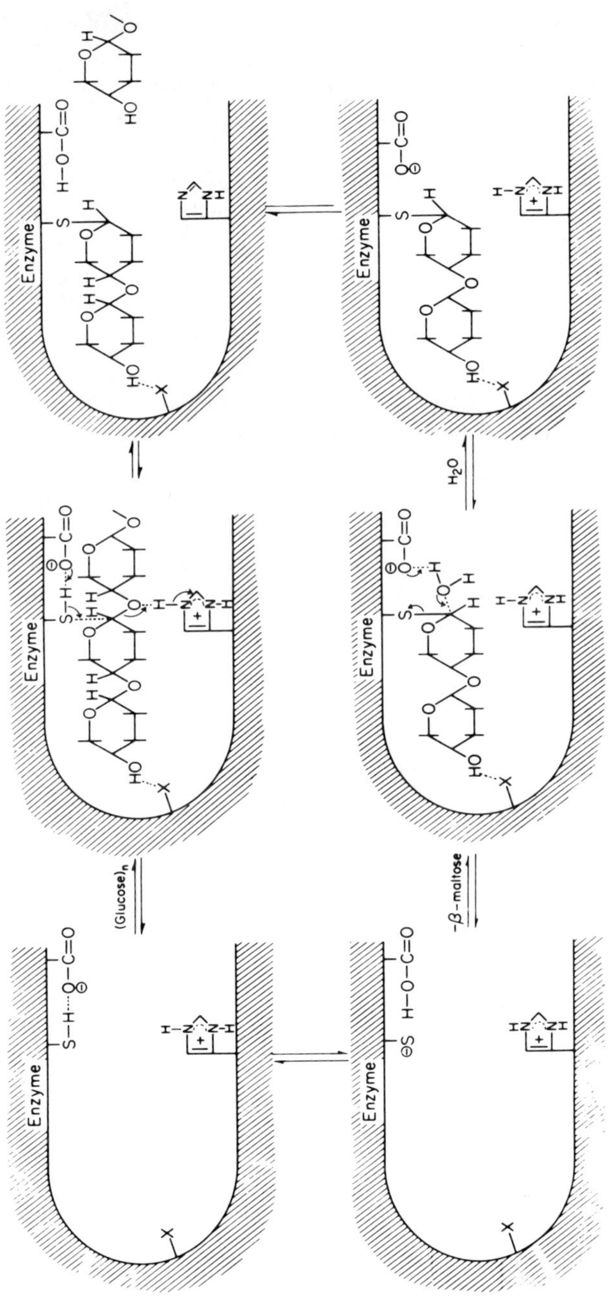

Figure 17 Proposed mechanism of β-amylase-catalyzed hydrolysis of amylose. A carboxylate group, an imidazolium group, and a sulfhydryl group are shown in the transforming locus. In the first transition-state intermediate, the carboxylate group acts as a general base and the imidazolium group as a general acid to facilitate attack of sulfhydryl group on C(1) of the susceptible bond. In the second transition state the carboxylate group serves as a general base to facilitate the attack of water on the maltosylenzyme intermediate.

β-glucose, so there is an inversion of configuration as in the case of β-amylase. The action of several exo- and endo-splitting glycosidases were examined; it was found that all exo-splitting enzymes caused an inversion of configuration about the anomeric C(1) position, while all the endo-splitting enzymes caused a retention of configuration [11].

The same rationale as presented in Fig. 16 for β-amylase may be used to explain the ability of the enzyme to split off only glucose units. In the case of this enzyme, the spatial distance between X and A and B would be such that when X is attached to the C(4) group, A and B would be juxtapositioned with respect to the first glucosidic bond.

Both the V_{max}/K_m versus pH and V_{max} versus pH profiles at 25°C are bell-shaped curves for glucoamylase with a pH optimum at 4.0 to 4.4. At 25°C, pK_1^E and pK_2^E were found to be 2.9 and 5.90, respectively, while pK_1^{EA} and pK_2^{EA} were ~1.9 and 5.9, respectively, for action of the enzyme on both maltose and panose [*O*-α-D-glucopyranosyl-(1→6)-*O*-α-D-glucosylpyranosyl-(1→4)-α-D-glucopyranose]. Values of ΔH_1 and ΔH_2 for ionization of the groups were found to be 0 and −0.8 kcal/mol, respectively [14]. The pK and ΔH values support the concept that two carboxyl groups, one as COO^- and the other as COOH, are involved in catalysis by the enzyme. This is to be contrasted with the possible involvement of an imidazole and a carboxylate group (and a sulfhydryl group in β-amylase) in the case of α- and β-amylases.

H. Pullulanase

Pullulanase is an endo-splitting enzyme that hydrolyzes the α-1,6-glucosidic bonds of pullulan, amylopectin, and glycogen and their α- and β-limit dextrins. Pullulanase requires that each of the two chains of amylopectin linked by an α-1,6-glucosidic bond contain at least two adjacent α-1,4-linked glucose units. In this regard, pullulan is an ideal substrate since it is a linear molecule composed of repeating maltotriose units (with two α-1,4 linkages) joined by α-1,6-glucosidic linkages. In other words, every third glucosidic bond is α-1,6. Pullulanase and similar debranching enzymes are very useful in making light beer (all the carbohydrate is converted to ethanol) and in preventing staleness of breads.

III. DEXTRANSUCRASE

The overall reaction in the biosynthesis of di-, oligo-, and polysaccharides is

$$\text{glycosyl-}O\text{-X} + \text{HO-acceptor} \rightleftharpoons \text{glucosyl-}O\text{-acceptor} + \text{HOX} \qquad (2)$$

The nature of X differentiates the three known pathways of condensation of saccharides: X may be a phosphate group, as in aldose-1-phosphate; a nucleoside pyrophosphate, as in nucleoside diphosphate glycosides (e.g., UDP-glycosides); or it may be a saccharide with one, two, or any number of monomer units. The respective enzymes involved in glycosyl transfer to the acceptor by the three pathways are called *phosphorylases*, *nucleoside diphosphate transglycosylases*, and *transglycosylases*, respectively.

In biosynthesis of disaccharides most of the reactions involve nucleoside diphosphate glycosides as donors. In biosynthesis of polysaccharides, all three pathways have been found. In the biosynthesis of glycogen and starch, the phosphorylase-catalyzed reaction is quite important.

The third type of reaction, catalyzed by the transglycosylases, is particularly important in microorganisms (Table 4). The energy for formation of the new glycosidic bonds comes from that present in the existing glycosidic bond of the donor (Table 5). It is clear from this table that while the C—N bond in UDP-glucose has a little more energy that the glycosidic bond of sucrose, sucrose should be a better glucosyl donor than glucose-1-phosphate and considerably better than maltose. It is easy to see why, in terms of energy, sucrose can serve as a glycosyl donor in the synthesis of dextran and levan.

Of the transglycosidases listed in Table 4, dextransucrase is of particular interest to the food scientists. With its α-1,6-glucosidic linkages, dextran can be used as a thickening agent for dietetic foods since humans do not have enzymes in the intestinal tract that hydrolyze these polymers at an appreciable rate. The dextrans are the starting material for the production of Sephadex, used so widely in gel filtration techniques, and are also important in blood transfusions. Dextran-

Table 4 Formation of Polysaccharides by Transglycosylation

Donor	Acceptor	Polymer	Other product	Enzyme
Sucrose	$(\text{Glucose})_n$	Amylodextrin (α-1,4)	D-Fructose	Amylosucrase (sucrose:1,4-α-D glucan 4-α-D-glucosyltransferase, EC 2.4.1.4)
Sucrose	$(\text{Glucose})_n$	Dextran (α-1,6)	D-Fructose	Dextransucrase (Sucrose:1,6-α-D-glucan 6-α-D-glucosyltransferase, EC 2.4.1.5)
Sucrose	$(\text{Fructose})_n$	Levan	D-Glucose	Levansucrase (Sucrose:2,6-β-D-fructan 6-β-D-fructosyltransferase, EC 2.4.1.10)
Maltose	$(\text{Glucose})_n$	Amylodextrin (α-1,4)	D-Glucose	Amylomaltase (1,4-α-D-glucan:1,4-α-D-glucan 4-α-D-glucosyl-transferase, EC 2.4.1.25)

Table 5 Energies of Hydrolysis of Glycosyl Donors and of Some Polysaccharides

Compound	ΔG° (cal/mol)
Glucose 1-phosphate	−5500
UDP-glucose	−7600
Sucrose	−6600
Maltose	−3000
Glycogen	−4300
Dextran	−2000
Levan	−4600

Source: Ref. (15), p. 382, by courtesy of Pergamon Press.

sucrase or a similar enzyme may be responsible for the "ropy" nature of some beers and other products.

The activity of dextransucrase can be assayed by following the rate of formation of the reducing sugar D-fructose or by measuring the increase in viscosity of the solution.

The only known donor for this enzyme is sucrose. A long list of other sugars and sugar derivatives have been tried, but they are inactive. A great variety of acceptor molecules are known, including glucose and maltose. However, the enzyme has a stronger preference for low-molecular-weight dextrans. The size of the final product depends on the nature of the acceptor and its concentration. In the presence of high concentrations of glucose, only the dimer, trimer, and very little polysaccharide are formed. It appears that the enzyme initiates chain lengthening on all acceptor molecules simultaneously, and the reaction is terminated whenever the enzyme runs out of donor molecules. The short-chain products formed cannot serve as donor molecules since the glucosidic bonds formed have a ΔG° value of −2000 cal/mol, the same as that of the glucosidic bonds of the higher polymers (Table 5).

Dextransucrase is not inhibited by F^-, CN^-, iodoacetate, Cu^{2+}, and Ag^+, nor is its activity affected by dialysis or EDTA. It has been reported that photooxidation causes a loss of activity, possibly indicating the involvement of an imidazole group in the active site.

IV. CELLULASE

Cellulose is the most abundant carbohydrate in nature (annual production of about 4×10^{10} tons) and is thought to be synthesized through the involvement of UDP-

and GDP-glucose as the donor and [O-β-D-glucopyranosyl-(1→4)]$_n$ as the acceptor. The chain lengths are variable. Because of the β-configuration around C(1), cellulose chains interact with each other by hydrogen bonding to form oriented, crystalline fibrillar micelles (Fig. 7), which have little water solubility. Due to this inertness to water, cellulose is relatively inert to action of hydrolytic enzymes.

The potential use of cellulase to convert waste paper, sawdust, and other cellulosic waste to glucose (with fermentation to ethanol), to tenderize plant materials, to break microbial cell walls for release of protein, and to elucidate the structure of cellulosic material should not be minimized. The cellulases of microorganisms are of tremendous importance in the degradation of cellulose waste products.

The enzymes that act on cellulose and derived products of cellulose can be divided into four groups [16]. (a) Endoglucanases [1,4(1,3;1,4)-β-D-glucan 4-glucanohydrolase; EC 3.2.1.4] are inactive against crystalline celluloses such as cotton and Avicel, but they hydrolyze amorphous celluloses (including amorphous regions of crystalline celluloses) and soluble substrates such as carboxymethyl cellulose and hydroxyethyl cellulose. Endoglucanase activity is characterized by random hydrolysis of β-glucosidic bond cleavage, resulting in a rapid decrease in viscosity relative to the rate of increase in reducing groups. The products include glucose, cellobiose, and cellodextrins of various sizes. (b) The second group of enzymes are the cellobiohydrolases (1,4-β-D-glucan cellobiohydrolase, EC 3.2.1.91), which are exo-splitting enzymes. They degrade amorphous cellulose by consecutive removal of cellobiose from the nonreducing ends of the substrate. When pure, they usually have little activity on cotton, but can degrade Avicel, a microcrystalline cellulose, by as much as 40%. The rate of increase in reducing groups in relation to decrease in viscosity is much higher than for the endoglucanases. Endoglucanases and cellobiohydrolases act synergistically in the hydrolysis of crystalline cellulose. (c) The third group of enzymes are the exoglucohydrolases (1,4-β-D-glucan glucobiohydrolase, EC 3.2.1.74), which hydrolyze consecutively the removal of glucose units from the nonreducing end of cellodextrins; the rate of hydrolysis decreases as the chain length of the substrate decreases. (d) The fourth group of enzymes are the β-glucosidases (β-D-glucoside glucohydrolase, EC 3.2.1.21), which cleave cellobiose to glucose and remove glucose from the nonreducing end of small cellodextrins. Unlike the exoglucohydrolases, the rate of β-glucosidase increases as the size of the substrate decreases, with cellobiose being hydrolyzed the fastest. The β-glucosidases and exoglucohydrolases have in common the substrates cellobiose to cellohexaose (six glucose units). They may readily be distinguished on the basis of their relative activity on the two substrates cellobiose and cellohexaose. The β-glucosidases hydrolyze cellobiose much more rapidly than cellohexaose, whereas the reverse is true for the exoglucohydrolases (Fig. 18).

Homogeneous *Myrothecium verrucaria* cellulase has a molecular weight of

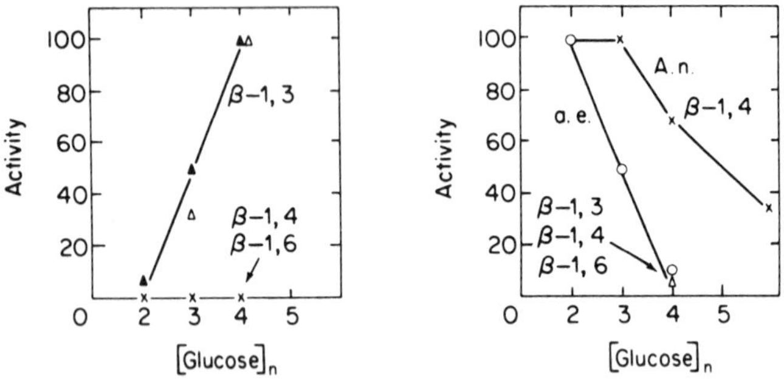

Figure 18 Relative activities of exoglucohydrolases and β-glucosidases in hydrolysis of substrates with varying degree of polymerization ($[\text{glucose}]_n$): left, exoglucohydrolases; right, β-glucosidases. Exoglucohydrolases from *Basidiomycete* sp. (▲, ×) and *Sporotrichum* (△, ×); β-glucosidases from *Aspergillus niger* (×) and almond emulsion (○). (From Ref. 17, p. 27, by courtesy of the National Research Council of Canada.)

63,000, a pH optimum of 5.5 to 6.0, and is stable for 5 minutes at 100°C at pH 7. The enzyme is inhibited by heavy metal ions, by sulfhydryl reagents, by oxidizing and reducing agents, and by glucose.

Cellulase activity can be determined by measuring the rate of reducing groups released from cellulose. The heterogeneity of the system (insoluble substrate) does not permit good kinetic data. It is preferable to use cellulose that has been carboxymethylated (carboxymethyl cellulose) until it is water soluble. The action of cellulase on carboxymethyl cellulose can be followed by a decrease in viscosity or by determining the rate of reducing-group formation. In combination, the two methods can distinguish between endo- and exo-splitting cellulases and sometimes among enzymes in either group. Many microorganisms—particularly fungi, as well as snails and termites—contain small amounts of cellulase.

V. β-FRUCTOFURANOSIDASE

Experimental work on the sucrose-degrading enzyme β-fructofuranosidase (β-D-fructofuranoside fructohydrolase, EC 3.2.1.26) presents a paradox. On the one hand, the ability of a substance extracted from yeast to hydrolyze sucrose was recognized as early as 1860 by Berthelot. The enzyme was a popular one to work with because its action can be followed continuously by the use of a polarimeter (see below). It was one of the enzymes treated in Sörensen's classic papers of 1909 [18] on the effect of pH on enzyme activity, and it was data on β-fructofuranosidase that led Michaelis and Menten in 1913 [19] to a mathematical derivation of the Michaelis–Menten equation. On the other hand, it is very difficult to obtain the enzyme in a satisfactory state of homogeneity [20]. The reason for this is the glycoprotein nature of the molecule.

The reaction catalyzed by the enzyme is

$$\underset{+66.5^\circ}{\text{sucrose}} + H_2O \rightarrow \underset{+52.2^\circ}{\text{glucose}} + \underset{-93^\circ}{\text{fructose}} \qquad (3)$$

where the numbers under the compounds represent the specific optical rotations. The rate of the reaction can be followed most conveniently in a polarimeter since there is a net molar change in rotation of 87°. Because of inversion of rotation (+66.5 to −20.5°) of the solution during reaction, the enzyme became known as invertase. The rate of glucose production can be followed by the glucose oxidase–peroxidase–chromogen method.

β-Fructofuranosidase is found widely distributed in nature and is present in plants, animals, and microorganisms. There are two types of enzymes, β-fructofuranosidase and α-glucosidase (α-D-glucoside glucohydrolase, EC 3.2.1.20), which can hydrolyze sucrose. The glycosidic bond in sucrose involves the reducing groups of both D-glucose and D-fructose. One type of enzyme hydrolyzes the glucosidic bond between C(2) and O (β-fructofuranosidase) and the other hydrolyzes the bond between C(1) and O (α-glucosidase).

The two types of enzymes can be distinguished by use of $H_2{}^{18}O$, but this is an inconvenient method since the two products would have to be separated and analyzed by mass spectrometry. The best way to distinguish between the two is by the use of the two trisaccharides raffinose and melezitose (Fig. 19). α-Glucosidase directs its specificity toward the glucose moiety of sucrose, including the C(6) position. Modification of C(6) of this moiety prevents α-glucosidase from hydrolyzing raffinose. In melezitose, the glucosyl residue attached to fructose is not modified and thus α-glucosidase can hydrolyze melezitose. The converse is true of β-fructofuranosidase. Its specificity is directed toward the fructose moiety of sucrose, and modification of this moiety, as in melezitose, prevents the compound from serving as a substrate, while raffinose is a substrate for β-fructofuranosidase.

The following properties have been described for yeast β-fructofuranosidase. K_m values for sucrose and raffinose are 0.016 *M* and 0.24 *M*, respectively. In order to see appreciable activity with raffinose, one must have a very high concentration of substrate. The pH optimum of the enzyme is in the range 4.5 to 5.5. Its activity appears to depend on two prototropic groups of p*K* ~3 and ~7. The enzyme is stable to around 35°C. The enzyme is inhibited by iodine (to only 50%) and by heavy metal ions (Hg^{2+}, Ag^{+}, Cu^{2+}). Iodoacetate and ferricyanide have very little effect on activity.

Exercise 1. Assuming that V_{max} is the same on both substrates, calculate the relative rates of activity of β-fructofuranosidase on raffinose and sucrose at 0.016 *M* concentrations.

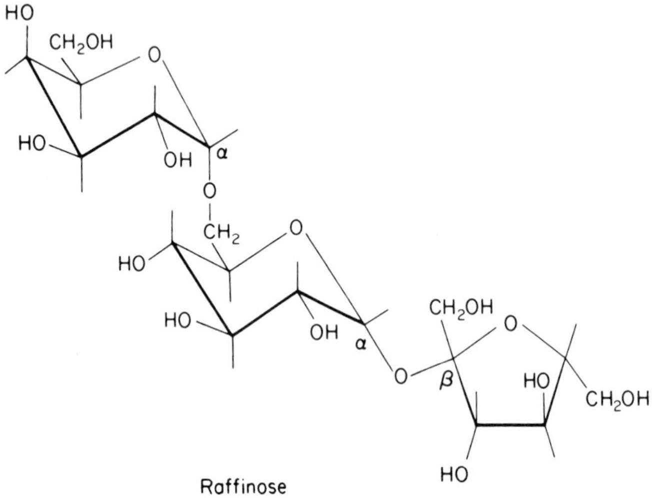

Raffinose

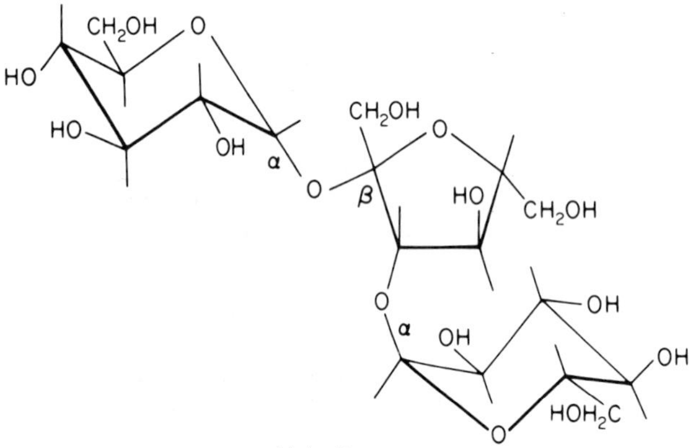

Melezitose

Figure 19 Structures of raffinose [*O*-α-D-galactopyranosyl-(1→6)-*O*-α-D-glucopyranosyl-(1→2)-*O*-β-D-fructofuranoside] and melezitose [*O*-α-D-glucopyranosyl-(1→3)-*O*-β-D-fructofuranosyl-(2→1)-*O*-α-D-glucopyranoside].

VI. β-GALACTOSIDASE

β-Galactosidase (β-D-galactoside galactohydrolase, EC 3.2.1.23), referred to frequently as lactase (trivial name), has been one of the most carefully studied glycosidases because of its ready inducibility in microorganisms. The nature of the lac operon has been well elucidated and researchers have been successful in separating the area containing this operon from the chromosome.

The enzyme catalyzes the hydrolysis of β-D-galactosides and α-L-arabinosides. Among the naturally occurring substrates, its action on lactose has been most studied. This enzyme is of interest to the food scientist for several reasons. Hydrolysis of lactose leads to glucose and galactose, which are much sweeter than lactose. Lactose has a limited solubility and tends to precipitate out in frozen dairy products, which gives a grainy texture to the product. Partial hydrolysis of the lactose prevents this. It is important to appreciate that most adults are often allergic to milk since they do not have β-galactosidase and therefore cannot metabolize lactose.

Several methods are available for following the activity of β-galactosidase. There is a change in specific rotation from +52.5° to +67.0° when lactose is hydrolyzed [Eq. (4)]. This is a small change but adequate to permit the use of a

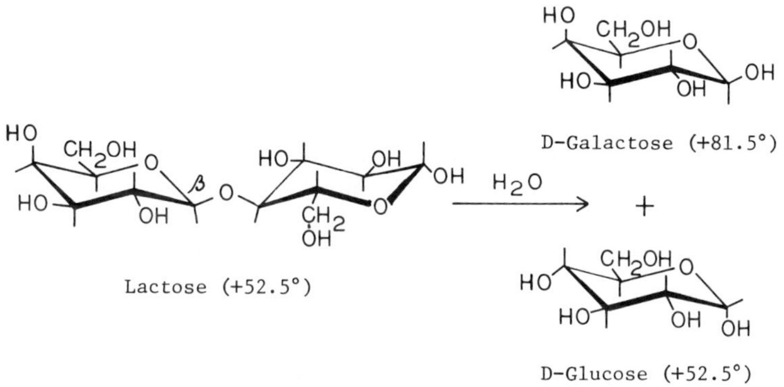

(4)

polarimeter for following the reaction. The rate of reaction can also be followed by determining the glucose produced by use of glucose oxidase–peroxidase–chromogen (Glucostat reagent). A reducing sugar method is not recommended because of the high initial blank value. (Lactose is a reducing sugar.) One of the best methods is to use a chromogenic substrate. A chromogenic substrate is one that gives a colored product on hydrolysis. As shown in Table 6, either *o*- or *p*-nitrophenyl galactoside may be used since the products, *o*- and *p*-nitrophenol, have an absorbance maximum near 400 nm (above pH 7).

o-Nitrophenyl β-D-galactoside is by far the best substrate for determining the activity of β-galactosidase. This is true not only because of ease in following the reaction (chromogenic substrate), but at less than saturating substrate concentra-

Table 6 Substrate Specificity of β-Galactosidase of *E. coli*, ML 309[a]

Substrate	K_m (mol/L)	V_{max} (μmol/min/mg enzyme)	V_{max}/K_m ($\times 10^{-4}$)
o-Nitrophenyl-β-D-galactoside	1.61×10^{-4}	178	110
p-Nitrophenyl-β-D-galactoside	5.31×10^{-5}	22.4	44
Methylsalicylate-β-D-galactoside	2.5×10^{-3}	4.2	0.17
Phenyl-β-D-galactoside	1.47×10^{-3}	10.4	0.71
α-Lactose	1.9×10^{-3}	6.55	0.35
allo-Lactose[b]	8.32×10^{-3}	12.8	0.15
Thio-(*o*-nitrophenyl)-β-D-galactoside[b]	1.20×10^{-4}[c]	0.00	0

Source: Ref. 21.
[a]The reactions were performed at pH 7.6, 20°C, 0.05 M Tris–HCl buffer, and 0.05 M NaCl.
[b]No NaCl.
[c]K_i value determined with lactose as substrate.

tions, it is hydrolyzed at about 300 times the rate of α-lactose (compare the V_{max}/K_m values in Table 6). V_{max}/K_m is a measure of the catalytic efficiency and specificity of an enzyme.

The effect of pH on V_{max} in the presence and absence of 0.05 *M* NaCl is shown in Fig. 20. In the absence of NaCl, the pH optimum is 7.7, with p*K* values of 6.67 and 8.90 at 20°C. In the presence of 0.05 *M* NaCl there is almost a threefold increase in activity and the pH optimum is shifted to 6.6 with p*K* values of 5.87 and 8.60. Potassium chloride causes a shift in pH optimum but does not increase activity. The ionizable group on the alkaline side has been identified as a sulfhydryl group based on inhibition by heavy metal ions and by *p*-mercuribenzoate. The other ionizable group has a ΔH_{ion} of 5.75 kcal/mol. From the ΔH_{ion} and pK, it appears that this is an imidazole group.

A proposed mechanism for action of the enzyme is shown in Fig. 21. The sulfhydryl group acts as a general acid to protonate the glycosidic oxygen atom, and the imidazole group acts as a nucleophile in that it attacks the nucleophilic center at C(1) of the glycone. A covalent intermediate involving a carbon–nitrogen bond is proposed. In removal of the galactosyl group, the sulfhydryl anion (S^-) acts as a general base to abstract a proton from water (shown as R′OH) which assists in the attack of OH^- at the C(1) position. There is no inversion of the anomeric carbon at either step in the reaction and the product retains the β-configuration around the anomeric carbon.

The number of active sites in β-galactosidase per molecular weight of 700,000 is temperature dependent. At 4 to 6°C, there are 1.9, whereas at 20 to 22°C, there are 4.7 experimentally determined active sites. It appears that the molecule is

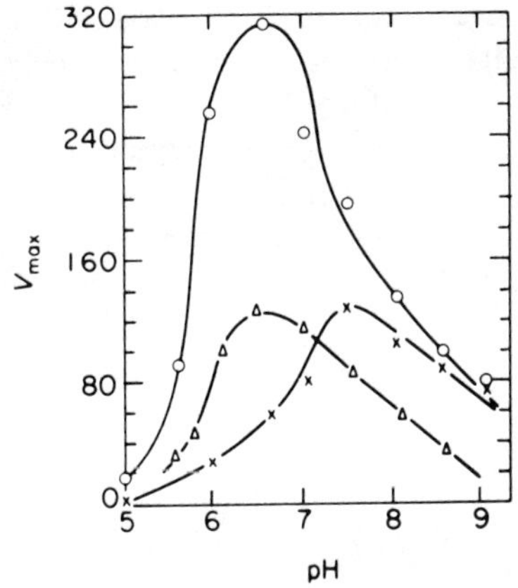

Figure 20 Effect of pH on V_{max} of β-galactosidase in the presence and absence of NaCl and KCl. ×, no added salt; ○, 0.05 *M* NaCl; △, 0.05 *M* KCl. (From Ref. 22, p. 381, by courtesy of Springer-Verlag.)

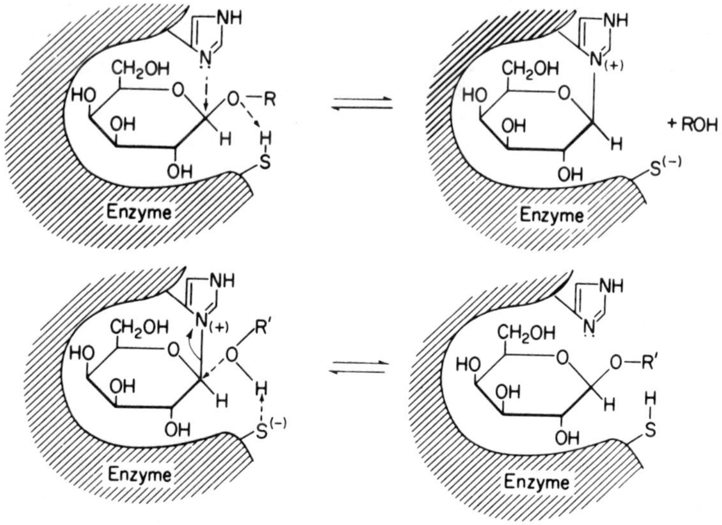

Figure 21 Proposed mechanism for β-galactosidase. (From Ref. 23, p. 427, by courtesy of Academic Press.)

composed of several subunits (possibly five) and that at lower temperatures the conformation of the molecule is such that not all the active sites are exposed to the substrate.

Exercise 2. How would you determine the number of active sites per molecule of β-galactosidase? Having determined the number of active sites at several temperatures, how would you use this information to determine the true activation energy for the hydrolysis of *o*-nitrophenyl β-D-galactoside, for example?

REVIEW QUESTIONS

See the end of Chapter 19.

REFERENCES

General References

Caputto, R., H. S. Barra, and F. A. Cumar, Carbohydrate metabolism, *Ann. Rev. Biochem.*, **36**, 211 (1967).

Fischer, E. H., and E. A. Stein, α-Amylases, in *The Enzymes* (P. D. Boyer, H. Lardy, and K. Myrbäck, eds.), Vol. 4, Academic Press, New York, 1960, p. 313.

Fogarty, W. M., and C. T. Kelly, eds., *Microbial Enzymes and Biotechnology*, 2nd Ed., Elsevier Applied Science, New York, 1990.

French, D., β-Amylases, in *The Enzymes* (P. D. Boyer, H. Lardy, and K. Myrbäck, eds.), Vol. 4, Academic Press, New York, 1960, p. 345.

Friedman, R. B., ed., *Biotechnology of Amylodextrin Oligosaccharides*, ACS Symposium Series 458, American Chemical Society, Washington, D.C., 1991.

Wallenfels, K., and O. P. Malhotra, β-Galactosidase, in *The Enzymes* (P. D. Boyer, H. Lardy, and K. Myrbäck, eds.), Vol. 4, Academic Press, New York, 1960, p. 409.

Cited References

1. D. R. Whitaker, *Arch. Biochem. Biophys.*, **53**, 439 (1954).
2. S. Nasuno and M. P. Starr, *J. Biol. Chem.*, **241**, 5298 (1966).
3. J. Robyt and D. French, *Arch. Biochem. Biophys.*, **100**, 451 (1963).
4. K. H. Meyer and P. Bernfeld, *Helv. Chim. Acta*, **23**, 875 (1940).
5. P. Bernfeld, *Adv. Enzymol.*, **12**, 379 (1951).
6. K. Myrbäck, *Z. Physiol. Chem.*, **159**, 1 (1926).
7. E. H. Fischer and E. A. Stein, α-Amylases, in *The Enzymes* (P. D. Boyer, H. Lardy, and K. Myrbäck, eds.), Vol. 4, Academic Press, New York, 1960, p. 313.
8. D. E. Koshland, Jr., Group Transfer as an Enzymatic Substitution Mechanism, in *The Mechanism of Enzyme Action* (W. D. McElroy and B. Glass, eds.), John Hopkins University Press, Baltimore, 1954, p. 608; *Discussions Faraday Soc.*, **20**, 142 (1956).
9. J. Wakim, M. Robinson, and J. A. Thoma, *Carbohydr. Res.*, **10**, 487 (1969).
10. G. Semenza, H.-Ch. Curtius, O. Raunhardt, P. Hore, and M. Müller, *Carbohydr. Res.*, **10**, 417 (1969).

11. D. E. Eveleigh and A. S. Perlin, *Carbohydr. Res.*, **10**, 87 (1969).
12. K. Myrbäck, *Arkiv Kemi*, **2**, 417 (1951).
13. J. A. Thoma and D. E. Koshland, Jr., *J. Am. Chem. Soc.*, **82**, 3329 (1960).
14. K. Hiromi, K. Takahashi, Z. Hamauzu, and S. Ono, *J. Biochem. (Tokyo)*, **59**, 469 (1966).
15. P. Bernfeld, The Biogenesis of Carbohydrates, in *Biogenesis of Natural Compounds* (P. Bernfeld, ed.), Pergamon Press, Elmsford, N.Y., 1967, p. 315.
16. M. P. Coughlan, Cellulose degradation by fungi, in *Microbial Enzymes and Biotechnology*, 2nd Ed. (W. M. Fogarty and C. T. Kelly, eds.), Elsevier Applied Science, New York, 1990, p. 1.
17. E. T. Reese, A. H. Maguire, and F. W. Parrish, *Can. J. Biochem.*, **46**, 25 (1968).
18. S. P. L. Sörensen, *Biochem. Z.*, **21**, 131, 201 (1909).
19. L. Michaelis and M. L. Menten, *Biochem. Z.*, **49**, 333 (1913).
20. N. P. Neumann and J O. Lampen, *Biochemistry*, **6**, 468 (1967); **8**, 3552 (1969).
21. K. Wallenfels, J. Lehman, and O. P. Malhotra, *Biochem. Z.*, **333**, 209 (1960).
22. K. Wallenfels, O. P. Malhotra, and D. Dabich, *Biochem. Z.*, **333**, 377 (1960).
23. K. Wallenfels and O. P. Malhotra, β-Galactosidase, in *The Enzymes* (P. D. Boyer, H. Lardy, and K. Myrbäck, eds.), Vol. 4, Academic Press, New York, 1960, p. 409.

16
Pectic Enzymes

Two of the three types of enzymes that act on pectic substances do not hydrolyze glycosidic bonds of the substrate; therefore, it is best to discuss these enzymes under a separate category from the glycoside hydrolases. This should not obscure the fact the substrate is a glycoside. The pectic substances include the galacturonans (polymers of galacturonic acid), the rhamnogalacturons [mixed polymers of rhamnose (2 to 4%) and galacturonic acid], arabinans (polymers of arabinose), galactans (polymers of galactose), and arabinogalactans (mixed polymers of arabinose and galactose). The pectic enzymes recognize the β-1,4-D-galacturonic acid units of the galacturans and the rhamnogalacturons specifically as substrates (Fig. 1).

A number of names are associated with the substances composed primarily of α-1,4-D-galacturonopyranose units. *Pectic substance* is a group designation for the colloidal carbohydrates of plants that contain a large proportion of anhydrogalacturonic acid units. *Protopectin* is the water-insoluble parent pectic substance in plants. Its exact structure is not known since it must be partially degraded in order to be extracted from plant cells. *Pectin* is the general term for pectic substances that form characteristic sugar–acid gels (Fig. 1). Pectin has about 75% of its carboxyl groups esterified with methanol. *Pectic acid* is a compound in which essentially all of the carboxyl groups of the anhydrogalacturonic acid units are free. *Pectinic acid* is a group of compounds, including pectin, which contain more than a negligible number of methyl ester groups.

The pectic substances are constituents of cell walls and of intercellular layers of all higher plants, along with cellulose and hemicelluloses. These substances are

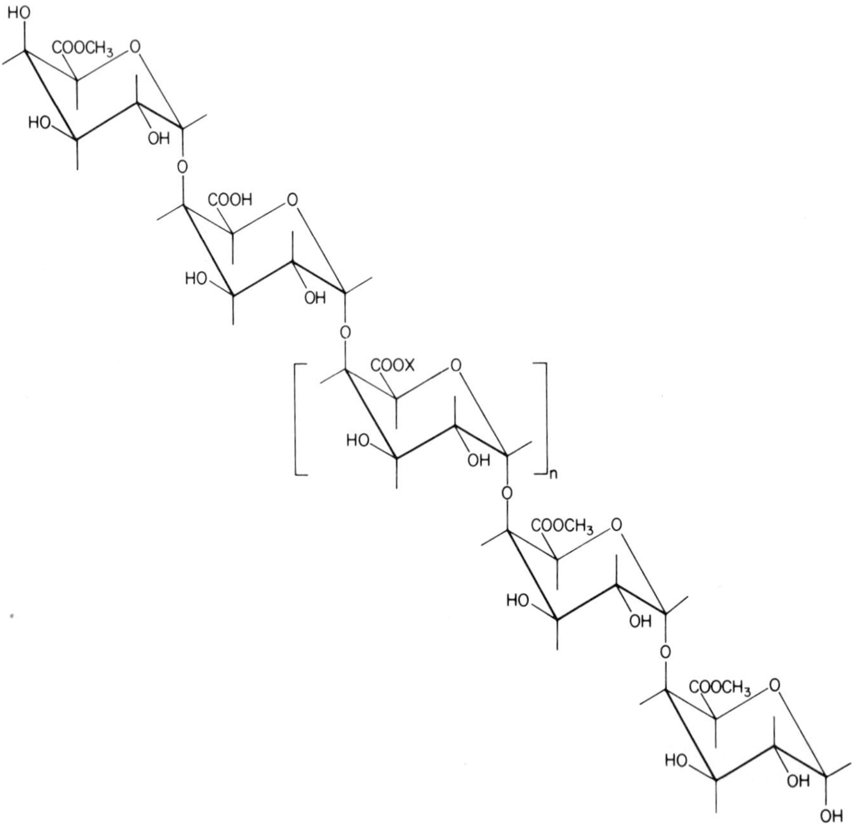

Figure 1 Structure of pectin. X is either H or CH_3, with ~75% of the carboxyl groups being esterified with methanol. The secondary conformation of the polymer is not shown.

also found in plant juices and saps and contribute in a substantial way to the texture of fruits such as tomatoes and citrus.

The pectic enzymes that degrade the pectic substances have been found in higher plants and in microorganisms but, other than the snail, are not found in animals. These enzymes are of great importance to food scientists. They are useful for treatment of fruit juices and beverages to facilitate filtration and clarification and to increase juice yields and in the production of low-methoxyl pectins and galacturonic acids. They are deteriorative enzymes in that they cause excessive softening of many fruits and vegetables and they cause "cloud" separation in such products as tomato and citrus juices.

There are three types of pectic enzymes. These three types of enzymes catalyze three different types of reactions and belong to two of the six major classes of enzymes. The pectic enzyme types are: pectinesterase, the "polygalacturonases" and the pectate lyases.

I. PECTINESTERASE

Pectinesterase (pectin pectylhydrolase, EC 3.1.1.11) removes methoxyl groups from methylated pectic substances (pectin) and therefore belongs to the sub-subdivision of enzymes that hydrolyze carboxylic acid esters. The enzyme has often been referred to as pectase, pectin methoxylase, pectin demethoxylase, pectolipase, and pectin methylesterase.

Activity of the enzyme on pectin [Eq. (1)] can be followed most conveniently and continuously in a pH Stat at pH 7.5 since a proton is released when the ester

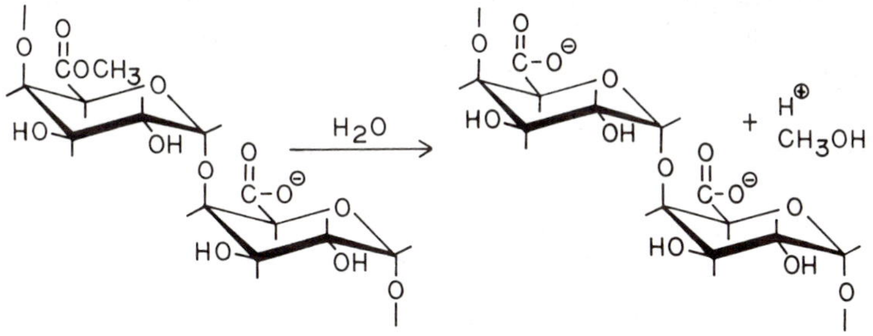

(1)

bond is hydrolyzed [Eq. (1)]. The rate of methanol production and the Ca^{2+} precipitability of the pectic acid formed have also been used to determine enzyme activity.

Pectinesterases are generally thought to initiate attack on pectin at a position adjacent to a preexisting free carboxyl group. Only 65 to 75% of the carboxyl groups of pectin are esterified, so there are a number of points along the chain where the enzyme may start removal of the methoxyl groups. Totally methylated polygalacturonic acid is not acted upon by a pectinesterase that requires a free carboxyl group for initiation of attack.

One of the best studied of the pectinesterases is the enzyme from orange albedo. It has been obtained in reasonably pure form [1] and its activity tested on a wide variety of compounds. The enzyme is not absolutely specific for the methyl ester of pectin in that the ethyl ester is hydrolyzed at 3 to 13% of the rate of the methyl ester. Derivatives in which the alcohol moiety is glycol, glycerol, or butanol are not hydrolyzed. Orange pectinesterase is specific for esters of polygalacturonic acid in that a great variety of other carboxylic acid esters were hydrolyzed at less than 0.001 of the rate of pectin.

II. THE POLYGALACTURONASES

The polygalacturonases (poly-α-1,4-galacturonide glycanohydrolase, EC 3.2.1.15) hydrolyze glycosidic linkages in pectic substances with water [Eq. (2)]. The

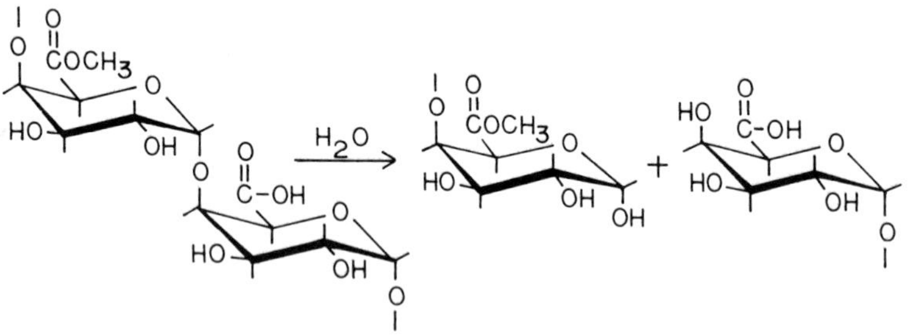

(2)

Commission on Enzymes did not distinguish between the polygalacturonases and polymethylgalacturonases (see below) but classified them all as polygalacturonases (trivial name). We shall use "polygalacturonases" as a group name and "polygalacturonases" as a subgroup distinct from the subgroup polymethylgalacturonases.

The polygalacturonases are subdivided on the basis of the method of attack on the substrate and the nature of the substrate (Table 1). The pH optima for most of the polymethylgalacturonases and polygalacturonases are in the range 4.5 to 6.0. Both types of endo-splitting enzymes hydrolyze large substrates at a maxi-

Table 1 Subdivision of the Polygalacturonases

1. Random mechanism of hydrolysis
   a. Protopectinases, protopectin hydrolyzed (solubilized) to pectin.
   b. Endo-polymethylgalacturonases. Pectin hydrolyzed in preference to pectic acid. Products are methyl oligogalacturonates.
   c. Endo-polygalacturonases (EC 3.2.1.15). Pectic acid hydrolyzed in preference to pectin. Products are oligogalacturonates.
2. Terminal mechanism of hydrolysis
   a. Exo-polymethylgalacturonases. Pectin hydrolyzed in preference to pectic acid. Product is methyl monogalacturonate.
   b. Exo-polygalacturonases. Pectic acid hydrolyzed in preference to pectin. Product is monogalacturonate.
   c. Oligogalacturonate hydrolases. Di- and trigalacturonates are hydrolyzed to monogalacturonate.
   d. Δ4:5 Unsaturated oligogalacturonate hydrolases. Δ4:5 (Galacturonate)$_n$ hydrolyzed to unsaturated monogalacturonate and saturated galacturonate$_{n-1}$.

Source: Adapted from Refs. 2 and 3.

mum rate. The relative rates of hydrolysis by tomato endopolygalacturonase acting on pectic acid and tetra-, tri-, and digalacturonic acids have been reported to be 100, 7, 1.5, and 1, respectively [3,4].

Endo-polymethylgalacturonase and endo-polygalacturonase may readily be distinguished from each other on the basis of their relative activities on fully methylated pectin and on pectic acid (Fig. 2). Endo-polymethylgalacturonase has maximum activity on pectin and no activity on pectic acid; the converse is true of endo-polygalacturonase. Use of pectin, with 65 to 70% methylation, does not give as clear-cut results. The same experimental technique would also distinguish between exo-polymethylgalacturonase and exo-polygalacturonase. While exo- and endo-polymethylgalacturonases continue to be reported in the literature, some major researchers question whether these enzymes really exist. Only the use of 100% methylated pectin can determine this.

Several methods are available to distinguish between the exo- and endo-splitting enzymes. One of the most useful methods is to compare the rate of decrease in viscosity with rate of hydrolysis as measured by increase in reducing groups. The results are the same as described for α- and β-amylases (Chapter 15, Figs. 11 and 12). An endo-splitting enzyme causes ~50% reduction in viscosity when only 3 to 5% of the glycosidic bonds have been hydrolyzed, while with an exo-splitting enzyme ~10 to 15% hydrolysis of glycosidic bonds is needed to produce a 50% reduction in viscosity.

A second very important method used to distinguish between an endo- and an exo-splitting enzyme is the nature of products formed near the beginning of the

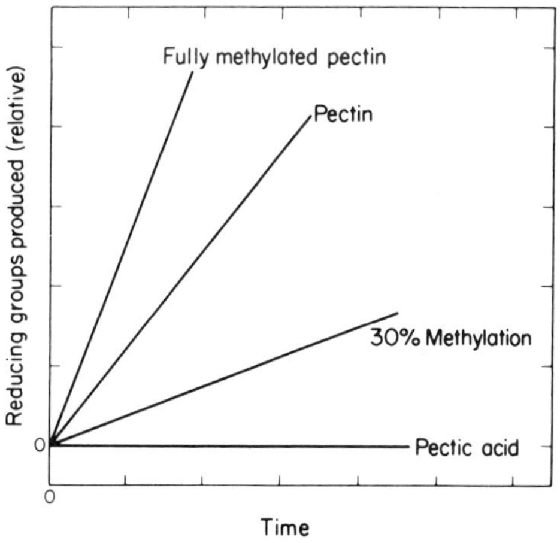

Figure 2 Schematic representation of the initial velocity of hydrolysis of several pectic substances by an endo-polymethylgalacturonase as measured by reducing group formation.

reaction. This is conveniently determined by column ion-exchange chromatography and gel filtration. No low-molecular-weight products will be found near the beginning of reaction with an endo-splitting enzyme; with an exo-splitting enzyme low-molecular-weight products will be present at the very beginning (Figs. 3 and 4). The data in Figure 4 indicate that the exo-splitting enzyme hydrolyzes digalacturonide units from the chain. The mono-, tri-, and tetragalacturonic units come from the fragments left after nearly complete hydrolysis of a molecule.

The relative activities of an enzyme on different-size substrates may be observed by following the rate of hydrolysis of a polymeric molecule to near completion or by starting with different-size substrates. For an enzyme that shows no preference for size of substrate the rate will remain constant to complete hydrolysis. On the other hand, the rate will decrease as the size of the substrate becomes smaller when the enzyme has a preference for large substrates (Table 2). Tomato polygalacturonase has a higher level of activity on larger substrates (Fig. 5).

The substrate is hydrolyzed from the reducing end by tomato polygalacturonase [8]. When tetragalacturonic acid was used as substrate, the relative rates of hydrolysis of the glycosidic bonds were 1 $\gg$ 2 $\gg$ 3 (numbered from the reducing end of the substrate).

The oligogalacturonate hydrolases are exo-splitting enzymes that have maximum activity on the trigalacturonates produced by the action of the endo-polygalacturonases on pectic acid and the action of the Δ4:5 unsaturated oligogalacturonate hydrolases (see Table 1). The final product is monogalacturonate

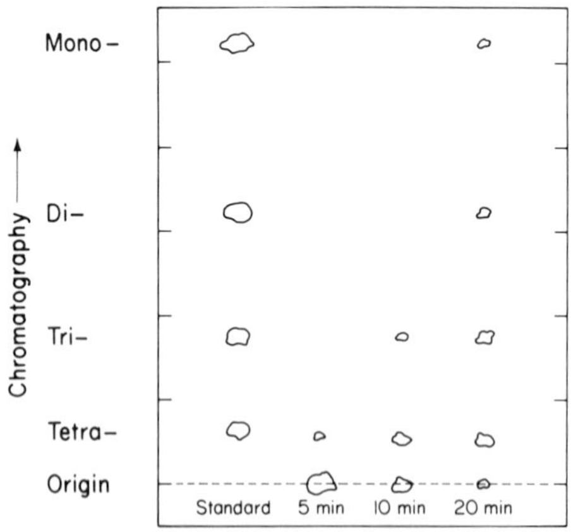

Figure 3 Schematic representation of relative rates of formation of products from the action of an endo-splitting enzyme on a pectic substance. Separation is by paper chromatography.

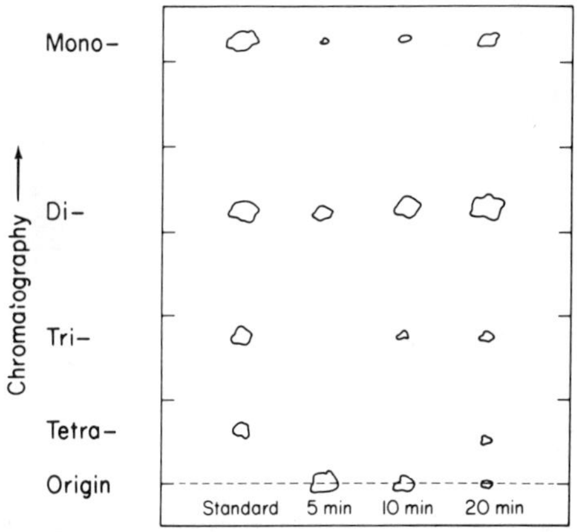

Figure 4 Schematic representation of relative rates of formation of products from the action of an exo-splitting enzyme on a pectic substance. Separation is by paper chromatography.

Table 2 Effect of Size of Substrate on Relative Rates for Some Polygalacturonases and Pectate Lyases

	Enzymes					
Substrate N[a]	A[b]	B[c]	C[d]	D[e]	E[f]	F[g]
>300	100	100	100	100	100	100
12	52					
8			103	38	36	17
7			78	35	15	13
6	12		41	27	11	7
5		12	32	15	3	2
4	2	2	20	7	0.6	0.6
3	2	0.2	2	1	0.02	0.008
2	0	0				

[a]*N*, number of galacturonate units.
[b]Polygalacturonase of *Erwinia carotovora*. From Ref. 4.
[c]Polygalacturonase of *Trichosporon penicillatum* SNO-3. From Ref. 5.
[d]Pectate lyase-1 of *Bacillus polymyxa*. From Ref. 6.
[e]Pectate lyase-2 of *B. polymyxa*. From Ref. 6.
[f]Pectate lyase-3 of *B. polymyxa*. From Ref. 6.
[g]Pectate lyase-4 of *B. polymyxa*. From Ref. 6.

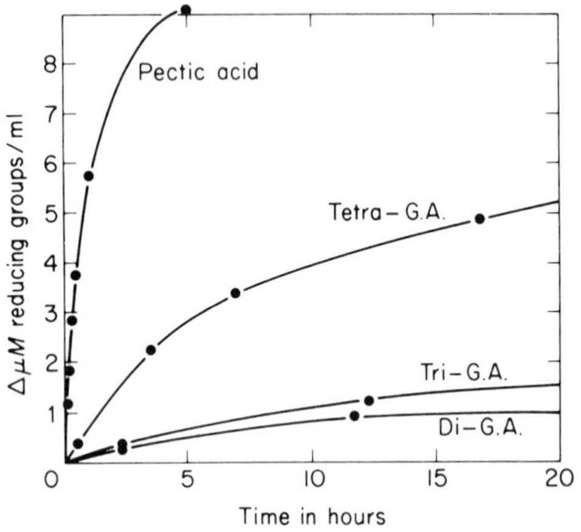

Figure 5 Initial rates of hydrolysis of pectic acid, tetra-, tri-, and digalacturonic acids by tomato polygalacturonase at pH 4.5 (From Ref. 7, p. 453, by courtesy of the Institute of Food Technologists.)

(Table 1). The Δ4:5 unsaturated oligogalacturonate hydrolases act on the Δ4:5 unsaturated oligogalacturonates produced by the action of endo-polygalacturonate lyase on pectic acid (Table 3). The exo-splitting enzymes remove the unsaturated monogalacturonate unit from the Δ4:5 unsaturated oligogalacturonates, giving unsaturated monogalacturonate and a saturated oligogalacturonate. The oligogalacturonates are hydrolyzed further by the oligogalacturonate hydrolases.

The role of NaCl, needed for maximum activity of the polygalacturonases, is

Table 3 Subdivision of the Pectate Lyases

3. Random mechanism of *trans*-eliminative degradation
   a. Endo-polymethylgalacturonate lyases (endopectin lyases; EC 4.2.2.10). Pectin split, not pectic acid. The products are unsaturated methyl oligogalacturonates.
   b. Endo-polygalacturonate lyases (endopectate lyases; EC 4.2.2.2). Pectic acid split, not pectin. The products are unsaturated oligogalacturonates.
4. Terminal mechanism of *trans*-eliminative degradation
   a. Exo-polymethylgalacturonate lyases (exopectin lyases). Pectin split, not pectic acid. This enzyme may not exist.
   b. Exo-polygalacturonate lyases (exopectate lyases; EC 4.2.2.9). Pectic acid split, not pectin.
   c. Oligogalacturonate lyases (EC 4.2.2.6). Unsaturated oligogalacturonate split by *trans*-elimination to unsaturated monogalacturonate.

Source: Adapted from Ref. 10.

not clear, but it appears to function in part to prevent inhibition of the enzyme by the products. Some of the polygalacturonases require Ca^{2+}. Whether Ca^{2+} performs a role in binding and/or catalysis, in maintaining the conformation of the enzyme, or whether it masks the carboxylate groups of the substrate is not clear.

Activity of the polygalacturonases may be followed by measuring (a) the rate of decrease in viscosity of the reaction mixture, (b) the rate of formation of reducing groups, (c) the decrease in optical rotation, or (d) the decrease in precipitability by calcium ions or nonpolar solvents. Only the reducing group and decrease in optical rotation measurements can be correlated directly with the number of glycosidic bonds hydrolyzed. An endo- and exo-splitting enzyme may be distinguished by comparing the rate of viscosity reduction with the rate of formation of reducing groups. One of the assays alone is not adequate to make the distinction.

Two precautions are necessary in working with the polygalacturonases. An enzyme preparation must be purified sufficiently to remove all pectinesterase before testing it for polymethylgalacturonase activity by use of 100% methylated pectin. Otherwise, the hydrolysis observed may be due to polygalacturonase acting on the pectic acid produced by action of pectinesterase. Contamination of a polygalacturonase preparation with pectinesterase does not affect the results obtained since pectic acid is the substrate.

The second concern is with the presence of a third type of depolymerizing enzyme, pectate lyase, discussed below. This group of enzymes also gives the same changes in viscosity, reducing groups, precipitability, and optical rotation as do the polygalacturonases. The amount of pectate lyase present can be determined by the change in absorbance at 235 nm, discussed below.

III. PECTATE LYASES

The Commission on Enzymes recommended that the name *trans*-eliminase not be used to describe this group of enzymes, although it is the name that occurs most frequently, especially in the older literature.

The pectate lyases [poly(1,4-α-D-galacturonide) lyase, EC 4.2.2.2] split the glycosidic bond by *trans* elimination of hydrogen from the C-4 and C-5 positions of the aglycone portion of the substrate [Eq. (3)]. Note that water is not

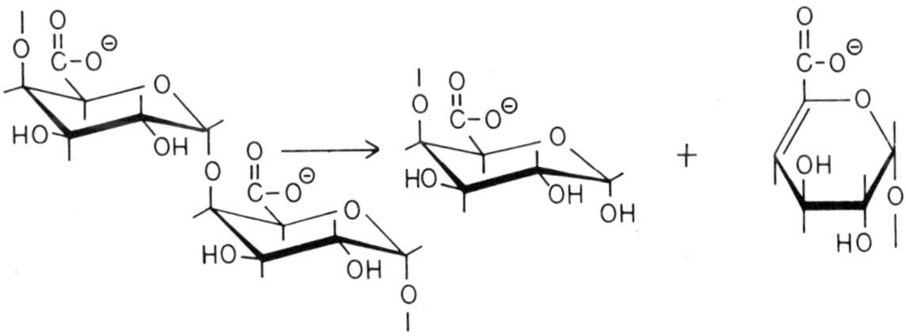

(3)

involved as a substrate. Therefore, these enzymes belong to the "lyase" group of enzymes (EC 4).

Activity of the pectate lyases can be measured by the same methods as described for the polygalacturonases. However, it is more specific and convenient to follow the rate of splitting of the glycosidic bond by observing the increase in absorbance at 235 nm due to formation of the double bond. The molar extinction coefficient of the double bond is 4800 M^{-1} cm^{-1} [9].

The pectate lyases are subdivided on the basis of whether they prefer pectin or pectic acid as substrate and whether they are exo- or endo-splitting enzymes (Table 3). All four types of pectate lyases described in the classification are known. Methods for determining whether the enzyme is an endo- or exo-splitting enzyme and whether it has specificity for pectin or pectic acid are the same as described for the polygalacturonases.

The pectate lyases were unknown before 1960 and all enzymes acting on pectin or pectic acid were designated as either pectinesterases or polygalacturonases. The literature prior to 1960 concerning the polygalacturonases may be interpreted on the basis of the following: (a) pectate lyases have only been found in microorganisms and not higher plants; (b) the pH optima of the pectate lyases are in the region 8.5 to 9.5, while those of the polygalacturonases are in the pH region 5 to 6.5; and (c) all the pectate lyases appear to require Ca^{2+}, whereas not all the polygalacturonases do.

Some microorganisms produce only pectate lyases, whereas others produce both pectate lyases and polygalacturonases. It is easy to distinguish between these possibilities by correlating the number of reducing groups formed with the number of double bonds formed (as measured spectrophotometrically at 235 nm). When only pectate lyase is present, the number of reducing groups formed are exactly identical to the number of double bonds formed in both the crude and purified preparations. When polygalacturonase and pectate lyases are present together in a crude preparation, the number of reducing groups formed will be larger than the number of double bonds formed.

A requirement for Ca^{2+} appears to be a characteristic of the pectate lyases. Other divalent cations, such as Sr^{2+}, Mn^{2+}, and Mg^{2+}, also stimulate activity but not to the extent of Ca^{2+}; in some cases Ca^{2+} appears to be the only cation that is permitted. The essentiality of Ca^{2+} for the enzyme from *Clostridium multifermentans* is shown in Fig. 6. The enzyme has activity on polygalacturonic acid in the absence of added Ca^{2+} because of Ca^{2+} contamination of the substrate used. Addition of 0.001 *M* $CaCl_2$ increased the activity a little but not markedly. Addition of 5×10^{-4} *M* EDTA to the reaction mixture completely inhibited all enzyme activity which could be restored on addition of 0.001 *M* $CaCl_2$.

In the cases that have been investigated, the exo-pectate lyases have been shown to attack the substrate from the *reducing end* of the chain just as do the exo-polygalacturonases [9]. This specificity of the exo-polygalacturonases and pectate

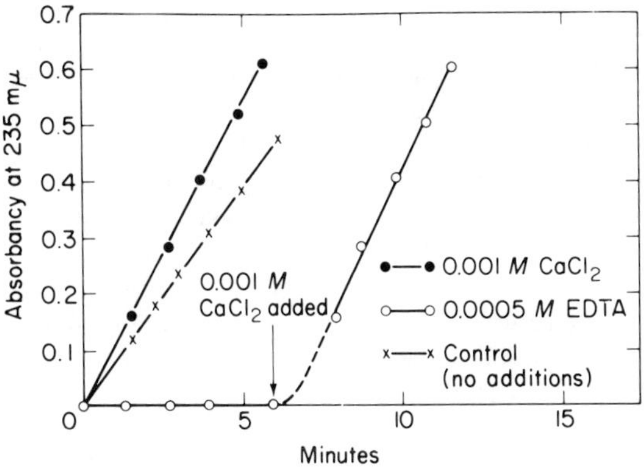

Figure 6 Effect of calcium chloride and ethylenediaminetetraacetic acid (EDTA) on activity of exo-polygalacturonate lyase from *Clostridium multifermentans*. Reaction contained 0.46% polygalacturonic acid in 0.033 *M* Tris buffer, pH 8.0. (From Ref. 7, p. 569, by courtesy of the American Chemical Society.)

lyases for the reducing end of the polymer chain should be contrasted with that of the exo-amylases (β-amylase and glucoamylase) and the exo-β-1,4-glucanases, which have specificity for the nonreducing end of the polymer chain.

Exercise 1. Can you think of features of the active sites of the enzymes, exo-pectate lyase and β-amylase, which might account for one having specificity for the reducing end of the polymer and the other for the nonreducing end? What experimental approach might you use to prove your hypotheses?

REVIEW QUESTIONS

See the end of Chapter 19.

REFERENCES

General References

Whitaker, J. R., Microbial proteolytic enzymes, in *Microbial Enzymes and Biotechnology*, 2nd Ed. (W. M. Fogarty and C. T. Kelly, eds.), Elsevier Applied Science, New York, 1990, p. 133.

Whitaker, J. R., Pectic substances, pectic enzymes and haze formation in fruit juices, *Enzyme Microb. Technol.*, **6**, 341 (1984).

Cited References

1. L. R. MacDonald, R. Jang, E. F. Jansen, and H. Lineweaver, *Arch. Biochem. Biophys.*, **28**, 260 (1950).
2. A. L. Demain and H. J. Phaff, *Wallerstein Lab. Commun.*, **20**, 119 (1957).
3. J. R. Whitaker, Microbial pectolytic enzymes, in *Microbial Enzymes and Biotechnology*, 2nd Ed. (W. M. Fogarty and C. T. Kelly, eds.), Elsevier Applied Science, New York, 1990, p. 133.
4. S. Nasuno and M. P. Starr, *J. Biol. Chem.*, **241**, 5298 (1966).
5. T. Sakai, M. Okushima, and M. Sawada, *Agric. Biol. Chem.*, **46**, 2223 (1982).
6. C. W. Nagel and T. M. Wilson, *Appl. Microbiol.*, **20**, 374 (1970).
7. B. S. Luh, S. J. Leonard, and H. J. Phaff, *Food Res.*, **21**, 448 (1956).
8. D. S. Patel and H. J. Phaff, *Food Res.*, **23**, 693 (1958).
9. J. D. Macmillan, H. J. Phaff, and R. H. Vaughn, *Biochemistry*, **3**, 572 (1964).
10. D. F. Bateman and R. L. Millar, *Ann. Rev. Phytopathol.*, **4**, 119 (1966).
11. J. D. Macmillan and R. H. Vaughn, *Biochemistry*, **3**, 564 (1964).

17

The Esterases

The esterases are ubiquitous. Perhaps no other group of enzymes is so widely distributed in nature, and most tissues contain a great variety of enzymes with esterase activity. As a group, the esterases are involved in the splitting of ester linkages by the addition of water as the second substrate:

$$\underset{\text{ester}}{R\text{—}O\text{—}R'} + H_2O \rightarrow \underset{\text{acid}}{R\text{—}H} + \underset{\text{alcohol}}{R'\text{—}OH} \quad (1)$$

The alcohol may be monohydric or polyhydric, aliphatic or aromatic. The acid may be an organic or an inorganic acid. Usually, the esterase is specific for either the acid or the alcohol moiety but not both.

The esterases may be subdivided into several groups primarily on the basis of specificity for the acid involved in the ester substrate: (a) carboxylic ester hydrolases, (b) thiolester hydrolases, (c) phosphoric monoester hydrolases, (d) phosphoric diester hydrolases, and (e) sulfuric ester hydrolases. Only the thiolester hydrolases are named on the basis of the alcohol involved. This group of hydrolases involves a number of enzymes that split acyl CoA (and a few other thiol alcohol derivatives) by the addition of water.

$$R\text{—}\overset{\overset{\displaystyle O}{\|}}{C}\text{—}S\text{—}CoA + H_2O \rightarrow R\text{—}\overset{\overset{\displaystyle O}{\|}}{C}\text{—}OH + CoASH \quad (2)$$

We shall not treat the thiolester hydrolases (EC 3.1.2) further.

I. CARBOXYLIC ESTER HYDROLASES

The carboxylic ester hydrolases (EC 3.1.1) are a diverse group of enzymes that have in common the splitting of carboxylic acid esters by the addition of water:

$$\underset{\text{carboxylic ester}}{R{-}\overset{\overset{\displaystyle O}{\|}}{C}{-}O{-}R'} + H_2O \rightarrow \underset{\text{carboxylic acid}}{R{-}\overset{\overset{\displaystyle O}{\|}}{C}{-}OH} + \underset{\text{alcohol}}{R'{-}OH} \tag{3}$$

For convenience in the following discussion we divide the carboxylic ester hydrolases into two groups, the nonspecific carboxylic ester hydrolases and the specific carboxylic ester hydrolases.

A. Nonspecific Carboxylic Ester Hydrolases

1. Esterases

Three different groups of nonspecific carboxylic ester hydrolases are found in tissues. According to the Commission on Enzymes, classification is on the basis of substrate specificity, but the differences extend beyond substrate specificity (Table 1).

The three types of esterases can readily be distinguished on the basis of several

Table 1 Types of Nonspecific Carboxylic Ester Hydrolases

Property	Type		
Systematic name	Carboxylic-ester hydrolase (EC 3.1.1.1)	Aryl-ester hydrolase (EC 3.1.1.2)	Acetic-ester acetylhydrolase (EC 3.1.1.6)
Recommended trivial name	Carboxylesterase	Arylesterase	Acetylesterase
Other names	B-esterase, aliesterase	A-esterase	C-esterase
Substrate	Carboxylic esters of aliphatic and aromatic alcohols	A phenyl acetate	An acetic ester
Ethyl acetate	+	—	+
Ethyl butyrate	+	—	—
Tributyrin	+	—	—
Phenyl acetate	+	+	+
DFP inhibition	+	—[a]	—
EDTA inhibition	—	+	—
Ca^{2+} activation	—	+	—

[a]Some arylesterases hydrolyze DFP (diisopropyl fluorophosphate).

of the criteria shown in Table 1. All three types hydrolyze phenyl acetate, only carboxylesterases and acetylesterases hydrolyze ethyl acetate, and only carboxylesterases hydrolyze ethyl butyrate. Carboxylesterases also hydrolyze triglycerides as indicated by their hydrolysis of tributyrin. However, they may be distinguished from lipase as they hydrolyze soluble substrates, while lipase acts only at water–lipid interfaces (Section I.A.2 below). Carboxylesterases are inhibited by low concentrations of organophosphorus compounds such as diisopropyl fluorophosphate (DFP), while arylesterases and acetylesterases are not. Some arylesterases can even hydrolyze organophosphorus compounds. The arylesterases appear to contain an essential Ca^{2+} since they are inhibited by EDTA, but the activity can be restored by addition of Ca^{2+}.

Very little is known about the mechanism of action of this group of esterases. Since carboxylesterases are inhibited by organophosphorus compounds it seems reasonable to suggest that the enzymes contain a seryl hydroxyl group in the active site. It has been suggested that the mechanism of action may be similar to that of chymotrypsin [1].

2. Lipase (Triacylglycerol Acylhydrolase, EC 3.1.1.3)

When ingested as part of the diet, the triglycerides (triacylglycerols) are hydrolyzed to glycerol, monoglycerides, and fatty acids (as salts) primarily by pancreatic lipase in the small intestine. Lipases are found in pancreas, blood plasma, saliva, pancreatic juice, milk, in a number of triglyceride-producing plants (soybean, castor bean, peanut, etc.), and in molds and bacteria. The lipases of milk are responsible for hydrolytic rancidity of improperly pasteurized milk, of milk that has been held without pasteurization, and of cream and butter.

Pancreatic lipase has been the most studied of the lipases and has been obtained in homogeneous form [2]. It has a molecular weight near 42,000 [3]. There is probably only one lipase secreted by the pancreas, but two esterases are secreted by the intestinal mucosa. Much of the data on lipase action are impossible to interpret because methyl butyrate, or even *p*-nitrophenyl acetate, was used as a substrate. Use of *p*-nitrophenyl acetate is particularly bad since nearly all esterases can hydrolyze this compound (Table 1) as can the proteolytic enzymes, chymotrypsin and trypsin, also found in pancreatic secretions. Both carboxylesterase and lipase hydrolyze methyl butyrate.

Pancreatic lipase only acts at a lipid–water interface [4] and has very little activity on soluble triacetin (Fig. 1). At substrate concentrations below saturation (<0.328 *M*) triacetin is in true solution and purified lipase has essentially no activity at all on the dissolved substrate. As soon as the solution becomes saturated with triacetin so that an emulsion is formed, the activity increases markedly. It should be noted that a velocity–substrate saturation curve for lipase is obtained but only after the solution has become saturated with triacetin. The saturation curve should not be interpreted in the same way as the usual velocity–substrate curves.

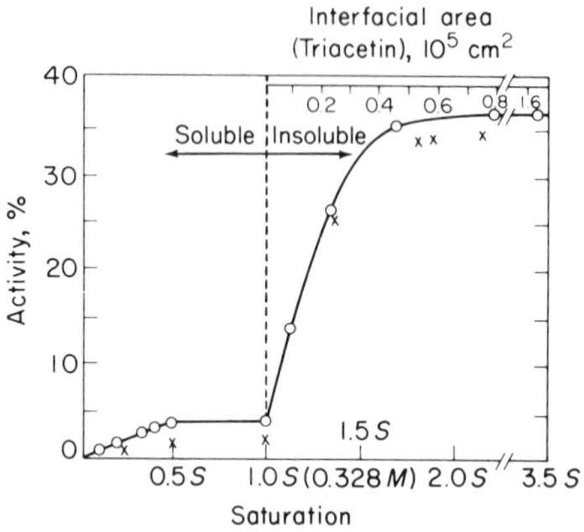

Figure 1 Activity of porcine pancreatic lipase on triacetin. The substrate concentration (x-axis) is expressed as a function of concentration of ester needed to saturate the solution; the percent activity is with respect to maximal activity on a triolein emulsion. ○, lipase preparation containing some esterase activity; ×, electrophoretically purified lipase. (From Ref. 4, p. 517, by courtesy of Elsevier Publishing Co.)

Starting with a triglyceride, the relative rates of hydrolysis by pancreatic lipase are triglycerides > diglycerides ≫ monoglycerides (Fig. 2). Figure 2 has a third dimension, that of time. Aliquots were removed from the reaction at various times and assayed for the amounts of tri-, di-, and monoglycerides and glycerol present. There was a rapid decrease in triglyceride with a rapid increase in diglyceride and a somewhat slower increase in monoglyceride content. Glycerol content was quite low until the later stages of hydrolysis, where its rate of formation paralleled the rate of hydrolysis of monoglyceride.

Several independent methods have shown that pancreatic lipase acts at the 1 and 3 positions of a triglyceride and that the 2-monoglyceride accumulates. This has been demonstrated, for example, by use of three triglycerides: 1,3-dipalmityl-2-oleyl glyceride (POP), 2,3-dipalmityl-1-oleyl glyceride (OPP), and 2,3-dioleyl-1-palmityl glyceride (POO) (Table 2). If the ester bonds are split randomly regardless of position, the relative percent liberation of oleic acid should be the same for POP and OPP and one-half the relative percent liberation for POO. However, if hydrolysis at the 1 and 3 positions is favored, the relative percent liberation of oleic acid will be the same for OPP and POO and the relative percent liberation of oleic acid from POP will be much lower. The latter was found to be the case, thus demonstrating the specificity of enzyme for positions 1 and 3 on the triglyceride. Slow hydrolysis at ester position 2 is due to acyl migration from the 2 to the 1 position at pH 8 followed by enzymatic hydrolysis [7]. Milk lipase also shows a

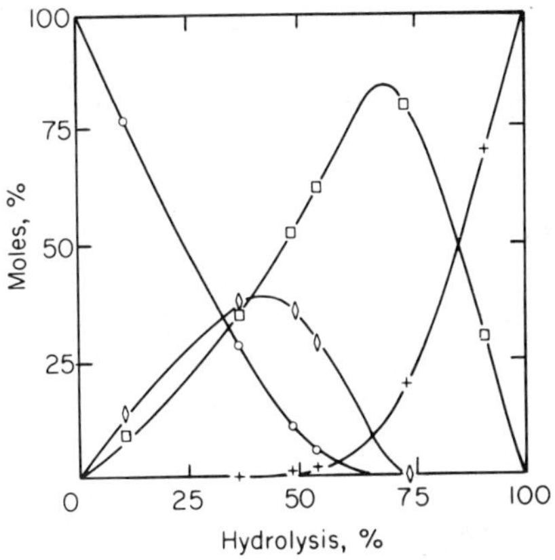

Figure 2 Sequence of events in the hydrolysis of a triglyceride by porcine pancreatic lipase. Experiments performed in the presence of taurocholate and Ca^{2+}. ○, triglyceride; ◇, diglycerides; □, monoglycerides; +, glycerol. (From Ref. 5, p. 107, by courtesy of Elsevier Publishing Co.)

strong preference for triglycerides and does not hydrolyze the 2-monoglyceride [8]. There are lipases which have specificity for the 2-position (β) fatty acid.

Not all lipases have a marked specificity for position of the ester bond hydrolyzed. Ricinus lipase [9] and heparin-induced serum lipoprotein lipase [10] hydrolyze all three ester bonds at about the same rates. Positional specificity is very important in the industrial use of lipases for transesterification reactions. Through use of highly specific lipases, lipids with selected chemical and physical properties can be synthesized, such as those found in cocoa butter.

Porcine pancreatic lipase is not absolutely specific for the glycerol component of a triglyceride since there is only a threefold difference in the relative rates

Table 2 Demonstration of the Specificity for Position of Hydrolysis by Porcine Pancreatic Lipase

	Oleic acid content (mol/100 mol substrate)	
Substrate	Free fatty acid	Monoglyceride
1,3-Dipalmityl-2-oleyl glyceride (POP)	5	87–89
2,3-Dipalmityl-l-oleyl glyceride (OPP)	46–51	4
2,3-Dioleyl-l-palmityl glyceride (POO)	53–58	76–86

Source: Ref. 6.

of hydrolysis of monobutyrin and methyl butyrate. What about specificity for the acid part of the ester? Some typical results are shown in Table 3. Although there appear to be some quantitative differences among the different lipases, they all hydrolyze tributyrin at the fastest rate. The lipases have maximum specificity for the four-carbon chain acid, but this specificity is only relative. It is of interest that human gastric lipase, which has a pH optimum of about 5.5 to 7, compared with pH 8 to 9 for the other lipases, has the same relative specificities on the triglycerides. Interpretation of the apparent lower activities of the lipases on triglycerides composed of fatty acids with more than 10 carbons is complicated in that these homogeneous triglycerides are solids at the temperature of the experiment.

Since the lipases act best at a lipid–water interface, any condition that increases the surface area of the substrate–water mixture would be expected to increase the observed activity of the enzyme (until it is saturated with substrate). For this reason, lipase action is much more rapid in homogenized (but unpasteurized) milk. In the small intestine the action of pancreatic lipase is aided by bile salts, which emulsify the lipids. Addition of taurocholate and detergents as emulsifying agents in in vitro experiments increases lipase activity.

The nature of the emulsifying agent can change the pH optimum of the lipase and its relative velocities on triglycerides. For example, adipose tissue lipase had a pH optimum on tricaproin of 7.5, 6.8, and 6.6, respectively, when the emulsifying agent was 1% Amino-O (a cationic detergent), 1% Triton WR-1339 (a nonionic detergent), or 0.5% Triton X-220 (an anionic detergent) [13].

Calcium ion increases the rate of hydrolysis of triglycerides, particularly

Table 3 Comparison of Relative Velocities of Several Lipases Acting on a Variety of Triglycerides

		Lipases			
Substrate	Carbon number of acid	Hog pancreas[a]	Human milk[a]	Bovine pancreas[a]	Human gastric[b]
Triacetin	2	0.33	2.4	1.9	—
Tripropionin	3	20.7	18.9	18.9	27
Tributyrin	4	100	100	100	100
Tricaproin	6	36.6	58	36.8	39
Tricaprylin	8	36.6	57.5	40.6	—
Tricaprin	10	31.9	51.5	37	13
Trilaurin	12	31.1	8.1	23.5	2
Trimyristin	14	11.0	0.49	12.7	—
Tristearin	18	0.47	0.37	1.3	0.8

[a]Ref. 11.
[b]Ref. 12.

toward the end of hydrolysis. This is not an effect on the enzyme but rather is due to removal of the free fatty acids as insoluble calcium salts.

Exercise 1. Two possible explanations for increase in rate on removing the free fatty acids as calcium salts are: (a) the fatty acids act as competitive inhibitors of the enzyme, and (b) removal of fatty acids shifts the equilibrium of the reaction further toward complete hydrolysis. Design an experiment(s) that would differentiate between these possibilities.

Although a great deal of work has been done on the substrate specificity of porcine pancreatic lipase, almost nothing has been done in a definitive way in elucidating its mechanism of action. This is due largely to the difficulty of studying the kinetics of an enzyme in a heterogeneous system (emulsion). Pancreatic lipase contains essential serine, histidine, and aspartic acid residues in the active site. It is suggested that the mechanism of hydrolysis of the ester bonds of lipids resembles the mechanism of hydrolysis of peptide bonds by the serine proteases (see Chapter 19). Most likely an acylenzyme intermediate is formed. The acylenzyme intermediate is then hydrolyzed by water to give the free fatty acid and regenerate the enzyme or the acyl group can be transferred to a nucleophilic group, such as the OH of glycerol to give a transesterification. The pH dependence of porcine pancreatic lipase activity on tributyrin gives a bell-shaped curve, with a pH optimum at 8.0 and a dependence on pK_{a1} of 5.8 and pK_{a2} of 10.1.

A number of methods are used to follow the hydrolysis of triglycerides by lipases: (a) removal of aliquots of reaction, extraction of the free acids into chloroform after acidification, and titration of the chloroform layer with standard base, (b) continuous monitoring of the reaction in a pH Stat at pH 8 where the protons liberated ($RCOOH \rightleftharpoons RCOO^- + H^+$) on hydrolysis of the ester are titrated, (c) measurement of CO_2 liberation from $NaHCO_3$ by a Warburg apparatus (liberation of protons causes formation of H_2CO_3 which decomposes to CO_2), (d) reduction in surface tension of the reaction due to release of fatty acids, (e) changes in turbidity of the emulsion on hydrolysis as measured spectrophotometrically, (f) use of the chromogenic substrate, α-naphthyl acetate, and (g) use of fluorometric substrates such as 4-methylumbelliferone heptanoate.

Exercise 2. Write balanced structural equations showing the basis of assay for each of the seven methods listed above. In some cases diagrams would also be appropriate in understanding the assay.

Despite the effect that detergents have on the reaction, it is recommended that a nonionic detergent be used to stabilize the emulsion during activity assays.

B. Specific Carboxylic Ester Hydrolases

1. Phospholipases

The hydrolysis of lecithin and cephalin are of interest because of their occurrence in the diet, their metabolism in vivo, and because of the need for four specific enzymes to hydrolyze the ester bonds present (Fig. 3). The positions where the

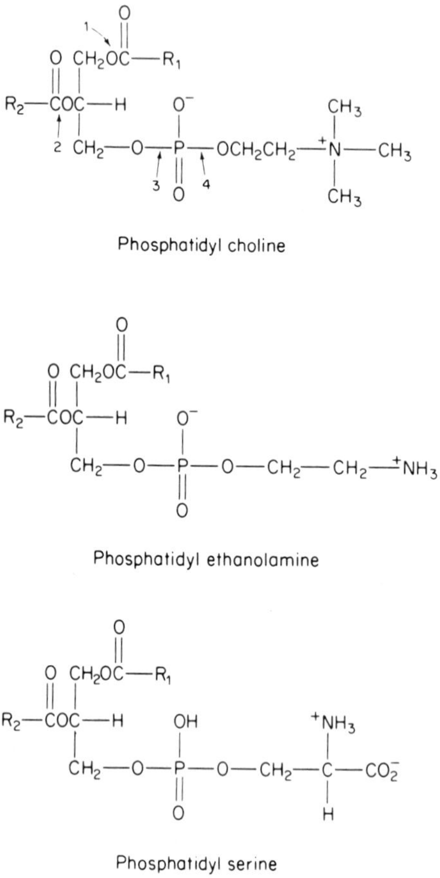

Figure 3 Structures of α-lecithin (phosphatidylcholine) and two α-cephalins (phosphatidylethanolamine and phosphatidylserine).

bonds are hydrolyzed are shown by the arrows. The enzymes involved in hydrolysis at positions 3 and 4 are phosphodiesterases and will be treated under the phosphatases (Section II.B).

Hydrolyses at positions 1 and 2 are catalyzed by separate enzymes. The enzyme that splits the bond at position 1 is generally referred to as phospholipase B (2-lysophosphatidylcholine acylhydrolase, EC 3.1.1.5) but has also been called lysophospholipase, lysolecithinase, and lecithinase B. The enzyme has been found in animal tissues such as the liver and pancreas, in molds, and in barley malt. The plant and animal enzymes have pH optima at 6 to 7, while the pH optimum of the enzyme from *Penicillium notatin* is at pH 4. Calcium ion, which is required for activity of phospholipase A (see below), has no effect on the activity of phospholipase B. Phospholipase B, but not phospholipase A, is inhibited by cyanide. It is generally agreed that phospholipase B is active only on lysolecithin (and lysocephalin), that is, α-lecithin in which the fatty acid in the 2-position is removed. Therefore, in the hydrolysis of lecithin the action of phospholipase A must precede that of phospholipase B.

Phospholipase A (phosphatidylcholine acylhydrolase, EC 3.1.1.4) hydrolyzes the ester bond at position 2 in α-lecithin and α-cephalin. Phospholipase A is widely distributed in nature. Its activity is greatly enhanced by Ca^{2+}, which appears to act not only in removing the free fatty acids formed as insoluble calcium salts but may also be involved in binding the substrate to the enzyme. The specific activity of phospholipase A is nearly independent of the length of the fatty acid in position 2. Phospholipase A is quite stable at 100°C and pH 5.9. Above pH 7.0 and at 100°C it loses its activity rapidly.

Unlike the situation with lipase, it is not necessary to use a detergent for emulsifying the substrate for the phospholipases. The substrates are excellent emulsifying agents.

Exercise 3. Based on the reactions catalyzed by phospholipases A and B, design a set of experiments that would permit their individual determination even in the presence of each other.

2. Chlorophyllase

Chlorophyllase (chlorophyll chlorophyllido-hydrolase, EC 3.1.1.14) hydrolyzes the phytyl group from chlorophyll (Fig. 4) to give phytol and chlorophyllide. Chlorophyllase probably occurs in all plant and microbial tissues that contain chlorophyll. Despite a number of studies designed to elucidate its biological role, it is not yet clear whether it functions in vivo to remove the phytyl group from chlorophyll or to add the phytyl group to chlorophyllide as the final step in the biosynthesis of chlorophyll. The enzyme is quite unusual in that it has appreciable activity in the presence of acetone concentrations as high as 70%. The enzyme is quite important in determining the stability of the green color of vegetables prior to and during processing.

3. Pectinesterase

Pectinesterase (pectin pectylhydrolase, EC 3.1.1.11), which belongs to the carboxylic ester hydrolases, removes the methoxyl groups esterified at C(6) of anhydrogalacturonic acid units of pectin. For this reason it was discussed along with the other pectin- and pectic acid–degrading enzymes in Chapter 16.

4. Acetylcholinesterase

Acetylcholine functions as an intercellular agent in both the transmission and conduction of nerve impulses. It is liberated at the synapse between the nerve cell and the endplate region of muscle fibers. There it stimulates the muscle fibers to

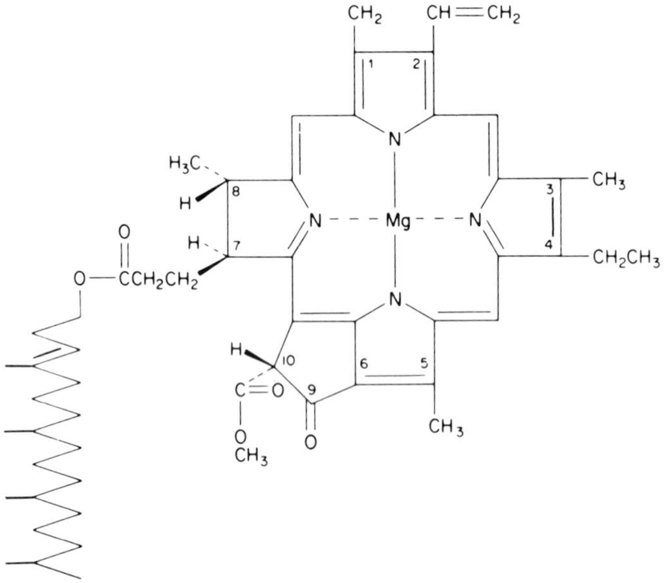

Figure 4 Structure of chlorophyll *a*.

twitch. The role of acetylcholinesterase (acetylcholine acetylhydrolase, EC 3.1.1.7), found in all nerve tissues of all animals, is to hydrolyze acetylcholine immediately after it has transmitted the impulse from the nerve cell to the motor neuron:

$$
\begin{aligned}
&(CH_3)_3N^+CH_2CH_2O\overset{\overset{\displaystyle O}{\|}}{C}CH_3 + H_2O \\
&\quad \rightarrow (CH_3)_3N^+CH_2CH_2OH + CH_3COO^- + H^+
\end{aligned}
\tag{4}
$$

Organophosphorus compounds such as DFP, and carbamates such as Sevin and a number of other compounds, are toxic because of their irreversible inhibition of acetylcholinesterase. This is also their mechanism of action on insects.

Much is known about the binding of substrate and its hydrolysis by acetylcholinesterase. The bonds involved in binding of substrate to enzyme are (a) electrostatic interaction between the positive charge of the substrate [Eq. (4)] and an anionic group at the active site of the enzyme, (b) van der Waals attractive forces between the methyl and methylene groups of the substrate and a nonpolar region of the enzyme, and (c) interaction between the basic group of the transforming site and the electrophilic carbonyl carbon atom of the ester.

Organophosphorus compounds inhibit acetylcholinesterase by alkylation of a single seryl hydroxyl group just as in α-chymotrypsin (Chapter 9). From the pH-activity profile, inhibition by photooxidation, and other data, an imidazole group is also involved in the transformation process. Therefore, the binding and hydrolysis of acetylcholine by acetylcholinesterase may be described as in Fig. 5.

The general features of the mechanisms of hydrolysis are identical for acetylcholinesterase and chymotrypsin (Chapter 19) except in one feature. The binding of substrate in the active site of acetylcholinesterase involves in part an electrostatic interaction, while the combining of substrate and α-chymotrypsin involves primarily hydrophobic interaction. In this regard, trypsin and acetylcholinesterase resemble each other. The imidazole acts as a general base to stretch the hydrogen-oxygen bond from the seryl-OH group, thereby facilitating nucleophilic attack at the carbonyl group of the substrate. In deacetylation, the imidazole removes a proton from a water molecule, thus facilitating deacetylation by the hydroxide ion formed.

II. THE PHOSPHORIC HYDROLASES

The phosphoric hydrolases catalyze the hydrolysis of ester bonds in which the acid moiety of the ester is supplied by phosphoric acid. The phosphoric mono- and diesters are substrates for two types of enzymes, the phosphoric monoester hydrolases and phosphoric diester hydrolases, respectively.

A. Phosphoric Monoester Hydrolases

There are two distinct groups of the phosphoric monoester hydrolases (EC 3.1.3). One group of enzymes has a rather broad specificity in that it has primary specificity for the monophosphate group only and the alcohol moiety of the substrate is of secondary importance. This group will be referred to as the nonspecific phosphatases. The second group of phosphoric monoester hydrolases has primary specificity for both the monophosphate moiety and the alcohol moiety and will be referred to as the specific phosphatases.

The general overall reaction catalyzed by the phosphoric monoester hydrolases is the same and involves hydrolysis of the P—O bond as shown by the use of $H_2{}^{18}O$:

$$ROPO_3^{2-} + H_2{}^{18}O \rightarrow ROH + H^{18}OPO_3^{2-} \qquad (5)$$

The oxygen-18 is incorporated into the inorganic phosphate rather than into the alcohol, as shown by mass spectrometry on the isolated products.

1. Nonspecific Phosphatases

The nonspecific phosphatases can be further divided, on the basis of difference in pH optimum, into two groups, the alkaline phosphatases [orthophosphoric-monoester phosphohydrolase (alkaline optimum), EC 3.1.3.1], and the acid

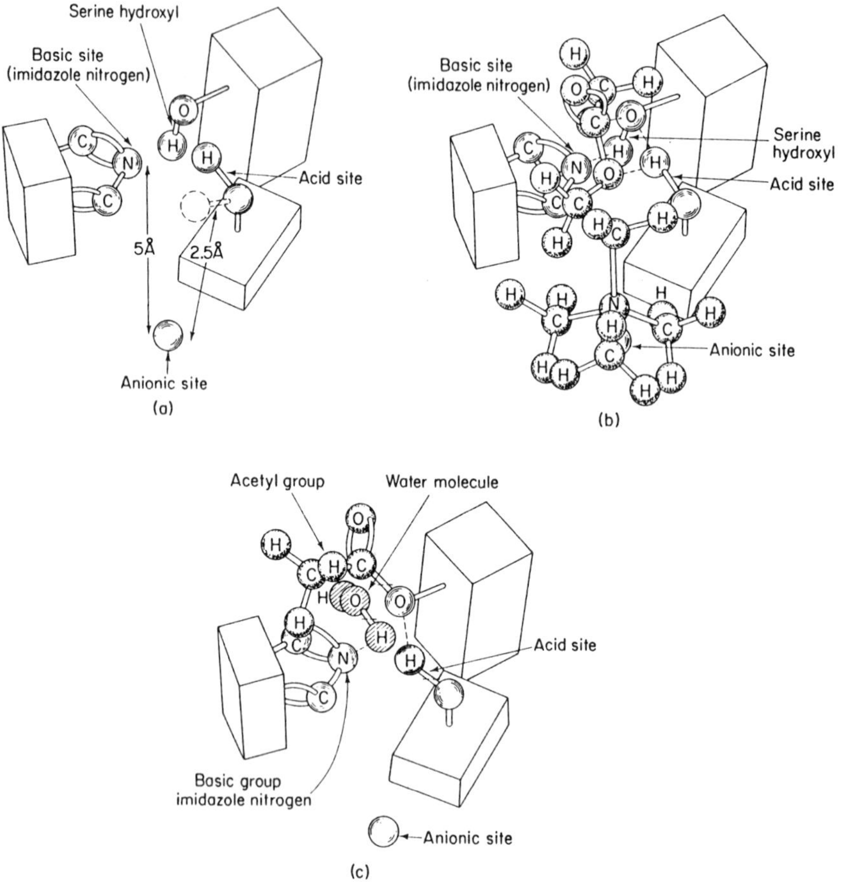

Figure 5 Schematic representation of active site of acetylcholinesterase and mechanism of action. (a) shows catalytic groups (imidazole, seryl hydroxyl, and carboxyl group) and a component of the binding site (anionic site); (b) shows the enzyme·substrate complex, and (c) shows the acetylenzyme intermediate formed and the water molecule involved in deacetylation of enzyme. White balls, groups on enzyme; stippled balls, substrate or derivative; diagonally lined balls, water molecule. The blocks were used in model building and represent the peptide chain of enzyme. (From Ref. 14, p. 1459, 1460, by courtesy of the American Chemical Society.)

phosphatases [orthophosphoric-monoester phosphohydrolase (acid optimum), EC 3.1.3.2]. For the acid phosphatases the pH optimum is between 3.5 and 5, depending on the particular source of enzyme. For the alkaline phosphatases, the pH optimum is between 9 and 10, again depending on the source of enzyme.

The nonspecific phosphatases are found in all tissues of an organism. The alkaline phosphatases are present in particularly high concentrations in the lactating mammary gland and in the kidney. Normal human serum contains alkaline as well as acid phosphatases in small but measurable concentrations that vary only within narrow limits. These levels are often changed markedly in certain pathological diseases, and determination of phosphatase activity is a rapid diagnostic tool in the clinical laboratory.

Table 4 shows the relative substrate specificities of three acid phosphatases: yeast α-glycerophosphomonoesterase, erythrocyte phosphomonoesterase, and acid prostatic phosphomonoesterase. Several conclusions may be drawn from the data of Table 4. The specificities of the enzymes decrease in the order yeast > erythrocyte > prostate. The yeast enzyme is rather specific in that it hydrolyzes

Table 4 Relative Substrate Specificities of Acid Phosphatases from Three Sources

Substrate (0.01 M)	Relative activity		
	Yeast[a]	Erythrocyte[b]	Prostate[c]
Phenyl phosphate	1	100	100
α-Glycerophosphate	100	33	69
β-Glycerophosphate	1	2	84
1-Propanediol phosphate	76	17	53
3-Phosphoglycerate	0	4	Very small
Glucose 1-phosphate	0	1	19
Glucose 6-phosphate	0	0	Small
Ribose 5-phosphate	0	8	41
Riboflavin 5-phosphate	0	40	35
Adenosine phosphate (2′ + 3′)	0	1	95
Cytidine phosphate (2′ + 3′)	0	2	94
Adenosine 5′-phosphate	0	3	60
Adenosine triphosphate	0	0	0
Sodium pyrophosphate	1	1	0
Diphenyl phosphate	0	0	0

Source: Ref. 15, p. 201, by courtesy of Academic Press.
[a]At pH 6.5.
[b]At pH 6.0.
[c]At pH 5.5

only α-glycerophosphate and 1-propanediol phosphate at appreciable rates. It is clear why the enzyme is referred to as yeast α-glycerophosphomonoesterase. Most acid phosphatases have maximum activity on phenyl phosphate, as typified by erythrocyte and prostate phosphatases. Both yeast and erythrocyte phosphatases show position specificity, as demonstrated by the relative rates of hydrolysis of α- and β- glycerophosphate. Prostate phosphatase does not show this discrimination. Prostate phosphatase can hydrolyze rapidly several nucleoside phosphates, whereas the other two enzymes cannot. All three enzymes have a strict requirement for a phosphoric monoester, however, as shown by the failure of all three to hydrolyze adenosine triphosphate, sodium pyrophosphate, and diphenyl phosphate.

Most of the acid phosphatases are not inhibited by metal chelating agents but are inhibited by orthophosphate. There are some acid phosphatases which are activated by Mg^{2+}.

In contrast to the acid phosphatases, the alkaline phosphatases are usually inhibited by chelating agents such as EDTA, α,α′-dipyridyl, and *o*-phenanthroline. They are also often inhibited by orthophosphate, arsenate, pyrophosphate, borate, and carbonate. Several of the alkaline phosphatases contain Zn(II), while others appear to contain Fe(II), Mn(II), or Mg(II).

The best studied of the alkaline phosphatases is that from *E. coli*. It has a molecular weight of 80,000, contains four Zn(II) [two active site and two structural Zn(II)] and forms a covalent intermediate that has two phosphate groups attached to seryl hydroxyl groups. The enzyme is inhibited by DFP. In the presence of 6 *M* urea and thioglycolic acid, the enzyme is split into two inactive units of molecular weight 40,000. On removal of urea by dialysis (or gel filtration) and on standing in air to reform the disulfide bonds, the enzyme regains its activity.

The general mechanism for *E. coli* alkaline phosphatase can be written as in Eq. (6), where a covalent phosphorylenzyme intermediate is formed. Direct

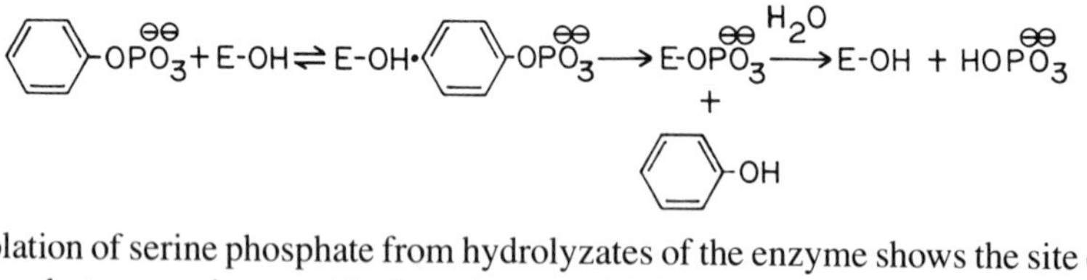

(6)

isolation of serine phosphate from hydrolyzates of the enzyme shows the site of the phosphate group is a seryl hydroxyl group. Additional evidence that a phosphorylenzyme intermediate is formed is obtained by comparison of the Cd(II) and Zn(II) enzymes. Both are active. With the Zn(II) enzyme, the rate-determining step is formation of the phosphorylenzyme, while for the Cd(II) enzyme, the rate-determining step is hydrolysis of the phosphorylenzyme [16]. Since the rate-determining steps are different, the activities of the two enzymes are affected differently by variation in experimental parameters.

Water is a reactant in the reaction described in Eq. (6) as it is in all hydrolase-catalyzed reactions. It is appropriate to ask if there is a binding site for the water molecule. An answer to this may be obtained by comparing the reactivity of water and methanol in such reactions. If there is no specific binding site for water, then water and methanol will react with the phosphorylenzyme intermediate at rates dependent on their relative nucleophilicity and concentration. Evidence for a specific water-binding site has been indicated for myosin ATPase [myosin ATP phosphohydrolase (actin-translocating), EC 3.6.1.32], 5′-nucleotidase (5′-ribonucleotide phosphohydrolase, EC 3.1.3.5), *E. coli* alkaline phosphatase, and hexosediphosphatase (D-fructose-1,6-bisphosphate 1-phosphohydrolase, EC 3.1.3.11). For example, the relative reactivity of water and methanol in the hydrolysis of ATP by myosin ATPase was found to be 1×10^3, while the predicted value based on relative nucleophilicity and concentration was 2.5 [17]. On the other hand, attempts to show a specific water-binding locus in α-chymotrypsin have not been successful.

The broad specificity of *E. coli* alkaline phosphatase is shown in Table 5. The only real specificity is toward the orthophosphate moiety and the need for an O—P bond in C—O—P. The enzyme does not hydrolyze creatine phosphate, which contains a N—P bond, nor does it hydrolyze metaphosphates and diphosphates.

Several methods are available for assay of the nonspecific acid and alkaline phosphoric monoester hydrolases. The most important of these methods are given.

1. Use of the chromogenic substrate *p*-nitrophenyl phosphate and measurement of rate of formation of *p*-nitrophenol spectrophotometrically (below pH 7 at 340 nm and above pH 7 at 403 nm).
2. Use of 4-methylumbelliferone phosphate as substrate, which gives a fluorescent product, 4-methylumbelliferone, with λ_{ex} = 326 nm.
3. Measurement of rate of formation of orthophosphate by use of the Fiske–

Table 5 Relative Activities of *E. coli* Alkaline Phosphatase on a Variety of Phosphate Compounds

Substrate	Relative rate	Substate	Relative rate
p-Nitrophenyl phosphate	1.0	Uridine phosphate	1.3
β-Glycerophosphate	0.9	Riboflavin 5′-phosphate	0.7
Glucose 1-phosphate	0.9	L-Histidinol phosphate	0.9
Adenosine 3′-phosphate	1.0	Creatine phosphate	0
Adenosine 5′-phosphate	0.8	Adenosine triphosphate	0
Cytidine phosphate	1.2	Sodium pyrophosphate	0
Guanosine phosphate	1.0		

Source: Ref. 18, p. 478, by courtesy of Elsevier Publishing Co.

SubraRow method (or equivalent method). This method is used where phenyl phosphate, glycerophosphate, or other nonchromogenic compounds are the substrates.

4. Use of naphthyl phosphates and measurement of the rate of release of naphthol by combining with dyes such as the diazotized derivatives of naphthylamine, *o*-dianisidine, *p*-nitroaniline, and benzidine.

2. Specific Phosphoric Monoesterases

In contrast to the nonspecific acid and alkaline phosphoric monoester hydrolases, there are a number of specific phosphoric monoesterases in which both the alcohol and the orthophosphate moiety are of importance in determining the specificity. These enzymes usually have a pH optimum between 6 and 8. Among this group of enzymes are glucose-6-phosphatase (D-glucose-6-phosphate phosphohydrolase, EC 3.1.3.9), glucose-1-phosphatase (D-glucose-1-phosphate phosphohydrolase, EC 3.1.3.10), hexosediphosphatase (D-fructose-1,6-bisphosphate 1-phosphohydrolase, EC 3.1.3.11), 5′-nucleotidase (5′-ribonucleotide phosphohydrolase, EC 3.1.3.5), and 3′-nucleotidase (3′-ribonucleotide phosphohydrolase, EC 3.1.3.6). The last two enzymes are responsible for hydrolysis of the phosphate group from 5′- and 3′-ribonucleotides formed from phosphoric diester hydrolase activity on RNA.

Exercise 4. Design an experiment which would permit you to tell whether an enzyme that hydrolyzes glucose 1-phosphate is a glucose-1-phosphatase or is a nonspecific acid or alkaline phosphoric monoester hydrolase.

B. Phosphoric Diester Hydrolases

The phosphoric diester hydrolases are a group of enzymes involved in the hydrolysis of *one* of the two ester bonds formed with orthophosphate:

$$RO-\overset{\overset{O}{\|}}{\underset{\underset{OH}{|}}{P}}-OR' + H_2O \xrightarrow[\text{diester hydrolase}]{\text{phosphoric}} \underset{R'OH}{\overset{ROPO_3H^{\ominus}}{+}} \xrightarrow[\text{monoester hydrolase}]{\text{phosphoric}} ROH + HOPO_3^{\ominus\ominus} \qquad (7)$$

Theoretically, it should be possible to distinguish this group of enzymes from the phosphoric monoester hydrolases by the use of the two substrates, phenyl phosphate and diphenyl phosphate. However, not all of the enzymes in each group will act on one or the other of these two substrates. The most important of this group of enzymes are those involved in the hydrolysis of nucleic acids (see Chapter 18) and the enzymes that hydrolyze specific phosphoric diester bonds in phospholipids.

Phospholipase C (phosphatidylcholine cholinephosphohydrolase, EC 3.1.4.3) hydrolyzes phosphatidylcholine (lecithin) as shown in Eq. (8). Phospholipase D (phosphatidylcholine phosphatidohydrolase, EC 3.1.4.4) hydrolyzes phosphatidylcholine (lecithin) as shown in Eq. (9). Phospholipase C hydrolyzes the ester

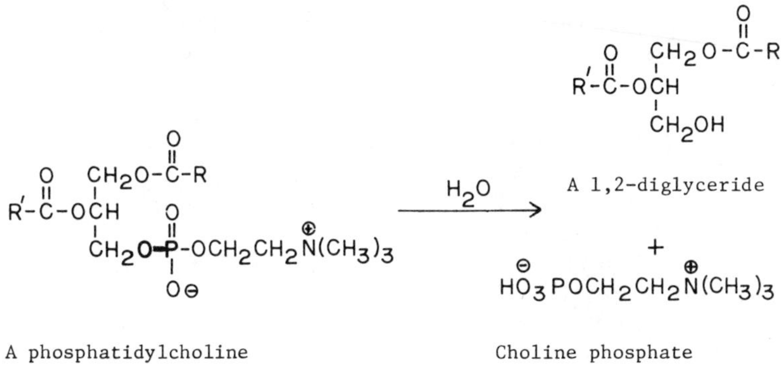

(8)

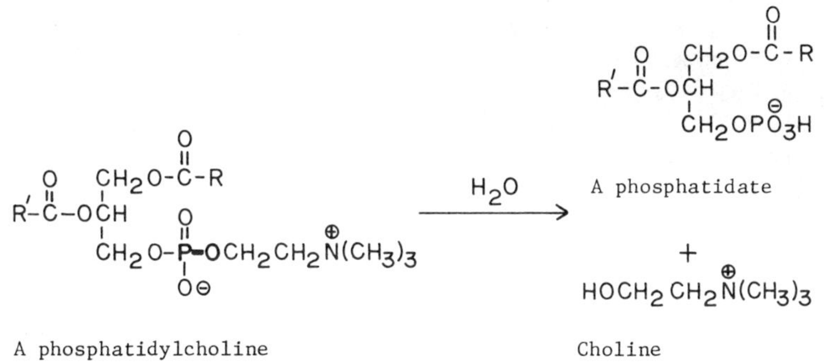

(9)

linkage formed between phosphate and the hydroxyl group of the glycerol moiety, while phospholipase D hydrolyzes the ester linkage formed between phosphate and the hydroxyl group of choline. Neither enzyme can hydrolyze the products formed by the other enzyme.

REVIEW QUESTIONS

See the end of Chapter 19.

REFERENCES

General References

Borgström, B., and H. L. Brockman, eds., *Lipases*, Elsevier, Amsterdam, 1984.

Brockerhoff, H., and R. G. Jensen, *Lipolytic Enzymes*, Academic Press, Orlando, Fla., 1974.

Chandan, R. C., and K. M. Shahani, Milk lipases: a review, *J. Dairy Sci.*, **47**, 471 (1964).

Desnuelle, P., Pancreatic lipase, *Adv. Enzymol.*, **23**, 129 (1961).

Desnuelle, P., and J. P. Savary, Specificities of lipases, *J. Lipid Res.*, **4**, 369 (1963).

Fishman, W. H., consulting ed., *The Phosphohydrolases: Their Biology, Biochemistry and Clinical Enzymology*, *Ann. New York Acad. Sci.*, **166**(2), 365–819 (1969).

Godtfredsen, S. E., Microbial lipases, in *Microbial Enzymes and Biotechnology*, 2nd Ed. (W. M. Fogarty and C. T. Kelly, eds.), Elsevier Applied Science, New York, 1990, p. 255.

Goldfine, H., Lipid chemistry and metabolism, *Ann. Rev. Biochem.*, **37**, 303 (1968).

Khorana, H. G., Phosphodiesterases, in *The Enzymes* (P. D. Boyer, H. Lardy, and K. Myrbäck, eds.), Vol. 5, Academic Press, New York, 1961, p. 79.

Morton, R. K., Phosphatases, in *Comprehensive Biochemistry* (M. Florkin and E. H. Stotz, eds.), Vol. 16, Elsevier, Amsterdam, 1965, p. 55.

Olivecrona, T., and G. Bengtsson-Olivecrona, Lipase, in *Food Enzymology* (P. F. Fox, ed.), Vol. 1, Elsevier Applied Science, New York, 1991, p. 62.

Oosterbaan, R. A., and H. S. Jansz, Cholinesterases, esterases and lipases, in *Comprehensive Biochemistry* (M. Florkin and E. H. Stotz, eds.), Vol. 16, Elsevier, New York, 1965, p. 1.

Schmidt, G., Nonspecific acid phosphomonoesterases, in *The Enzymes* (P. D. Boyer, H. Lardy, and K. Myrbäck, eds.), Vol. 5, Academic Press, New York, 1961, p. 37.

Stadtman, T. C., Alkaline phosphatases, in *The Enzymes* (P. D. Boyer, H. Lardy, and K. Myrbäck, eds.), Vol. 5, Academic Press, New York, 1961, p. 55.

van Deenen, L. L. M., and G. H. de Haas, Phosphoglycerides and phospholipases, *Ann. Rev. Biochem.*, **35**, 157 (1966).

Cited References

1. R. A. Oosterbaan and H. S. Jansz, Cholinesterases, esterases and lipases, in *Comprehensive Biochemistry* (M. Florkin and E. H. Stotz, eds.), Vol. 16, Elsevier, Amsterdam, 1965, p. 1.
2. P. Desnuelle, *Adv. Enzymol.*, **23**, 129 (1961).
3. W. K. Downey and P. Andrews, *Biochem. J.*, **94**, 642 (1965).
4. L. Sarda and P. Desnuelle, *Biochim. Biophys. Acta*, **30**, 513 (1958).
5. M. J. Constantin, L. Pásero, and P. Desnuelle, *Biochim. Biophys. Acta*, **43**, 103 (1960).
6. P. Savary and P. Desnuelle, *Biochim. Biophys. Acta*, **21**, 349 (1956).
7. B. Entressangles, H. Sari, and P. Desnuelle, *Biochim. Biophys. Acta*, **125**, 597 (1966).
8. R. G. Jensen, J. Sampugna, R. M. Parry, Jr., and K. M. Shahani, *J. Dairy Sci.*, **46**, 907 (1963).
9. P. Savary, J. Flanzy, and P. Desnuelle, *Bull. Soc. Chim. Biol.*, **40**, 637 (1958).
10. E. D. Korn, On the fatty acid and positional specificities of lipoprotein lipase, in *6th Intern. Conf. Biochem. Lipids* (P. Desnuelle, ed.), Pergamon Press, Elmsford, N.Y., 1961, p. 231.
11. F. Schönheyder and K. Volqvartz, *Enzymologia*, **11**, 178 (1944).
12. F. Schönheyder and K. Volqvartz, *Acta Physiol. Scand.*, **10**, 62 (1945); **11**, 349 (1946).
13. D. P. Wallach, *Can. J. Biochem.*, **45**, 133 (1967).
14. R. M. Krupka and K. J. Laidler, *J. Am. Chem. Soc.*, **83**, 1458 (1961).
15. K. K. Tsuboi and P. B. Hudson, *Arch. Biochem. Biophys.*, **61**, 197 (1956).
16. M. L. Applebury, B. P. Johnson, and J. E. Coleman, *Federation Proc.*, **29**, 531Abs. (1970).
17. D. E. Koshland, Jr., and E. B. Herr, Jr., *J. Biol. Chem.*, **228**, 1021 (1957).
18. A. Garen and C. Levinthal, *Biochim. Biophys. Acta*, **38**, 470 (1960).

18
The Nucleases and Biotechnology

I. GENERAL DEGRADATION OF NUCLEIC ACIDS IN THE GASTROINTESTINAL TRACT

Not all the enzymes involved in the degradation of nucleic acids belong to the hydrolases; therefore, it appears best to discuss this topic as a unit and separately from the esterases.

Enzymes that act on nucleic acids are important not only in the metabolism of growing plants and animals, they are important in the digestion of human food. The digestion products are sometimes associated with gout and other diseases. The nucleases of the gastrointestinal tract are rather broad in specificity and act on single-stranded RNA or DNA, while the restriction endonucleases are highly specific and act on double-stranded DNA only. Nucleic acid–synthesizing enzymes, the ligases, are also very specific.

Small amounts of nucleic acids are ingested in the diet and are degraded by enzymes secreted by the pancreas and the small intestine. How important these enzymes really are in the overall metabolism of an animal is not known with certainty. Ruminants such as the cow can tolerate up to 5% nucleic acids in the diet as a result of the high concentration of ribonucleases (RNases) and deoxyribonucleases (DNases) present in the stomach because of the microflora. Humans have a low concentration of these enzymes in the small intestine and can usually tolerate no more than 1% nucleic acids in the diet. Another reason for questioning the importance of nucleic acid digestion in the gastrointestinal tract is that the major part of the purine and pyrimidine bases are synthesized in vivo from smaller molecules, with simultaneous incorporation of the sugar moiety of the nucleoside rather than utilizing the bases that arise from the diet.

With such complex molecules as DNA and RNA (Figs. 1 and 2) a whole host of different enzymes must be involved in their degradation to purine and pyrimidine bases, ribose or deoxyribose, and orthophosphate. Let us examine each of these different types of enzymes with emphasis on what we can learn about the factors that determine enzyme specificity in general and nucleic acid degradation in particular.

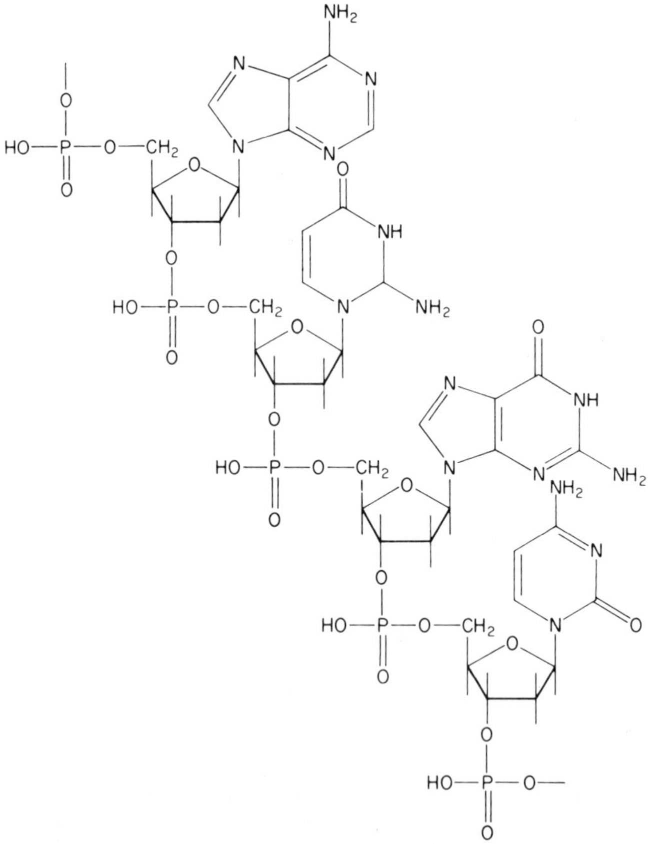

Figure 1 Segment of a DNA chain with the base sequence (top to bottom) of adenine, thymine, guanine, and cytosine. No secondary structure of the polymer is shown.

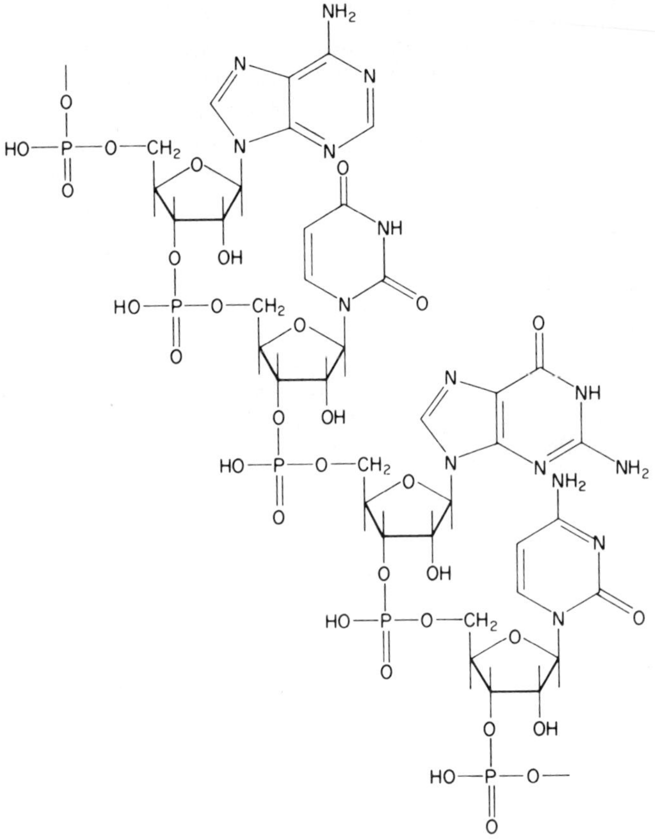

Figure 2 Segment of a RNA chain with the base sequence (top to bottom) of adenine, uracil, guanine, and cytosine. No secondary structure of the polymer is shown.

A. Phosphoric Diester Hydrolases Acting on Nucleic Acids

The first group of enzymes to attack single-stranded DNA and RNA convert the polymer to nucleotides by hydrolysis of the O-P bonds of the backbone of the polymer. Since these compounds are phosphoric diesters, the enzymes involved are the *phosphoric diester hydrolases*. What are some of the factors that determine whether an enzyme will hydrolyze one of the phosphoric diester bonds?

1. The Sugar Moiety

The major difference in the primary structures of single-stranded DNA and RNA is in the sugar moiety. In the case of DNA, the 2′-OH group is replaced by H (Figs. 1 and 2). This difference is of importance to an enzyme. Three different groups of enzymes can be distinguished on this basis: (a) those that act exclusively on RNA (ribonucleases), (b) those that act exclusively on DNA (deoxyribonucleases), and (c) those that act on both RNA and DNA (i.e., they have no preference for the nature of the sugar moiety).

2. The Phosphate-Sugar Linkage

There are two types of O—P linkages in DNA and RNA (Figs. 1 and 2), the 3′-linkage and the 5′-linkage. The enzymes are absolutely specific for one or the other of these bonds. Those enzymes that hydrolyze the 3′-linkage are referred to as 5′-formers, while those that hydrolyze the 5′-linkage are called 3′-formers in that 5′- and 3′-nucleotides are formed in the two cases, respectively.

3. Method of Cleavage of the 5′-Linkage

Two possible mechanisms of splitting the 5′-linkage have been found for the enzymes that act on RNA. The first involves direct addition of water to the P—O bond as shown in Eq. (1), where R′ and R″ represent continuation of the RNA chain

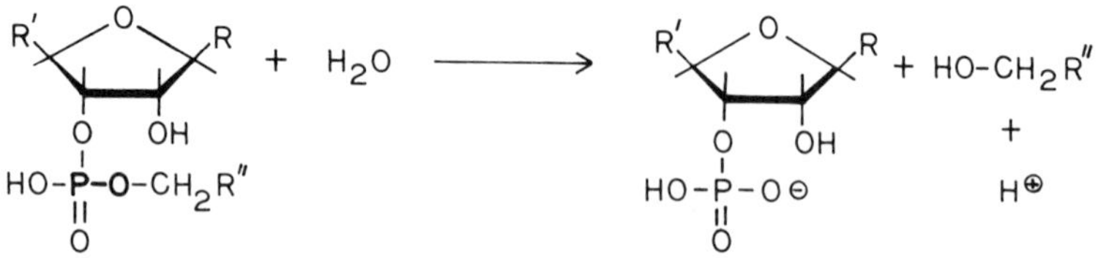

(1)

and R is the purine or pyrimidine base. These would be ribonucleases that form 3′-nucleotides.

A second, and apparently more widely catalyzed reaction is an initial depolymerization step which involves a nucleophilic attack of the oxygen of the 2′-OH group on the phosphorus to give a 2′,3′-cyclic intermediate [Eq. (2)]. Some

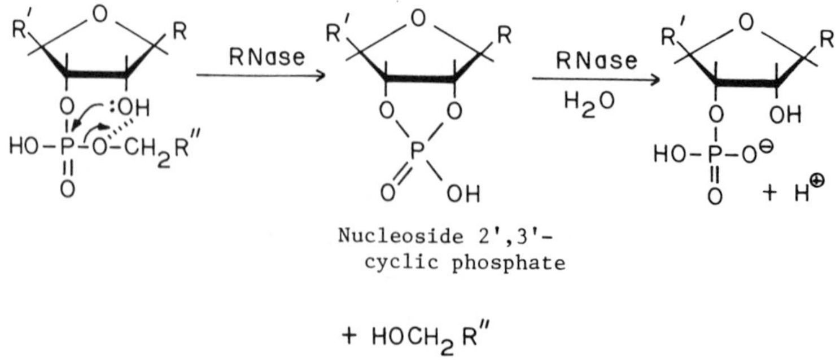

Nucleoside 2′,3′-cyclic phosphate

(2)

ribonucleases cannot perform the second step, which involves hydrolysis, or they do so at a very slow rate, so that the 2′,3′-cyclic phosphate accumulates. Other ribonucleases catalyze hydrolysis of the 2′,3′-cyclic phosphate at a rapid rate. Theoretically, both the 2′- and 3′-nucleotides could be formed; actually, only the 3′-nucleotide is found. The first reaction shown in Eq. (2) is not a hydrolytic reaction but rather involves the transfer of a phosphate group. Therefore, that group of ribonucleases that split the 5′-linkage by the mechanism shown in Eq. (2) belong to the general subdivision of the nucleotidyltransferases. Their specific name depends on preference for either a purine or a pyrimidine base.

4. Nature of the Purine or Pyrimidine Base

Although there appears to be some degree of specificity among the deoxyribonucleases with respect to the nature of the base, it is much less strict in the case of the ribonucleases. At least eight distinct types of ribonucleases have been recognized on this basis. Ribonuclease T_2 (EC 3.1.27.1) splits the 5′-linkage by formation of the 2′,3′-cyclic intermediate with both purine and pyrimidine nucleotide residues (i.e., it has not specificity for the base). Pancreatic ribonuclease (EC 3.1.27.5) splits the 5′-linkage by formation of a 2′,3′-cyclic intermediate only with pyrimidine nucleotide residues. Ribonuclease T_1 (EC 3.1.27.3) from *Aspergillus oryzae* and some other organisms splits the 5′-linkage by formation of a 2′,3′-cyclic intermediate with only guanylate nucleotides. *Bacillus subtilis* ribonuclease (EC 3.1.27.2) gives only endonucleolytic cleavage to 2′,3′-cyclic nucleotides. Ribonuclease V (EC 3.1.27.8) hydrolyzes poly(A) preferentially, but also poly(U). Ribonuclease F (EC 3.1.27.7) cleaves only between a cytosine and an adenine moiety.

5. Location of Attack

Like other depolymerizing enzymes that act on polysaccharides and proteins, the enzymes acting on nucleic acids can be divided into two groups on the basis of whether they split the terminal bonds (exo-enzymes) or those in the interior of the molecule (endo-enzymes).

6. Hydrolysis of Ribonucleoside Cyclic Phosphates

Some of the phosphoric diester hydrolases cannot split the phosphoric diester bonds found in nucleic acids. However, they can hydrolyze either the ribonucleoside 2′,3′-phosphates or the ribonucleoside 3′,5′-phosphates.

B. Assay Methods for Ribonucleases and Deoxyribonucleases

Several general methods are available for measuring the activity of the DNases and RNases. These may be divided into those that use single-stranded DNA or RNA as substrate and those that use a cyclic phosphate compound.

1. Use of Single-Stranded DNA or RNA as Substrate for Specific Enzymes

As shown by the general reaction in Eq. (1), a proton is liberated from ionization of the phosphate group formed (pK ~ 3). Therefore, the reaction may be monitored continuously in a pH Stat or in a Warburg apparatus in the presence of bicarbonate buffer. Because of a spectral change produced by initial formation of a 2′,3′-cyclic phosphate, RNase activity on RNA can be assayed continuously in a spectrophotometer at 300 nm. Another group of assay methods depends on changes in the precipitability of DNA or RNA with acidified tertiary butanol as the reaction proceeds. The change in absorbance of the supernatant liquid is read at 260 nm. Conversely, one can use a ^{32}P-labeled substrate, which makes the assay even more sensitive.

2. Use of Cyclic Phosphates

The ribonucleases can be assayed by the use of 2′,3′-cytidine. The reaction is monitored continuously in a pH Stat or in a spectrophotometer at 280 nm.

C. Nucleotidases

The products formed from action of the phosphoric diester hydrolases on nucleic acids are 3′- and 5′-nucleotides. These products are further hydrolyzed by two specific types of phosphoric monoester hydrolases: (a) 3′-nucleotidase (3′-ribonucleotide phosphohydrolase, EC 3.1.3.6) and (b) 5′-nucleotidase (5′-ribonucleotide phosphohydrolase, EC 3.1.3.5). These enzymes act on all the 3′- and 5′-nucleotides regardless of the specific nature of the base. The products of this action are the nucleosides (base + sugar) and inorganic phosphate, as shown in Eq. (3).

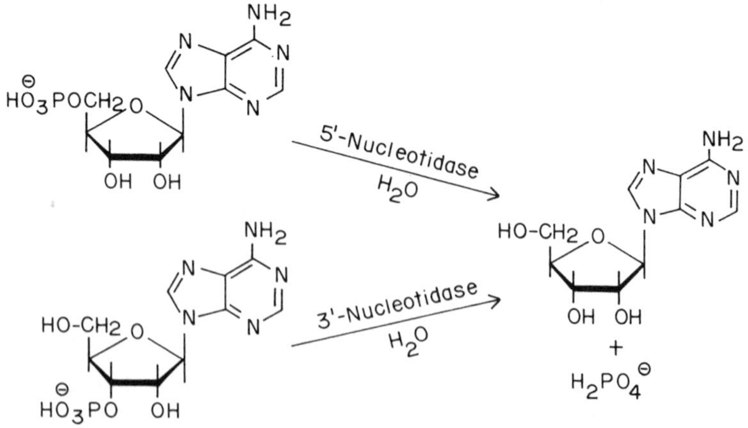

(3)

D. Nucleosidases

As shown for the specific example *N*-ribosyl uracil, the linkage between ribose and the base in a nucleoside involves a C—N bond [Eq. (4)]. The group of enzymes

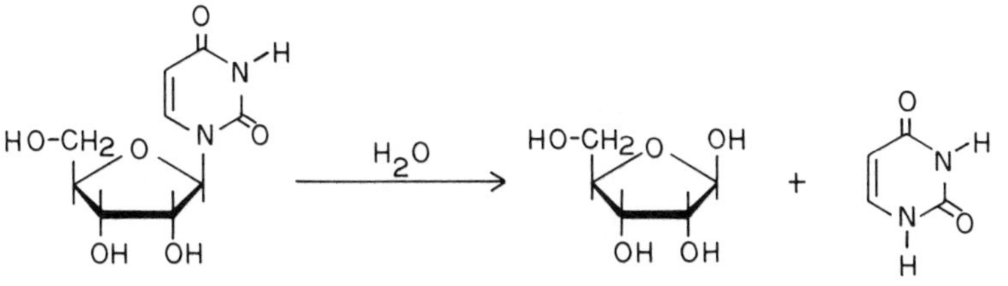

(4)

N-Ribosyl uracil D-Ribose Uracil

involved are those that hydrolyze *N*-glycosyl compounds. An enzyme specific for *N*-ribosyl purines (*N*-D-ribosyl-purine ribohydrolase, EC 3.2.2.1) has been described. Eighteen of these nucleosidases are listed in *Enzyme Nomenclature* [7].

II. RECOMBINANT DNA TECHNOLOGY

Recombinant DNA technology involves the removal of a fragment of double-stranded DNA from a chromosome of a donor and the insertion of the fragment into a chromosome of a host organism, as shown in Fig. 3. The plasmid containing the foreign DNA fragment is introduced into host cells where the plasmid is caused to undergo replication and multiplication. Proteins are expressed by the genes contained in the foreign DNA fragment. Human insulin, bovine chymosin, and other proteins can be manufactured by the host cell. Alternatively, DNA biochemically synthesized in the laboratory by DNA polymerase can be inserted in place of the foreign DNA to cause production of new proteins or of antisense RNA that inhibits the synthesis of one or more enzymes. Recombinant DNA technology is possible because of these highly specific types of enzymes. These are the DNA restriction endonucleases, the DNA ligases, and the polymerases [1].

A. DNA Restriction Endonucleases

The type II restriction endonucleases (EC 3.1.21.4) recognize specific base sequences along double-stranded DNA, and they cleave both strands of the duplex. They are Mg^{2+}-requiring enzymes. More than 600 restriction endonucleases are known, having first been discovered in 1968. More than 100 restriction endonucleases are supplied in pure form by biochemical companies. Restriction endonucleases are found in many bacteria, where they function to hydrolyze foreign DNA introduced by phage infection, conjugation, or transformation. Type

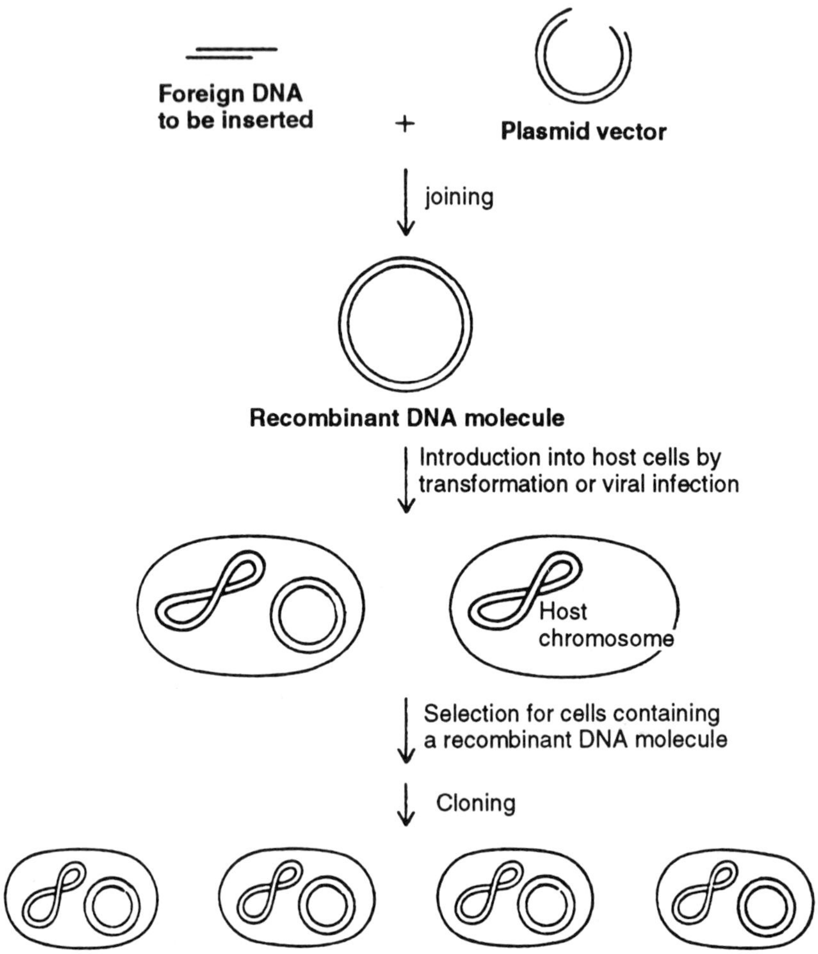

Figure 3 Schematic diagram of insertion of a foreign DNA fragment (donor) into a plasmid vector (host) and cloning of DNA molecules in host cells.

I restriction endonucleases (EC 3.1.21.3) have an absolute requirement for ATP and simultaneously hydrolyze it as they hydrolyze phosphodiester bonds. Type III restriction endonucleases (EC 3.1.21.5) have an absolute requirement for ATP but do not hydrolyze it.

The restriction endonucleases are used to specifically cut out a fragment of double-stranded DNA from a plasmid of a donor or host (Fig. 4). In a plasmid

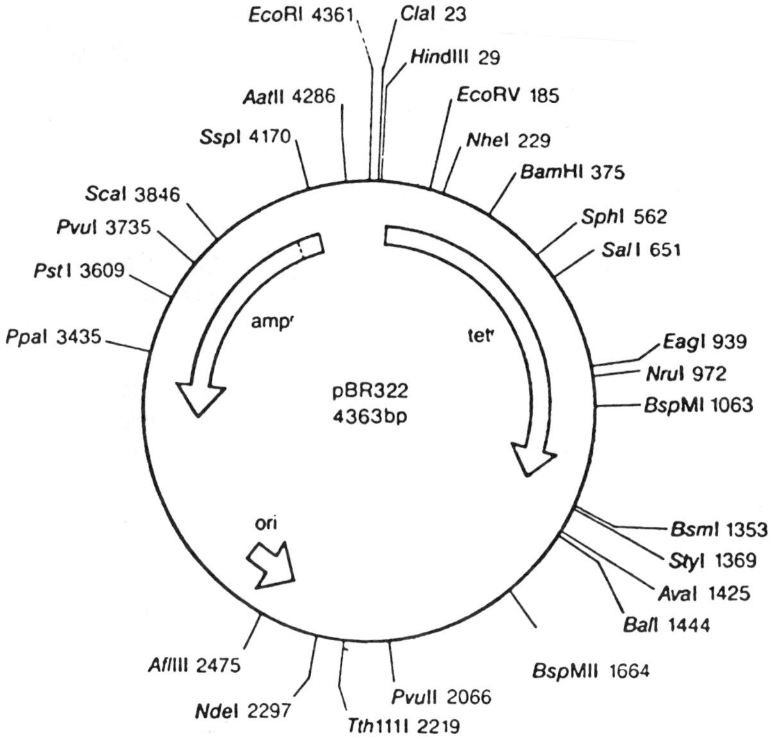

Figure 4 Some restriction endonucleases hydrolysis sites on *Escherichia coli* plasmid pBR322.

such as pBR322 (Fig. 4), a desired portion of the chain can be excised by use of two restriction endonucleases. For example, use of restriction endonucleases EcoRV and Sal I permits a segment to be excised between base pair 185 and base pair 651 of pBR322. A restriction endonuclease recognizes a sequence of base pairs as shown in Fig. 5 for BamH I, Kpn I, Hae III, and Hha I. Hae III produces a split with a blunt end, while BamH I, Kpn I, and Hha I hydrolyze the phosphodiester bonds at staggered locations, producing "sticky" ends. The restriction enzymes have high, absolute specificity, giving reproducible cleavage of phosphodiester bonds in DNA substrates (Fig. 6). Therefore, they are indispensable tools for sequencing long strands of DNA, isolating genes, analyzing chromosome structures, and creating new DNA molecules (above) that can be cloned (recombinant DNA technology).

Eco RI restriction endonuclease is one of the best studied of the enzymes [2,3].

It is a small protein of 276 amino acid residues (31,065 MW). The active enzyme, composed of two identical subunits, hydrolyzes the phosphodiester bonds between adjacent deoxyguanylic (G) and deoxyadenylic (A) acid residues of duplex 5′-GAATTC-3′ as shown generically in Fig. 6.

B. DNA Ligases

DNA ligases catalyze the joining of two DNA chain segments, or the closure of a single DNA chain, by condensation of the 5′-phosphoryl group with the adjacent 3′-hydroxyl group, coupled with hydrolysis of the pyrophosphate group of NAD^+ (*E. coli* enzyme; EC 6.5.1.2) or ATP (T4 enzyme; EC 6.5.1.1). The sequence of reactions involved are shown in Eq. (5) [4,5]. In the first reaction, the ε-amino group of a lysyl residue of the ligase, in a nucleophilic attack on the adenylyl phosphorus of NAD^+, is phosphorylated to give the adenylated ligase with the elimination of NMN (nicotinamide mononucleotide). In the second reaction, the adenylyl group is transferred from the enzyme to the 5′-phosphoryl terminus of the nicked DNA to form a pyrophosphate bond. In the third reaction, the 3′-OH group, in a nucleophilic attack on the 5′-phosphoryl group, forms a phosphodiester bond, with elimination of AMP, joining the two ends of the nicked chain.

E. coli DNA ligase is a single-chain protein of 74,000 MW, somewhat elongated in shape. The T4 DNA ligase is also a single-chain protein of about 68,000 MW and elongated in shape. The enzymes are not specific for the base at the nicked end.

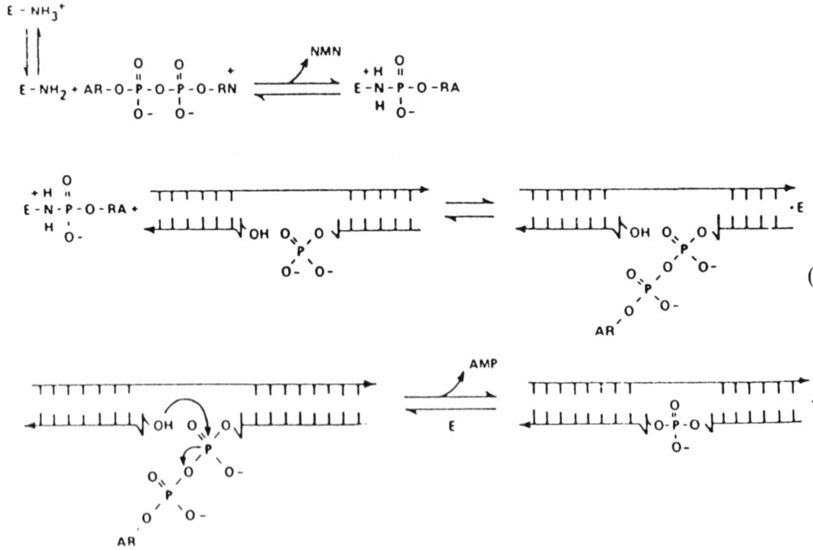

(5)

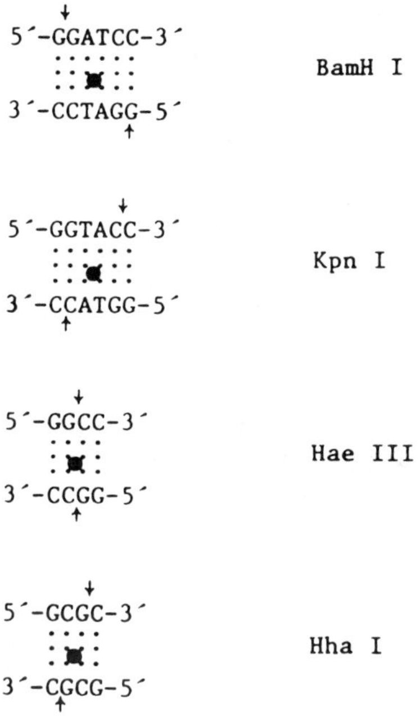

Figure 5 Recognition and cleavage sites (arrows) of double-stranded nucleotide sequences by some restriction endonucleases. The closed circle indicates axis of twofold rotational symmetry of the two complementary strands.

C. The Polymerases

Both RNA and DNA polymerases occur; both types have been isolated and studied. The polymerases catalyze the addition of nucleoside triphosphates to the 3′-OH terminals of single strands of preexisting polyribonucleotides (primers) with release of pyrophosphate [Eq. (6)]. For DNA synthesis (for example), all four deoxyribonucleoside 5′-triphosphates (dATP, dCTP, dGTP, dTTP) must be present, as well as a primer molecule [Eq. (6), top], Mg^{2+}, and a DNA template. A

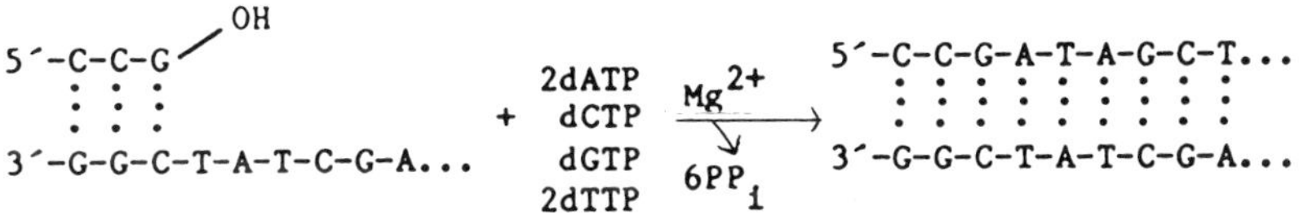

$(\text{Deoxyribonucleotide})_n$ (top) DNA template (bottom) — Nucleotides — $(\text{Deoxyribonucleotide})_{n+6}$ (top) DNA template (bottom)

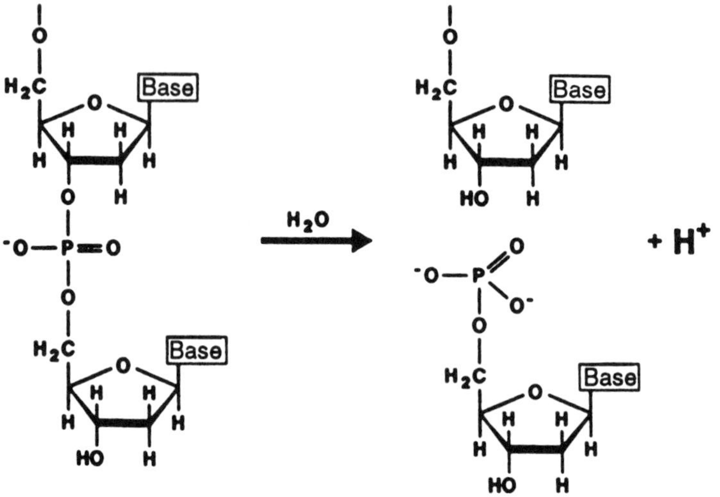

Figure 6 Restriction endonuclease-catalyzed hydrolysis of the 3′-O—P bond between two nucleotides.

proposed mechanism of DNA polymerase must consider that the enzyme binds at least four compounds simultaneously (a deoxyribonucleoside 5′-triphosphate, Mg^{2+}, a deoxyribonucleotide primer, and a DNA template) (Fig. 7). By binding the four reactants, the enzyme converts an impossible five-component system into a unimolecular reaction, thereby contributing to the enormous efficiency of the catalyzed reaction. The enzyme makes life possible!

As shown in Fig. 7, the enzyme catalyzes the nucleophilic attack of the 3′-OH terminus of the primer molecule on the α-phosphorus of the deoxyribonucleotide triphosphate to form a new phosphodiester bond with release of pyrophosphate. The DNA chain elongation proceeds in the 5′→3′ direction at a rate of about 10 nucleotides per second per molecule of DNA polymerase I. The reaction is probably processive, in that many nucleotide units are added without release of the enzyme from template. The PCR (polymerase chain reaction) reaction is now routine in most recombinant DNA technology laboratories.

Several DNA polymerases have been isolated and studied. They have similar general properties, but differ somewhat in specificity. DNA polymerase I, from wild-type *E. coli*, is a single-chain protein of 103,000 MW. The protein folds into three domains to give an approximately spherical overall shape [6]. It acts best to fill in extended single-stranded regions near double-helical regions of DNA. DNA polymerases II and III, produced by the *polB* and *dnaA* genes of *E. coli*, act optimally on double-stranded DNA templates that have short gaps.

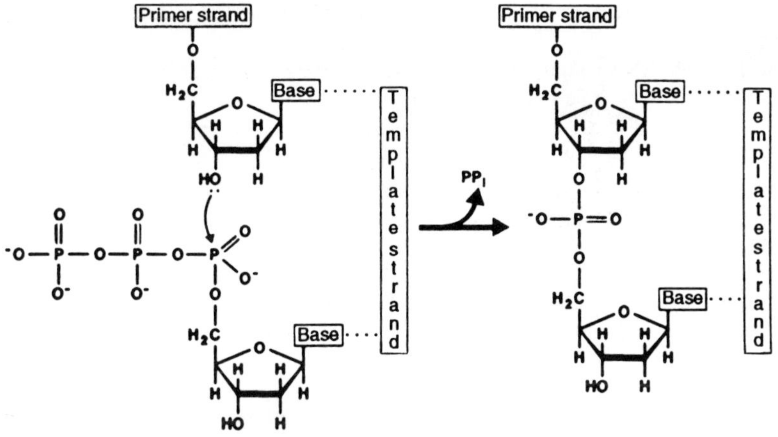

Figure 7 Proposed mechanism for DNA polymerase I–catalyzed reaction.

III. OTHER NUCLEASES

There are a number of other nucleases that are important in the in vivo biosynthesis of RNA and DNA and their modification and degradation. A list of these enzymes and their functions can be found in Ref. 1.

REVIEW QUESTIONS

See the end of Chapter 19.

REFERENCES

General References

Oxender, D. L., and C. F. Fox, eds., *Protein Engineering*, Alan R. Liss, New York, 1987.

Whitaker, J. R., and P. E. Sonnet, eds., *Biocatalysis in Agricultural Biotechnology*, ACS Symposium Series 389, American Chemical Society, Washington, D.C., 1989.

Cited References

1. J. R. Whitaker, Characteristics of some enzymes used in genetic engineering, in *Biocatalysis in Agricultural Biotechnology* (J. R. Whitaker and P. E. Sonnet, eds.), ACS Symposium Series 389, American Chemical Society, Washington, D.C., 1989, p. 44.
2. P. Modrich and D. Zabel, *J. Biol. Chem.*, **251**, 5866 (1976).

3. L. Jen-Jacobsen, M. Kurpiewski, D. Lesser, J. Grable, H. W. Boyer, J. M. Rosenberg, and P. J. Greene, *J. Biol. Chem.*, **258**, 14638 (1983).
4. I. R. Lehman, DNA joining enzymes (ligases), in *The Enzymes*, 3rd Ed. (P. D. Boyer, ed.), Vol. 10, Academic Press, New York, 1974, p. 237.
5. M. J. Engler and C. C. Richardson, DNA ligases, in *The Enzymes*, 3rd Ed. (P. D. Boyer, ed.), Vol. 15B, Academic Press, New York, 1983, p. 3.
6. T. A. Steitz and C. M. Joyce, Exploring DNA polymerase I of *E. coli* using genetics and x-ray crystallography, in *Protein Engineering* (D. L. Oxender and C. F. Fox, eds.), Alan R. Liss, New York, 1987, p. 227.
7. *Enzyme Nomenclature*, recommendation (1984) of the Nomenclature Committee of the International Union of Biochemistry, published for the International Union of Biochemistry, Academic Press, San Diego, Calif., 1984.

19

The Proteolytic Enzymes

Proteins are among the major constituents of organisms. The nature of these polymers and how they are synthesized in vivo are discussed in Chapter 2. In vivo, the proteins must be turned over. This is the responsibility of a group of hydrolytic enzymes, the proteolytic enzymes (peptide hydrolases). Proteolytic enzymes are of particular importance to the food scientist. Pepsin, trypsin, chymotrypsin, and the carboxypeptidases and aminopeptidases are responsible for hydrolysis of ingested proteins to amino acids in the digestive tract of humans. Proteolytic enzymes are responsible for the clotting of blood (see Chapter 20, Section II.E). There are proteolytic enzymes in the phagocytic cells of blood which are responsible for hydrolyzing foreign proteins. The lysosomes of cells contain a number of proteolytic enzymes, the cathepsins, which are responsible for the cellular catabolism of proteins. Some highly specific proteases are responsible for conversion of preproteins to proproteins and then to proteins.

The proteolytic enzymes are by far the most important group of enzymes in the food processing industry. They are used in the production of cheeses, chill-proofing of beer, tenderization of meat, and modification of the properties of the proteins of cereals in bread and cereal manufacture. The common reaction catalyzed by the proteolytic enzymes is the hydrolysis of peptide bonds of a protein:

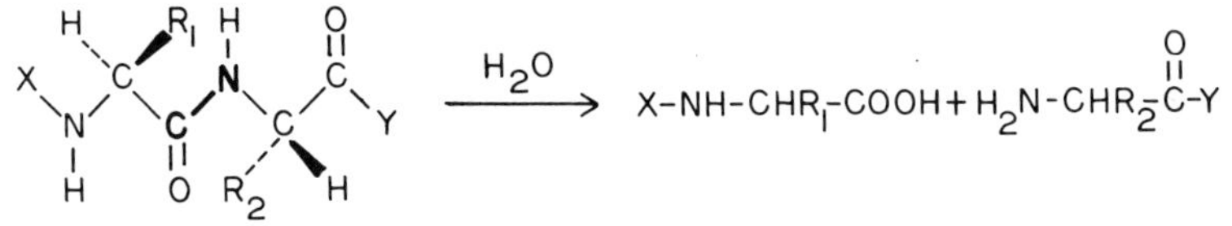

(1)

I. SPECIFICITY REQUIREMENTS

Let us consider the requirements that must be met in order to obtain hydrolysis of the peptide bond by a specific proteolytic enzyme.

A. Nature of the R_1 and R_2 Groups

A specific requirement in terms of the nature of R_1 and/or R_2 [Eq. (1)] is the major distinguishing feature among the proteolytic enzymes. For example, α-chymotrypsin hydrolyzes the peptide bond at an appreciable rate only when R_1 is the side chain of a tyrosyl, phenylalanyl, or tryptophanyl residue. Trypsin hydrolyzes the peptide bond only when R_1 is the side chain of an arginyl or lysyl residue. On the other hand, pepsin and the carboxypeptidases have specificities directed in part toward the nature of the R_2 side chain. Both of these enzymes hydrolyze the peptide bond at a maximum rate if R_2 is the side chain of a phenylalanyl residue. For carboxypeptidase, the Y constituent must be an —OH group.

B. Configuration of the Amino Acids

Not only are the enzymes specific for the nature of the R_1 and/or R_2 side chains, these side chains must be contributed by L-amino acids. The amino acids of proteins are of the L-configuration. Sometimes the D-amino acid residues may bind into the active site, but the peptide bond is not hydrolyzed.

C. Size of Compound

For several of the proteolytic enzymes, the size of the substrate does not appear to be important. For example, the best synthetic amide substrates for α-chymotrypsin and trypsin are α-*N*-acetyl-L-tyrosinamide and α-*N*-benzoyl-L-argininamide, respectively (Fig. 1). These substrates contain only one amino acid; however, the requirements for the nature of the R_1 group and of L-configuration of the amino acid have been met. Some of the other proteolytic enzymes have stricter requirements as to the size of substrate. This appears to be generally true of the group of proteolytic enzymes called the "aspartic proteases."

D. Nature of X and Y

Whether the enzyme requires that X and Y [Eq. (1)] be a hydrogen and hydroxyl group, respectively, or that they be derivatized permits us to distinguish between the endopeptidases and the exopeptidases. In the case of the endopeptidases, which hydrolyze peptide bonds randomly in the interior of the protein (if the requirement for R_1 and/or R_2 is met), maximum activity is obtained when X and Y are derivatized. X may be an acyl group (acetyl, benzoyl, benzyloxycarbonyl, etc.) and Y an amide or ester group, or the groups may be other amino acid residues, as in proteins.

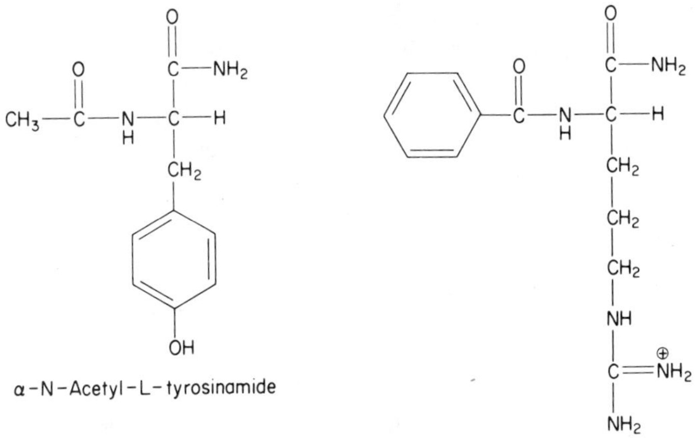

α-N-Acetyl-L-tyrosinamide

α-N-Benzoyl-L-argininamide

Figure 1 Structures of α-*N*-acetyl-L-tyrosinamide and α-*N*-benzoyl-L-argininamide, which are substrates for chymotrypsin and trypsin, respectively.

The carboxypeptidases (peptidyl-amino-acid hydrolases, EC 3.4.16–18) require that Y be a hydroxyl group. The carboxyl group is involved in binding and properly orienting the substrate in the active site of the enzyme (see Section V). For best activity by the carboxypeptidases, X should not be a hydrogen. The carboxypeptidases have their major specificity toward the R_2 side chain.

The aminopeptidases (α-amino-acyl-peptide hydrolases, EC 3.4.1) require that X be a hydrogen and prefer that Y not be a hydroxyl group. Specificity of these enzymes is directed mainly toward the R_1 residue. Leucine aminopeptidase is the best studied of this group of enzymes.

The dipeptidases (dipeptide hydrolases, EC 3.4.3) and tripeptidases require that X and Y be hydrogen and hydroxyl groups, respectively. The tripeptidases will hydrolyze only tripeptides, while some of the dipeptidases act on both di- and tripeptides. The specificity of these enzymes appears to be directed more toward the N-terminal rather than the C-terminal end of the peptide chain.

E. Requirement for Peptide Bond

Most of the proteolytic enzymes hydrolyze other types of bonds than just the peptide bond. The peptide bond may be replaced with an amide ($—NH_2$), ester (—COOR), thiolester (—COSR), or a hydroxamate (—CONHOH) group. The ester analogs of the amides shown in Fig. 1 are hydrolyzed at 200 to 1000 times the rate of the amides (because E_a for hydrolysis of esters is much lower than for amides). In the case of α-chymotrypsin, trypsin, and some of the other proteolytic enzymes, it appears to be sufficient for the enzyme that it have a substrate which binds properly in the active site so that the susceptible bond is oriented correctly with respect to the catalytic groups. The nature of this susceptible bond, within limits, does not matter.

Pepsin, as well as some of the other acid proteases, is more discriminating in the type of bond hydrolyzed. Even though the specificity requirements are met with respect to R_2, substitution of a peptide bond with an ester bond prevents the compound from serving as a substrate.

II. SUBDIVISION OF THE PROTEOLYTIC ENZYMES

As suggested by Hartley in 1960 [1], it is convenient to divide the proteolytic enzymes into four groups on the basis of their mechanism of action. The four groups are (a) the serine proteases, (b) the sulfhydryl proteases (the names *thiol proteases* and *cysteine proteases* are also used), (c) the metal-containing proteases, and (d) the aspartic proteases (the names *carboxyl proteases* and *acidic proteases* are also used). The names indicate one of the key catalytic groups in the active site.

III. THE SERINE PROTEASES

A. General Properties

The serine proteases include the chymotrypsin family (α, δ, and π-chymotrypsins and chymotrypsins B, EC 3.4.21.1, and C, EC 3.4.21.2), trypsin (EC 3.4.21.4), elastase (pancreatoelastase, EC 3.4.21.36), thrombin (EC 3.4.21.5), subtilisin (EC 3.4.21.14), and α-lytic protease from *Sorangium* sp. All these enzymes have in common the characteristic that they are inhibited by diisopropylphosphorofluoridate (DFP), which reacts with the hydroxyl group of a specific seryl residue in the active site of the enzyme. The enzymes are all endopeptidases. In addition, they all contain an imidazole group and an aspartyl carboxyl group in the active site. The amino acid sequences around the essential seryl and histidyl residues for several of the enzymes are similar (Table 1). Even some nonproteolytic enzymes such as phosphoglucomutase and alkaline phosphatase have essential seryl residues in their active sites. The active-site His and Ser residues are in positions 57 and 195 for bovine chymotrypsin A and 64 and 221 for subtilisin. Otherwise, these two serine proteases have very little sequence homology. Note that the enzymes are arranged in decreasing homology around the active-site His and Ser residues.

The substrate specificities of the individual members of this group are often quite different (Table 2). The general features of catalysis by this group of enzymes are identical; only the specific groups involved in binding of substrate are different, which results in the different specificities. The cause of this difference in specificity among α-chymotrypsin, trypsin, and elastase is shown in Fig. 2. In

Table 1 Amino Acid Sequence Around the Essential Seryl and Histidyl Residues of Several Serine Enzymes

Enzyme	Sequence
	Around Seryl Residue
Trypsin, bovine	-Asp-Ser-Cys-Gln-Gly-Asp-Ser*-Gly-Gly-Pro-Val-Val-Cys-Ser-
Trypsin, pig	-Asn-Ser-Cys-Gln-Gly-Asp-Ser*-Gly-Gly-Pro-Val-Val-Cys-Gly-
Chymotrypsin A, bovine	-Ser-Ser-Cys-Met-Gly-Asp-Ser*-Gly-Gly-Pro-Leu-Val-Cys-Lys-
Chymotrypsin B, bovine	-Ser-Ser-Cys-Met-Gly-Asp-Ser*-Gly-Gly-Pro-Leu-Val-Cys-Gln-
Elastase, pig	-Ser-Gly-Cys-Gln-Gly-Asp-Ser*-Gly-Gly-Pro-Leu-His-Cys-Leu-
Thrombin, bovine	-Asp-Ala-Cys-Glu-Gly-Asp-Ser*-Gly-Gly-Pro-Phe-Val-
α-Lytic protease, *Sorangium*	-Cys-Met-Gly-Arg-Gly-Asp-Ser*-Gly-Gly-Ser-Trp-Ile-Thr-Ser-
Subtilisin, *B. subtilis* Carlsberg	-Ala-Thr-Leu-Asn-Gly-Thr-Ser*-Met-Ala-Ser-Pro-His-Val-Ala-
Alkaline phosphatase, *E. coli*	-Pro-Asp-Tyr-Val-Thr-Asp-Ser*-Ala-Ala-Ser-Ala-
Phosphoglucomutase, rabbit	-Asx-Leu-Gly-Val-Thr-Ala-Ser*-His-Asx-Gly-Glx-Ser-Ala-Gly-
	Around Histidyl Residue
Trypsin, bovine	-Val-Val-Ser-Ala-Ala-His*-Cys-Tyr-Lys-Ser-Gly-Ile-Gln-
Chymotrypsin A, bovine	-Val-Val-Thr-Ala-Ala-His*-Cys-Gly-Val-Thr-Thr-Ser-Asp-
Chymotrypsin B, bovine	-Val-Val-Thr-Ala-Ala-His*-Cys-Gly-Val-Thr-Thr-Ser-Asp-
Elastase, pig	-Val-Met-Thr-Ala-Ala-His*-Cys-Val-Asp-Arg-Glu-Leu-Thr-
Thrombin, bovine	-Ala-Ala-Thr-Ala-Ala-His*-Cys-Lev-Lev-Tyr-Pro-
α-Lytic protease, *Sorangium*	-Phe-Val-Thr-Ala-Gly-His*-Cys-Gly-Thr-Val-Asn-Ala-Thr-
Subtilisin, *B. subtilis* Carlsberg	-Thr-Asp-Gly-Asn-Gly-His*-Gly-Thr-His-Val-Ala-Gly-Thr-

Source: Ref. 2, p. 49, by courtesy of National Biomedical Research Foundation.

Table 2 Substrate Specificities of Several Serine Proteases

Enzyme	Specificity
α-Chymotrypsin, bovine	Tyrosine, phenylalanine, tryptophan
Chymotrypsin B, bovine	Tyrosine, phenylalanine, tryptophan
Trypsin, bovine	Lysine, arginine
Thrombin, bovine	Lysine, arginine
Elastase, bovine	Alanine
α-Lytic protease, *Sorangium*	Tyrosine, phenylalanine, tryptophan
Subtilisin, *B. subtilis* Carlsberg	Tyrosine, phenylalanine, tryptophan

α-chymotrypsin and trypsin the binding pocket is quite open because of the Gly-216 and Gly-226 residues. Thus large side chains of amino acid residues can fit into the pockets. The two differ, however, in that trypsin has Asp-189 at the bottom of the pocket, whereas α-chymotrypsin has Ser-189. The negative charge of the carboxyl group of Asp-189 forms an electrostatic bond with the positive charge of an ϵ-amino group of lysine, or the guanidinium group of arginine, in a substrate. The binding pocket of elastase is quite different in that it is largely filled with the bulky side chains of Val-216 and Thr-226. Only a small side chain such as —CH_3 can fit. Therefore, elastase has specificity for alanine-containing substrates.

It is not too surprising that the chymotrypsins, trypsin, and elastase should be similar in their mechanism of action. They all have their origin in the pancreas. Prothrombin is found in the blood and is converted to thrombin during the blood clotting process. It is interesting that the microbial enzymes subtilisin and α-lytic

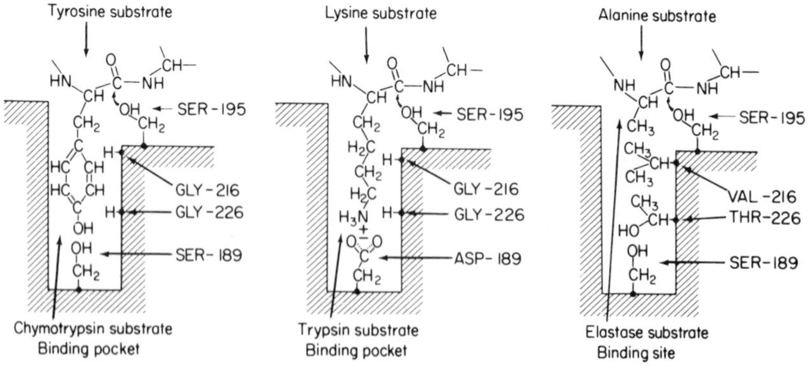

Figure 2 Proposed substrate binding loci in α-chymotrypsin, trypsin, and elastase. (From Ref. 3, p. 53, by courtesy of Walter de Gruyter.)

protease should resemble α-chymotrypsin so closely. Could it be that there are only a few possible arrangements of amino acids in a protein that can give rise to an active site capable of hydrolyzing proteins?

Exercise 1. Which of these proteolytic enzymes would you suspect have evolved from (a) a common ancestral gene, and (b) by convergent evolution from different ancestral genes?

B. α-Chymotrypsin

Rather than catalog the many properties of the serine proteases that have been described for the individual enzymes, we shall restrict ourselves to a consideration of a few important features of α-chymotrypsin. Many of these features will fit in a general way the other serine proteases.

Bovine pancreas produces two inactive proteins, chymotrypsinogens A and B. The two differ markedly in their amino acid compositions, and this is reflected in their different isoelectric points (8.5 and 4.5 for A and B, respectively). When converted to the active enzymes, they still differ markedly in their physical properties, but their specificity requirements are the same. Therefore, they are assigned the same Enzyme Commission number (EC 3.4.21.1). Porcine pancreas produces, in addition, another zymogen, chymotrypsinogen C, which is activated by trypsin to chymotrypsin C. This enzyme has specificity for leucyl side chains rather than for the aromatic residues. Therefore, it is assigned a different Enzyme Commission number (EC 3.4.21.2). The process of activation of chymotrypsinogen A to α-chymotrypsin has been carefully worked out and is shown in Chapter 20, Section II.C.2, as an example of limited specific hydrolysis.

The primary sequence of α-chymotrypsin is shown in Fig. 3 and the secondary and tertiary structures of α-chymotrypsin are shown in Fig. 4. For the sake of simplicity only the amino acid residues involved in the active site, the terminal residues, and the disulfide bonds are specifically identified in Fig. 4. The locus where the substrate is bound is not shown specifically but involves Gly-216, Gly-226, and Ser-189 (see Fig. 2 for position).

Five specific amino acid residues have been shown to be important for the catalytic activity of α-chymotrypsin. These are the N-terminal isoleucyl-16 residue, the aspartyl-194 residue, the aspartyl-102 residue, the seryl-195 residue, and the histidyl-57 residue. Thus the catalytic locus is the result of cooperative action among five groups widely separated in terms of the primary structure of the protein. Note also that the amino acid residues in the binding locus are not in sequence (Fig. 2).

The N-terminal isoleucyl-16 and aspartyl-194 residues are involved in maintaining the proper conformation of the active site through the electrostatic bond formed between the NH_3^+ and COO^- groups of the isoleucyl and the aspartyl residues. Evidence for this include (a) loss of activity of the enzyme when the NH_3^+

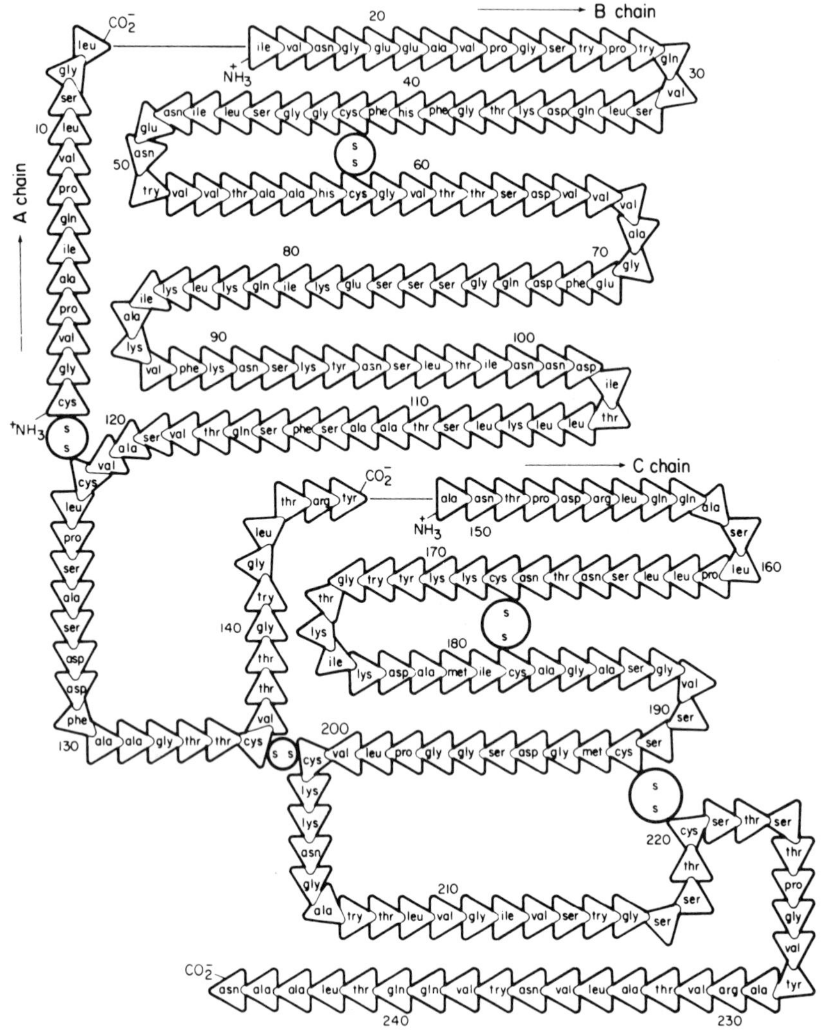

Figure 3 Amino acid sequence of α-chymotrypsin. The numbering system is for chymotrypsinogen A. The discontinuities at residues 13-16 and 146-149 arise from the loss of two dipeptides on activation (see Chapter 20, Section II.C.2). Note the positions of the A (13aa, upper left), B (131aa, upper middle and upper right), and C (198aa, bottom middle and bottom right) chains. (Courtesy of the Medical Research Council, Cambridge, England.)

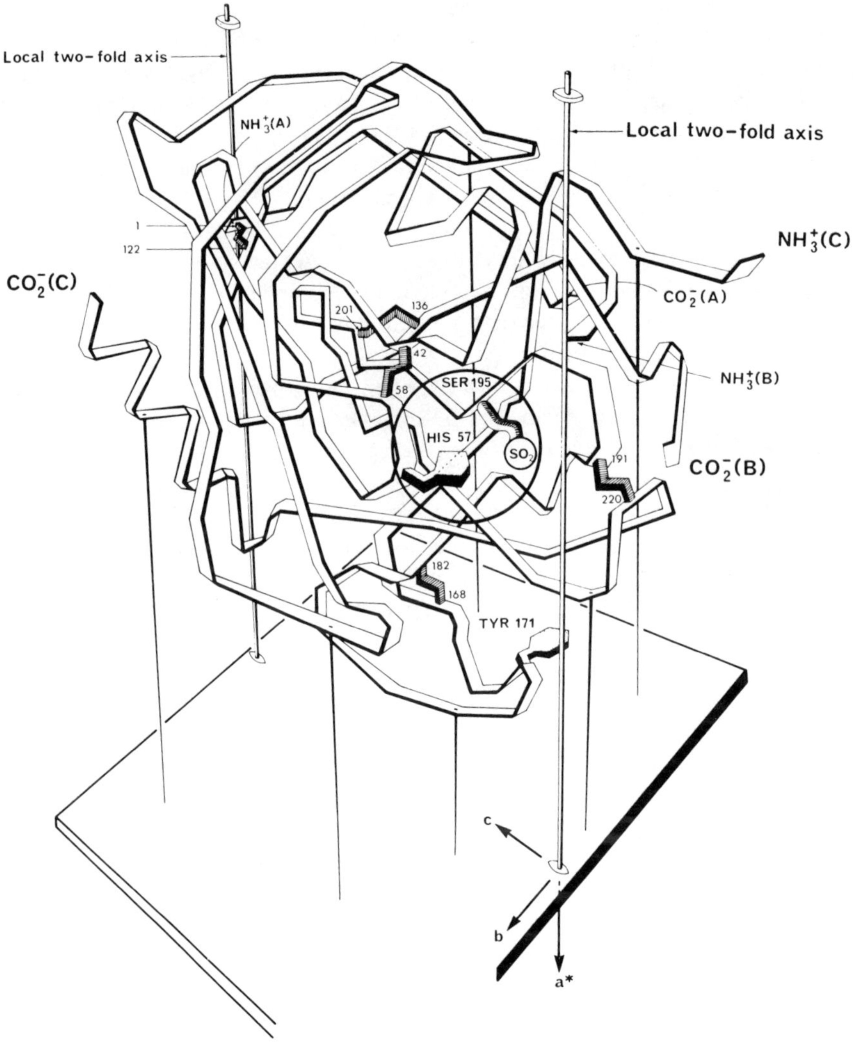

Figure 4 Schematic illustration of secondary and tertiary structures of α-chymotrypsin. The disulfide bridges, side chains of His-57, Tyr-171, and the sulfonated Ser-195 are shown. (Courtesy of the Medical Research Council, Cambridge, England.)

group is selectively acylated; (b) loss of activity of the enzyme when the pH of the medium is adjusted to above pH 9, where NH_3^+ is converted to NH_2; and (c) the spatial arrangements of these two groups as indicated from x-ray diffraction data.

The seryl-195, histidyl-57, and aspartyl-102 residues are involved in catalyzing the hydrolysis of the susceptible bond of the substrate. Evidence for involvement of the seryl-195 residue in activity of the enzyme includes (a) loss of activity of the enzyme when this residue is specifically modified and (b) the finding of covalent attachment of the acyl portion of substrate and pseudosubstrates to this residue. The hydroxyl group of seryl-195 is a peculiarly reactive group since there are 26 other seryl hydroxyl groups in α-chymotrypsin and none of these 26 groups are modified as is seryl-195. The special reactivity of this one seryl hydroxyl group must be due to its special environment.

Evidence for involvement of histidyl-57 in the active site of chymotrypsin includes (a) loss of activity on specific modification of this group, (b) kinetic studies showing the involvement of the group with a pK_a value of 6.5 to 7.0, (c) ΔH_{ion} of 7.5 kcal/mol, (d) the spatial location of this residue with respect to seryl-195 residue as shown by x-ray diffraction studies (see Fig. 4), and (e) by substitution of other amino acid residues by recombinant DNA technology.

Based on substantial evidence, the pathway of hydrolysis of a substrate by α-chymotrypsin can be described by the general equation

$$\mathrm{E} + \mathrm{A} \underset{k_{-1}}{\overset{k_1}{\rightleftharpoons}} \mathrm{E{\cdot}A} \underset{k_{-2}}{\overset{k_2}{\rightleftharpoons}} \underset{\substack{+\\ \mathrm{P_1}}}{\mathrm{EA'}} \underset{k_{-3}}{\overset{k_3,\mathrm{H_2O}}{\rightleftharpoons}} \mathrm{EP_2} \underset{k_{-4}}{\overset{k_4}{\rightleftharpoons}} \mathrm{E} + \mathrm{P_2} \tag{2}$$

E·A and E·P_2 are Michaelis–Menten adsorptive complexes and EA′ is a covalent acylenzyme intermediate. For hydrolysis of the ester substrate acetyl-L-tryptophan ethyl ester, the rate-limiting step is that controlled by k_3 (deacylation), while for the amide substrate acetyl-L-tryptophanamide, the rate-limiting step is that controlled by k_2 (acylation). The pH dependencies of these two steps are shown in Fig. 5(a) and (b). Therefore, both the rates of acylation (k_2) and deacylation (k_3) are dependent on a prototropic group with a pK_a value of 6.7 to 7. This is undoubtedly the imidazole group of histidyl-57. The effect of pH on k_{cat}/K_m is a bell-shaped curve with pK_a values of 6.7 and 9 [Fig. 5(c)] and may be interpreted as an influence of pH on K_m, k_{cat}, or both. The data of Fig. 5(a) and (b) show that the prototropic group with pK_a of 6.7 is involved in k_{cat} (k_{cat} equals k_2 for acetyl-L-tryptophanamide and k_3 for acetyl-L-tryptophan ethyl ester). By difference, the group with a pK of 9 must be involved with K_m. A number of independent studies have shown that the enzyme undergoes a conformational change in the alkaline region [Fig. 5(d)] and that the pK_a of the group involved in this conformational change is 8.5 to 9. As indicated above, this group is the -NH_3^+ of the N-terminal isoleucyl-16 residue, which interacts electrostatically with the carboxyl group of aspartyl-194 [Eq. (3)] to maintain the conformation of the active site. The conformational form at high pH can no longer bind substrate or competitive inhibitor [Fig. 5(d)].

$$\text{Asp194-COO}^- \cdots\cdot \text{H}_3^+\text{N-Ile16} \overset{\text{pK } 8.5}{\rightleftharpoons} \text{Asp194-COO}^- + \text{H}_2\text{N-Ile16} + \text{H}^+ \quad (3)$$

A mechanism of action of chymotrypsin is proposed in Eq. (4) [6,7]. In the

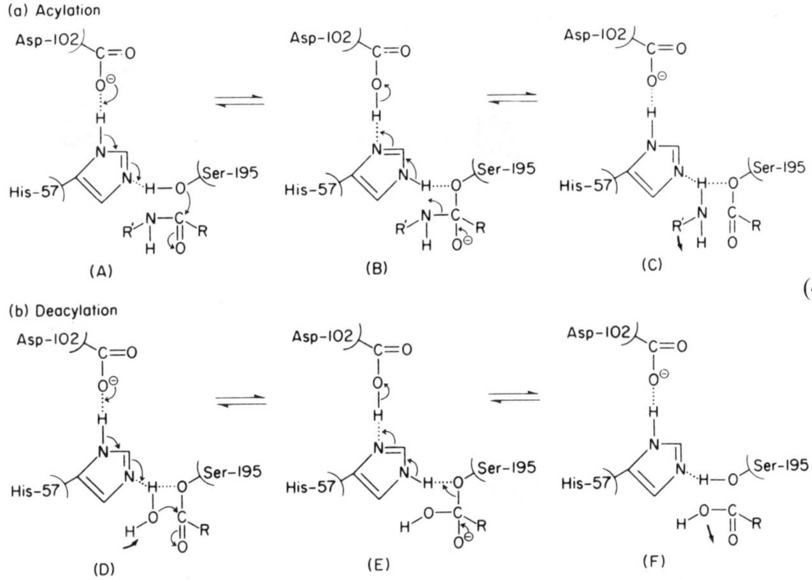

(4)

reaction a stereospecific adsorptive complex is formed between the substrate and enzyme (E·A), structure B. Then the imidazole group of histidyl-57 acts as a general base to stretch the H—O bond of the hydroxyl group of the seryl-195 residue, thus facilitating the nucleophilic attack of the oxygen of the seryl group at the carbonyl of the substrate. This leads to an expulsion of the R′ residue and formation of an acylenzyme (structure C). In the deacylation process, the imidazole group serves as a general base to extract a proton from water to facilitate the attack of the hydroxide ion at the carbonyl group of the acylenzyme (structure D). In the second transition-state intermediate (structure E), imidazole acts as a general acid in deacylation of the acylenzyme. This leads to release of the second product and regeneration of the enzyme (structure F). The action of histidyl-57 as a general base (acylation step) and general acid (deacylation step) is facilitated by hydrogen bonding involving the aspartyl-102 residue. It should be noted that all these changes can take place without movement of any of the functional groups on the enzyme.

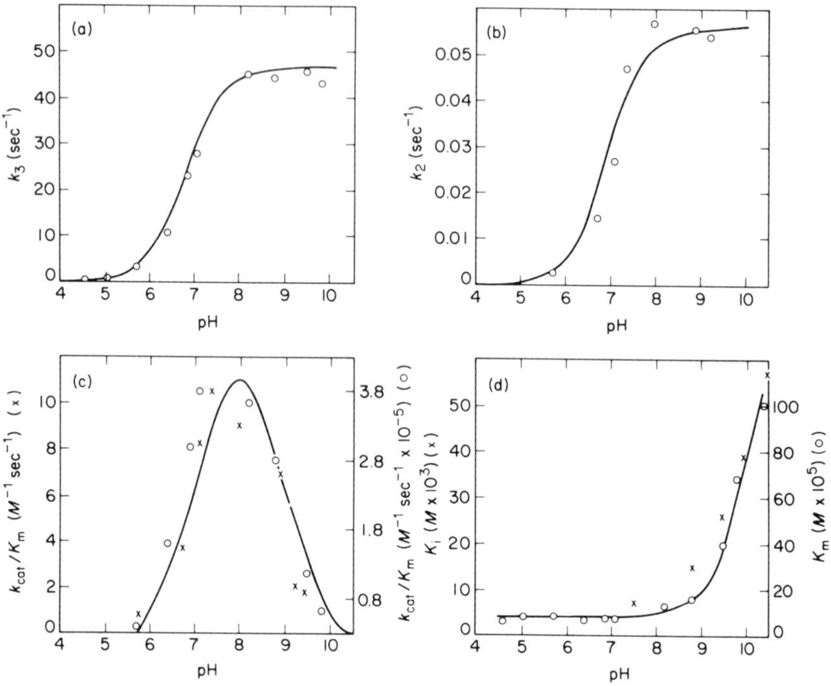

Figure 5 Effect of pH on rate parameters for α-chymotrypsin-catalyzed hydrolysis. (a) k_3–pH profile for α-chymotrypsin-catalyzed hydrolysis of *N*-acetyl-L-tryptophan ester ester. Data from Ref. 4. Theoretical curve drawn for p*K* of 6.77 and k_3(lim) of 46.6 sec^{-1}. (b) k_2–pH profile for α-chymotrypsin-catalyzed hydrolysis of *N*-acetyl-L-tryptophanamide. Data at pH 7.09 and below from Ref. 4, above pH 7.09 from Ref. 5. Theoretical line drawn for pK_a of 6.88 and k_2(lim) of 0.056 sec^{-1}. (c) k_{cat}/K_m–pH profile for α-chymotrypsin-catalyzed hydrolysis. ○, *N*-acetyl-L-tryptophan ethyl ester, data from Ref. 4; ×, *N*-acetyl-L-tryptophanamide, data at pH 7.09 and below from Ref. 4, above pH 7.09 from Ref. 5. Semiempirical curve drawn on basis of pK_{a1} and pK_{a2} of 6.7 and 9.2, respectively. (d) K_m–pH and K_i–pH profiles for α-chymotrypsin. ×, K_i for benzamide; data from Ref. 5. ○, K_m for *N*-acetyl-L-tryptophan ethyl ester, data from Ref. 4.

IV. THE SULFHYDRYL PROTEASES

The sulfhydryl proteases (also called the thiol proteases and cysteine proteases) have in common the ability to hydrolyze the peptide bonds of proteins and their inhibition by sulfhydryl reagents. The group of enzymes includes the higher plant enzymes, papain (EC 3.4.22.2), ficin (EC 3.4.22.3), and bromelain (EC 3.4.22.4), and the microbial enzyme *Streptococcus* protease (*Streptococcus* cysteine proteinase, EC 3.4.22.10). A great many other proteolytic enzymes are also inhibited by sulfhydryl reagents and may eventually be shown to have similar kinetic properties to the best understood of these enzymes, papain.

As with the serine proteases the similarity among the sulfhydryl proteases extends to the amino acid sequence around the essential cysteinyl and histidyl residues of the active site (Table 3). The active site Cys and His residues are at positions 25 and 159, respectively, in papain.

Among this group of enzymes, considerable work has been done on papain (from papaya), ficin (from fig), and bromelain (from pineapple). They have in common a broad pH optimum between pH 6 and 7.5 and their activity appears to be dependent on two prototropic groups with pK_a values near 4 and 8.5. The enzymes are quite heat stable up to 60 to 80°C at neutral pH.

The enzymes have rather broad substrate specificity. Papain and ficin hydrolyze *N*-benzoyl and *N*-carbobenzoxy derivatives of substrates containing L-arginine, L-lysine, glycine, and L-citrulline with about equal efficiency. The positive charge on the arginyl and lysyl residues is not necessary for binding of the substrate to the active site of the enzyme since substrates containing glycyl and L-citrullyl residues are equally well bound. The enzymes specifically recognize the aromatic substituents rather than the amino acid residue as can be shown by using acetyl- or formyl derivatives, which are not hydrolyzed and by using L-phenylalanyl-X derivatives. Therefore, the specificity is for the second amino acid residue from the scissle peptide bond.

In the following discussion, papain will be used as the prototype sulfhydryl protease. The complete amino acid sequence is known for papain (Fig. 6). It is composed of a single polypeptide chain of 212 amino acids and has a molecular weight of 23,900. The tertiary structure, as determined by x-ray diffraction, has been elucidated (Fig. 7). The molecule has a deep cleft running diagonally down the center so that it almost appears to be two molecules. Only three segments of the chain cross this cleft and the essential sulfhydryl group on cysteinyl residue 25 lies just to one side of this cleft (see the center circle). (The essential sulfhydryl group is easy to place in the molecule since it is the only one.) The substrate is bound stereospecifically to groups in the cleft so that the susceptible bond is positioned adjacent to the sulfhydryl group.

Evidence for involvement of the sulfhydryl group in the active site includes (a) loss of activity when the sulfhydryl group is modified or cysteine is replaced by another amino acid, (b) the spectrophotometric observation and isolation of an acylenzyme intermediate that has all the properties of a thiolester, and (c) kinetic studies which indicate involvement of a group with a pK_a value of 8.3 and a ΔH_{ion} value of 5.1 kcal/mol.

A second group involved in the active site of papain is either a carboxyl group, a histidyl group, or both. Based on values of pK_a 4 and ΔH_{ion} 0 kcal/mol, this group would appear to be a carboxyl group. In the x-ray diffraction pattern (Fig. 7) the nearest carboxyl group is 7.5 Å away from the essential sulfhydryl group. This

Table 3 Amino Acid Sequences Around the Essential Sulfhydryl Group and Histidine in Several Sulfhydryl Proteolytic Enzymes

Bromelain, stem	-Asn-Gln-Asp-Pro-Cys-Gly-Ala-Cys*-Trp-
Papain[a]	-Pro-Val-Lys-Asn-Gln-Gly-Ser-Cys-Gly-Ser-Cys*-Trp-
Ficin	-Pro-Ile-Arg-Gln-Gln-Gly-Gln-Cys-Gly-Ser-Cys*-Trp-
Proteinase, streptococcal[b]	-Ser-Phe-Val-Gly-Gln-Ala-Ala-Thr-Gly-His-Cys*-Val-
Bromelain, stem	-His*-Ala-Val-Thr-Ala-Ile-Gly-Tyr-
Papain[a]	-Val-Gly-Pro-Cys-Gly-Asn-Lys-Val-Asp-His*-Ala-Val-Ala-Ala-Val-Gly-Tyr-
Ficin	-Thr-Gly-Pro-Cys-Gly-Thr-Ser-Leu-Asp-His*-Ala-Val-Ala-Leu-
Proteinase, streptococcal[b]	-Gln-Ala-Ala-Thr-Gly-His*-Cys-Val-Ala-Thr-Ala-Thr-

Source: Ref. 8, except as noted.
[a]Cys* is amino acid residue 25 and His* is amino acid residue 159.
[b]Ref. 9.

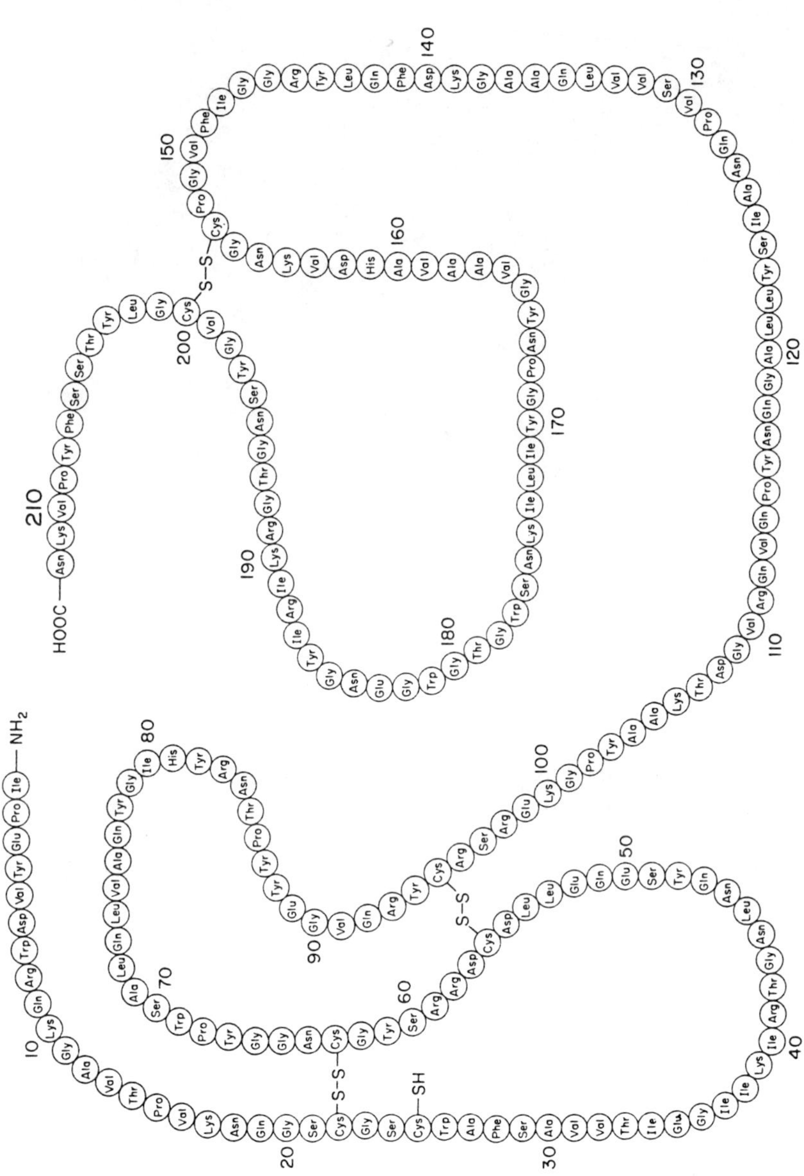

Figure 6 Primary sequence of papain. (From Ref. 10, p. 286, by courtesy of the Biochemical Society. Residue 64 changed from Asp to Asn on the basis of data of Ref. 11.)

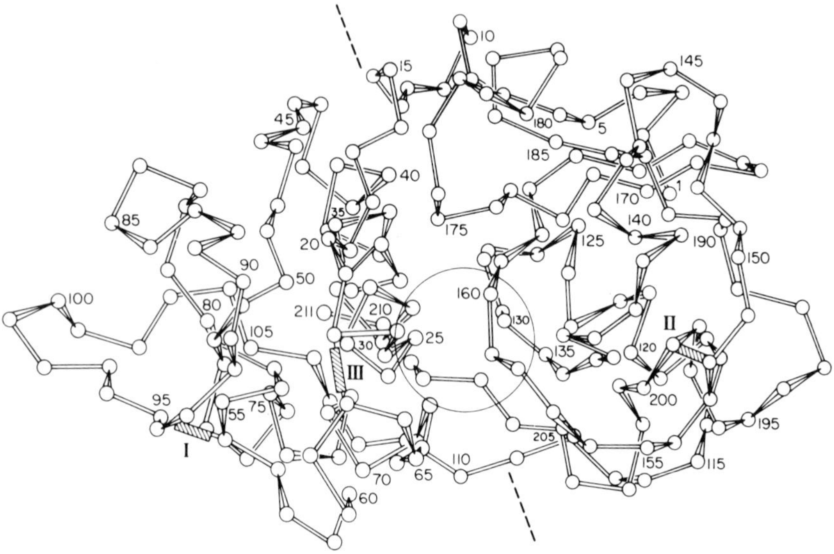

Figure 7 Secondary and tertiary structures of papain as determined by x-ray crystallography. The circles represent the α-carbon atoms of the 212 residues. [From Ref. 12, p. 932, by courtesy of Macmillan (Journals) Ltd.]

distance is too far, by about 3 Å, to permit the carboxyl group (Asp-158) to be a catalytic group unless there is a conformational change in the active site on binding substrate. The carbonyl group of Asn-175 is a better candidate.

An alternative explanation is that the carbonyl group of Asn-175 is in the active site of papain but is hydrogen bonded to the imidazole group of His-159, thereby facilitating involvement of His-159 in the hydrolysis. This is substantiated by (a) the x-ray diffraction pattern of the resting enzyme, in which a histidyl residue (No. 159) is found to be 4.8 Å away from the essential sulfhydryl group, and (b) by chemical modification with the bifunctional reagent, dibromoacetone. This reagent reacts first with the very reactive sulfhydryl group, but then the other bromo group reacts with histidyl-159 [Eq. (5)]. Because of the small size of dibromo-

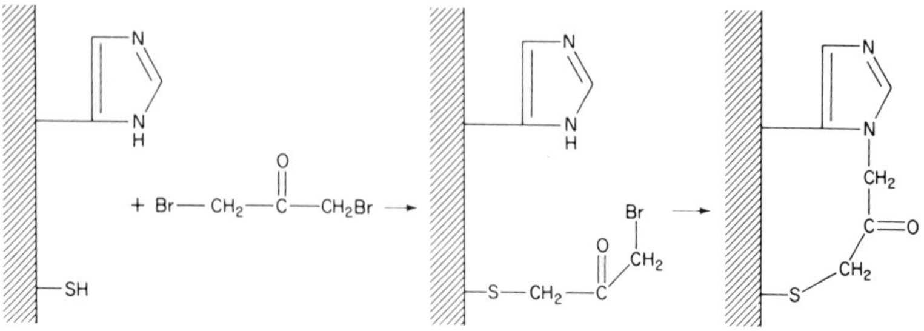

(5)

acetone, the histidyl residue most certainly must be close to the sulfhydryl group. Based on kinetic evidence and on the isolation of a thiolester intermediate, the reaction pathway for papain can be described by the general reaction [Eq. (6)].

$$ESH + R\text{-}\overset{\overset{O}{\|}}{C}\text{-}OR' \underset{k_{-1}}{\overset{k_1}{\rightleftharpoons}} ESH{\cdot}R\overset{\overset{O}{\|}}{C}\text{-}OR' \underset{k_{-2}}{\overset{k_2}{\rightleftharpoons}} E\text{-}S\text{-}\overset{\overset{O}{\|}}{C}\text{-}R + R'OH$$

$$E\text{-}S\text{-}\overset{\overset{O}{\|}}{C}\text{-}R \underset{k_{-3}}{\overset{k_3,\ H_2O}{\rightleftharpoons}} ESH{\cdot}RCOOH \tag{6}$$

$$H^{\oplus} + RCOO^{\ominus} \underset{k_{-5}}{\overset{k_5}{\leftleftharpoons}} ESH + RCOOH \underset{k_{-4}}{\overset{k_4}{\leftleftharpoons}} ESH{\cdot}RCOOH$$

The effects of pH on K_m, k_2, k_3, and k_{cat}/K_m are shown in Fig. 8 for papain. The effect of pH on the various steps in the reaction are consistent with the involvement of two prototropic groups of pK 4 and 8.5 in the acylation reaction and of a single prototropic group of pK 4 in the deacylation of the intermediate thiolester. Binding of substrate to the enzyme appears to be essentially pH independent above pH 4. The detailed mechanism for papain-catalyzed hydrolysis is thought to be identical to that for α-chymotrypsin [Eq. (4)] except that the Ser-195 -OH group is replaced by the Cys-25 -SH group. (His-159 and Asn-175 of papain replace His-57 and Asp-102 of α-chymotrypsin).

V. METAL-CONTAINING PROTEOLYTIC ENZYMES

It is of interest that almost all the proteolytic enzymes that belong to this group are exopeptidases. Carboxypeptidases A (peptidyl-L-amino-acid hydrolase, EC 3.4.17.1) and B (peptidyl-L-lysine hydrolase, EC 3.4.17.2), glycyl-glycine dipeptidase (dipeptide hydrolase, EC 3.4.13.11) from rat muscle, carnosinase (acts on β-alanyl-L-histidine and related compounds; amino-acyl-histidine hydrolase, EC 3.4.13.3), and cytosol aminopeptidase [α-aminoacyl-peptide hydrolase (cytosol), EC 3.4.11.1] require Zn^{2+}, while prolidase (dipeptidase that hydrolyzes dipeptides in which proline or hydroxyproline is the carboxyl terminal residue; aminoacyl-L-proline hydrolase, EC 3.4.13.9), and iminodipeptidase (dipeptidase that hydrolyzes dipeptides in which proline or hydroxyproline is the N-terminal residue; L-prolyl-amino-acid hydrolase, EC 3.4.13.8-9) require Mn^{2+}. These are only a few representatives from the large number of exopeptidases found in animal tissues. All of these metal-containing enzymes are inhibited by metal-chelating agents. Whether the metal ion, which appears to be divalent in all cases, performs a similar function in all these enzymes is not known. Thermolysin (EC 3.4.24.4) is a metalloendopeptidase, with much similarity of active site, specificity, and mechanism to carboxypeptidase A. Other metalloendopeptidases include the collagenases and the hemorrhagic proteinases of snake venoms.

Carboxypeptidase A will be described as a prototype metal-containing proteolytic enzyme. As described in Chapter 20, bovine procarboxypeptidase A is

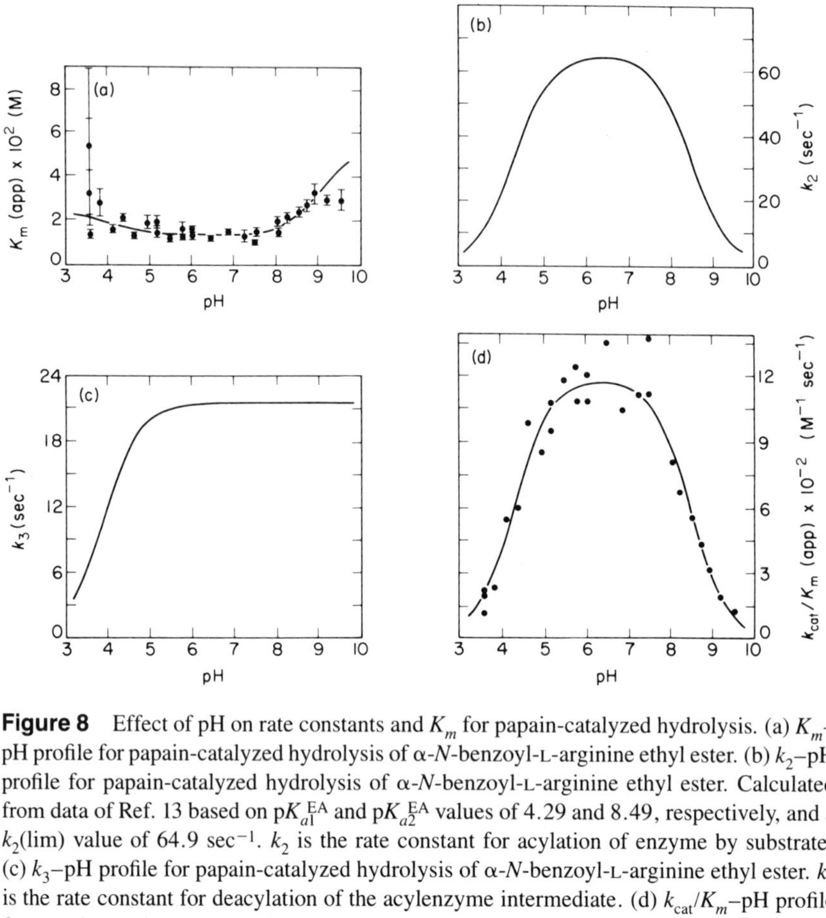

Figure 8 Effect of pH on rate constants and K_m for papain-catalyzed hydrolysis. (a) K_m–pH profile for papain-catalyzed hydrolysis of α-*N*-benzoyl-L-arginine ethyl ester. (b) k_2–pH profile for papain-catalyzed hydrolysis of α-*N*-benzoyl-L-arginine ethyl ester. Calculated from data of Ref. 13 based on pK_{a1}^{EA} and pK_{a2}^{EA} values of 4.29 and 8.49, respectively, and a k_2(lim) value of 64.9 sec^{-1}. k_2 is the rate constant for acylation of enzyme by substrate. (c) k_3–pH profile for papain-catalyzed hydrolysis of α-*N*-benzoyl-L-arginine ethyl ester. k_3 is the rate constant for deacylation of the acylenzyme intermediate. (d) k_{cat}/K_m–pH profile for papain-catalyzed hydrolysis of α-*N*-benzoyl-L-arginine ethyl ester. [(a), (c), and (d) reprinted from Ref. 13, p. 2731, by courtesy of the American Chemical Society.]

excreted by the pancreas as an 80,000 molecular weight compound of three polypeptide chains and is activated to carboxypeptidase A of 34,500 molecular weight through the action of trypsin. The pancreas also produces another zymogen which, on activation, gives carboxypeptidase B. The two enzymes are remarkably similar in the general details of mechanism of action. Both enzymes require that the carboxyl group of the C-terminal amino acid residue be free and they both contain Zn(II) in the active site. However, they are quite different in their substrate specificities. Carboxypeptidase B requires that the C-terminal amino acid residue be arginine or lysine, while carboxypeptidase A will act on any peptide in which the C-terminal amino acid residue is not arginine, lysine, or proline.

Extensive data are available for carboxypeptidase A with respect to its substrate specificity, amino acid sequence, tertiary structure, and mechanism of action. The two substrates normally used in the assay of this enzyme are benzyloxycarbonylglycyl-L-phenylalanine (CGP) and benzoylglycyl-β-L-phenyllactic acid (HPLA; Fig. 9). Chloroacetyl-L-phenylalanine is hydrolyzed by carboxypeptidase A but at a slower rate than is benzyloxycarbonylglycyl-L-phenylalanine.

The Zn(II) can be removed from carboxypeptidase A by a strong chelating agent such as 1,10-phenanthroline. The metal-free protein has no enzymatic activity but it still binds peptide substrates. The Zn(II) can be replaced by several other divalent metal ions, including Co^{2+}, Ni^{2+}, and Pb^{2+}. The enzyme is active but the ratio of HPLA to CGP activity is not that of the original enzyme.

Data obtained by chemical modifications indicate that one (or two) tyrosyl residue(s) and an imidazole residue are involved in the active site of carboxypeptidase A. Evidence for the involvement of an imidazole group also comes from kinetic data which indicate that a group with a pK_a value of 6.5 to 7 is involved in activity of the enzyme. For a peptide substrate such as CGP, the pH–activity profile is a bell-shaped curve with a pH optimum of 7.8.

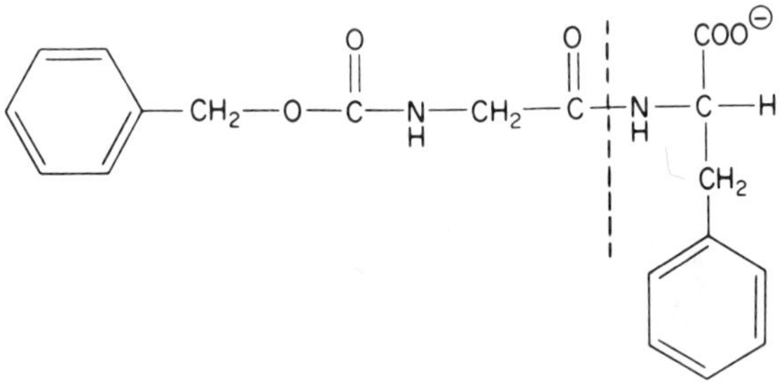

Benzyloxycarbonylglycyl-L-phenylalanine (CGP)

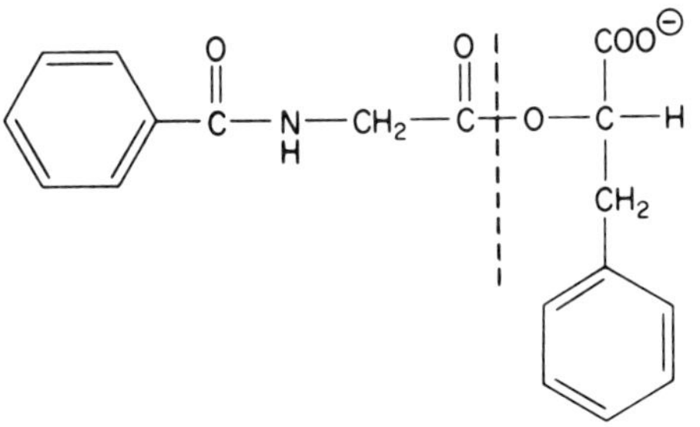

Benzoylglycyl-β-L-phenyl-lactic acid (HPLA)

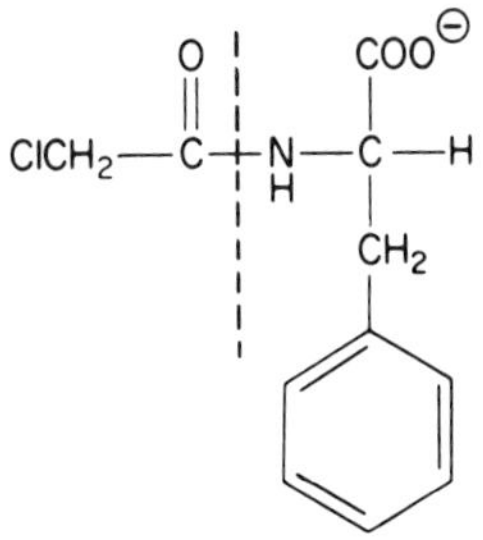

Chloroacetyl L-phenylalanine

Figure 9 Structures of some substrates of carboxypeptidase A.

The complete amino acid sequence of bovine carboxypeptidase A, which contains 300, 305, or 307 amino acid residues (γ, β, and α forms, respectively; see Chapter 20), has been completed [14] and its tertiary structure has been completely elucidated by x-ray diffraction studies (Fig. 10). The molecule is composed of a single polypeptide in which alanyl and asparagine residues are N- and C-terminal, respectively. Some 55 residues in carboxypeptidase A are involved in a twisted β-pleated sheet structure. The sheet structure is made up of four pairs of parallel and three pairs of antiparallel chains which form one side of the pocket which extends into the molecule and forms the active site. The helical regions of the molecule, which involve about 92 amino acids (30%), are mainly α-helices. The regions involving the other 160 amino acid residues are random coils.

The region around the active site is not a cleft, as found in a number of other

Figure 10 Secondary and tertiary structures of carboxypeptidase A as determined by x-ray crystallography. The small circles along the protein chain represent the α-carbon atoms of the 307 residues. The active site is indicated by the circle in the center of the molecule. (From Ref. 15, p. 61, by courtesy of Brookhaven National Laboratory.)

enzymes (see Fig. 7 for papain), but rather, appears to be a hole in the enzyme. It is postulated that the R group of the carboxyl terminal end of a substrate can protrude into the hole. A side view of this hydrophobic hole is shown schematically in Fig. 11 with the hydrophobic side chain of the C-terminal amino acid bound. Other points of binding to the enzyme surface include an electrostatic bond between the negative charge of the essential carboxyl group of the substrate and the positive charge of Arg-145 of the enzyme, hydrogen bonds between the two NH groups of the susceptible and nonsusceptible bonds of the substrate and Tyr-248 of the enzyme, and dipolar bonds between the Zn(II) and the oxygen of the carbonyl group of susceptible bond of substrate and between Glu-270 of the enzyme and the carbonyl carbon of the susceptible bond of substrate. The Zn(II) is coordinated to

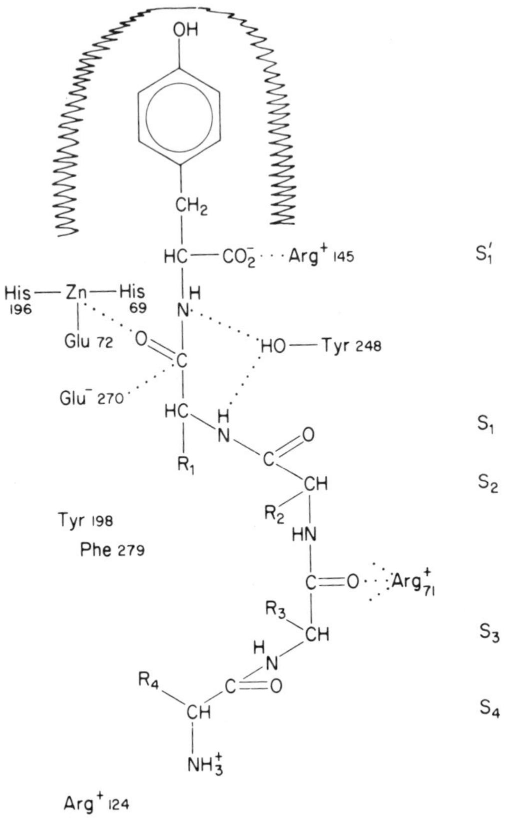

Figure 11 Drawing of a substrate bound in the active site of carboxypeptidase A. The NH and CO groups of the sensitive peptide bond are in a plane perpendicular to that of the drawing. (From Ref. 15, p. 68, by courtesy of Brookhaven National Laboratory.)

His-196, His-69, and Glu-72. Depending on the size and nature of the substrate, it may also interact with Arg-71, Tyr-198, and Phe-279.

Lipscomb et al. [15] proposed the following mechanism for the action of carboxypeptidase A.

> The most probable steps in the mechanisms that we slightly favor at present for polypeptide substrates, not necessarily in the order given, are:
>
> (a) Water is displaced as the C-terminal side-chain of the substrate moves into the (hydrophobic) pocket, the carboxylate group of the substrate saltlinks to Arg-145, the carbonyl groups of S_3 and perhaps of S_4 bind to Arg-71, and side-chains particularly of S_3 but also of S_2 and S_4 associate with Tyr-198 and Phe-279. (Residues S_2, S_3 and S_4 are amino acid residues 1, 2, and 3 removed from the terminal amino acid residue of substrate.) The oxygen of the carbonyl group of the susceptible peptide (bond) probably coordinates to Zn, but the alternative in which this carbonyl is directed away from Zn has also been discussed (in text). Large conformational changes associated with the binding step are the movement of the guanidinium group of Arg-145 by about 2 Å to bind the terminal carboxyl group of the substrate, the coordinated movement of the OH of Tyr-248 by 12 Å to donate a hydrogen bond to the NH of the (strained) susceptible peptide bond and to receive a hydrogen bond from the penultimate peptide bond, and the movement of the carboxylate group of Glu-270 about 2 Å away from the region of the susceptible peptide bond in order to preserve normal van der Waals contacts in the binding step.
>
> (b) Tyr-248 donates H^+ to the NH of the peptide bond, forming a phenoxide ion (of Tyr-248) which is stabilized by the hydrogen bond donated from the penultimate peptide bond, and in the case of the 'Zn-carbonyl' mechanism, Glu-270 forms an anhydride with the carbon of the polarized $C^{\delta+} \cdots O^{\delta-} \cdots$ Zn carbonyl bond of the susceptible peptide (or Glu-270 promotes the attack of water on this carbon atom).
>
> (c) After the peptide bond is split, a water molecule can cleave the anhydride (if formed) in a step which may be facilitated by the nearness of the phenoxide ion of Tyr-248.
>
> (d) The remainder of the polypeptide chain moves away, the conformational changes are reversed, the C-terminal residue leaves the pocket, and the water molecules re-enter. (From Ref. 15, pp. 84–85, by courtesy of Brookhaven National Laboratory.)

Whether or not the details of the mechanism are correct in all aspects, the complexity of the enzyme action is not likely to be any less complicated than described above. For the substrate, benzyloxycarbonylglycyl-L-phenylalanine, k_{cat} is 106 sec^{-1} at pH 7.50 and 25.0°C, which means that one molecule of enzyme converts 106 molecules substrate to product in 1 sec. Truly an enzyme is a very efficient catalyst!

VI. THE ASPARTIC PROTEASES

The group name *aspartic protease* indicates that the carboxyl groups of two aspartic acid residues are the catalytic groups in the active sites. This group of enzymes are also referred to as *carboxyl proteases* and *acid proteases* since the pH optima are around pH 2 to 4. The best studied of this group of enzymes is pepsin. Chymosin (previous name, rennin) is the most useful commercially because of its wide use in cheese production. A large number of microbial proteases also belong in this group. Among this group are three fungal proteases from *Endothia parasitica*, *Mucor pusillus* and *M. miehei*, which are good chymosin substitutes. This group of proteases is inhibited by pepstatin.

Chymosin is the major proteolytic enzyme of the fourth stomach of the suckling calf and is excreted as a zymogen, prochymosin. As the calf grows older and its feed changes from milk to grass and grain, the amount of chymosin produced decreases and the amount of pepsin increases. Chymosin has maximal activity on proteins near pH 3.5. However, in the clotting of milk, the pH of the medium is usually around pH 5.5 to 6.5. The substrate specificity of chymosin has not been studied extensively but it appears to be quite similar to that for pepsin (Fig. 12). Chymosin is most stable in the region around pH 5. At pH 3.5 it undergoes autolysis and above pH 6 it is rapidly denatured. It has a molecular weight of 30,000.

Calf prochymosin has been genetically engineered into microorganisms by several research groups and activated to chymosin. This ensures a continuing commercial supply of this valuable milk-clotting enzyme. This research has permitted replacement of selected amino acid residues by site-directed mutagenesis. Replacement of Tyr-77 with Phe or Val-113 with Phe gives significant increases in milk-clotting activity relative to general proteolysis.

Pepsinogen is excreted by the chief cells of the mucosa of the stomach lining and is converted, by an extensive proteolysis process (see Chapter 20, Section II.B), into pepsin by an autocatalytic process in the presence of a high concentration of HCl in the stomach (~0.01 *N*). Pepsin probably occurs in the gastric

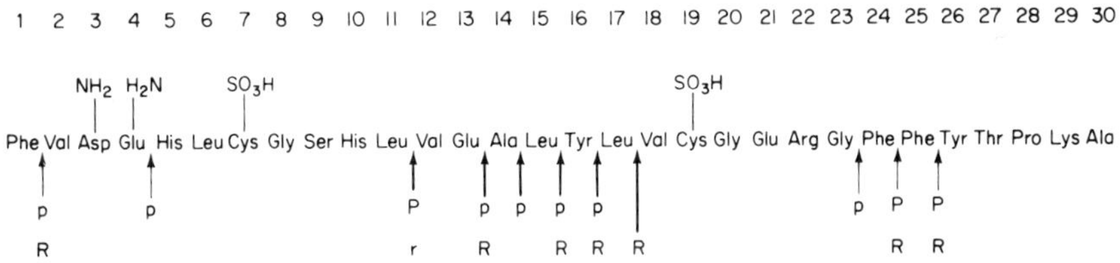

Figure 12 Specificity of hydrolysis of peptide bonds of oxidized B-chain of insulin by pepsin (P) and by chymosin (R). The capital letters represent points of most rapid hydrolysis, the lowercase letters points of slower hydrolysis. (Data for pepsin from Ref. 16, for chymosin from Ref. 17.)

juice of all vertebrates and has been purified from beef, swine, sheep, chicken, and from several fish.

Pepsin is composed of a single polypeptide chain of 321 amino acids and has a molecular weight of 35,500. Its tertiary structure is stabilized in part by three disulfide bridges and a phosphate linkage (see Chapter 20, Section II.B). A phosphate group, attached to the hydroxyl group of a seryl residue, can be removed without loss of enzymatic activity. The enzyme is quite stable from pH 2 to about 5, but above pH 5 it rapidly loses activity due to denaturation.

The pH optimum of pepsin on proteins is about pH 2, but on synthetic substrates it is around pH 4.0. Pepsin has its primary specificity toward the amino acid residue that furnishes the NH group to the susceptible peptide bond [R_2 in Eq. (1)] and prefers that this residue be phenylalanyl, tyrosyl, or tryptophanyl. Pepsin and other aspartic proteases readily catalyze transpeptidation in the presence of suitable reactants.

The relative substrate specificities of pepsin and rennin on the oxidized B-chain of insulin are shown in Fig. 12. There is a preference for hydrolysis at the aromatic amino acid residues. Chymosin is more selective in its action than pepsin since it fails to hydrolyze three peptide bonds hydrolyzed by pepsin.

Based on kinetic studies on synthetic substrates which indicate a bell-shaped pH–activity profile dependent on pK values of ~2 and ~5, and on inactivation of the enzyme with diazo compounds which react with carboxyl groups, it has been concluded that two carboxyl groups are involved in the active site of pepsin. A proposed mechanism, involving nucleophilic catalysis, for the action of pepsin is shown in Eq. (7) [18]. According to this proposal, the free enzyme has two

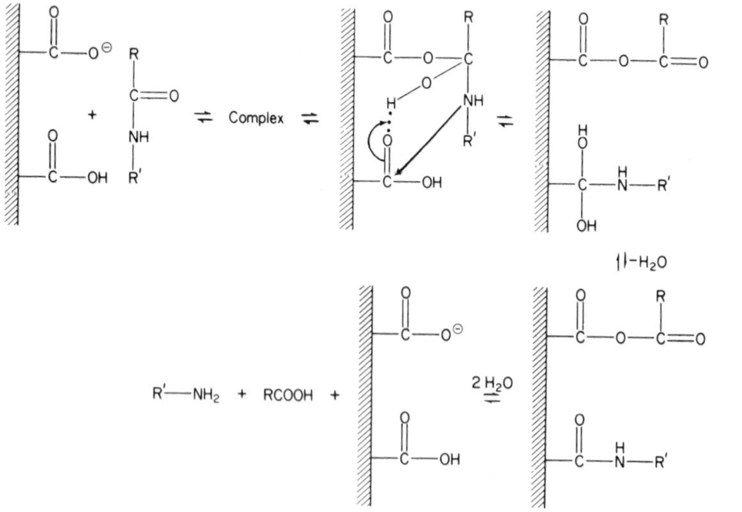

(7)

carboxyl groups, one in the protonated form and the other in the ionized form, in the transforming locus of the active site. The enzyme·substrate adsorptive complex is formed, followed by a nucleophilic attack of the carboxylate group on the carbonyl group of the peptide bond. This leads to formation of a covalent tetrahedral intermediate. The carbonyl oxygen of the protonated carboxyl group then extracts a proton from the hydroxyl group, facilitating an electrophilic attack of the carbonyl carbon on the NH group of the peptide bond. The result is the formation of an aminoacylenzyme intermediate which then reacts with water to give the products of the reaction. There is reasonably good evidence for the aminoenzyme intermediate but no compelling evidence yet for the acylenzyme intermediate. Extensive studies have failed to detect a covalently bound enzyme·substrate intermediate, as required for the mechanism detailed in Eq. (7). This led other researchers to suggest a general acid/general base-catalyzed mechanism [19, Eq. (8)]. The general acid/general base mechanism is favored by

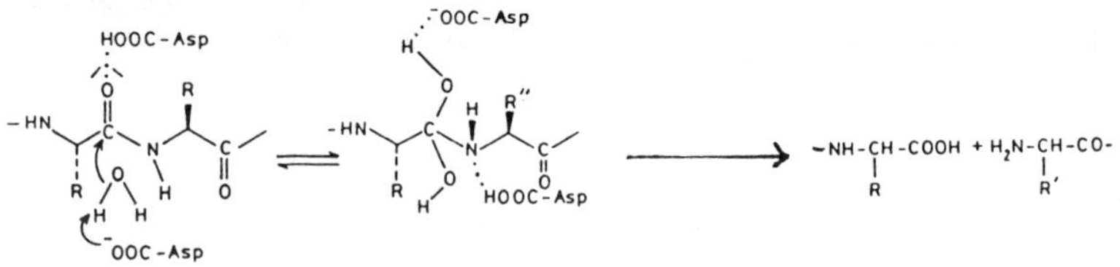

(8)

^{13}C-NMR data. Further research is needed on the mechanism of hydrolysis by the aspartic proteases.

VII. ASSAY OF PROTEOLYTIC ACTIVITY

Many methods are available for determining the activity of proteolytic enzymes that use proteins and synthetic compounds as substrates.

A. Proteins as Substrates

With proteins, a widely used method is the change in the trichloroacetic acid (TCA) solubility of a protein when it is subjected to the action of a proteolytic enzyme. The two most commonly used proteins are casein and acid- or urea-denatured hemoglobin. As a proteolytic enzyme acts on a protein the amount of TCA-soluble peptides produced is proportional to the amount of enzyme and time of action. The amount of TCA-soluble products formed can be determined by measuring the absorbance of the supernatant liquid at 280 nm or by use of color reactions involving tyrosine (Lowry method) or the peptide bonds (biuret method) in the soluble peptides. The method is fast and precise but does not give the number of peptide bonds hydrolyzed. For this reason the method should not be used for detailed fundamental studies.

The number of peptide bonds hydrolyzed in a protein can be determined by use of the ninhydrin reagent, which reacts stoichiometrically with the free amino groups to give a purple color with maximum absorbance at 570 nm. From a standard curve prepared with leucine, the absorbance can be related to the number of peptide bonds hydrolyzed. One problem with this method is the high background color due to preexisting amino groups on the protein. The use of reductive alkylation to eliminate the preexisting amino groups as well as the use of trinitrobenzene sulfonic acid for determining the number of peptide bonds hydrolyzed is recommended [20]. Increased sensitivity is possible using fluorescent-derivatizing reagents, such as fluorescamine.

B. Synthetic Substrates

Use of synthetic substrates is a necessity in studies designated to elucidate the specificity and mechanism of action of an enzyme.

Most synthetic substrates for proteolytic enzymes contain a susceptible ester, amide, or peptide bond. Hydrolysis of a nitrophenyl ester (chromogenic substrate) can be followed at 400 nm (above pH 7) or 340 nm (below pH 7) as shown in Eq. (9). Frequently, hydrolysis of other esters or amides can be followed spectrophoto-

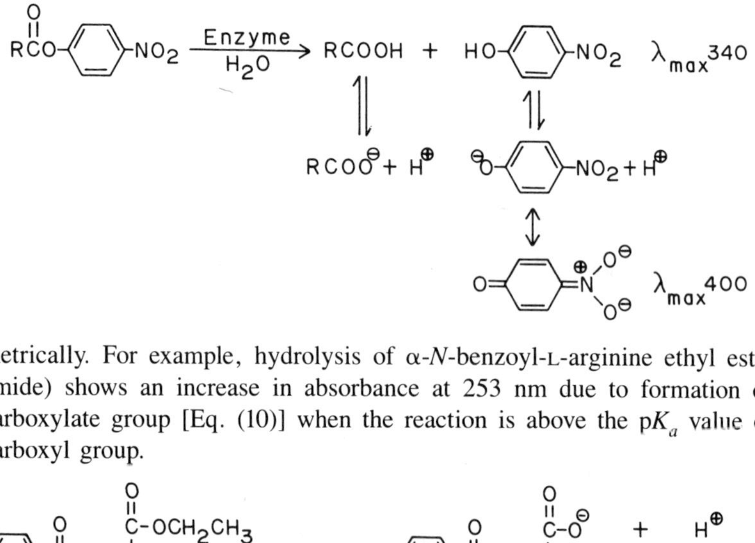

(9)

metrically. For example, hydrolysis of α-*N*-benzoyl-L-arginine ethyl ester (or amide) shows an increase in absorbance at 253 nm due to formation of the carboxylate group [Eq. (10)] when the reaction is above the pK_a value of the carboxyl group.

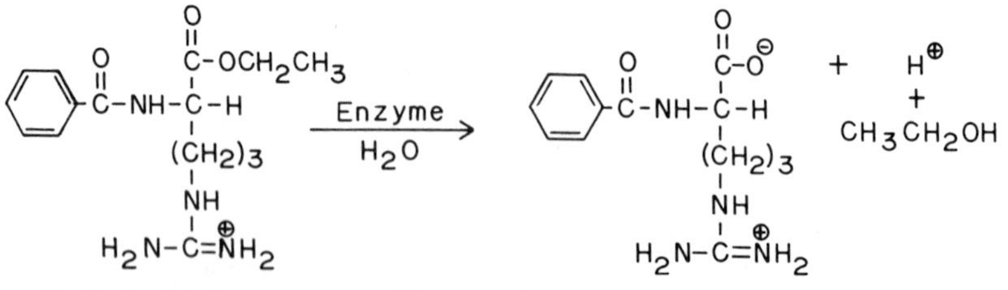

(10)

Hydrolysis of an ester at a pH above the pK value of the carboxyl group formed results in liberation of a proton. This can be followed conveniently in a pH Stat which automatically adds base to neutralize the proton produced. At pH values either below the pK_a of the carboxyl group formed or above the pK_a of the amino group (or ammonia) formed, hydrolysis of an amide substrate can also be followed in a pH Stat.

The rate of hydrolysis of an ester can also be followed conveniently by the alkaline hydroxylamine–$FeCl_3$ reaction by removing aliquots of the reaction periodically and determining how much ester is left. Hydroxylamine reacts with the remaining ester in alkaline solution to give a hydroxamate, which in turn gives a red color with Fe^{3+} in an acid solution.

$$R-\overset{\overset{\displaystyle O}{\|}}{C}-O-R' + H_2N-OH \xrightarrow{OH^-} R-\overset{\overset{\displaystyle O}{\|}}{C}-\overset{H}{N}OH + R'OH \qquad (11)$$

$$\xrightarrow{Fe^{3+},\ H^+} \text{red color read at 540 nm}$$

REVIEW QUESTIONS

1. Experimentally, how would you distinguish between the following enzymes?
   (a) Trypsin and chymotrypsin
   (b) Papain and trypsin
   (c) Acid phosphatase and alkaline phosphatase
   (d) Lipase and a carboxylesterase
   (e) DNase and RNase
   (f) A β-glucosidase and an exo-β-1,4-glucanase
   (g) β-Amylase and glucoamylase
   (h) Pectate lyase and polygalacturonase
   (i) An endo-polygalacturonase and an exo-polygalacturonase
   (j) An endo-polygalacturonase and an endo-polymethylgalacturonase

2. An enzyme, isolated from a certain bacterium, was found to produce rapidly a large amount of reducing groups when allowed to act on a 1% starch solution. Was the enzyme an α- or a β-amylase? How do you know?

3. An enzyme was found to hydrolyze *p*-nitrophenyl-α-D-glucoside.
   (a) Write the structure of this compound.
   (b) Discuss parts of the substrate for which the enzyme may have specificity, and indicate experimentally how you would verify this.
   (c) How would you measure activity of the enzyme on this substrate?

4. Give a method for measuring the activity of each of the enzymes listed below. Give the substrate(s) and product(s) involved. Indicate what you measure and how you would measure it in following the reactions.
   (a) Lipase
   (b) Alkaline phosphatase
   (c) α-Amylase

(d) β-Amylase
(e) Leucine aminopeptidase
(f) Trypsin
(g) Pectinesterase
(h) α-Glucosidase
(i) β-Fructofuranosidase
(j) β-Galactosidase

5. By use of the appropriate substrate structure, point out the features of the molecule that are essential if the enzyme listed is to convert the substrate to product.
(a) Pepsin
(b) Carboxypeptidase A
(c) α-Amylase
(d) Polygalacturonase
(e) Lipase
(f) Phosphatase

6. Give at least one area in which the following enzymes are of importance to the food scientist.
(a) Alkaline phosphatase
(b) Papain
(c) α-Amylase
(d) Glucoamylase
(e) Pectinesterase
(f) β-Fructofuranosidase
(g) Chymosin
(h) Lipase

7. Summarize the general features of a compound that determine whether it will serve as a substrate for a proteolytic enzyme.

8. How would you determine whether the softening of tomatoes on ripening is due to action of
(a) polymethylgalacturonase?
(b) polymethylgalacturonic acid lyase?
(c) both?
(d) neither?

9. How would you perform the following?
(a) Determination of the activity of α-amylase in a solution that also contains β-amylase.
(b) Determination of the activity of chymotrypsin in a solution that also contains trypsin.
(c) Determination of the activity of β-fructofuranosidase in a preparation that also contains α-glucosidase.

(d) Determination of whether an enzyme required α- or β-maltose as substrate since both are present in a solution of maltose.

(e) Determination of whether an enzyme preparation that hydrolyzed *p*-nitrophenyl acetate contains a single or several different types of enzymes.

10. Experimentally, how would you determine whether a proteolytic enzyme was an endo- or an exo-splitting enzyme?

11. Pectic enzymes cause softening of tomatoes, for example. What practical steps might be taken to ensure that these enzymes do not cause excessive softening during storage and initial steps of processing?

REFERENCES

General References

Bender, M. L., and F. J. Kézdy, Mechanism of action of proteolytic enzymes, *Ann. Rev. Biochem.*, **34**, 49 (1965).

Cunningham, L., The structure and mechanism of action of proteolytic enzymes, in *Comprehensive Biochemistry* (M. Florkin and E. H. Stotz, eds.), Vol. 16, Elsevier, Amsterdam, 1965, p. 85.

Desnuelle, P., H. Neurath, and M. Ottesen, eds., *Structure-Function Relationships of Proteolytic Enzymes*, Academic Press, New York, 1970.

A discussion on the structures and functions of proteolytic enzymes, *Phil. Trans. Roy. Soc. London*, **B257**, 63–266 (1970).

Hydrolysis: peptide bonds, in *The Enzymes*, 3rd Ed. (P. D. Boyer, ed.), Vol. III, Academic Press, New York, 1971.

Neurath, H., and R. A. Bradshaw, Evolution of proteolytic function, *Accounts Chem. Res.*, **3**, 249 (1970).

Outtrup, H., and C. O. L. Boyce, Microbial proteinases and biotechnology, in *Microbial Enzymes and Biotechnology*, 2nd Ed. (W. M. Fogarty and C. T. Kelly, eds.), Elsevier Applied Science, New York, 1990, p. 227.

Page, M. I., and A. Williams, eds., *Enzyme Mechanisms*, Royal Society of Chemistry, London, 1987.

Proteolytic enzymes, *Methods in Enzymology*, (S. Colowick and N. O. Kaplan, series eds.; G. E. Perlmann and L. Lorand, volume eds.), Vol. 19, Academic Press, New York, 1970.

Cited References

1. B. S. Hartley, *Ann. Rev. Biochem.*, **29**, 45 (1960).
2. M. O. Dayhoff, *Atlas of Protein Sequence and Structure 1969*, Vol. 4, National Biomedical Research Foundation, Silver Spring, Md., 1969, p. 49.
3. D. Shotton, The molecular architecture of the serine proteinases, in *Proc. Intern. Res. Conf. on Proteinase Inhibitors* (H. Fritz and H. Tschesche, eds.), Walter de Gruyter, Berlin, 1971, p. 47.

4. M. L. Bender, G. E. Clement, F. J. Kézdy, and H. D'A. Heck, *J. Am. Chem. Soc.*, **86**, 3680 (1964).
5. M. L. Bender, M. J. Gibian, and D. J. Whelan, *Proc. Natl. Acad. Sci. USA*, **56**, 833 (1966).
6. L. Polgár and M. L. Bender, *Proc. Natl. Acad. Sci. USA*, **64**, 1335 (1969).
7. D. M. Blow, J. J. Birktoft, and B. S. Hartley, *Nature*, **221**, 337 (1969).
8. S. S. Husain and G. Lowe, *Biochem. J.*, **117**, 341 (1970).
9. T.-Y. Liu, W. H. Stein, S. Moore, and S. D. Elliott, *J. Biol. Chem.*, **240**, 1143 (1965).
10. S. S. Husain and G. Lowe, *Biochem. J.*, **114**, 279 (1969).
11. S. S. Husain and G. Lowe, *Biochem. J.*, **116**, 689 (1970).
12. J. Drenth, J. N. Jansonius, R. Koekoek, H. M. Swen, and B. G. Wolthers, *Nature*, **218**, 929 (1968).
13. J. R. Whitaker and M. L. Bender, *J. Am. Chem. Soc.*, **87**, 2728 (1965).
14. R. A. Bradshaw, L. H. Ericsson, K. A. Walsh, and H. Neurath, *Proc. Natl. Acad. Sci. USA*, **63**, 1389 (1970).
15. W. N. Lipscomb, J. A. Hartsuck, G. N. Reeke, Jr., F. A. Quiocho, P. H. Bethge, M. L. Ludwig, T. A. Steitz, H. Muirhead, and J. C. Coppola, *Brookhaven Symp. Biol.*, **21**, 24 (1969).
16. F. Sanger and H. Tuppy, *Biochem. J.*, **49**, 481 (1951).
17. V. Bang-Jensen, B. Foltmann, and W. Rombauts, *Compt. Rend. Trav. Lab. Carlsberg*, **34**, 326 (1964).
18. E. Zeffren and E. T. Kaiser, *J. Am. Chem. Soc.*, **89**, 4204 (1967).
19. G. Fisher, Acyl group transfer-aspartic proteinases, in *Enzyme Mechanisms* (M. I. Page and A. Williams, eds.), Royal Society of Chemistry, London, 1987, p. 229.
20. Y. Lin, G. E. Means, and R. E. Feeney, *Anal. Biochem.*, **32**, 436 (1969).

20
Ordinary and Limited Proteolysis

Proteolysis is so important to humans in digestion and assimilation of food, in inter- and intracellular maintenance and turnover of proteins, in the transport and activation of preproproteins, in blood coagulation, and in the manufacture of certain foods that we shall devote some space to this subject in our treatment of enzymes. In some cases, such as digestion, broad specificity is desirable, while in conversion of preproproteins and proproteins to biologically active proteins, highly specific proteases are required. Specific information on the enzymatic properties of proteolytic enzymes is provided in Chapter 19.

I. DIGESTION OF PROTEINS

A. Digestion in Gastrointestinal Tract

When food is ingested it is masticated in the mouth and mixed with saliva containing salivary α-amylase. Digestion of carbohydrates begins in the mouth and continues until prevented by acidity in the stomach. In the stomach, food is mixed with secretions from the stomach lining which contain hydrochloric acid and the proteolytic enzyme pepsin. The acidic conditions (pH 1 to 2) denature proteins, thus exposing peptide bonds to hydrolysis by pepsin. Pepsin hydrolyzes most rapidly those peptide bonds in which the amino acid residue furnishing the amino group to the peptide bond is an aromatic amino acid. Pepsin is an endopeptidase, and extensive but not complete hydrolysis of protein occurs in the stomach.

After a period of time determined largely by the nature and size of the meal, food is passed from the stomach into the small intestine, where it is mixed with secretions from the pancreas, the gallbladder, and the intestinal mucosa. All these secretions have a pH near 8, so that the pH of the partially digested food soon becomes alkaline. Action of pepsin soon stops, due to its rapid denaturation by the alkaline conditions after the food enters the small intestine, and because it has no activity at alkaline pH values.

Enzymes capable of digesting all the polymeric compounds found in food are present in the small intestine. These include the proteolytic enzymes chymotrypsin, trypsin, elastase, carboxypeptidases A and B, and leucine aminopeptidase; the nucleic acid–degrading enzymes ribonuclease and deoxyribonuclease; the carbohydrate degrading enzyme α-amylase; and the lipid-degrading enzyme lipase. These enzymes are excreted by the pancreas through the pancreatic duct into the small intestine. Chymotrypsin, trypsin, and elastase are endopeptidases. The carboxypeptidases and leucine aminopeptidase are exopeptidases which hydrolyze peptide bonds at the carboxyl and amino terminal ends of the peptide chain, respectively. Humans cannot digest cellulose, pectins, lignins, and tannins of plants, since they do not have the required enzymes. Cellulose and pectins are of major importance as essential fiber. There may be a small amount of hydrolysis due to microbial enzymes in the large intestine.

The gallbladder excretes bile, but no enzymes, which aids in emulsifying the lipids. Emulsification is essential for action of lipase. Secretions from the intestinal mucosa contain a number of di- and tripeptidases and carbohydrases (such as maltase), which function to hydrolyze the smaller peptide and carbohydrate units to monomeric units.

The monomer units, as well as other somewhat larger fragments in some cases, are actively transported across the intestinal mucosa and eventually pass into the bloodstream, where they are transported to the liver and other organs. There they are resynthesized into the specific proteins of that organ or cell, into carbohydrates and lipids, or stored in reserve pools, or used to supply the energy requirements of the organism.

B. Intracellular Digestion

At the cellular level the amount of polymeric materials is controlled by a balance between the rate of biosynthesis (anabolism) and degradation (catabolism). It would be extremely wasteful in terms of energy if the organism depended only on a balance between the two rates to control the level of polymeric materials. At the cellular level the organism also controls the rate of biosynthesis by feedback inhibition and by repression at the gene level. Much but not all of the catabolic activity takes place inside the lysosomes. These are small subcellular organelles which contain a great variety of hydrolytic enzymes. Enzymes for hydrolysis of all types of polymeric compounds are present, so that the lysosomes have often been described as small stomachs. Therefore, humans have two digestive systems, the large gastrointestinal tract and the lysosomes. The lysosomal enzymes include cathepsins A, B, C, and D, which are involved in degradation of proteins and peptides.

C. Specificity of Proteolytic Enzymes Involved

Each of the proteolytic enzymes hydrolyzes only a few of the total peptide bonds in a protein. Trypsin has specificity toward those peptide bonds in which the carbonyl group is contributed by a lysyl or arginyl residue and chymotrypsin has specificity toward those peptide bonds in which the carbonyl group is furnished by a tyrosyl, phenylalanyl, or tryptophanyl residue. Elastase has specificity toward alanyl residues. Carboxypeptidase A hydrolyzes peptide bonds at the carboxyl terminal end of the peptide chain if the carboxyl terminal residue is not lysine, arginine, or proline. Carboxypeptidase B hydrolyzes peptide bonds at the carboxyl terminal end of the chain if the carboxyl terminal amino acid is lysine or arginine. Leucine aminopeptidase acts from the amino terminal end; it can hydrolyze peptide bonds involving any amino acid residue except proline. There is a prolidase that hydrolyzes peptide bonds containing proline.

D. The Denaturated State of Proteins and Digestion

There is more involved in hydrolysis of a protein by a proteolytic enzyme than just the presence of a peptide bond which satisfies the specificity requirements of that enzyme. The peptide bond must be exposed to the proteolytic enzyme in such a manner that the enzyme can combine with the substrate in a correct stereospecific fashion so that the susceptible bond is positioned correctly with respect to the catalytic groups in the active site of the enzyme involved in hydrolysis.

Generally, a proteolytic enzyme hydrolyzes a peptide bond in a protein molecule only when that portion of the molecule is in a denatured state. A solution of a protein may contain at least two forms, a native form and a reversibly denatured form, and only the reversibly denatured form is subject to proteolysis (Fig. 1). Irreversibly denatured protein is also subject to proteolysis. Any condition, such as pH, binding of ligands, buffer concentration, and ionic strength, which favors the native form over the reversibly denatured form protects a protein against proteolysis. For example, α-amylase contains one Ca^{2+} per molecule (molecular weight 50,000). In the presence of this Ca^{2+} the enzyme is quite resistant to proteolysis, while in the absence of Ca^{2+} it is susceptible to proteolysis. It is not reasonable to believe that a small ion of atomic weight 40 could protect a large protein of molecular weight 50,000 simply by masking all peptide bonds for which a proteolytic enzyme has specificity. The explanation is that the

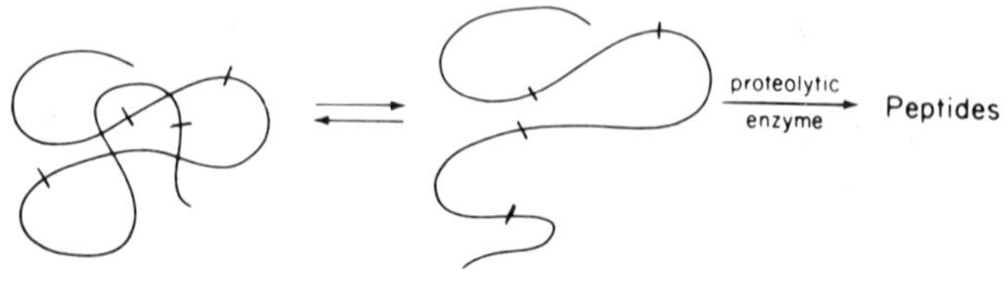

Figure 1 Equilibrium between native and reversibly denatured forms of a protein, showing proteolysis of the reversibly denatured form. The potentially susceptible bonds are shown by the short lines perpendicular to polypeptide chain.

Ca^{2+} maintains the enzyme in the native form and by this means eliminates or greatly slows down the rate of proteolysis.

There are wide differences among proteins in their susceptibility to proteolysis. These differences appear to correlate well with the conformational stability of the native form of the protein in solution as measured by other methods.

II. LIMITED PROTEOLYSIS

In most cases a protein molecule becomes more susceptible to further proteolysis after initial hydrolysis of one or a few peptide bonds. However, there are some very important exceptions to this. In these exceptions, after one, or at most a few peptide bonds have been hydrolyzed, the remainder of the molecule becomes more resistant to proteolysis. We shall refer to this as limited proteolysis, which should not be confused with substrate specificity requirements of the proteolytic enzymes. Two explanations for this behavior are valid. The first holds that hydrolysis of a peptide bond occurs in a particularly susceptible region of the molecule but that similar bonds in the remainder of the protein molecule are resistant to proteolysis. This could be the result of a random coil region between two folded regions or domains. The second explanation is that there are several particularly susceptible bonds but that after hydrolysis of one of these bonds the remainder of the protein molecule changes conformation so that remaining peptide bonds are no longer susceptible to proteolysis.

A large number of examples of limited proteolysis are known. These include conversion of proenzymes (zymogens) to enzymes (pepsinogen to pepsin, chymotrypsinogen to chymotrypsin, trypsinogen to trypsin, prochymosin to chymosin, proelastase to elastase, the conversion of a number of proenzymes to enzymes in the very complex blood clotting mechanism), the conversion of prohormones to hormones, proinsulin to insulin, procollagen to collagen, and removal of signal peptide from preproteins or preproproteins. Also, the coagulation of milk produced by chymosin (older name, rennin) action is due to limited proteolysis of κ-casein.

A. Conversion of Proinsulin to Insulin

One of the most interesting examples of limited proteolysis is the conversion of proinsulin to insulin [1]. The protein hormone insulin has been known for a long time because a deficiency of this hormone is responsible for diabetes mellitus (sugar diabetes). Insulin has a molecular weight of 6000 and is composed of two polypeptide chains held together by two interchain disulfide bonds. The A-chain has 21 amino acid residues and the B-chain has 30 amino acid residues (Fig. 2). Proinsulin, a protein of molecular weight 9100 composed of a single polypeptide chain, is synthesized by the islets of Langerhans, which are attached to the pancreas. Proinsulin is activated to insulin by hydrolysis of two specific peptide bonds by a trypsinlike enzyme either before or after it is excreted into the blood. This conversion is shown schematically in Fig. 2, where peptide bonds Arg-63–Gly-1 and Ala-30–Arg-31 are hydrolyzed. Thirty-three of the original 84 amino

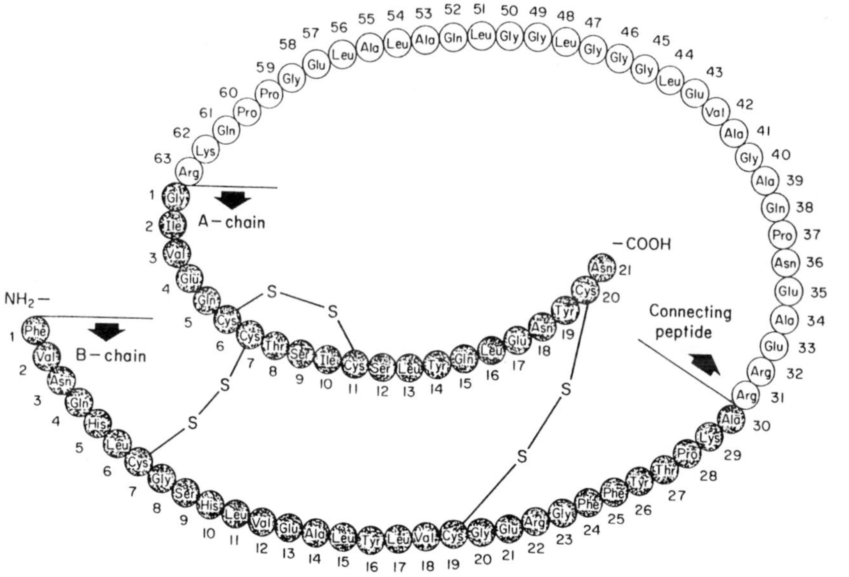

Figure 2 Primary structure of porcine proinsulin. On hydrolysis of peptide bonds Ala-30–Arg-31 and Arg-63–Gly-1 insulin, in dark, is produced. Gln-35 is corrected to Glu-35 according to Ref. 2. (From Ref. 1, p. 166, by courtesy of the American Association for the Advancement of Science.)

acids are removed in conversion of proinsulin to insulin. A person who suffers from diabetes may be deficient in insulin because insufficient proinsulin is produced or because of insufficiency of the activating enzyme. This example is of historical importance also, since it demonstrated to the chemist how nature solved the dilemma of correctly aligning the four —SH groups to properly form the two disulfide bonds. Chemists had previously synthesized the A and B polypeptide chains but were unable to align the —SH groups properly for oxidation to —S—S— bonds.

B. Conversion of Pepsinogen to Pepsin

The inactive protein pepsinogen is produced by the chief cells of the stomach wall and secreted into the stomach, where in the presence of hydrochloric acid, it is converted to the active enzyme pepsin. Conversion of pepsinogen to pepsin is catalyzed by pepsin and results in hydrolysis of a number of peptide fragments from the N-terminal end of pepsinogen (Fig. 3). The large peptide labeled B remains attached by noncovalent bonds to pepsin at higher pH values and is inhibitory. It readily dissociates from pepsin at pH 1 to 2.

Pepsinogen has a molecular weight of 42,000, contains three disulfide bonds, and has an isoelectric point of 3.7. It is quite stable at pH 7 to 9, but at low pH it is rapidly converted to pepsin. Six peptide bonds are hydrolyzed in the initial conversion of pepsinogen to pepsin (Fig. 3). Pepsin continues to act on the last large peptide, B, to produce three more bond cleavages (bonds cleaved indicated by lower case p). Pepsin is produced only when the bond (double-underlined p in Fig. 3) is hydrolyzed. Pepsin is quite resistant to further proteolysis in acid solution, but at pH 5 to 7 it is rapidly inactivated, in contrast to pepsinogen. During conversion of pepsinogen to pepsin the molecular weight decreases from 42,000 to 35,500 and the isoelectric point changes from 3.7 to less than 1. The decrease in isoelectric point is due to loss of 9 of the 10 lysyl residues, 2 of the 3 histidyl residues, and 2 of the 4 arginyl residues during activation.

C. Biosynthesis, Storage, and Activation of Trypsinogen and Chymotrypsinogen

Biosynthesis of trypsinogen and chymotrypsinogen and their conversion to active enzymes are interesting illustrations not only of limited proteolysis but also of mechanisms that an organism uses to protect itself from digestion by proteolytic enzymes. The zymogens are synthesized in large quantities (~10 g total per day) in the pancreas by the usual biosynthetic mechanism involving DNA, messenger RNA, transfer RNA, amino acid–activating enzymes, and ribosomes (see Chapter 2). The zymogens, while still attached to the rough endoplasmic reticulum, are packaged for storage into zymogen granules by participation of the Golgi appa-

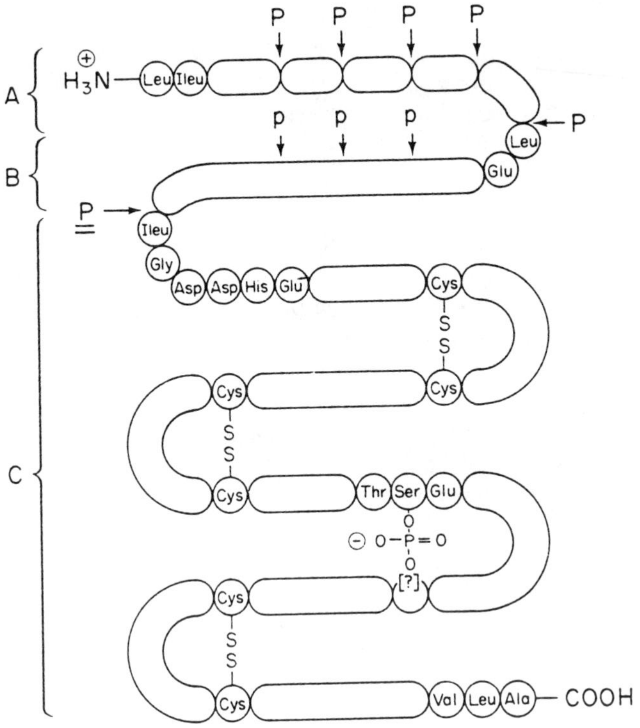

Figure 3 Schematic representation of the structure of pepsinogen and its conversion to pepsin. The major points of hydrolysis are marked with P and result in release of several peptides (A), pepsin inhibitor (B), and pepsin (C). Hydrolysis of the bond P is essential for activation. (From Ref. 3, p. 69, by courtesy of Academic Press.)

ratus. The presence of food in the small intestine stimulates release of these zymogens into the pancreatic duct, where they are transported to the small intestine and activated to the enzymes.

Three mechanisms protect the pancreas against digestion by the potentially high concentrations of proteolytic enzymes: (a) the enzymes are produced in an inactive form (zymogens), (b) the zymogens are stored inside granules to protect the remainder of the organ against hydrolysis if some were inadvertently activated, and (c) the pancreas produces high concentrations of protein inhibitors which react specifically to inactivate any enzyme molecules inadvertently formed. In the stomach and small intestine the organism is protected against self-digestion by the thick coating of complex carbohydrate lining the walls.

1. Conversion of Trypsinogen to Trypsin

Trypsinogen is converted to trypsin by cleavage of a single peptide bond located between Lys-6 and Ile-7. The conversion is catalyzed by trypsin or by enterokinase excreted by the small intestinal mucosa. There is a change in molecular weight from around 23,800 to 23,100 and a change in isoelectric point from 9.3 to 10.5. The increase in isoelectric point is due to the loss of four acidic aspartyl residues and only one basic lysyl residue (Fig. 4). Ser-183 and His-46 residues are involved in the active site of the enzyme. Since the zymogen has no activity, it is reasonable to conclude that these two residues are not properly positioned in the zymogen but that due to a conformational change produced by splitting of the single peptide bond, they are brought into juxtaposition with respect to each other and with the binding site. This, indeed, has been shown by x-ray crystallography.

It is interesting that activation of trypsinogen to trypsin in the presence of Ca^{2+} leads to nearly 100% conversion to active enzyme. In the absence of Ca^{2+} only about 50% active enzyme is obtained. This is because Ca^{2+} stabilizes the structure of trypsin and protects it against further proteolysis.

2. Conversion of Chymotrypsinogen to Chymotrypsin

The bovine pancreas produces two chymotrypsinogens, A and B. The two differ markedly in amino acid content and in electrophoretic mobilities (pI of A is 9.1, pI of B is 5.2). On activation, the two zymogens give two different series of enzymes, but they have similar substrate specificity in that they hydrolyze peptide

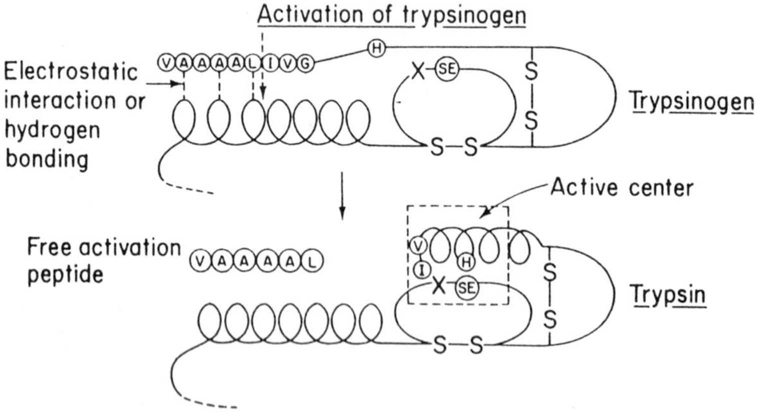

Figure 4 Schematic representation of conversion of bovine trypsinogen to trypsin. The following symbols are used: A, Asp; G, Gly; H, His; I, Ile; SE, Ser; V, Val; X, specificity site. (From Ref. 4, p. 799, by courtesy of the Federation of American Societies for Experimental Biology.)

bonds in which tyrosine, phenylalanine, and tryptophan furnish the carbonyl group. We shall only discuss the conversion of chymotrypsinogen A to enzyme.

Chymotrypsinogen A is a single polypeptide chain of molecular weight 25,000 containing 245 amino acid residues and 5 disulfide bonds. Conversion of chymotrypsinogen A to α-chymotrypsin is a complicated process, although hydrolysis of a single polypeptide bond leads to active enzyme. When chymotrypsinogen A is treated with catalytic amounts of trypsin, the Arg-15–Ile-16 peptide bond is hydrolyzed, and there is a conformational change that brings Ser-195 and His-57 (Fig. 5) in proximity in the active site to give π-chymotrypsin, which is fully active. π-Chymotrypsin consists of two polypeptide chains (Fig. 6) held together by interchain disulfide bonds. π-Chymotrypsin then hydrolyzes peptide bond Leu-13–Ser-14 to release the dipeptide Ser-14–Arg-15 to form δ-chymotrypsin, which is fully active, contains two less amino acids than the zymogen, and is composed of two polypeptide chains. Continued autolysis of the π- and δ-chymotrypsins leads to hydrolysis of peptide bonds Tyr-145–Thr-146 and Asn-147–Ala-148, releasing a second dipeptide, Thr-146–Asn-147, and formation of α-chymotrypsin, the most stable form of chymotrypsin. α-Chymotrypsin is composed of three polypeptide chains held together by two interchain disulfide bonds and three intrachain disulfide bonds. Two neochymotrypsinogens, with no activity, can be formed by action of δ-chymotrypsin on chymotrypsinogen A. The three-dimensional structure of α-chymotrypsin, as determined by x-ray crystallography, is shown in Fig. 5. All the carboxyl- and amino-terminal ends produced in conversion from the zymogen are on the right side of the molecule. The A, B, and C chains contain 13, 131, and 97 amino acid residues, respectively.

D. Conversion of Procarboxypeptidase A to Carboxypeptidase A

Conversion of bovine procarboxypeptidase A to carboxypeptidase A is complex and is further complicated in that genetically controlled variants of the proenzyme are usually present. Bovine procarboxypeptidase A has a molecular weight of 87,000 and consists of three polypeptide chains, I, III, and III (Fig. 7). The use of circles to indicate the three polypeptides does not imply that they are cyclic peptides. During isolation of the proenzyme, a two-polypeptide unit is also obtained, and when the complex is treated with depolymerizing agents such as urea, pH 10.5, or succinic anhydride, individual polypeptide units are obtained.

Let us focus our attention on conversion of the three-polypeptide unit to active enzyme. Treatment of the proenzyme with trypsin leads initially to hydrolysis of one or more peptide bonds in subunit II and its conversion to an enzyme (ATEEase) having activity on acetyl-L-tyrosine ethyl ester (ATEE) similar to that of α-chymotrypsin. Continued treatment with trypsin hydrolyzes a peptide bond in subunit I giving carboxypeptidase A activity. This enzyme is referred to as

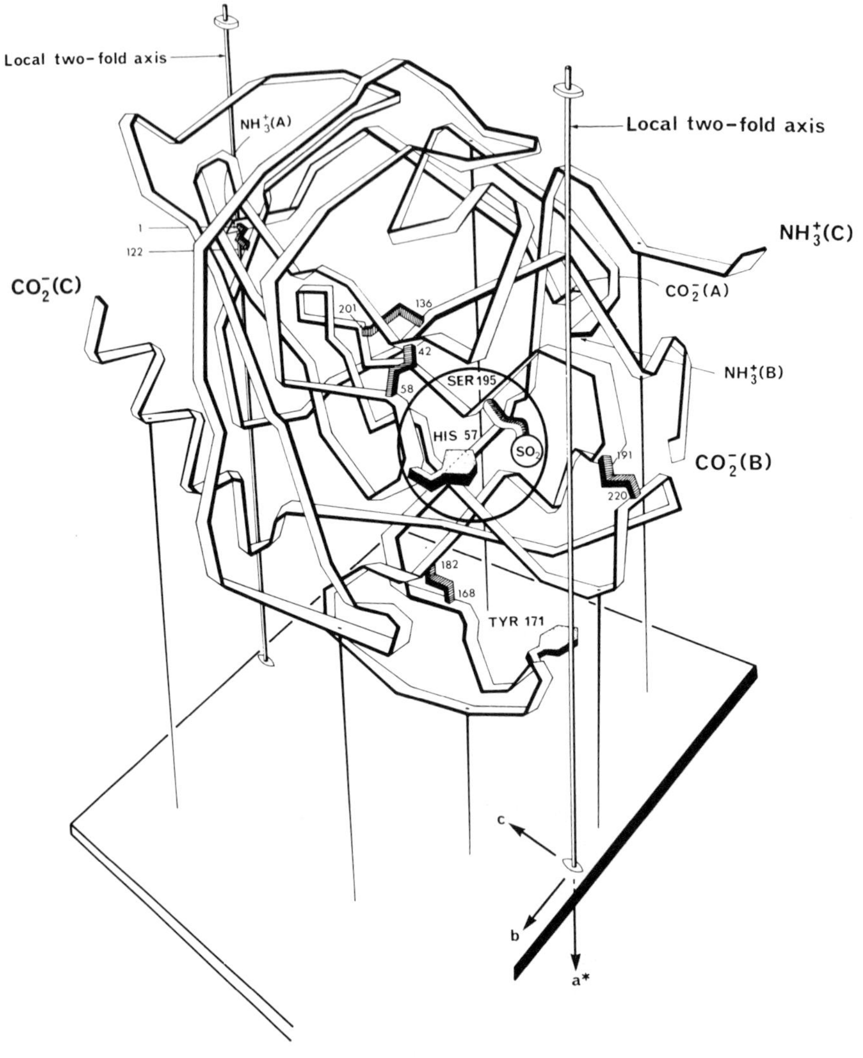

Figure 5 Schematic drawing representing the conformation of the three polypeptide chains of α-chymotrypsin. Location of disulfide bonds (black and striped), amino and carboxyl ends of chains, Tyr-171, His-57, and the sulfonated Ser-195 (SO_2) are shown. (Personal communication, D. M. Blow, Medical Research Council, Cambridge.)

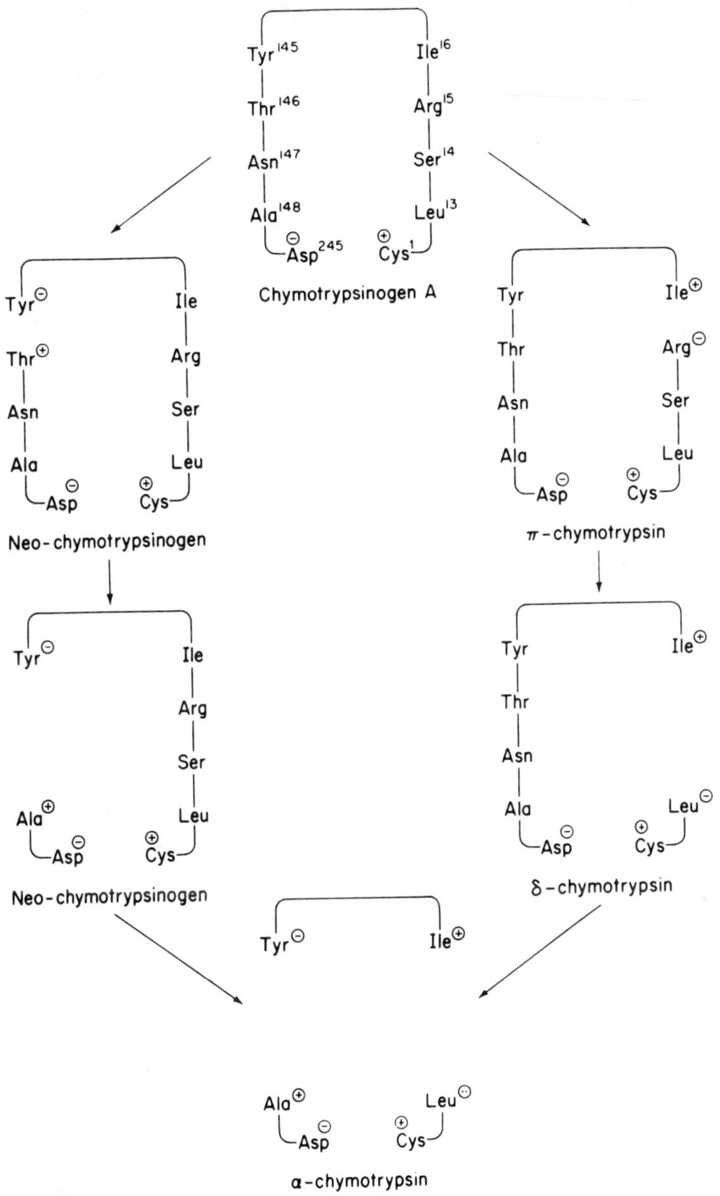

Figure 6 Schematic representation of conversion of bovine chymotrypsinogen A to π-, δ-, and α-chymotrypsins, and to neochymotrypsinogens. Hydrolysis of the Arg-15–Ile-16 bond by trypsin results in activation (From Ref. 5.)

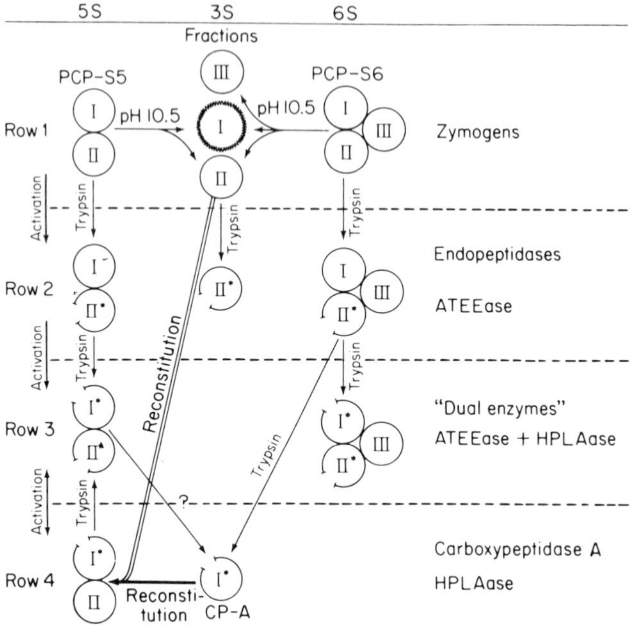

Figure 7 Schematic representation of the conversion of procarboxypeptidase A to carboxypeptidase A. See the text for details. (From Ref. 6, p. 883, by courtesy of the American Chemical Society.)

HPLAase (Fig. 7) since hippuryl-L-β-phenyllactic acid is frequently used in its assay. The unit, still close to 87,000 in molecular weight, has two activities; it has been called a "dual enzyme." More extensive trypsin treatment leads to destruction of both subunits II and III with retention of only activated subunit I. This is carboxypeptidase A, with a molecular weight of 34,500.

During conversion of subunit I to active enzyme, several different peptide bonds at the N-terminal end are hydrolyzed to give at least four different enzymes (Fig. 8), named carboxypeptidases A_α, A_β, A_γ, and A_δ. A_α contains 307 amino acid residues and has an N-terminal alanyl residue. A_β and A_γ contain 305 and 300 residues, respectively, and seryl and asparaginyl are the N-terminal residues, respectively. A_γ and A_δ both have 300 residues with an N-terminal asparaginyl residue; however, they differ in physical properties. In addition, three genetically controlled amino acid replacements occur at positions 179, 228, and 305. These amino acid replacements are genetically linked since only two allotypic forms, Ile-179–Ala-228–Val-305 and Val-179–Glu-228–Leu-305, are formed. The different carboxypeptidases A have been isolated by use of ion-exchange chromatography [8].

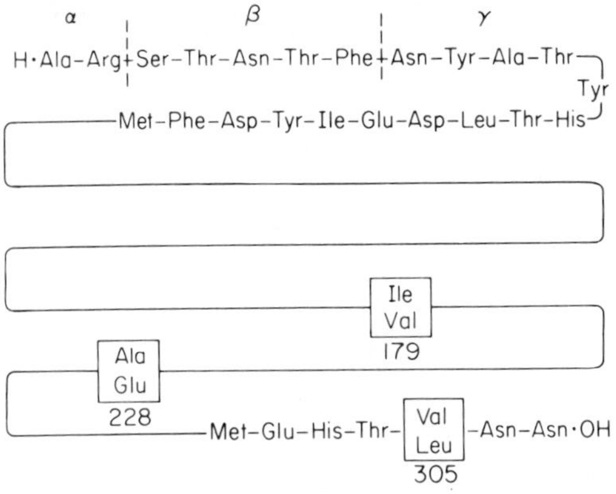

Figure 8 Schematic representation of carboxypeptidases A_α, A_β, and A_γ obtained from activation of subunit I of procarboxypeptidase A. The allotypic amino acid replacements are shown at positions 179, 228, and 305. (From Ref. 7, p. 5030, by courtesy of the American Chemical Society.)

E. Proenzyme Activation in the Blood Clotting Mechanism

The blood clotting mechanism is the most complicated process of limited proteolysis known. Conversion of the soluble protein fibrinogen to the insoluble protein fibrin is the process that actually produces clotting, but a number of steps precede this final step (Fig. 9).

The sequence of events is started by conversion of the Hageman factor into the activated Hageman factor [9]. This step is catalyzed at the surface of a fresh wound. In order, each of the proenzymes is converted to an active proteolytic enzyme which catalyzes the next step. The final proteolytic enzyme in the sequence, thrombin, hydrolyzes four peptide bonds near the N-terminal end of four of the polypeptide chains in fibrinogen to produce fibrin. A strong, irreversible clot is formed through a transpeptidation reaction involving the γ-carboxyl group of glutamyl residues and the amino groups of the newly formed N-terminal residues of fibrin. This reaction is catalyzed by a specific glutamyl transpeptidase.

At least nine specific proteins, each genetically controlled, are involved in the clotting process. With such a complex mechanism, it is a wonder that there are not more diseases associated with abnormalities in the clotting mechanism. In addition, Ca^{2+} and vitamin K are required.

The advantages of such a complicated process include multipoint control and tremendous amplification. To illustrate the latter, suppose that each preceding enzyme molecule causes the activation of 100 molecules of proenzyme to enzyme.

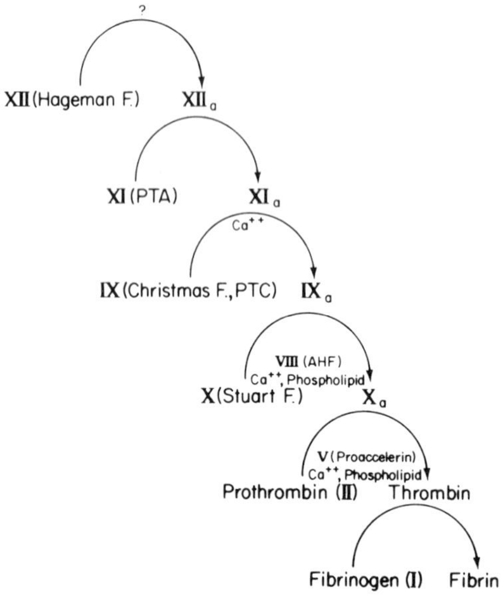

Figure 9 Proposed process of interrelated proenzyme–enzyme transformations leading to clotting of blood. PTA, plasma thromboplastin antecedent; PTC, plasma thromboplastin component, a synonym of Christmas factor; AHF, antihemophilic factor. (From Ref. 9, p. 14, by courtesy of J. and A. Churchill Ltd.)

This would lead to a 10^{12}-fold amplication at the stage of conversion of fibrinogen to fibrin.

F. Clotting of Milk

Clotting of milk for production of cheeses and the ability of chymosin to bring about clot formation have been known for a long time. However, it was only recently that the primary event associated with this process was elucidated for chymosin. The casein micelle is an aggregate of a number of different caseins. The major types of protein molecules involved are α_s-casein, β-casein, and κ-casein. These different caseins contain phosphate groups that complex Ca^{2+}. The micelle is stabilized by κ-casein. Addition of chymosin to milk at pH 5 to 6 results in a rapid hydrolysis of a single phenylalanyl–methionyl bond in κ-casein. The hydrolysis of the peptide bond in κ-casein destroys the stability of the micelle, which then dissociates and reaggregates to form a clot (precipitate) (Fig. 10).

The initial step is the only enzyme-catalyzed step; the others are nonenzymatic. The activation energy for the first step is about 6 kcal/mol, that for the dissociation and/or reaggregation step is about 40 kcal/mol.

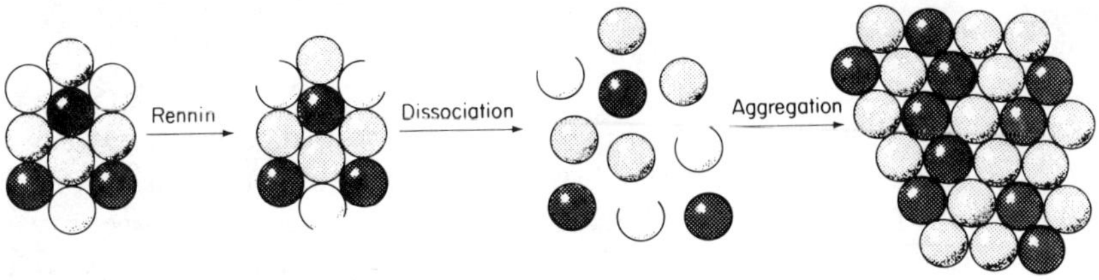

Figure 10 Schematic representation of events in clotting of milk. The α_s-, β-, and κ-caseins are shown by striped, stippled, and white balls, respectively. The involvement of Ca^{2+}, attached to phosphate groups of the caseins, is not shown.

Not all proteolytic enzymes are equally good in clotting milk for cheese production. Most proteolytic enzymes, in addition to causing initial clot formation, continue to act on peptide bonds in all the caseins. This general proteolysis leads to formation of bitter peptides, a weakening of curd strength, and eventual dissolution of the clot if proteolysis is extensive. The acceptability of a proteolytic enzyme for cheese production is determined largely by (a) the ratio of its milk-clotting activity to general proteolysis, and (b) its inability to form bitter peptides even after several months' storage of cheese. The higher the ratio of milk-clotting activity to general proteolysis, the better. Only a few other enzymes even approach the acceptability of chymosin in the production of cheeses. These other enzymes include porcine pepsin and three fungal proteinases from *Endothia parasitica*, *Mucor pusillus*, and *M. miehei*. The gene for calf chymosin has now been genetically engineered into microorganisms, so there should be an ample supply of this enzyme.

G. Other Examples of Limited Proteolysis

The examples given do not cover all aspects of the topic of limited proteolysis. Example of more specific endoproteinases include endoproteinase Glu-C (*Staphylococcus aureus* V8), endoproteinase Lys-C (*Lactobacter enzymogenes*) and endoproteinase Arg-C (mouse submaxillary gland). Enterokinase of the small intestine hydrolyzes a unique Lys–Ile peptide bond in trypsinogen as the zymogen enters the duodenum to give trypsin. Trypsin in turn initiates activation of chymotrypsin, procarboxypeptidases A and B, and prolipase. There are specific proteases that activate prohormones [10] and that remove the signal peptide from preproteins. The pancreatic zymogens are synthesized as proproteins, for example A suggested group name for many of these processing proteases is *convertases* (or *maturases*) [10].

REVIEW QUESTIONS

1. A number of examples of limited proteolysis are described in this chapter. Is there a central theme(s) involved in the process, and if so, what is it?

2. Speculate as to why such a complex blood-clotting mechanism has evolved in humans as indicated by Fig. 9. Wouldn't it be safer to have had only one or two zymogens involved? What is the role of vitamin K in blood clotting? How do you distinguish between the *intrinsic* and *extrinsic* blood-clotting systems?

3. It would seem rather wasteful that the islets of Langerhans synthesize a protein of molecular weight 9100 that is decreased to molecular weight 6000 in formation of active insulin. Speculate as to why human evolution led to this method.

4. Chymosin is produced as the zymogen prochymosin. Read and summarize the important features of conversion of prochymosin to chymosin.

5. It will be noted that the names of certain zymogens are formed by adding the prefix *pro* to the enzyme name, as in *prochymosin* and *prothrombin*, while the names of other zymogens are formed by adding the suffix *ogen*, as in *chymotrypsinogen* and *trypsinogen*. Why the difference in nomenclature? Which is preferred?

6. Experimentally, how might you obtain data to support the hypothesis that the native form of a protein molecule is resistant to proteolysis and that only a reversibly or irreversibly denatured form undergoes proteolysis?

7. The conversions of pepsinogen and trypsinogen to pepsin and trypsin, respectively, are referred to as autocatalytic processes. What does this mean? How would a plot of concentration of enzyme formed versus time look?

8. Based on the assumption that 10 g total of trypsinogen and chymotrypsinogen is produced per day by the pancreas, calculate the amount of energy needed for this synthesis. What is the immediate source of this energy?

REFERENCES

General References

Bond, J. S., and P. Butler, Intracellular proteases, *Ann. Rev. Biochem.*, **56**, 333 (1987).

Desnuelle, P., Chymotrypsin, in *The Enzymes* (P. D. Boyer, H. Lardy, and K. Myrbäck, eds.), Vol. IV, Academic Press, New York, 1960, p. 93.

Desnuelle, P., and M. Rovey, The proteins of the exocrine pancreas, in *Advan. Protein Chem.*, **16**, 139 (1961).

Linderstrøm-Lang, K., Structure and enzymatic break-down of proteins, *Cold Spring Harbor Symp. Quant. Biol.*, **14**, 117 (1950).

Neurath, H., Mechanism of zymogen activation, *Federation Proc.*, **23**, 1 (1964).

Ottesen, M., Induction of biological activity by limited proteolysis, *Ann. Rev. Biochem.*, **36**, 55 (1967).

Pontremoli, S., and E. Melloni, Extralysosomal protein degradation, *Ann. Rev. Biochem.*, **55**, 455 (1986).

Cited References

1. R. E. Chance, R. M. Ellis, and W. W. Bromer, *Science*, **161**, 165 (1968).
2. W. W. Bromer, *BioScience*, **20**, 171 (1970).
3. F. A. Bovey and S. S. Yanari, Pepsin, in *The Enzymes* (P. D. Boyer, H. Lardy, and K. Myrbäck, eds.), Vol. IV, Academic Press, New York, 1960, p. 63.
4. H. Neurath and G. H. Dixon, *Federation Proc.*, **16**, 791 (1957).
5. P. Desnuelle, Chymotrypsin, in *The Enzymes* (P. D. Boyer, H. Lardy, and K. Myrbäck, eds.), Vol. IV, Academic Press, New York, 1960, p. 93.
6. J. R. Brown, M. Yamasaki, and H. Neurath, *Biochemistry*, **2**, 877 (1963).
7. P. H. Pétra and H. Neurath, *Biochemistry*, **8**, 5029 (1969).
8. P. H. Pétra and H. Neurath, *Biochemistry*, **8**, 2466 (1969).
9. E. W. Davie, C. Hougie, and R. L. Lundblad, Mechanism of blood coagulation, in *Recent Advances in Blood Coagulation* (L. Poller, ed.), Churchill, London, 1969, p. 13.
10. D. F. Steiner, S. P. Smeekens, S. Ohagi, and S. J. Chan, *J. Biol. Chem.*, 1992. Minireview Compendium, American Society for Biochemistry and Molecular Biology, Bethesda, Md.

21

Introduction to the Oxidoreductases

The oxidoreductases catalyze the oxidation or reduction of substrates. Oxidation or reduction of a substrate can occur in a number of ways, as shown in Eqs. (1) to (7), where the distinction is made on the basis of electron acceptor (B, O_2, or H_2O_2) and products formed.

A. $AH_2 + B \rightleftharpoons A + BH_2$ (1)

B. $AH_2 + O_2 \rightarrow A + H_2O_2$ (2)

C. $2AH_2 + O_2 \rightarrow 2A + 2H_2O$ (3)

D. $A + H_2O + B \rightarrow AO + BH_2$ (4)

E. $A + H_2O_2 \rightarrow AO + H_2O$ (5)

F. $A + O_2 \rightarrow AO_2$ (6)

G. $A + O_2 + BH_2 \rightarrow AO + B + H_2O$ (7)

By far the largest number of oxidoreductases (Table 1; 611 enzymes total) belong to type A reactions in which enzymes catalyze oxidation of the substrate by removal of hydrogens and/or electrons through participation of an acceptor B, such as NAD^+, $NADP^+$, ferricytochrome, oxidized lipoic acid, and so on. Oxidation of substrates in type B and C reactions also involves removal of hydrogen and/or electrons. However, these reactions differ from those of type A reactions in several important ways.

1. Since molecular oxygen is involved, these reactions are observed only

Table 1 The Oxidoreductases

Type of donor[a]	EC No.	Number of enzymes	Examples
CH—OH group	1:1	244	Glucose oxidase, lactate dehydrogenase
Aldehyde or oxo group	1:2	61	Formate dehydrogenase, aldehyde dehydrogenase
CH—CH group	1:3	54	Succinic dehydrogenase, dihydrouracil dehydrogenase
CH—NH_2 group	1:4	38	D-Amino acid oxidase, alanine dehydrogenase
CH—NH group	1:5	33	Dihydrofolate reductase, sarcosine oxidase
NADH or NADPH	1:6	33	Nitrate reductase, cytochrome b_5 reductase
Other N-containing groups	1:7	11	Nitrate reductase (cytochrome)
Sulfur group	1:8	16	Sulfite reductase (NADPH)
Heme group	1:9	4	Cytochrome-*c* oxidase
Diphenols and related groups	1:10	9	L-Ascorbate oxidase
H_2O_2	1:11	10	Glutathione peroxidase
Hydrogen group	1:12	2	Hydrogen dehydrogenase
Single donors with incorporation of molecular O_2	1:13	44	Lipoxygenase, catechol 1,2-dioxygenase
Paired donor with incorporation of molecular O_2	1:14	36	Lysine hydroxylase, phenol hydrogenase
Superoxide radical as acceptor	1:15	1	Superoxide dismutase
Oxidizing metal ions	1:16	2	Ferroxidase
—CH_2 groups	1:17	5	Pteridine oxidase
Reduced ferredoxin	1:18	6	Nitrogenase
Reduced flavodoxin	1:19	1	Nitrogenase (flavodoxin)
Others	1:97	1	Chlorate reductase

[a]Ref. 1. The EC system is not related to the Eqs. (1)–(7) subdivision.

under aerobic conditions, while type A reactions occur under either anaerobic or aerobic conditions.

2. Type B and C reactions are really three substrate reactions in that a cofactor (not shown) is involved. For example, in the glucose oxidase–catalyzed reaction, the overall reaction is given by

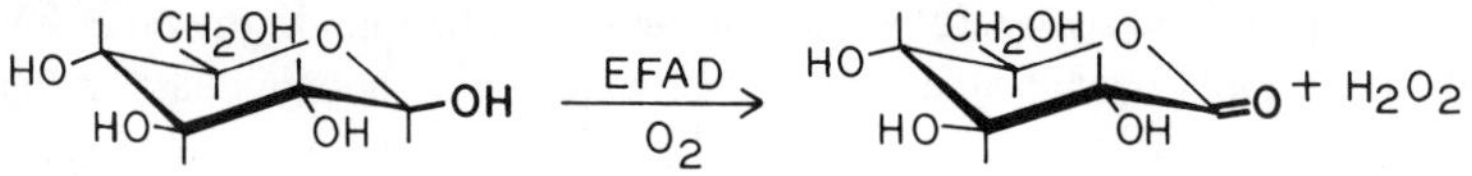

(8)

O_2, β-D-glucose, and FAD are all reactants. In type A reactions two substrates, AH_2 and B, are involved.

3. In type B and C reactions the cofactor is regenerated when the cycle of events is completed [and does not appear in the overall equation as written; Eqs. (2) and (3)], while in type A reactions the cofactor is used up [i.e., it is not regenerated unless a second enzyme system is coupled with it (see Chapter 12, Section I.A)].

4. In general, type A reactions are readily reversible, whereas type B and C reactions are not.

Reactions described by Eqs. (4) to (7) involve oxidation of substrate by incorporation of one or more oxygen atoms into the substrate. The reactions can readily be distinguished on the basis of the source of oxygen atoms, which can be determined by use of ^{18}O-labeled $H_2{}^{18}O$, $H_2{}^{18}O_2$, or $^{18}O_2$. In type D reactions [Eq. (4)], oxygen comes from water and there is need for another reactant, B, such as NAD^+, $NADP^+$, or O_2. In type E reactions, the oxygen incorporated into the substrate comes from H_2O_2. Typical enzymes of this type of reaction are peroxidase and catalase.

Type F and G reactions are quite distinctive from those described in Eqs. (1) to (5) since the oxygen incorporated into the substrate comes from O_2. In type F reactions both atoms of O_2 are incorporated into the substrate, while in type G reactions one atom is incorporated into the substrate and the other oxygen atom ends up in water as a result of participation of a reduced cofactor, BH_2, such as NADH or NADPH. In the latter case, use of $^{18}O_2$ gives both $A^{18}O$ and $H_2{}^{18}O$ as products.

The enzymes that catalyze reactions of type F and G are called "oxygenases" since the reactions are similar to those known to occur by chemical and photochemical processes. According to this proposal, the oxygenases would be subdivided into two subclasses, the *dioxygenases* (type F) and the *monooxygenases* (type G). Examples of dioxygenases include catechol 1,2-oxygenase [catechol:oxygen 1,2-oxidoreductase; EC 1.13.1.1; Eq. (9)] and lipoxygenase (Chapter 27).

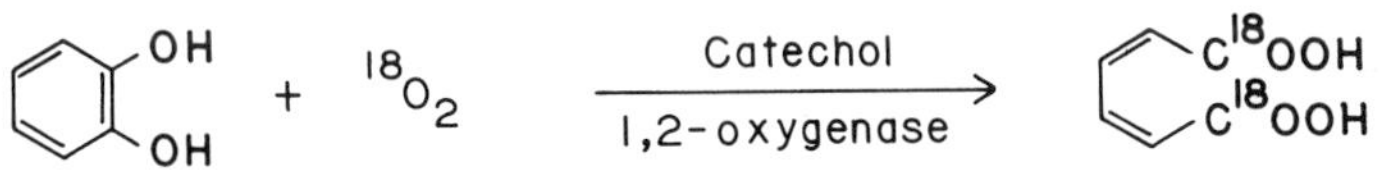

(9)

Catechol *cis,cis*-Muconic acid

The monooxygenases are further subdivided into two groups, *internal monooxygenases* and *external monooxygenases*, depending on whether BH_2 has to be added to the reaction mixture in order that there be catalysis. Those polyphenol oxidases that catalyze both hydroxylation of monophenols and dehydrogenation of *o*-diphenols are examples of internal monooxygenases. The required BH_2 (an *o*-diphenol) is produced by hydroxylation of the monophenol. On the other hand, phenylalanine hydroxylase is an example of an external monooxygenase in that it requires that tetrahydrobiopterin (THBP) be added to the system, which is oxidized to dihydrobiopterin [DHBP; Eq. (10)].

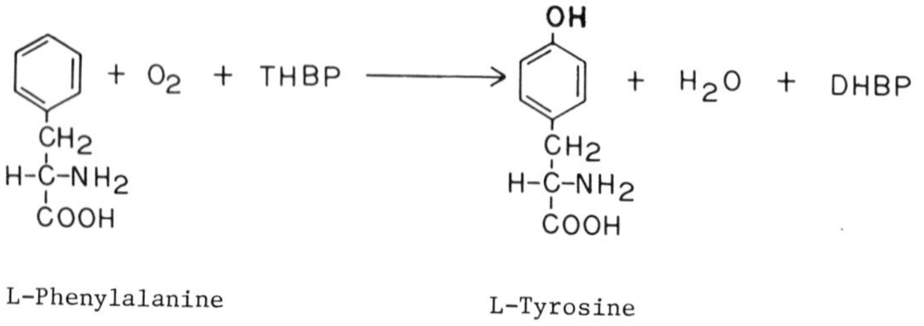

(10)

In the following chapters we have selected lactate dehydrogenase, glucose oxidase, polyphenol oxidase (*o*-diphenol activity), xanthine oxidase, catalase and peroxidase, lipoxygenase, and polyphenol oxidase (hydroxylation reaction) to illustrate type A to G reactions, respectively, as well as to emphasize the general features of enzymes described prior to Chapter 14.

REFERENCES

General References

Boyer, P. D., H. Lardy, and K. Myrbäck, eds., *The Enzymes*, Vols. 7 and 8, Academic Press, New York, 1963.

Crueger, A., Glucose transforming enzymes, in *Microbial Enzymes and Biotechnology*, 2nd Ed. (W. M. Fogarty and C. T. Kelly, eds.), Elsevier Applied Science, New York, 1990.

Hayaishi, O., in *The Enzymes* (P. D. Boyer, H. Lardy, and K. Myrbäck, eds.), Vol. 8, Academic Press, New York, 1963, p. 353.

Hayaishi, O., in *Biological Oxidations* (T. P. Singer, ed.), Interscience, New York, 1968, p. 581.

Hayaishi, O., *Ann. Rev. Biochem.*, **38**, 21 (1969).

Hayaishi, O., *Ann. N.Y. Acad. Sci.*, **158**(1), 318 (1969).

Hayaishi, O., and M. Nozaki, *Science*, **164**, 389 (1969).

Mason, H. S., *Ann. Rev. Biochem.*, **34**, 595 (1965).

Page, M. I., and A. Williams, eds., *Enzyme Mechanisms*, Royal Society of Chemistry, London, 1987.

Robinson, D. S., and N. A. M. Eskin, eds., *Oxidative Enzymes in Foods*, Elsevier Applied Science, New York, 1991.

Cited Reference

1. *Enzyme Nomenclature*, recommendations (1984) of the Nomenclature Committee of the International Union of Biochemistry, published for the International Union of Biochemistry, Academic Press, San Diego, Calif., 1984.

22

Lactate Dehydrogenase

Lactate dehydrogenase [(S)-L-lactate:NAD$^+$ oxidoreductase, EC 1.1.1.27] is one of the key enzymes in the glycolytic system and is ubiquitous to all animal cells. The presence of the enzyme in animal cells was described by Thunberg in 1920 [1].

I. REACTION CATALYZED

The reaction catalyzed in both directions by the enzyme is shown in Eq. (1). Enzyme activity can be determined by starting either with L(+)-lactate and NAD$^+$

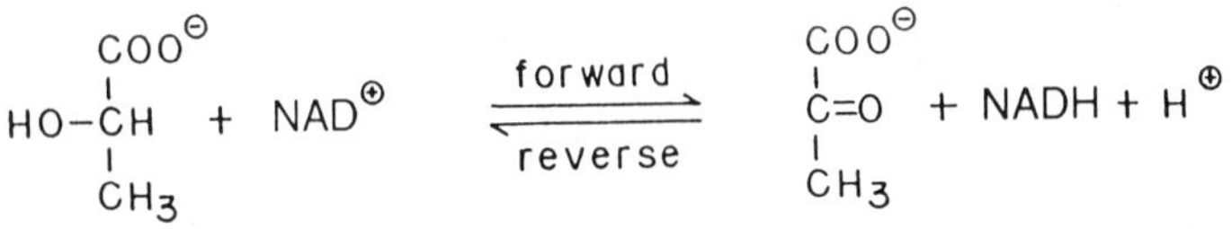

(1)

L(+)-Lactate Pyruvate

or with pyruvate and NADH. The method of choice is to follow the rate of change in absorbance at 340 nm due to the formation or disappearance of NADH. The assay is a particularly sensitive one since $\Delta\epsilon_m = 6.22 \times 10^3\ M^{-1}\ \text{cm}^{-1}$ for NADH versus NAD$^+$.

Exercise 1. Calculate the molar concentration of NADH that will give an absorbance of 0.450.

The relative concentrations of lactate and pyruvate and NAD$^+$ and NADH at equilibrium are dependent on the pH value at which the reaction is performed. At both pH 7 and 10 the forward reaction [as written in Eq. (1)] is favored. The equilibrium constant for the reaction is given by

$$K_{eq} = \frac{(\text{pyruvate})(\text{NADH})(\text{H}^+)}{(\text{lactate})(\text{NAD}^+)} \tag{2}$$

K_{eq} has been determined to be 2.76×10^{-12} M at 25°C [2]. The ratio of lactate to pyruvate at equilibrium can be calculated. For example, when reactions are performed at pH 6.0 and 10.0 at 25°C and with initial concentrations of 1.0×10^{-2} M lactate and 1×10^{-3} M NAD^+, the ratio of lactate to pyruvate is 1800 and 24, respectively.

Exercise 2. Calculate the ratio of lactate to pyruvate at equilibrium under the conditions used above except that the reaction is at pH 8.0.

II. SUBSTRATE SPECIFICITY

Most lactate dehydrogenases are specific for L(+)-lactate. D-(−)-Lactate is not oxidized nor is it inhibitory. In some organisms, such as *Lactobacillus plantarium*, there are two lactate dehydrogenases, one specific for L(+)-lactate and the other specific for D(−)-lactate (D-lactate dehydrogenase; D-lactate:NAD oxidoreductase, EC 1.1.1.28). Pyruvate serves as the best substrate for the reverse reaction, although α-keto acids from C_4 to C_9 and α,γ-diketo acids from C_5 to C_{11} are reduced slowly by lactate dehydrogenase. The α,γ-diketo acids are reduced at the α-position only.

III. COENZYME SPECIFICITY

NAD^+ is some 170 times more effective than $NADP^+$ as a coenzyme in the oxidation of lactate, and NADH is on the order of 200 times more effective than NADPH as a coenzyme in the reduction of pyruvate by muscle lactate dehydrogenase. A number of analogs of NAD^+ will serve as coenzymes in the reaction; however, the rate of the reaction will be changed.

The dehydrogenases are stereospecific in their addition and removal of hydrogen from the C(4) portion of the nicotinamide ring of the coenzyme (Chapter 12, Section II.A). Most of the lactate dehydrogenases for which the stereochemistry has been elucidated belong to the A-group; that is, they add and remove hydrogen stereospecifically from the A-position of C(4) of the nicotinamide ring of NAD^+.

IV. LACTATE DEHYDROGENASE AS A PROTEIN

Lactate dehydrogenases from many animals and from different organs of the same animal have been purified and a number of their properties have been studied. One of the major reasons for this intensive effort is that lactate dehydrogenase occurs in multiple molecular forms (isozymes) and the relative amounts of these forms serve as a diagnostic tool for analyzing for several diseased states.

The molecular weights of lactate dehydrogenases isolated from a great variety of animals and organs are invariant at 150,000. All lactate dehydrogenases are composed of four polypeptide chains of weight 37,500, and only the tetramer is enzymatically active.

A. Multiple Molecular Forms (Isozymes)

Five multiple molecular forms of lactate dehydrogenase are found in most tissues and the amount of each of these forms varies from one tissue to the other, even in the normal organism (Table 1). Heart muscle is particularly rich in the H_4 form, while skeletal muscle contains mostly the M_4 form. Extensive studies have shown that only two types of polypeptide chains are found in most lactate dehydrogenases and that the biosynthesis of each chain is under the control of a separate gene. These two types of polypeptide chains are designed H and M and the five isozymes are the result of the tetrameric molecule containing zero to four polypeptide chains of each type (Table 1). In many tissues, the relative amounts of isozymes are those expected from a random combination mechanism based on the relative amounts of the two types of polypeptide chains present in a cell or tissue.

B. Differences and Similarities Among the Multiple Molecular Forms

The two types of polypeptide chains have different properties which are carried over to the tetramers formed from their combination. They are different in charge density, which permits their ready separation by electrophoresis (Fig. 1) or by ion-

Table 1 Approximate Concentrations of Lactate Dehydrogenase Isozymes in Three Organs of the Rat

	Tissue concentration (pmol/g)[b]		
Isozyme[a]	Renal medulla	Heart muscle	Skeletal muscle
H_4	750	2340	52.1
H_3M	640	970	50.0
H_2M_2	337	532	31.5
HM_3	383	77.0	214
M_4	630	39.0	1720

Source: Ref. 3, p. 63, by courtesy of the American Association for Advancement of Science.

[a]H_4 indicates there are four H-type polypeptide chains in the active molecule, H_3M indicates three H type and one M type, and so on.

[b]A picomole (pmol) is 10^{-12} mol.

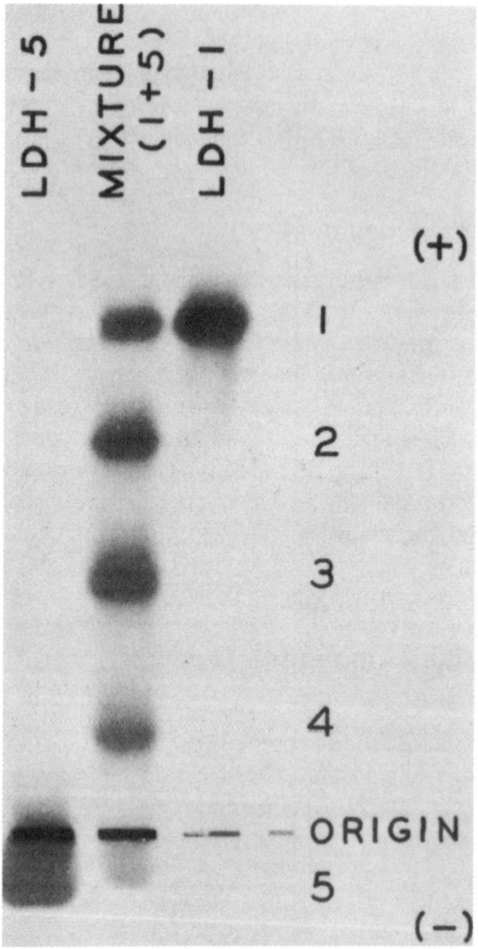

Figure 1 Electrophoretic separation of isozymes of lactate dehydrogenase. Preparations M_4 and H_4 were combined in equal quantities, frozen in 0.1 *M* sodium phosphate buffer, pH 7.0, containing 1.0 *M* NaCl, and thawed. The electrophoretic results on the mixture are shown in the center. (From Ref. 4, p. 1329, by courtesy of the American Association for the Advancement of Science.)

exchange chromatography. This is a result of their difference in amino acid composition [5]. Amino acid composition alone does not indicate the extent of the differences in primary sequence between the two types of polypeptide chains. These differences can readily be determined by comparing the number and common identity of tryptic peptides from each type [6] by separation on cellulose paper by high-voltage electrophoresis–chromatography or by high-performance liquid chromatography. Both types give the same number of tryptic peptides (Table 2). Sixteen of the 36 peptides are different between chicken H_4 and chicken M_4, while only 1 to 2 of the 36 peptides are different between chicken H_4 and turkey H_4. The differences between chicken H_4 and beef H_4 are larger than those between chicken H_4 and turkey H_4, as might be expected on the basis of evolution.

Although the H_4 and M_4 isozymes differ in amino acid composition and in amino acid sequence, the amino acid sequence around the essential sulfhydryl group is nearly identical among all the enzymes examined. An example is that of the chicken H_4 and M_4 isozymes (Table 3). The only change in the 12-amino acid sequence is a homologous exchange of threonine and serine. In the evolutionary process the areas essential for activity of the enzyme have been conserved. The sequences of amino acids around the active sites of triose phosphate dehydrogenase and horse liver and yeast alcohol dehydrogenases are different from those of lactate dehydrogenase. The secondary and tertiary structures of H and M polypeptide chains must be essentially identical since hybrid tetramers which are fully active can be formed from recombination of the two polypeptide chains from different organs as well as different organisms [5].

The isozymes of lactate dehydrogenase differ in stability to denaturation by heat and urea. For example, chicken H_4 is completely stable at 60°C for 10 min,

Table 2 Fingerprint Patterns of H_4 and M_4 Lactate Dehydrogenases

Lactate dehydrogenase type	Number of ninhydrin-positive spots	Number of peptides[a]	
		Identical	Different
Chicken H_4	34–38		
Chicken M_4	35–38		
Turkey H_4	33–35		
Beef H_4	33–35		
Chicken H_4 + chicken M_4	50–53	20	16
Chicken H_4 + turkey H_4	35–40	34–35	1–2
Chicken H_4 + beef H_4	45–47	26	10

Source: Ref. 6.

[a]This is the number of peptides in one type of lactate dehydrogenase that are different from another type.

Table 3 Amino Acid Sequence Around the Essential —SH of Lactate Dehydrogenase and Comparison with Sequence Around Essential —SH of Other Dehydrogenases

Chicken H_4[a]	-Val-Ile-Ser-Gly-Gly-Cys*-Asn-Leu-Asp-Thr-Ala-Arg-
Chicken M_4[a]	-Val-Ile-Ser-Gly-Gly-Cys*-Asn-Leu-Asp-Ser-Ala-Arg-
Chicken triose phosphate dehydrogenase[b]	-Val-Ser-Asn-Ala-Ser-Cys*-Thr-Thr-Asn-Cys-Leu-Ala-
Horse liver alcohol dehydrogenase[c]	-Val-Ala-Thr-Gly-Ile-Cys*-Arg-Ser-Asp-Asp-His-Val-
Yeast alcohol dehydrogenase[d]	-Tyr-Ser-Gly-Val-Cys*-His-Thr-Asp-Leu-His-Ala-

[a]Ref. 7.
[b]Ref. 8. Triose phosphate dehydrogenase has a molecular weight of 145,000, is composed of four identical polypeptide chains of weight 36,300, and has four active sites per molecule.
[c]Ref. 9. Horse liver alcohol dehydrogenase has a molecular weight of 84,000, is composed of two polypeptide chains, and has two active sites per molecule.
[d]Ref. 9. Yeast alcohol dehydrogenase has a molecular weight of 150,000, is composed of four polypeptide chains, and has four active sites per molecule.

Table 4 Immunological and Catalytic Properties of Crystalline Chicken Lactate Dehydrogenases

Enzyme form	Inhibition: By anti-H_4 antibody (%)	Inhibition: By anti-M_4 antibody (%)	$DeNADH_L/NADH_H$[a]	Inhibition by 3×10^{-4} M oxalate (%)
H_4	100	0	3.2	84
H_2M_2	46	51	1.97	61
HM_3	18	80	1.40	46
M_4	0	100	0.55	33

Source: Ref. 10, p. 528, by courtesy of the American Chemical Society.
[a]Ratio of rates of reaction with hypoxanthine analog of reduced NAD (DeNADH) at pyruvate concentration of 3×10^{-4} M and NADH at pyruvate concentration of 1×10^{-2} M.

while chicken M_4 is completely inactivated in 2 min at 60°C. H_2M_2 and HM_3 isozymes show an intermediate stability proportional to the amount of M polypeptide they contain. The isozymes also differ in their immunological properties, inhibition by oxalate, and the effectiveness of their utilization of NADH and NADH analogs. Again, the response of H_2M_2 and HM_3 isozymes is intermediate between that of the H_4 and M_4 isozymes (Table 4).

V. BIOLOGICAL SIGNIFICANCE OF MULTIPLE MOLECULAR FORMS

There is a marked difference between the isozymes H_4 and M_4 with respect to the influence of pyruvate concentration on activity (Fig. 2). Based on this difference, it has been suggested that the two are designed to perform different functional roles in an organism [12]. According to this suggestion, H_4-type lactate dehydrogenase functions in cells with a high degree of aerobic metabolism, while the M-type enzyme functions in cells with low oxygen content or high rates of glycolysis. Therefore, the H-type enzyme would be found in muscles designed to perform sustained activity, such as the heart, while the M-type would be found in voluntary muscles designed for sudden activity, such as skeletal muscle.

VI. MECHANISM OF ACTION

Lactate dehydrogenase contains four active sites per molecule; that is, each polypeptide subunit contains the equivalent of one active site. However, the subunits have activity only when associated in the tetramer. There is an essential

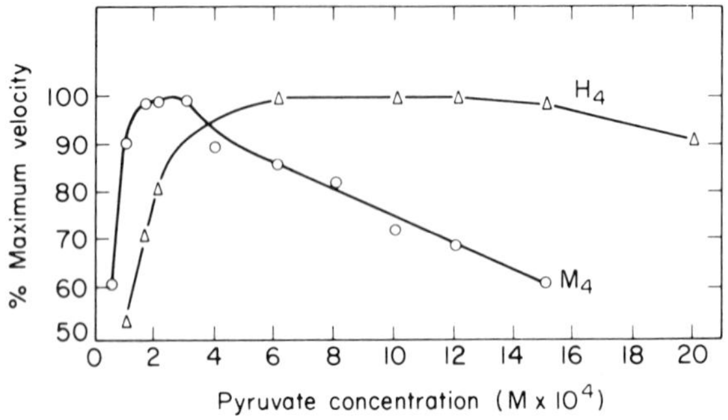

Figure 2 Effect of pyruvate concentration on the velocities of reactions catalyzed by human H_4 and M_4 lactate dehydrogenase isozymes at pH 7. (From Ref. 11, p. 2289, by courtesy of the American Society of Biological Chemists, Inc.)

sulfhydryl group for each active site. There are no cooperative effects among the active sites; therefore, the enzyme shows normal Michaelis–Menten kinetics.

Since the reaction catalyzed by lactate dehydrogenase involves two substrates, it is natural to ask whether the two substrates bind (a) independently of each other (randomly), (b) in an ordered sequence, or (c) only one substrate is on the enzyme at a time (ping-pong mechanism). Extensive kinetic data [2,13–17] indicate that there is an ordered sequence of addition of substrates to and release of products from the enzyme:

$$\begin{array}{ccccccccc} & \text{NAD}^+ & & \text{L} & & \text{P} & & \text{NADH} & \\ & \downarrow & & \downarrow & & \uparrow & & \uparrow & \\ \hline \text{E} & & \text{E}\cdot\text{NAD}^+ & & \begin{pmatrix}\text{E}\cdot\text{NAD}^+\cdot\text{L} \\ \text{E}\cdot\text{NADH}\cdot\text{P}\end{pmatrix} & & \text{E}\cdot\text{NADH} & & \text{E} \end{array} \qquad (3)$$

A. Ordered Bi Bi Mechanism

Some of the evidence for an ordered bi bi mechanism include:

1. When the reciprocal of velocity versus the reciprocal of lactate (or pyruvate) concentration at various NAD^+ (or NADH) concentrations is plotted, a series of converging lines results (Fig. 3). This indicates that the mechanism is a sequential one and rules out a ping-pong mechanism.

2. Binding studies indicate that NAD^+ and NADH can bind to the enzyme in the absence of lactate and pyruvate, respectively, but that lactate and pyruvate cannot bind to the enzyme in the absence of NAD^+ and NADH, respectively.

3. Product inhibition studies provided further evidence that the reaction occurs by an ordered mechanism (see Chapter 8, Section IV.B).

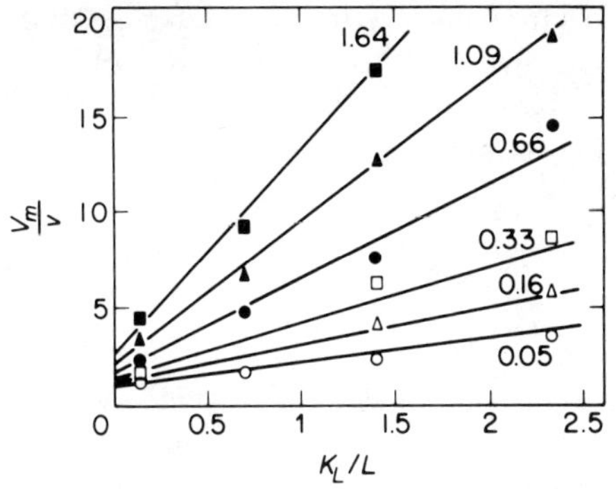

Figure 3 Effect of lactate concentration (L) on the initial reaction velocity at several concentrations of NAD$^+$. Reciprocal plots are shown as V_{max}/v versus K_L/L, where V_{max} and K_L are constants and are 1.43×10^4 min^{-1} and 8.8×10^{-4} *M*, respectively. The concentrations of NAD$^+$ shown on the graph are expressed as K_O/O, where K_O is 2.7×10^{-5} *M*. Reactions were at pH 6.80 and 29°C. (From Ref. 2, p. 198, by courtesy of the American Society of Biological Chemists, Inc.)

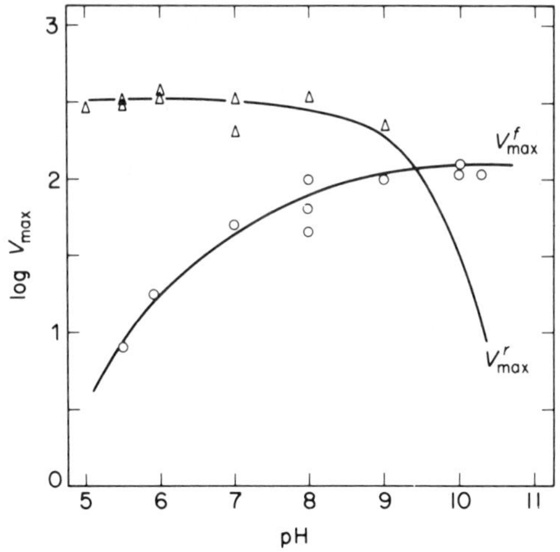

Figure 4 Effect of pH on some kinetic parameters of bovine heart lactate dehydrogenase (From Ref. 15.)

4. The order of release of the products from the enzyme is easy to determine because the rate-limiting step in the forward direction is the release of NADH from E·NADH and in the reverse direction is the release of NAD^+ from E·NAD^+. Release of product from the enzyme has been found to be rate determining for a number of other dehydrogenases.

B. Effect of pH

Vf_{max} [for the forward reaction; Eq. (1)] increases from zero at pH 5 to a maximum above pH 9 and is affected by a group of apparent value pK_a 7 (Fig. 4). Vr_{max} (for the reverse reaction) is maximal from pH 5 to 7 and then decreases at higher pH as affected by a group of apparent value pK_a 9. The pK_a 7 and pK_a 9 groups are an imidazole group and ammonium group, respectively. Apparently, these groups are important for the release of NADH and NAD^+, respectively, from the enzyme.

REVIEW QUESTIONS

See the end of Chapter 27.

REFERENCES

General References

Adams, M. J., Oxido-reductases: pyridine nucleotide-dependent enzymes, in *Enzyme Mechanisms* (M. I. Page and A. Williams, eds.), Royal Society of Chemistry, London, 1987, pp. 477–505.

Walsh, C., *Enzymatic Reaction Mechanisms*, W.H. Freeman, San Francisco, 1979, pp. 311–357.

Cited References

1. T. Thunberg, *Skand. Arch. Physiol.*, **40**, 1 (1920).
2. M. T. Hakala, A. J. Glaid, and G. W. Schwert, *J. Biol. Chem.*, **221**, 191 (1956).
3. T. Wuntch, R. F. Chen, and E. S. Vesell, *Science*, **167**, 63 (1970).
4. C. L. Markert, *Science*, **140**, 1329 (1963).
5. T. P. Fondy and N. O. Kaplan, *Ann. N.Y. Acad. Sci.*, **119**, 888 (1965).
6. A. Pesce, T. P. Fondy, F. Stolzenbach, F. Castillo, and N. O. Kaplan, *J. Biol. Chem.*, **242**, 2151 (1967).
7. T. P. Fondy, J. Everse, G. A. Driscoll, F. Castillo, F. E. Stolzenbach, and N. O. Kaplan, *J. Biol. Chem.*, **240**, 4219 (1965).
8. W. S. Allison, *Ann. N.Y. Acad. Sci.*, **151**(1), 180 (1968).
9. I. Harris, *Nature*, **203**, 30 (1964).
10. T. P. Fondy, A. Pesce, I. Freedberg, F. Stolzenberg, and N. O. Kaplan, *Biochemistry*, **3**, 522 (1964).

11. P. G. W. Plagemann, K. F. Gregory, and F. Wróblewski, *J. Biol. Chem.*, **235**, 2288 (1960).
12. N. O. Kaplan, in *Mechanisms of Action of Steroid Hormones* (C. A. Villee and L. L. Engel, eds.), Pergamon Press, Oxford, 1961, p. 247.
13. Y. Takenaka and G. W. Schwert, *J. Biol. Chem.*, **223**, 157 (1956).
14. A. D. Winer and G. W. Schwert, *J. Biol. Chem.*, **231**, 1065 (1958).
15. G. W. Schwert, B. R. Miller, and R. J. Peanasky, *J. Biol. Chem.*, **242**, 3245 (1967).
16. H. d'A. Heck, C. H. McMurray, and H. Gutfreund, *Biochem. J.*, **108**, 793 (1968).
17. R. S. Criddle, C. H. McMurray, and H. Gutfreund, *Nature*, **220**, 1091 (1968).

23

Glucose Oxidase

Glucose oxidase (β-D-glucose:oxygen 1-oxidoreductase; EC 1.1.3.4), discovered in 1928 by Müller [1] in *Aspergillus niger* and *Penicillium glaucum*, catalyzes the oxidation of β-D-glucose to δ-D-gluconolactone in the presence of molecular oxygen:

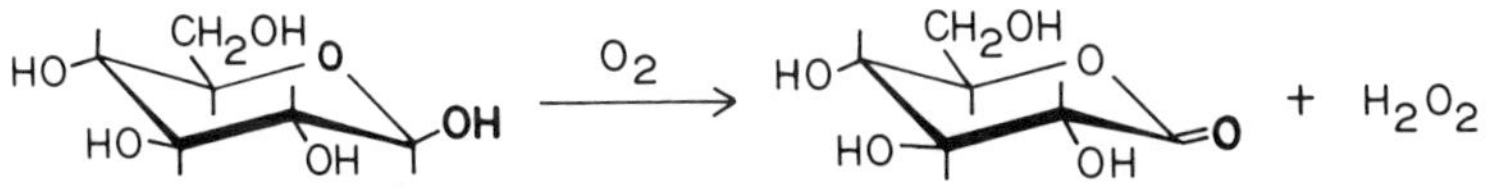

(1)

The enzyme has been called penicillium B, notatin, and glucose aerodehydrogenase, but the trivial name suggested by the Commission on Enzyme Nomenclature is glucose oxidase.

The enzyme, which has been found in other fungi, including *Aspergillus oryzae*, *Penicillium amagaskiense*, and *P. notatum*, has not been found in higher plants or in animals. The enzymes from different fungi are very similar in that all have molecular weights of 150,000, contain 2 mol of FAD (flavin adenine dinucleotide) per mole of enzyme, and have an isoelectric point near pH 4.2.

Other enzymes in animal tissues catalyze the oxidation of β-D-glucose (or derivatives) to δ-D-gluconolactone. They are readily differentiated from glucose oxidase, as they do not require molecular oxygen and hydrogen peroxide is not a product. Glucose-6-phosphate dehydrogenase (D-glucose-6-phosphate:NADP$^+$ 1-oxidoreductase; EC 1.1.1.49) is widely distributed in nature and catalyzes the NADP$^+$-dependent oxidation of glucose 6-phosphate to 6-phospho-δ-D-gluconolactone. Glucose dehydrogenase [β-D-glucose:NAD(P)$^+$ 1-oxidoreductase; EC

1.1.1.47] is found in ox liver and catalyzes the NAD^+-dependent oxidation of β-D-glucose to δ-D-gluconolactone.

I. SPECIFICITY

Glucose oxidase has rather high specificity for β-D-glucopyranose (Table 1). There is an absolute requirement for a hydroxyl group at C(1) and the activity is about 160 times higher if the hydroxyl group is in the β-position. Changes in the substrate at C(2) through C(6), except for L-glucose and 2-*O*-methyl-D-glucose, do not completely prevent the compounds from serving as substrates, but activity is much reduced.

II. MECHANISM OF ACTION

By 1948 it was known that glucose oxidase does not oxidize β-D-glucose by direct combination of molecular oxygen with the β-D-glucose. Evidence for this and for the mechanism of action include the following observations.

1. Glucose oxidase is not inhibited by HCN and CO, indicating that it does

Table 1 Substrate Specificity of Glucose Oxidase

Position modified	Compound	Changes from β-D-glucose	Relative rate
1	β-D-Glucose[a]		100
	α-D-Glucose[a]	Configuration of OH at C(1)	0.64
	1,5-Anhydro-D-glucitol[b]	Replacement of C(1) OH with H	0
2	2-Deoxy-D-glucose[c]	Replacement of C(2) OH with H	3.3
	D-Mannose[a]	Configuration of OH at C(2)	0.98
	2-*O*-Methyl-D-glucose[a]	Substitution of hydrogen of C(2) OH with methyl	0
3	3-Deoxy--glucose[b]	Replacement of C(3) OH with H	1
4	D-Galactose[b]	Configuration of OH at C(4)	0.5
	4-Deoxy-D-glucose[b]	Replacement of C(4) OH with H	2
5	5-Deoxy-D-glucose[b]	Replacement of C(5) OH with H (no pyranose ring can form)	0.05
	L-Glucose[b]	Configuration at C(5)	0
6	6-Deoxy-D-glucose[b]	Replacement of C(6) OH with H	10
	Xylose[a]	Replacement of C(6) with H	0.98

[a]Ref. 2.
[b]Ref. 3.
[c]Ref. 4.

not contain an essential metal ion. In general, oxidases contain metal ions as essential cofactors.

2. Other hydrogen acceptors, such as thionine, methylene blue, and quinone, can replace O_2. The ability to utilize such compounds is characteristic of dehydrogenases but not of oxidases.

3. Narcotics inhibit the oxidative capacity of the enzyme, which is in agreement with that observed for other dehydrogenases but not with oxidases.

4. The high substrate specificity of the enzyme is a general property of dehydrogenases but not of oxidases.

5. There is a direct correlation between riboflavin content and enzymatic activity. The highly purified enzyme is yellow and contains two molecules of FAD per molecule enzyme. Removal of FAD leads to complete loss of activity, which can be restored on addition of FAD.

6. Glucose oxidase in the oxidized form has an absorbance maximum at 280 nm, typical of all proteins, and absorbance maxima at 377 and 455 nm due to FAD (Fig. 1). On addition of glucose under anaerobic conditions the absorbance maxima at 377 and 455 nm disappear. When oxygen is admitted to the system, the two absorbance maxima reappear.

7. Glucose oxidase does not catalyze the complete reaction under anaerobic conditions and the reaction rate is dependent on O_2 concentration (Fig. 2). There is a saturation-like behavior with O_2. Evidence for the role of oxygen, and its source, can be obtained by experiments in which either $^{18}O_2$ or $H_2{}^{18}O$ is used. With $H_2{}^{18}O$, ^{18}O is found in the final product, D-gluconic acid, but not in H_2O_2. With $^{18}O_2$, ^{18}O is found only in H_2O_2.

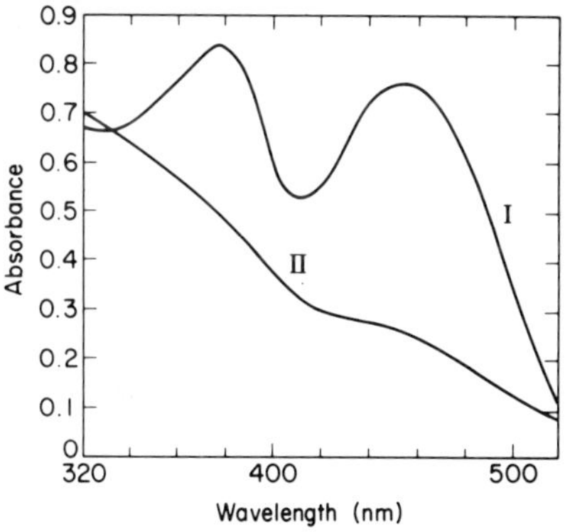

Figure 1 Absorbance spectra of a 0.608% solution of glucose oxidase in 0.03 *M* phosphate buffer, pH 6.5. I, without glucose; II, with glucose under anaerobic conditions. (From Ref. 2, p. 222, by courtesy of the Biochemical Society.)

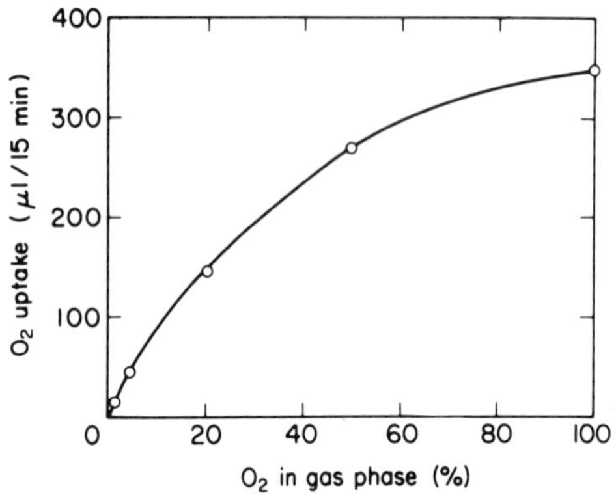

Figure 2 Effect of O_2 concentration on the activity of glucose oxidase at pH 5.6 and 39°C in the presence of catalase. (From Ref. 2, p. 226, by courtesy of the Biochemical Society.)

8. Plots of 1/rate versus 1/O_2 at several different glucose concentrations give a series of parallel lines (Fig. 3). These data indicate not only a dependence of rate on both O_2 and glucose concentrations but also that the mechanism is a Ping-Pong Bi Bi mechanism (see Chapter 6, Section III.C), which may be indicated schematically as

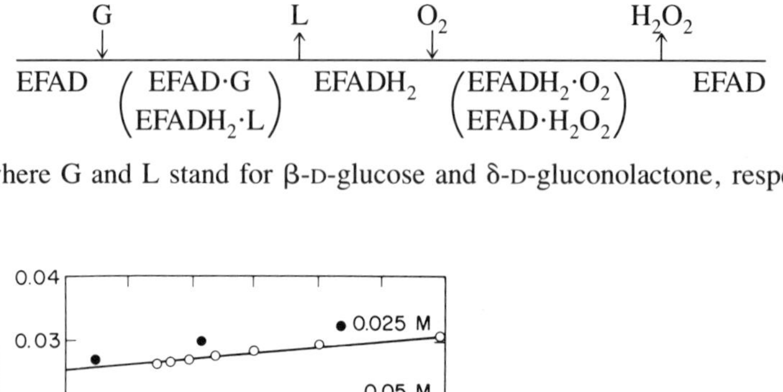

(2)

where G and L stand for β-D-glucose and δ-D-gluconolactone, respectively.

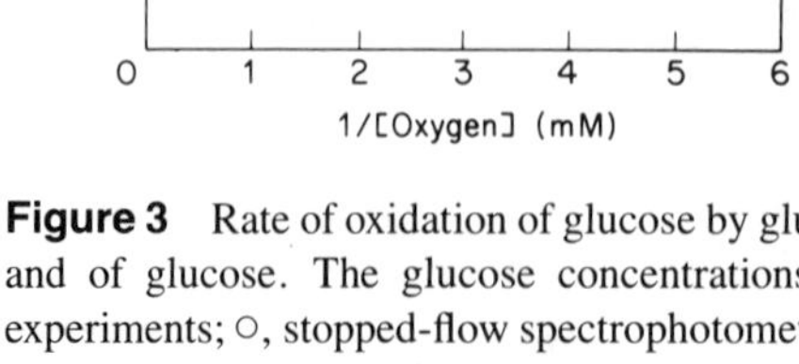

Figure 3 Rate of oxidation of glucose by glucose oxidase at various concentrations of O_2 and of glucose. The glucose concentrations are given on the graph. ●, Manometric experiments; ○, stopped-flow spectrophotometric experiments; 0°C, pH 5.6. (From Ref. 5, p. 3929, by courtesy of the American Society of Biological Chemists, Inc.)

Based on data presented above, a more detailed reaction can be written [Eq. (3)]. The oxidized form of the enzyme, EFAD, functions as a dehydrogenase

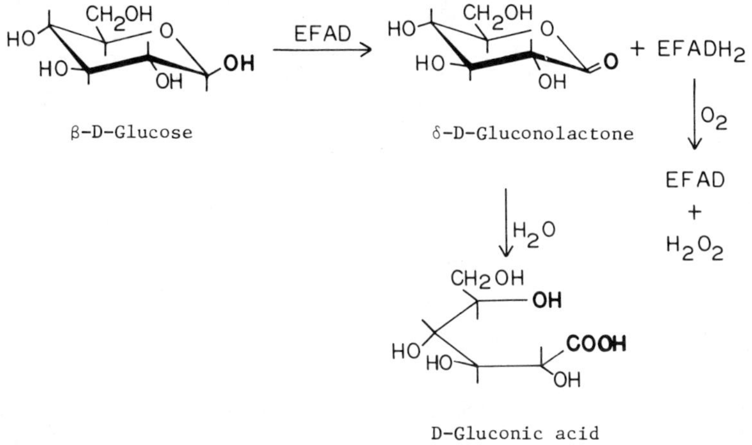

(3)

to extract two hydrogens from β-D-glucose to form the reduced enzyme, $EFADH_2$, and δ-D-gluconolactone. In a subsequent step, δ-D-gluconolactone is hydrolyzed, nonenzymatically, to D-gluconic acid and the reduced enzyme is reoxidized by molecular oxygen (enzymatic step).

III. EFFECT OF pH

A. Detailed Mechanism

Further details of the mechanism of action of glucose oxidase have been elucidated in part by the use of D-mannose and 2-deoxy-D-glucose, which are rather poor substrates (Table 1). Use of poor, and of different substrates in elucidation of enzyme mechanisms can be quite important, as different steps of the reaction may become rate determining. For example, consider the effect of pH on the maximum rates of oxidation of D-mannose, 2-deoxy-D-glucose, and β-D-glucose, respectively [Fig. 4(a) to (c)]. The pH–V_{max} profile for oxidation of D-mannose [Fig. 4(a)] is sigmoidal and dependent on a group with a pK of 5.00. The maximum activity of 15 M^{-1} sec^{-1} is very low and the second-order reaction rate constant depends on the concentration of both enzyme and D-mannose but is independent of O_2 concentration. V_{max} for oxidation of 2-deoxy-D-glucose at saturating concentrations of both 2-deoxy-D-glucose and O_2 is independent of pH [Fig. 4(b)]. The pH–V_{max} profile for oxidation of β-D-glucose under saturation conditions with respect to both substrates is bell-shaped and shows a dependence on prototropic groups of pK_a value 4.10 and 7.40, respectively [Fig. 4(c)]. Detailed studies using stopped-flow methods as well as normal turnover experiments in the absence and presence of oxygen led to formulation of the following

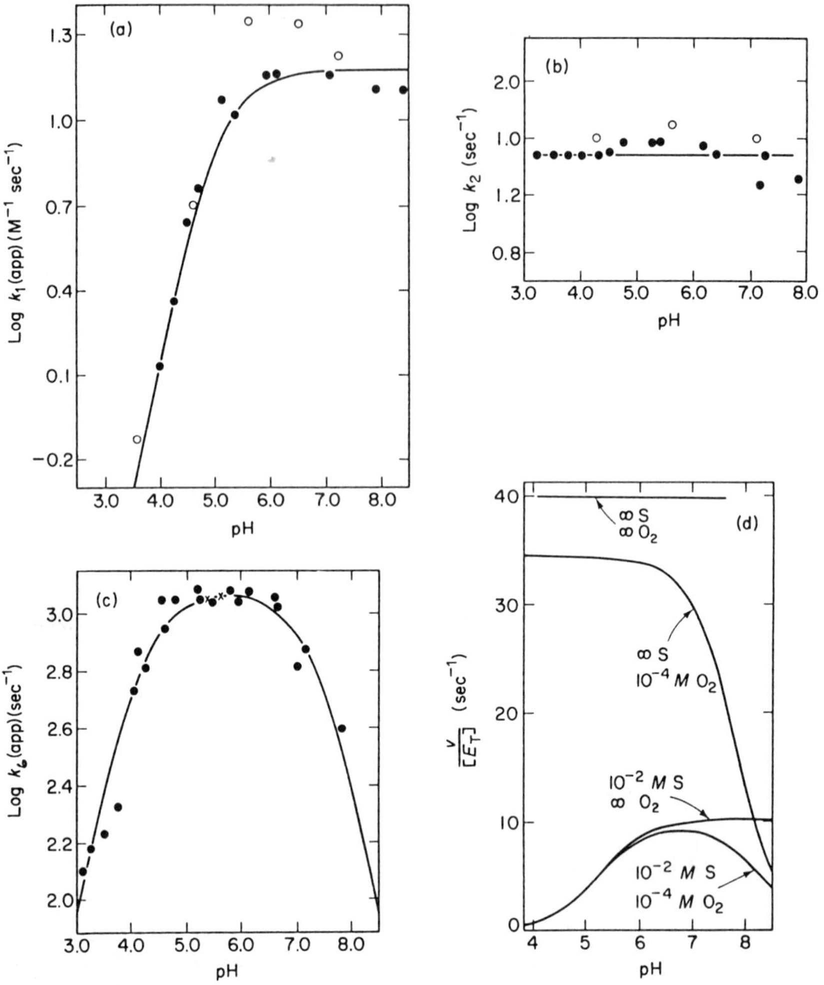

Figure 4 Effect of pH on the rate of oxidation of (a) D-mannose, (b) and (d) 2-deoxy-D-glucose, and (c) D-glucose by glucose oxidase at 25°C and ionic strength of 0.25. ○, Reductive half-reaction; ●, spectrophotometric turnover experiments; ×, stopped-flow turnover experiments. Plotted according to the Dixon method (see Chapter 10, Section II.C.3). Part (d) gives the pH dependence of the initial turnover velocity predicted for various concentrations of 2-deoxy-D-glucose and O_2 based on data of Table 2. (From Ref. 4, pp. 3628, 3630, and 3633, by courtesy of the American Society of Biological Chemists, Inc.)

rather complex scheme shown in Eq. (4) for the action of glucose oxidase [4]. For D-mannose, k_1 is the rate-determining step. This is in line with the rate constant

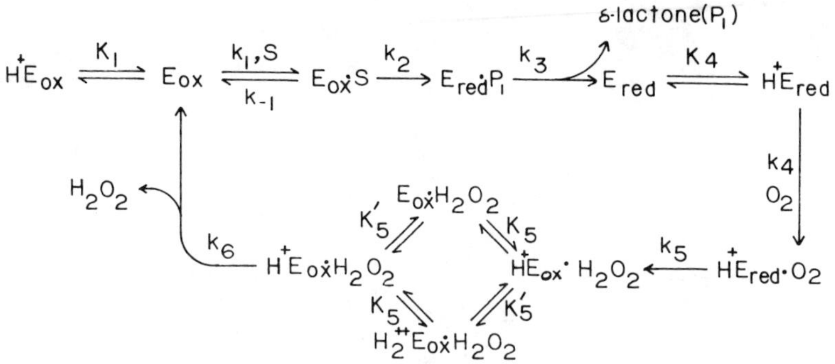

(4)

being of second order and independent of O_2 concentration. The pH dependence of k_1 reflects the involvement of a prototropic group on the free enzyme of pK 5.00, perhaps a carboxyl group, in binding of substrate. Only the unprotonated enzyme can combine with substrate. The rate of formation of the enzyme·sugar complex varies markedly with the nature of the sugar, varying from 15 to 12,600 M^{-1} sec^{-1} for D-mannose and D-glucose, respectively (Table 2).

With 2-deoxy-D-glucose, the rate-determining step is the conversion of $E_{ox}\cdot S$ to $E_{red}\cdot P_1$ and is indicated as k_2 in Eq. (4). The same results for k_2 are found in the presence and absence of O_2. Since the rate of conversion of $E_{ox}\cdot S$ to $E_{red}\cdot P_1$ is independent of pH [Fig. 4(b)], $E_{ox}\cdot S$ cannot undergo any proton association–dissociation that affects the rate.

With D-glucose, the rate-determining step is that designated as k_6 in Eq. (4) and Table 2. This step involves the dissociation of H_2O_2 from the $E_{ox}\cdot H_2O_2$ complex. The rate of dissociation of H_2O_2 from $E_{ox}\cdot H_2O_2$ is controlled by two prototropic groups, pK_5 and pK_5', of 4.10 and 7.40 [possibly a carboxyl and an imidazole group (?)]. H_2O_2 can only dissociate from the prototropic form $H^+E_{ox}\cdot H_2O_2$ and not from $E_{ox}\cdot H_2O_2$ or $H_2^{2+}E_{ox}\cdot H_2O_2$.

B. pH–Activity Profiles as Influenced by Substrate Concentration

The observed effect of pH on activity of an enzyme may be influenced by the degree of saturation of enzyme with substrate (Chapter 10). This is clearly the case for glucose oxidase.

1. With 2-deoxy-D-glucose, k_2 is the rate-determining step when the enzyme is saturated with both sugar and O_2 and the pH–activity profile is horizontal [Fig. 4(d)].
2. At saturating levels of 2-deoxy-D-glucose and less than saturating levels

Table 2 Kinetic and Acid Dissociation Constants for the Oxidation of Several Substrates by Glucose Oxidase[a]

Substrate	V_{max}[b] (M sec^{-1})	K_m,[b] sugar (M)	K_m,[b] O_2 (mM)	k_1 (M^{-1} sec^{-1})	k_2 (sec^{-1})	k_4 (M^{-1} sec^{-1})	k_6 (sec^{-1})	pK_1	pK_4	pK_5	pK'_5
D-Glucose	1150	0.11	0.48	12,600	3400	2.5×10^6	1200	5.00	6.90	4.10	7.40
2-Deoxy-D-glucose	50	0.04	0.03	1,400	40			5.35			
D-Mannose	22			15				5.00			

Source: Ref. 4, except as noted.
[a]Determined at 25°C and ionic strength of 0.25.
[b]Ref. 5.

of O_2 (10^{-4} *M*), the rate of oxidation is dependent on O_2 concentration as controlled by k_4. The pH–activity profile reflects the pH dependence of this step as controlled by ionization of a prototropic group of p*K* 6.90. Combination with O_2 requires that the group with p*K* 6.90 be protonated in order for the O_2 to combine with the enzyme and activity decreases above pH 6.

3. At less than saturating amounts of 2-deoxy-D-glucose (10^{-2} *M*) and saturating amounts of O_2, the pH–activity profile reflects the effect of pH on k_1 (p*K* 5.0) and a sigmoidal curve is obtained. Activity decreases below pH 6.5.

4. When less than saturating amounts of both substrates are used (10^{-2} *M* sugar and 10^{-4} *M* O_2), a bell-shaped curve is obtained which reflects the pH dependence of both k_1 and k_4.

The foregoing data make it very clear that interpretation of pH–activity curves must be done with great caution unless the complete sequence of kinetic events has been elucidated and the degree of saturation of enzyme with substrate is known.

IV. ASSAY OF GLUCOSE OXIDASE ACTIVITY

Glucose oxidase activity can be followed by several methods. The best two methods are given.

1. Since O_2 is one of the reactants, its rate of disappearance from the reaction can be followed continually by the use of a manometric technique (Warburg apparatus) or with an O_2-sensitive electrode. When initial rates are determined, the reaction can be run in the absence of catalase. Otherwise, catalase should be incorporated into the system to remove the H_2O_2 formed and prevent its detrimental effect on glucose oxidase.

In the presence of catalase there is a net loss of one O_2 per two D-glucose molecules oxidized rather than the 1:1 ratio found in the absence of catalase.

2. Activity of glucose oxidase can be determined by incorporation of peroxidase and a chromogen, such as *o*-dianisidine, into the reaction mixture. Peroxidase utilizes the H_2O_2 produced to oxidize the chromogen to a colored compound which can be determined spectrophotometrically (see Chapter 26 for a more complete description of this type of reaction).

REVIEW QUESTIONS

See the end of Chapter 27.

REFERENCES

General References

Barker, S. A., Carbohydrate oxidases, in *Oxidative Enzymes in Foods* (D. S. Robinson and N. A. M. Eskin, eds.), Elsevier Applied Science, New York, 1991, pp. 275–305.

Bright, H. J., and D. J. T. Porter, Flavoprotein oxidases, in *The Enzymes* (P. D. Boyer, ed.), Vol. XII, Academic Press, New York, 1975, pp. 421–505.

Crueger, A., Glucose transforming enzymes, in *Microbial Enzymes and Biotechnology*, 2nd Ed. (W. M. Fogarty and C. T. Kelly, eds.), Elsevier Applied Science, New York, 1990, pp. 177–226.

Cited References

1. D. Müller, *Biochem. Z.*, **199**, 136 (1928).
2. D. Keilin and E. F. Hartree, *Biochem. J.*, **42**, 221 (1948).
3. J. H. Pazur and K. Kleppe, *Biochemistry*, **3**, 578 (1964).
4. H. J. Bright and M. Appleby, *J. Biol. Chem.*, **244**, 3625 (1969).
5. Q. H. Gibson, B. E. P. Swoboda, and V. Massey, *J. Biol. Chem.*, **239**, 3927 (1964).

24
Polyphenol Oxidase

Polyphenol oxidase (1,2-benzenediol:oxygen oxidoreductase; EC 1.10.3.1) is frequently called tyrosinase, polyphenolase, phenolase, catechol oxidase, cresolase, or catecholase. The name *tyrosinase* came from tyrosine being used first as a substrate. It seems preferable to reserve the name *tyrosinase* for the mammalian enzyme, which has more limited specificity in that it acts only on tyrosine and dihydroxyphenylalanine. The enzymes from higher plants and fungi oxidize a great variety of monophenolic and *o*-diphenolic compounds.

In 1856, Schoenbein [1] described an enzyme in mushrooms which, in the presence of molecular oxygen, catalyzes the aerobic oxidation of certain compounds in plants. The enzyme is probably present in all plants, but it is in particularly high concentration in mushrooms, potato tubers, peaches, apples, bananas, avocados, tea leaves, coffee beans, and tobacco leaves. Polyphenol oxidase is present in animals, including humans, where it is important for skin, hair, and eye pigmentation. A number of important diseases in humans result from lack of sufficient activity or too much activity [2]. Polyphenol oxidase is also important in formation of the exoskeleton of insects.

There is another enzyme, laccase (benzenediol:oxygen oxidoreductase, EC 1.10.3.2) which catalyzes the hydroxylation of monophenols to *p*-diphenols and their further oxidation to *p*-benzoquinones. Laccase is most often found in fungi, which can be found on fruits, especially grapes, and vegetables. The two enzymes should not be confused.

Polyphenol oxidase is of great importance in determining the quality attributes of some fruits and vegetables. Its detrimental aspects caused by the browning of bruised and broken plant tissues are well known. Less often recognized is its beneficial role in the curing of tea, coffee, and tobacco [3], and the color of prunes and raisins.

I. REACTIONS CATALYZED; HYDROXYLATION REACTION

Unlike most enzymes, polyphenol oxidase can catalyze two quite different types of reactions, both of which involve phenolic compounds. These reactions involve hydroxylation of monophenols to give *o*-diphenols and the removal of hydrogens from *o*-diphenol to give an *o*-quinone.

Hydroxylation of monophenols is frequently referred to as cresolase activity since *p*-cresol is often used as substrate [Eq. (1)]. The oxygen for hydroxylation

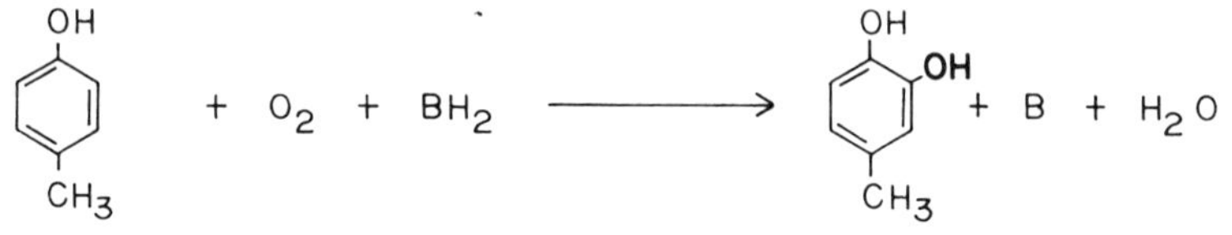

(1)

p-Cresol 4-Methylcatechol

comes directly from O_2, not from water [4]. When experiments are performed with $^{18}O_2$ in $H_2{}^{16}O$, the product contains oxygen-18; conversely, when experiments are performed with $^{16}O_2$ in $H_2{}^{18}O$, there is no oxygen-18 incorporation into the product. One atom of O_2 is incorporated into the phenol, the other into the water formed. Thus polyphenoloxidase acts as a monooxygenase in this reaction.

II. ASSAY OF HYDROXYLATION REACTION

The rate of hydroxylation may be followed by measuring O_2 uptake either in a Warburg respirometer or by use of an O_2-sensitive electrode. Both procedures are quite convenient and give a measure of activity of the enzyme; however, they also measure the O_2 uptake of oxidation of the *p*-diphenol and of subsequent nonenzyme-catalyzed reactions leading to melanin formation. A more specific method is to follow the reaction spectrophotometrically at a wavelength near 280 nm where only formation of *o*-diphenol is observed. Only initial rates should be used, so as to eliminate interference from later polymerization reactions. The reaction can also be followed by the rate of incorporation of tritium into the product from T_2O.

When a monophenol, such as *p*-cresol, is added alone as substrate there is a delay in the beginning of the reaction regardless of whether O_2 uptake, change in absorbance due to *o*-diphenol production, or benzoquinone formation is measured (Fig. 1, curve A). The lag period is due to the need for an *o*-diphenolic compound [indicated as BH_2 in Eq. (1)] and can be eliminated by addition of a small amount

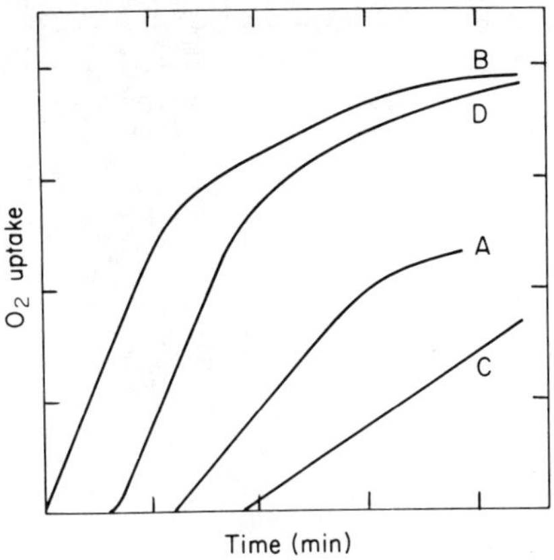

Figure 1 Effect of enzyme concentration and addition of *o*-diphenol on the lag period for the hydroxylation of monophenols by polyphenol oxidase. B, *o*-Diphenol added initially to reaction; A, C, and D, in the absence of added *o*-diphenol and at different enzyme concentrations (D > A > C).

of an *o*-diphenol at the beginning of the reaction (Fig. 1, curve B). In the absence of added *o*-diphenol there is a lag period until the enzyme can build up a sufficient concentration ($\sim 10^{-7}$ *M*) of the *o*-diphenol to permit the reaction to proceed. The length of this lag period in the absence of added *o*-diphenol is proportional to the amount of enzyme added (Fig. 1, curves A, C, and D).

III. MECHANISM OF HYDROXYLATION REACTION

Based on the data above and on knowledge that polyphenol oxidase is a copper-containing enzyme, one can write the reactions as shown in Eqs. (2) to (4) (where

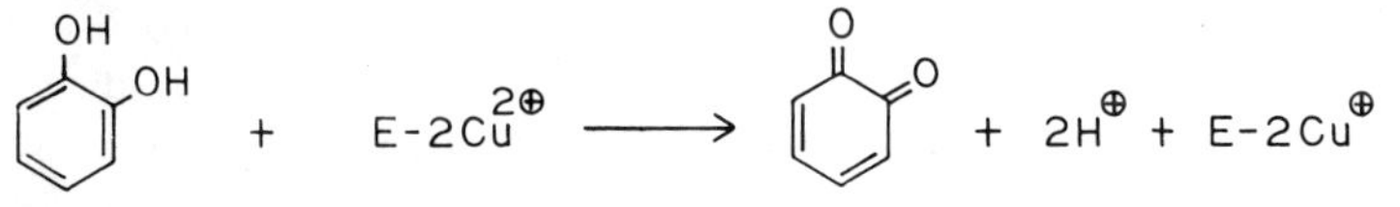

(2)

Catechol *o*-Benzoquinone

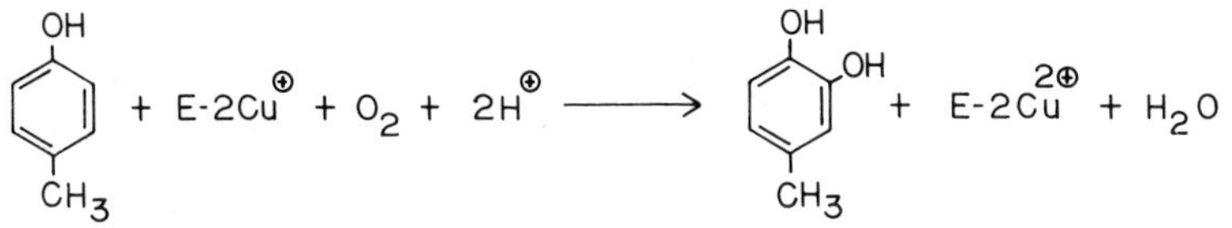

(3)

a small amount of catechol is added initially along with *o*-cresol). Apparently, in the absence of added *o*-diphenol, there is a small amount of enzyme form

E-$2Cu^+$ which can produce small amounts of *o*-diphenol via Eq. (3). The *o*-diphenol functions to reduce the enzyme form E-$2Cu^{2+}$ to E-$2Cu^+$ [Eq. (2)] which can combine with O_2, the second substrate [Eq. (3)], to hydroxylate the monophenol [Eq. (3)]. The overall reaction is

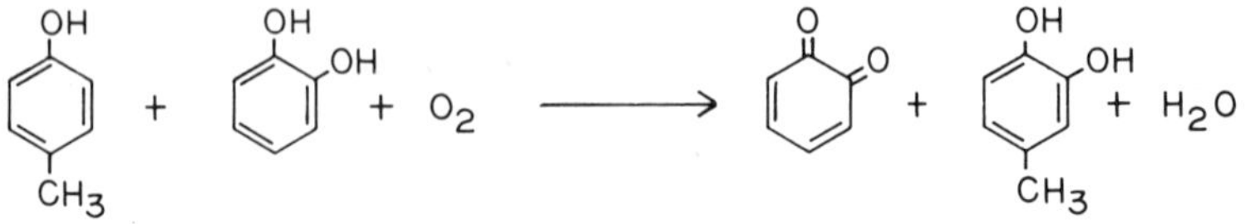

(4)

In the absence of added catechol, the overall reaction is

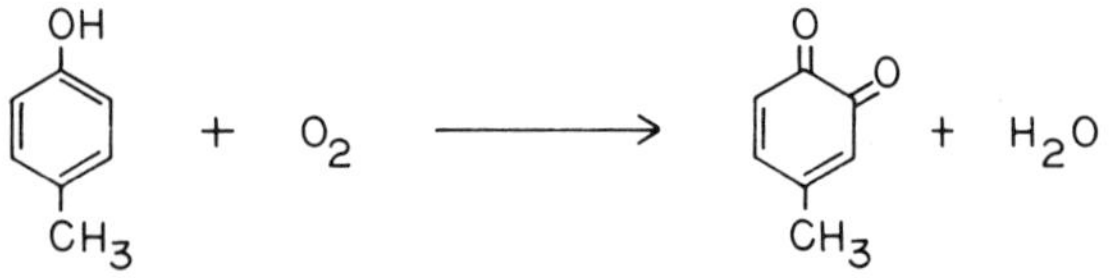

(5)

Exercise 1. If the reaction is found only to give 4-methylbenzoquinone and water as products as shown in Eq. (5), how could you prove that 4-methylcatechol is an intermediate in the reaction?

IV. OXIDATION OF *o*-DIPHENOLS (DEHYDROGENATION)

The second type of reaction catalyzed by polyphenol oxidase is the oxidation of an *o*-diphenol to a benzoquinone:

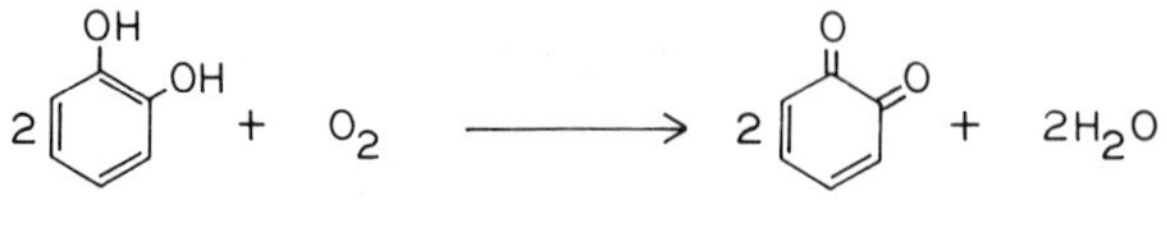

(6)

Catechol *o*-Benzoquinone

All polyphenol oxidases have activity on *o*-diphenols. Polyphenol oxidases from banana, tea leaf, tobacco leaf, and clingstone peach have been reported to have activity on *o*-diphenols exclusively and no ability to hydroxylate monophenols. Polyphenol oxidases from potato, apple, sugar beet leaf, and broad bean leaf have both types of activity.

V. MECHANISMS OF OXIDATION OF MONO- AND *o*-DIPHENOLS

The mechanisms of oxidation of mono- and *o*-diphenols by polyphenol oxidase shown in Fig. 2 are consistent with all experimental data [5]. The hydroxylation

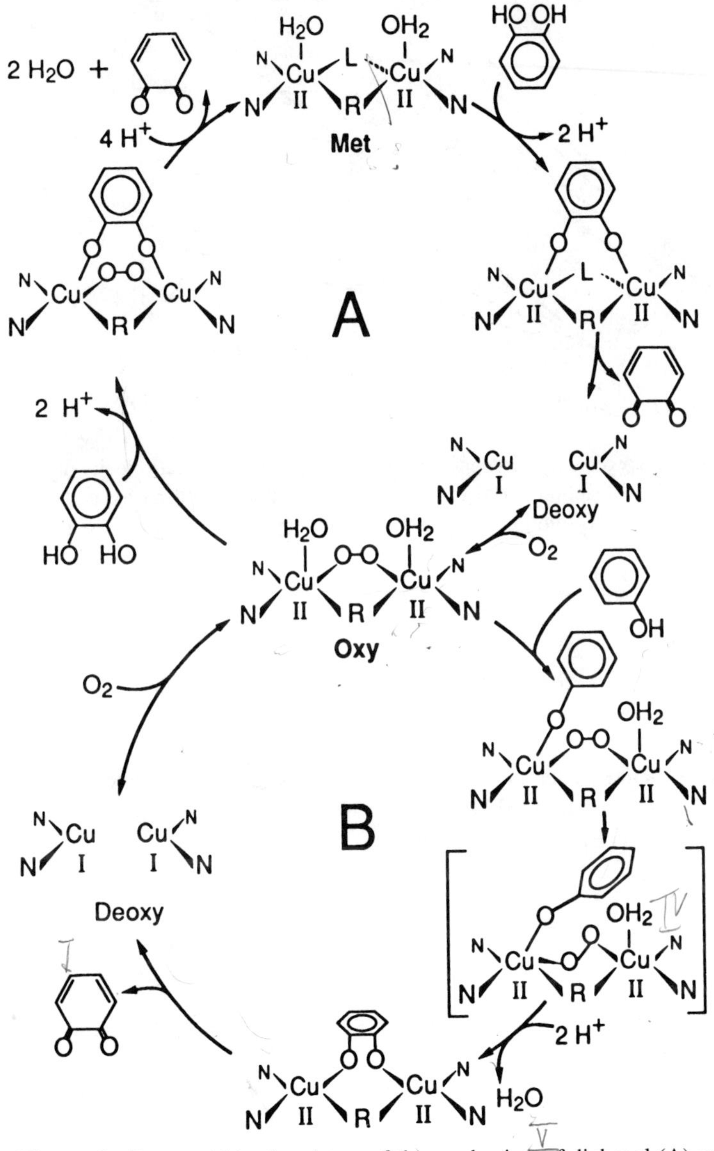

Figure 2 Proposed kinetic scheme of the mechanism of diphenol (A) and monophenol (B) activities of polyphenol oxidase. (From Ref. 5, p. 137, by courtesy of Martinus Nijhoff Publishers.)

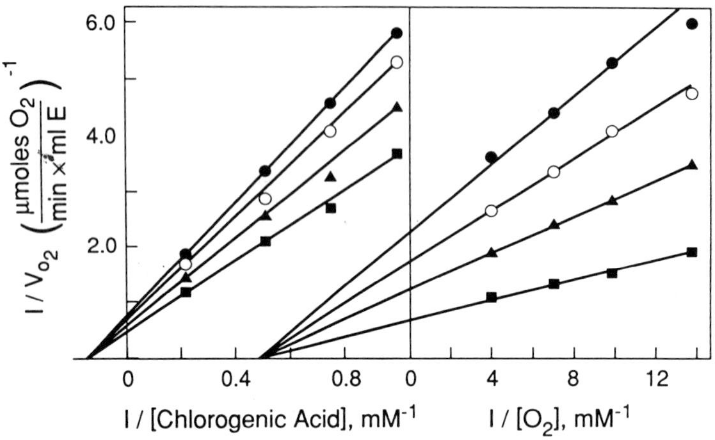

Figure 3 Effect of oxygen and chlorogenic acid concentrations on initial velocity of pear isoenzyme B–catalyzed reactions. The reactions were monitored with an oxygen electrode at pH 4.0 and 30°C. (From Ref. 7, p. 505, by courtesy of the American Society of Plant Physiologists.)

and dehydrogenation reactions differ from each other in detail. Not shown is a semibenzoquinone free-radical intermediate that is probably formed in the oxidation of *o*-diphenols to *o*-benzoquinones (see Section VI).

Investigations of the order of binding of O_2 and *o*-diphenol to the enzyme indicate that O_2 must bind first [6]. Therefore, the mechanism is thought to be an Ordered Sequential Bi Bi mechanism (see Fig. 3):

$$\begin{array}{ccccc} & O_2\downarrow & B\downarrow & P\uparrow & Q\uparrow \\ \hline \text{E-2 Cu}^{+} & \text{E-2 Cu}^{2+}\cdot O_2 & \text{E-2 Cu}^{2+}\cdot O_2\cdot B & \text{E-2 Cu}^{+}\cdot Q & \text{E-2 Cu}^{+} \\ & & \text{E-2 Cu}^{+}\cdot P\cdot Q & & \end{array} \tag{7}$$

Equation (7) shows the order of binding of the two substrates O_2 and P (an *o*-diphenol) and the release of products P (H_2O) and Q (benzoquinone).

The substrate most frequently used in the assay of activity of polyphenol oxidase on *o*-diphenols is catechol; thus this activity is sometimes referred to as catecholase activity. Catecholase activity of polyphenol oxidase can be measured independently of cresolase activity by the use of *o*-diphenols as substrates. However, measurement of activity is complicated by (a) rapid inactivation of the enzyme during reaction (Fig. 4), (b) further polymerization of the benzoquinone formed with a resultant change in absorbance (Fig. 4) and additional uptake of oxygen, and (c) inhibition of activity at high substrate concentrations.

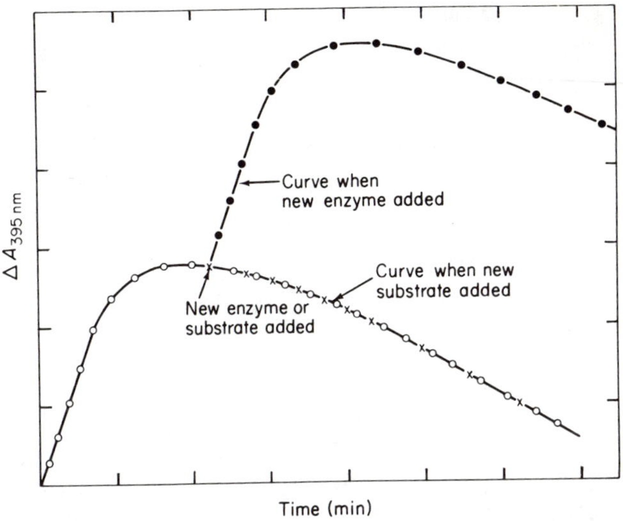

Figure 4 Activity curve (○) in the oxidation of catechol by polyphenol oxidase. At the time indicated new substrate (×) or new enzyme (●) was added to the reaction in similar experiments and the continued progress of the reaction followed.

VI. REACTION INACTIVATION OF POLYPHENOL OXIDASE

Oxidation of substrate stops long before all the substrate has been converted to product (Fig. 4). How much product will be formed before the reaction comes to completion is a function of the initial enzyme concentration added. That the reaction comes to a stop because of inactivation of the enzyme and not because of exhaustion of substrate can be shown rather easily by addition of new substrate or new enzyme to the reaction system (Fig. 4). Only when new enzyme is added is there an increase in absorbance until the enzyme has once again become inactivated. Because of rapid inactivation of enzyme, the initial rate is usually linear for only a very short time (~30 to 90 sec).

The cause of this inactivation of enzyme is not a result of instability of enzyme to the pH or temperature used nor is it an inactivation caused by catechol. Rather, inactivation is due to reaction of an intermediate product, *o*-semibenzoquinone free radical, with the imidazole group of one or more of the several histidyl groups that ligand the two copper ions in the active site [8]. As a result of the free-radical oxidation, the copper is released and the active site destroyed. This is called reaction inactivation (or suicide inactivation).

The decrease in absorbance at 395 nm shown in Fig. 4 is due to further

Table 1 4-Methyl-2,3-benzoquinone Formation and 4-Methylcatechol Disappearance in the Polyphenol Oxidase–Catalyzed Reaction

Time (min)	(1) Quinone formed (μmol)	(2) Substrate disappearing (μmol)	Ratio (1)/(2)
0.30	0.256	0.261	0.98
0.90	0.512	0.653	0.78
1.56	0.768	1.133	0.68
3.30	1.024	1.568	0.66
8.00	1.408	2.464	0.57

Source: Ref. 9, p. 787, by courtesy of Pergamon Press Ltd.

reactions (nonenzymatic), which eventually result in the formation of an insoluble brown pigment called melanin. These nonenzymatic reactions also consume O_2. Formation of *o*-benzoquinone from catechol consumes 1 gram-atom of O_2 per mole of *o*-diphenol [Eq. (6)]. However, the total process, including pigment formation, consumes 2.4 gram-atoms oxygen per mole of catechol oxidized. The effect of these further reactions on the ratio of benzoquinone formed to substrate disappearance is shown in Table 1. Only near the beginning of the reaction is the ratio close to unity.

VII. ASSAY OF ACTIVITY ON *o*-DIPHENOLS

Several methods are available for following the activity of polyphenol oxidase on *o*-diphenols such as catechol. These include manometric, polarographic, chronometric, and spectrophotometric methods. The manometric and polarographic methods measure the oxygen consumption of the system by use of a Warburg apparatus and an O_2-sensitive electrode, respectively. The chronometric method [10] determines the time at which the first color is observed when the reaction is performed in the presence of a specified amount of ascorbic acid [see Eqs. (8) and (9) in Section X].

Quite different results may be obtained by the various methods (Fig. 5). The manometric method gave the lowest results, which were linear with enzyme concentration only up to about 0.2 mL of enzyme. The low rates at low enzyme concentrations were due to the difficulty of getting true initial rates by this method, and the nonlinear response and flattening of the curve above 0.4 mL of enzyme were due to limitations placed on the rate of incorporation of O_2 into the solution by shaking. The chronometric and spectrophotometric methods gave about the same results at low enzyme concentrations, although the response in the

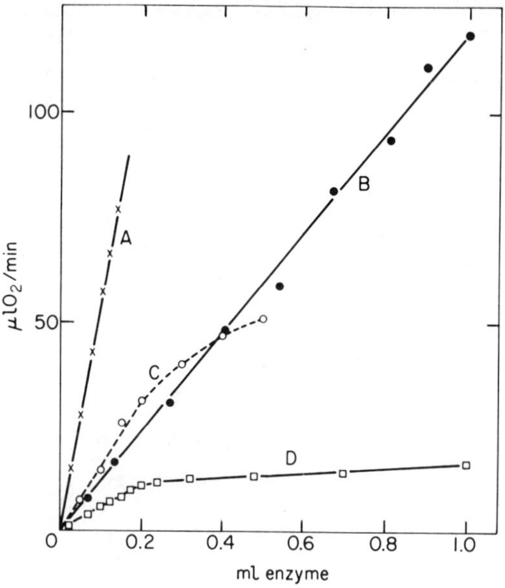

Figure 5 Dependence of initial rate of oxidation of 4-methylcatechol on polyphenol oxidase concentration as measured by four different methods; A, polarographic; B, spectrophotometric; C, chronometric, and D, manometric. Conditions were: pH 5.1, 25°C, 50 μg/mL apple enzyme, and 5×10^{-3} *M* substrate. The μL O_2/min consumed in the spectrophotometric and chronometric methods were calculated on the basis of one atom of oxygen consumed per molecule of substrate oxidized. (From Ref. 9, pp. 784 and 785, by courtesy of Pergamon Press Ltd.)

chronometric method was not linear. The polarographic method, which measures O_2 uptake, gave a linear response with enzyme concentration, but the rate of O_2 uptake was larger than that found by the manometric method, a surprising result. Initial rates obtained by the polarographic and by the spectrophotometric methods are preferred assay methods.

VIII. SUBSTRATE SPECIFICITY

Polyphenol oxidases from different sources differ markedly in their specific substrate requirements as shown in Table 2 for potato, peach, and broad bean leaf polyphenol oxidases. Notice particularly the marked differences among the three polyphenol oxidases in their relative activities on chlorogenic and caffeic acids, whereas they have similar relative activities on 3,4-dihydroxy-L-phenylalanine. Also, the activities on monophenols are much lower than on *o-diphenols*, which is

Table 2 Relative Substrate Specificities of Three Polyphenol Oxidases

Substrate	Activity relative to catechol		
	Potato[a]	Peach[b]	Broad bean leaf[c]
Di- or triphenolic compounds			
Catechol	100	100	100
4-Methylcatechol		51.5	200–225
d-Catechin		31.8	
Chlorogenic acid	140	22.2	8
Caffeic acid	76.5	0	12.5
Protocatechuic acid		16.3	0.11
3,4-Dihydroxy-L-phenylalanine	54.3	40.5	50
Dopamine		45.6	
Gallic acid		25.7	0.22
Pyrogallol			85–95
Monophenolic compounds			
p-Cresol	5.5	0	4
p-Coumaric acid	nil	0	0.05

[a]Ref. 11; pH 7.0.
[b]Ref 12. For isozyme A of clingstone peach at pH 6.8 and 30°C.
[c]Ref. 13.

typical for polyphenol oxidases. Peach and broad bean polyphenol oxidases differ in their relative activities on protocatechuic acid, gallic acid, and 4-methylcatechol.

IX. MULTIPLE MOLECULAR FORMS

A number of polyphenol oxidases catalyze both *o*-hydroxylation of monophenols and dehydrogenation of *o*-diphenols. However, the ratio of these two activities differs depending on the method of preparation. This was a particularly perplexing problem with mushroom polyphenol oxidase until it was shown conclusively that this enzyme exists in several multiple molecular forms [14] which do not have the same relative activities on *p*-cresol and catechol. Neither do they have the same amounts of copper (Table 3).

X. INHIBITION OF POLYPHENOL OXIDASE ACTIVITY

Browning of fruits caused by polyphenol oxidase can be prevented by exclusion of molecular oxygen (substrate limitation), by methylation of the phenols with *o*-methylase, by addition of reducing agents (ascorbate, bisulfite, thiols) which

Table 3 Activities and Copper Content of Four Multiple Molecular Forms of Mushroom (*Agaricus bispora*) Polyphenol Oxidase

		Specific activity (units/mg protein)			Copper content (μg/100 mg protein)		
Preparation	Component	*p*-Cresol	Catechol	Catechol/cresol	Cu^{2+}	Cu^{+}	Total
F	α	212	40	0.19			
	β	188	850	4.52	0.10	0.13	0.23
	γ	30	1275	42.5	0.04	0.19	0.23
	δ	55	850	15.5			
M	α	121			0.08	0.10	0.18
	β	243	940	3.87	0.07	0.17	0.24
	γ	51	1640	32.2	0.07	0.16	0.23
	δ	57	1280	22.5			0.32

Source: Ref. 15.

prevent the accumulation and polymerization of *o*-benzoquinone, by metal complexing agents such as sodium fluoride and azide which inactivate the enzyme by reacting with the essential copper, by heat treatment (thermal destruction of the enzyme), by lowering the pH to below pH 4.5, where most polyphenol oxidases have little activity (pH optima 6 to 6.5), or by the use of competitive inhibitors (sodium benzoate, 4-hexyl resorcinol, and a number of other compounds).

Addition of the reducing agent L-ascorbic acid to prevent enzymatic browning has been studied extensively. Ascorbic acid prevents browning by reducing the *o*-benzoquinone back to *o*-diphenol as rapidly as it is formed [Eqs. (8) and (9)]. No

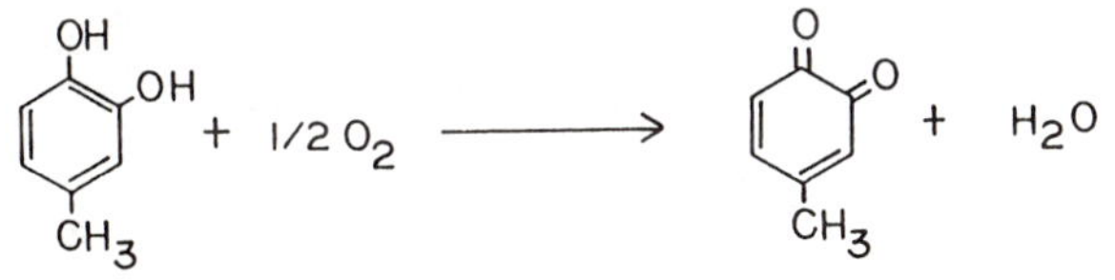

(8)

4-Methylcatechol   4-Methyl *o*-benzoquinone

browning occurs as long as ascorbic acid is present. Spectroscopically at 395 nm no activity would be measured; however, O_2 uptake measurements confirm that a reaction is taking place [Eq. (8)].

Although many of the phenolic and amino compounds in plants are not substrates for polyphenol oxidase, they can react rapidly with the *o*-benzoquinone produced (addition at the 4- or 5-position) to give compounds whose molar extinction coefficients are much higher than those of *o*-benzoquinone or its

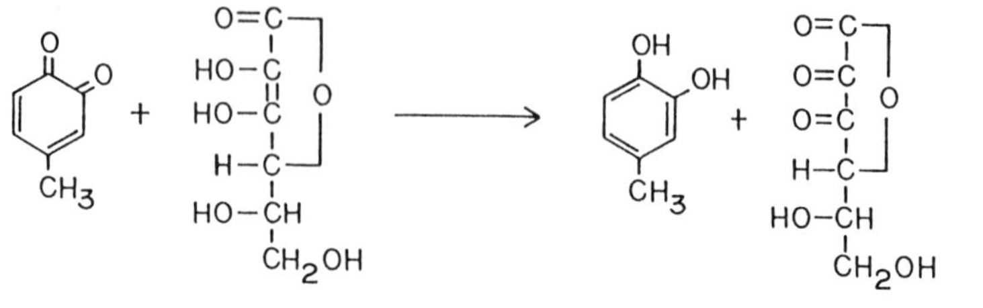

(9)

polymerization products alone. For this reason the products of polyphenol oxidase activity in fruits and vegetables are often very complex and highly colored.

Golan-Goldhirsh et al. [16] showed that ascorbic acid has a direct effect on polyphenol oxidase, especially in the presence of micromolar copper ions. Ascorbic acid and copper ions catalyze a free-radical reaction that oxidizes the imidazole group of the histidine residues that ligand the active-site coppers, in much the same way as expected for reaction inactivation (see Section VII), thereby inactivating the enzyme. This occurs rather rapidly at 20°C. Sodium bisulfite and thiol compounds directly inactivate polyphenol oxidase by reduction of the Cu^{2+} to Cu^{+}, which then dissociates from the active site, thereby inactivating the enzyme [17].

XI. PROTEIN NATURE OF ENZYME

Relatively little is known about the protein nature of the fruit and vegetable polyphenol oxidases. Mushroom (a fungus) polyphenol oxidase was studied intensively by Mason and colleagues at the University of Oregon. The molecular weight is 128,000, with four subunits, although the subunit molecular weights are not agreed upon. It appears that there are four active sites, each with two copper ions, based on the copper content of the isolated isozymes [15].

Neurospora crassa polyphenol oxidase is much easier to study because it is easily purified, unlike the ones from mushrooms, and occurs in only one isoform. It has 407 amino acid residues, with a molecular weight of 46,000, a single polypeptide chain, and one active site. Its complete amino acid sequence has been determined by the classical Edman degradation method and from the cDNA (gene) for the enzyme. The amino acid sequences for polyphenol oxidases from two other fungi (*Streptomyces glaucescens* and *S. antibioticus*), humans, and the rat have been determined by cDNA sequencing. The two *Streptomyces* polyphenol oxidases have ~24% exact homology with the *N. crassa* enzyme. There is little homology between *N. crassa* polyphenol oxidase and human and rat polyphenol oxidases; the latter enzymes are much larger, with molecular weights of 62,610 and 58,000 for the human and rat enzymes, respectively. Significantly, five out of the six histidines that ligand the active-site copper ions are conserved in all five enzymes. *N. crassa* polyphenol oxidase has much homology with one (copper binding) of the three subunits of spider hemocyanin, the O_2-carrying protein. The five histidines that bind the two copper ions responsible for binding O_2 are in the same relative positions as in the fungal, human, and rat polyphenol oxidases. (See Ref. 17 for a more thorough review.)

REVIEW QUESTIONS

See the end of Chapter 27.

REFERENCES

General References

Mayer, A. M., and E. Harel, Review: polyphenol oxidases in plants, *Phytochemistry*, **18**, 193 (1979).

Osuga, D., A. van der Schaaf, and J. R. Whitaker, Control of polyphenol oxidase, based on catalytic mechanism, in *Structure/Function Relationships in Food Proteins* (R. Yada, ed.), in press (1993).

Witkop, C. J., Jr., Inherited disorders of pigmentation, *Clin. Dermatol.*, **3**(1), 70 (1985).

Zawistowski, J., C. G. Biliaderis, and N. A. M. Eskin, Polyphenol oxidase, in *Oxidative Enzymes in Foods* (D. S. Robinson and N. A. M. Eskin, eds.), Elsevier Applied Science, New York, 1991, pp. 217–273.

Cited References

1. C. F. Schoenbein, *Phil Mag.*, **11**, 137 (1856).
2. C. J. Witkop, Jr., *Clin. Dermatol.*, **3**(1), 70 (1985).
3. W. G. C. Forsyth, *Ann. Rev. Plant Physiol.*, **15**, 443 (1964).
4. H. S. Mason, W. L. Fowlks, and E. Peterson, *J. Am. Chem. Soc.*, **77**, 2914 (1955).
5. K. Lerch, *Mol. Cell. Biochem.*, **52**, 125 (1983).
6. L. L. Ingraham, *J. Am. Chem. Soc.*, **79**, 666 (1957).
7. N. J. Rivas and J. R. Whitaker, *Plant Physiol.*, **52**, 501 (1973).
8. H. S. Mason, E. Spencer, and I. Yamazaki, *Biochem. Biophys. Res. Commun.*, **4**, 236 (1961).
9. A. M. Mayer, E. Harel, and R. Ben-Shaul, *Phytochemistry*, **5**, 783 (1966).
10. W. H. Miller and C. R. Dawson, *J. Am. Chem. Soc.*, **63**, 3375 (1941).
11. A. R. Macrae and R. G. Duggleby, *Phytochemistry*, **7**, 855 (1968).
12. T. C. Wong, B. S. Luh, and J. R. Whitaker, *Plant Physiol.*, **48**, 19 (1971).
13. D. A. Robb, T. Swain, and L. W. Mapson, *Phytochemistry*, **5**, 665 (1966).
14. J. L. Smith and R. C. Krueger, *J. Biol. Chem.*, **237**, 1121 (1962).

15. S. Bouchilloux, P. McMahill, and H. S. Mason, *J. Biol. Chem.*, **238**, 1699 (1963).
16. A. Golan-Goldhirsh, D. T. Osuga, A. O. Chen, and J. R. Whitaker, Effect of ascorbic acid and copper on proteins, in *The Bioorganic Chemistry of Enzymatic Catalysis: An Homage to Myron L. Bender* (V. T. O'Souza and J. Feder, eds.), CRC Press, Boca Raton, Fla., 1992, pp. 61–76.
17. D. Osuga, A. van der Schaaf, and J. R. Whitaker, Control of polyphenol oxidase activity, based on catalytic mechanism, in *Structure/Function Relationships in Food Proteins* (R. Yada, ed.), in press (1993).

25
Xanthine Oxidase

Just shortly after 1900, an enzyme was discovered in mammalian tissues (especially the liver) that catalyzes the oxidation of hypoxanthine and xanthine to uric acid. About this same time, an enzyme was found in milk that oxidizes aldehydes to carboxylic acids. However, it was not until 1935 that it was shown that these two enzymes are the same. Xanthine oxidase (xanthine:oxygen oxidoreductase; EC 1.1.3.22) is widely distributed in mammals, especially in cow's milk and calf liver. The enzyme from milk, being easy to purify, has been used in most studies of this enzyme.

I. REACTIONS CATALYZED; SUBSTRATE SPECIFICITY

The enzyme catalyzes the oxidation of hypoxanthine and xanthine to uric acid [Eq. (1)], which is a specific example of a type D reaction (Chapter 21). The heavy arrows show the reactions catalyzed by the enzyme; the dashed arrows show the net result of such catalysis.

Xanthine oxidase is a rather nonspecific enzyme and a great variety of compounds serve as substrates (electron donors). The rates of oxidation vary quite markedly among the different types of substrates, with hypoxanthine and xanthine being the best substrates (Table 1). In the case of the purines the reaction catalyzed is a hydroxylation of the substrate, while aldehydes are oxidized to carboxylic acids.

In Eq. (1), O_2 is the electron acceptor. However, the enzyme is about as

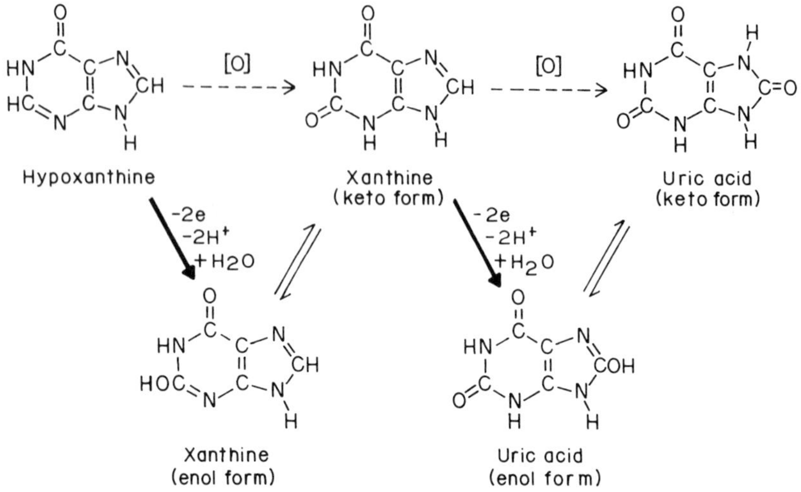

(1)

nonspecific for the electron acceptor as it is for the electron donor. A large number of dyes, such as methylene blue, 2,6-dichlorophenolindophenol, triphenyltetrazolium chloride, and phenazine methosulfate, as well as cytochrome *c* and ferricyanide, can serve as electron acceptors.

II. PROTEIN AND ACTIVE SITE PROPERTIES

Milk xanthine oxidase has a molecular weight of 283,000, is composed of two identical subunits, and contains (per mole of protein) 2 mol of FAD, 2 gram-atoms of molybdenum, 8 gram-atoms of nonheme iron, and 8 "labile sulfides" (as Fe_2S_2

Table 1 Substrate Specificity of Xanthine Oxidase from Milk

Substrate	Relative rates	Substate	Relative rates
Xanthine	140	Acetaldehyde	0.72
Hypoxanthine	100	Cinnamaldehyde	0.41
8-Hydroxypurine	6	Vanillin	0.01
6-Amino-2-hydroxypurine	4.5	Glyceraldehyde	0.014
6-Amino-8-hydroxypurine	7	*p*-Phthalaldehyde	0.013
Adenine	6	Decanaldehyde	0.0080
2,8-Dihydroxypurine	0.6	*n*-Butyraldehyde	0.0060
p-Hydroxybenzaldehyde	1.3	Octaldehyde	0.0002
Benzaldehyde	0.8		

Source: Ref. 1.

centers) which are readily released as H_2S from the protein by acidification or by boiling at pH 7. There are two active sites per molecule. The metal ions and flavin are bound quite firmly to the protein and are not released by dialysis. Denaturation of the protein by heat or by treatment with acids releases the metal ions and flavin and the activity is irreversibly lost. The flavin can be removed by $CaC1_2$ treatment; this hydrolyzes FAD to FMN, which being much less tightly bound, dissociates from the protein. The isoelectric point of the milk enzyme is 5.3 to 5.4 and the pH optimum is around 8.3.

III. INVOLVEMENT OF FAD IN ACTIVE SITE

That FAD is involved in the reaction can be shown spectrophotometrically (Fig. 1). The oxidized enzyme (E-FAD) has a characteristic absorption spectrum with a peak at 450 nm and a shoulder in the region of 360 nm. When xanthine is added to the enzyme solution under anaerobic conditions, both the 360-nm shoulder and the 450-nm peak disappear due to formation of E-$FADH_2$. The difference spectrum of oxidized and reduced enzyme is similar to the difference spectrum of FAD and $FADH_2$ at the same concentration (Fig. 1, curves C and D).

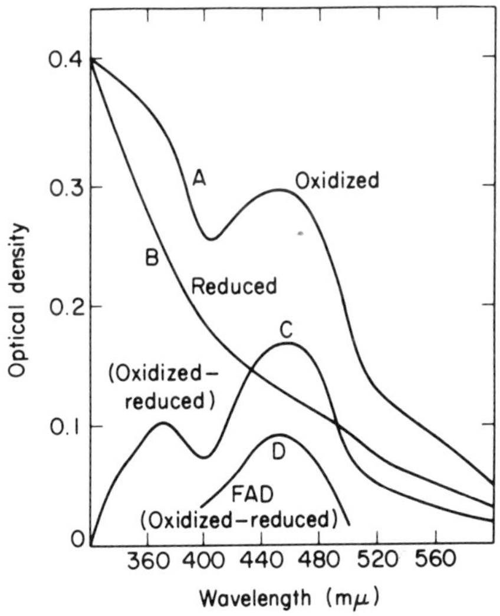

Figure 1 Absorption spectrum of xanthine oxidase. A, oxidized form; B, after complete reduction by reaction with xanthine under anaerobic conditions; C, the difference spectrum between oxidized and reduced enzyme; and D, the difference spectrum between FAD and $FADH_2$ at the same concentration. (From Ref. 2, p. 661, by courtesy of the Biochemical Society.)

IV. KINETICS OF REACTION

The experimental evidence is conclusive that the reaction occurs as two half-reactions. Addition of xanthine to the enzyme in the absence of O_2 results in reduction of flavin (Fig. 1). The rate of this reaction has been measured by rapid reaction techniques by following the rate of decrease in absorbance at 450 nm (or increase in absorbance at 610 nm due to semiquinone formation). When O_2 is then rapidly mixed with the system, the half-reaction involving reaction with O_2 can be followed by observing the increase in absorbance at 450 nm due to oxidation of the reduced flavin.

V. MECHANISM

Additional evidence that the mechanism of xanthine oxidase is a Ping-Pong Bi Bi mechanism comes from varying the concentration of one substrate while holding the other substrate concentration constant (Fig. 2). The data give parallel lines at different levels of the second substrate. By use of rapid reaction techniques it has been found that, kinetically, the mechanism of action of xanthine oxidase can be described by the scheme shown in Fig. 3 [4]. The rate constants k_1, k_2, k_3, and k_4 have been determined to be ~$5.0 \times 10^5\ M^{-1}\ sec^{-1}$, $10.5\ sec^{-1}$, $>4.0 \times 10^5\ M^{-1}\ sec^{-1}$, and $21.5\ sec^{-1}$, respectively. Note that k_1 and k_3 are for the rates of formation of the absorptive complexes, xanthine·E-FAD and O_2·E-$FADH_2$, respectively, while k_2 and k_4 are for the rates of reduction and oxidation of the enzyme, respectively; the latter steps are associated with oxidation of xanthine and reduction of O_2, respectively. The rates of reduction and oxidation of the enzyme are

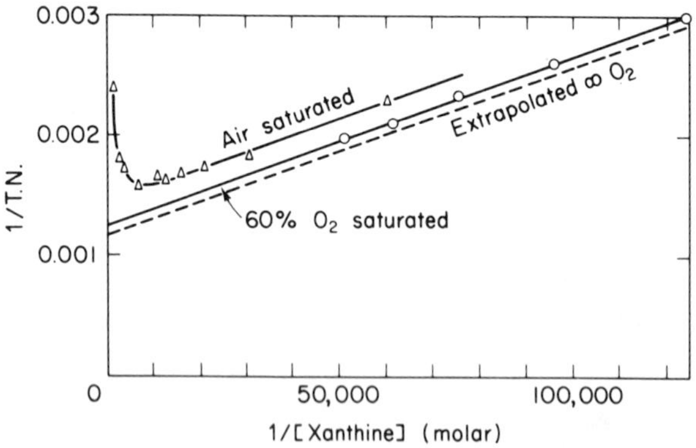

Figure 2 Effect of xanthine concentration on the rate of xanthine oxidation. T.N. is the turnover number measured at pH 8.3 and 25°C. (From Ref. 3, p. 1688, by courtesy of the American Society of Biological Chemists, Inc.)

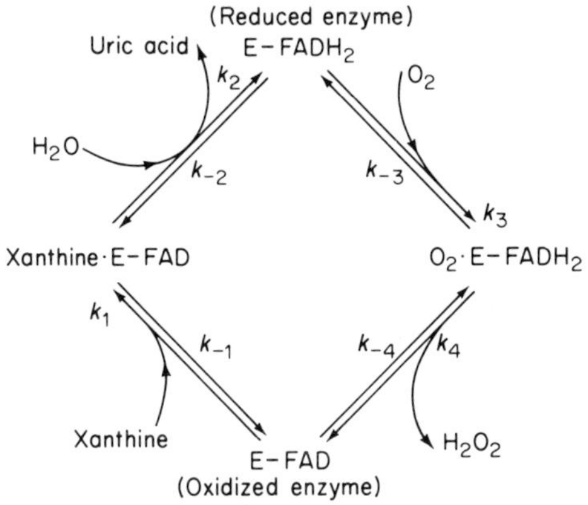

Figure 3 Mechanism of action of xanthine oxidase as observed kinetically. (From Ref. 4.)

nearly the same, so that both steps are rate determining in the overall reaction. The reaction is diagrammed by the Cleland method, as shown in

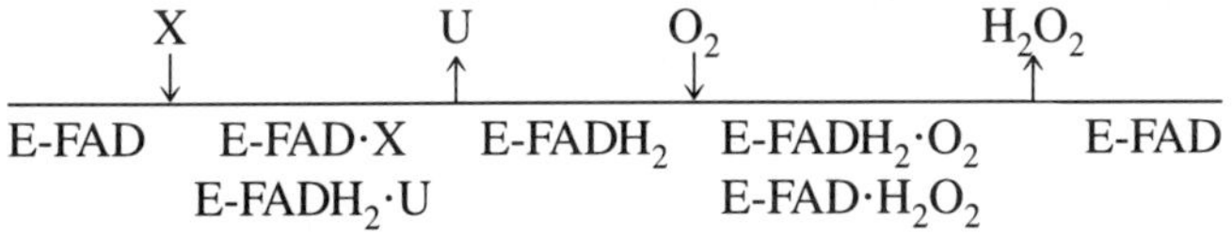

(2)

where X is xanthine and U is uric acid.

VI. ELECTRON PARAMAGNETIC RESONANCE SPECTROSCOPY AND MECHANISM OF ACTION

The mechanism shown in Fig. 3 is based on spectrophotometric changes measured at 450 nm (reduction and oxidation of flavin) and 290 nm (difference in xanthine and uric acid absorbances). When the reaction is examined by electron paramagnetic resonance (EPR) spectroscopy, the details of the mechanism are found to be much more complex and evidence is obtained for participation of molybdenum and iron as well as flavin (Fig. 4).

To demonstrate the rate of formation and decay of signals due to molybdenum, flavin, and iron, reactions were performed in which the xanthine concentration was essentially identical to that of the enzyme and oxygen was present in excess. Under such conditions, the enzyme can only go through the reaction one time and the xanthine is all converted to product. This is referred to as a "single turnover" experiment. A method of extremely rapid freezing was used to stop the reaction at various stages in the reaction and the EPR spectra were determined (Fig. 4). There

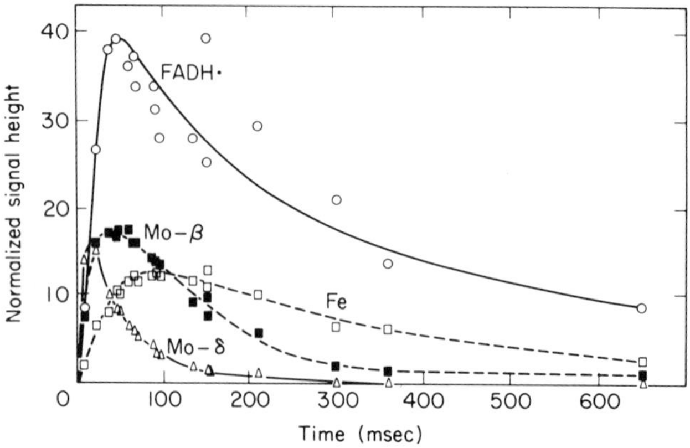

Figure 4 Electron paramagnetic resonance signals generated by xanthine oxidase in a "single turnover" experiment with xanthine as substrate. (From Ref. 5, p. 2668, by courtesy of the American Society of Biological Chemists, Inc.)

was an initial increase in the signals and then a slow decay. More important, however, was the rate of formation and disappearance of these signals. The Mo-δ signal, believed to be due to formation of Mo(V) from Mo(VI) of the oxidized enzyme, reached a maximum value at ~15 msec. The Mo-β, FADH·(FAD→FADH·), and Fe(II) [Fe(III)→Fe(II)] signals were maximal at about 40, 45, and 100 msec, respectively. These time courses indicate that the order of acceptance of electrons from the substrate proceeds as

$$\text{substrate}\rightarrow\text{Mo}\rightarrow\text{FAD}\rightarrow\text{Fe}\rightarrow O_2$$

Putting these data from EPR spectroscopy together with those from spectrophotometric analyses, the minimum scheme of Eq. (3) (excluding formation and

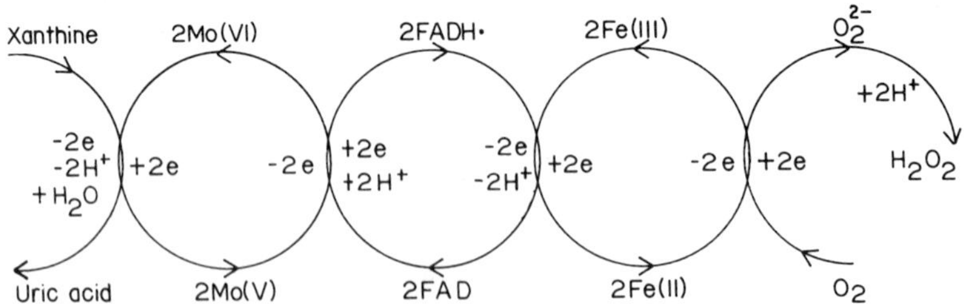

(3)

decomposition of enzyme·substrate complexes known to occur; Fig. 3) must be written for the sequence of events that occur in the enzyme-catalyzed oxidation of xanthine to uric acid. The overall reaction is

$$\text{xanthine} + H_2O + O_2 \rightarrow \text{uric acid} + H_2O_2$$

All of these reactions take place while all constituents are bound to the protein and do so at the rapid rate of 100 msec per turn through the cycle. The complexity of the mechanism has been a challenge and opportunity for the biochemist, chemist, and physicist.

VII. ENZYME INHIBITION

As might be expected for an enzyme as complex as xanthine oxidase, many substances inhibit the enzyme. Substances such as purines, pteridines, and other heterocyclic compounds inhibit by competing with the substrate for binding at the active site. Compounds such as arsenite, cyanide, and methanol appear to inhibit by reacting with the molybdenum. Other inhibitors include phosphate, imidazole, sodium and potassium chloride, benzoate, borate, copper, ascorbic acid, and dinitrophenol. The inhibition of xanthine oxidase activity by ascorbic acid, which has been noted repeatedly in milk, is probably due to an indirect effect caused by reduction of Cu^{2+} to Cu^{+} by ascorbic acid; Cu^{+} is a much more effective inhibitor of the enzyme that is Cu^{2+}.

VIII. ASSAY OF ACTIVITY

A number of methods are available for assaying xanthine oxidase activity. These include rates of methylene blue reduction, cytochrome *c* reduction, oxygen uptake, and uric acid formation. In the methylene blue method the reaction is performed under anaerobic conditions so that the methylene blue replaces O_2 as the terminal acceptor of electrons. The rate of reduction in blue color can then be followed continuously in a spectrophotometer. Rate of oxygen uptake is measured conveniently with an O_2-sensitive electrode. The rate of uric acid formation is measured conveniently in a spectrophotometer. At pH 8.3, 25°C and 290 nm, $\Delta\epsilon_m = 9.6 \times 10^3\ M^{-1}\ \text{cm}^{-1}$ for uric acid formation. In the oxygen uptake and uric acid formation methods, H_2O_2 is formed. Since H_2O_2 is detrimental to the enzyme, only initial rates should be used in assaying the enzyme. Alternatively, catalase can be incorporated into the system to remove H_2O_2. The assay with cytochrome *c* is carried out under aerobic conditions and the rate of increase in absorbance at 550 nm (reduced band of cytochrome *c*) is used to follow the assay.

Exercise 1. What would be the theoretical ratio of hypoxanthine converted to uric acid and oxygen uptake as measured by an O_2-sensitive electrode in the presence of catalase? (*Hint:* Write the reactions catalyzed by xanthine oxidase and by catalase.)

REVIEW QUESTIONS

See the end of Chapter 27.

REFERENCES

General References

Bray, R. C., The flavin and the other catalytic and redox centers of xanthine oxidase and related enzymes, in *Flavins and Flavoproteins* (V. Massey and C. H. Williams, eds.), Elsevier/North-Holland, Amsterdam, 1982, pp. 775–785.

Whitaker, J. R., Mechanisms of oxidoreductases important in food component modification, in *Chemical Changes in Food During Processing* (T. Richardson and J. W. Finley, eds.), AVI, Westport, Conn., 1985, pp. 121–176.

Cited References

1. V. H. Booth, *Biochem. J.*, **32**, 494 (1938).
2. D. B. Morell, *Biochem. J.*, **51**, 657 (1952).
3. V. Massey, P. E. Brumby, H. Komai, and G. Palmer, *J. Biol. Chem.*, **244**, 1682 (1969).
4. H. Gutfreund and J. M. Sturtevant, *Biochem. J.*, **73**, 1 (1959).
5. R. C. Bray, G. Palmer, and H. Beinert, *J. Biol. Chem.*, **239**, 2667 (1964).

26

Catalase and Peroxidase

Catalase (hydrogen-peroxide:hydrogen-peroxide oxidoreductase; EC 1.11.1.6) and peroxidase (donor:hydrogen-peroxide oxidoreductase; EC 1.11.1.7) differ markedly in their protein characteristics but have several common features in their mechanism of action. Catalase catalyzes a reaction in which one molecule of hydrogen peroxide acts as a donor and a second molecule acts as an acceptor of hydrogen atoms:

$$2H_2O_2 \xrightarrow{\text{catalase}} 2H_2O + O_2 \quad (1)$$

The products of the reaction are water and molecular oxygen. Peroxidase catalyzes a reaction in which hydrogen peroxide acts as the acceptor and another compound, AH_2, acts as the donor of hydrogen atoms:

$$H_2O_2 + AH_2 \rightarrow 2H_2O + A \quad (2)$$

Molecular oxygen is not a product of the reaction. The nature of AH_2 and A will be discussed later. The two types of activity [Eqs. (1) and (2)] often are referred to as "catalatic" and "peroxidatic" activities, respectively.

I. PEROXIDASE

In 1855, Schoenbein observed that extracts from some mushrooms and animal tissues caused the development of a blue color in guaiac solutions in the presence of air or with dilute solutions of hydrogen peroxide [1]. This activity is due to peroxidase. Peroxidase is widely distributed in higher plants, with especially high concentrations in fig sap and horseradish. It is also found in some animal tissues and in microorganisms. In 1943, Theorell crystallized horseradish peroxidase [2]; the enzyme has now been crystallized from several other sources.

A. Classes of Peroxidases

The peroxidases are classified as follows:

1. Iron-Containing Peroxidases

a. Ferriprotoporphyrin Peroxidases

The ferroprotoporphyrin group includes peroxidases from higher plants (horseradish, Japanese radish, turnip, fig sap), animals (tryptophan pyrrolase, iodine peroxidase of thyroid), and microorganisms (cytochrome *c* peroxidase of yeast). These peroxidases all contain ferriprotoporphyrin III as the prosthetic group which can be removed from the protein moiety on treatment with acidic acetone. The enzymes are brown in color when highly purified, due to the ferriprotoporphyrin III cofactor.

b. Verdoperoxidases

The verdoperoxidases are found in myelocytes (myeloperoxidase), milk (lactoperoxidase), and in other tissues. The prosthetic group of these enzymes is an iron porphyrin nucleus but is not ferriprotoporphyrin III. Highly purified myeloperoxidase and lactoperoxidase are green because of their absorbance maxima in the region 570 to 690 nm (as well as near 403 nm). The prosthetic group is not removed from the protein by treatment with acidic acetone or with silver sulfate.

2. Flavoprotein Peroxidases

Flavoprotein peroxidases have been purified from several streptococci, including *Streptococcus faecalis*, and from several animal tissues. The prosthetic group of these peroxidases is FAD.

B. Protein and Prosthetic Group

The following discussion of peroxidase is restricted to horseradish peroxidase. Horseradish peroxidase has a molecular weight of 40,200 and contains one ferriprotoporphyrin III (protohemin) group per molecule (Fig. 1). In protohemin, four of the six coordination bonds of iron interact with the pyrrole ring nitrogens. The other two coordination bonds, X and Y, are occupied by water molecules or OH^-, depending on the pH. In peroxidase, one of the two remaining coordination bonds (X or Y) is to a carboxyl group of the protein while the other is coordinated to an amino group or to a water molecule.

Because of protohemin, peroxidase has absorption maxima not only at 275 nm (tyrosine and tryptophan residues of the protein) but also at 403, 497, and 641.5

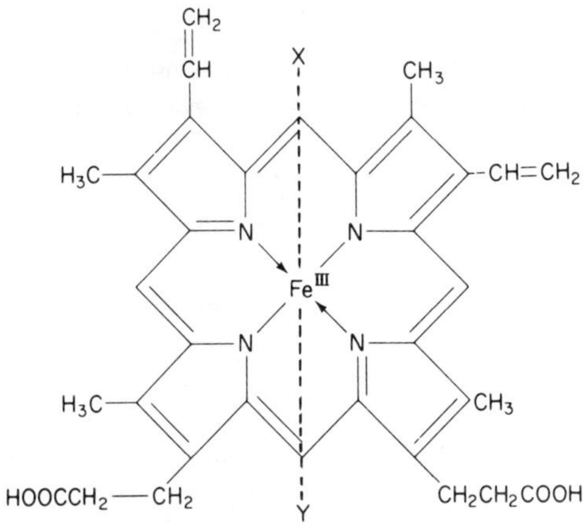

Figure 1 Structure of ferriprotoporphyrin III (protohemin).

nm (Fig. 2). Completely pure horseradish peroxidase has a ratio of absorption at 403 to 275 nm of 3.04, which is called the RZ (for *Reinheitszahl*) value. It has been recommended that the purity of peroxidase preparations be expressed in terms of RZ values. There are several disadvantages to this method of expressing the purity of hemoproteins.

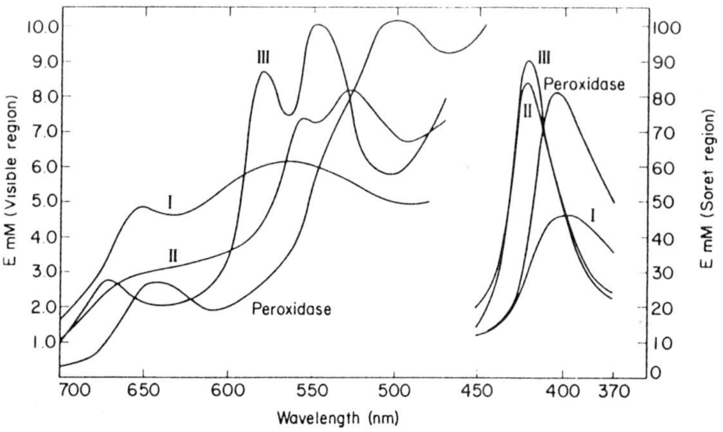

Figure 2 Spectra of peroxidase and its peroxide-induced compounds I, II, and III. The scale for the Soret bands is one-tenth of that for the visible region. (From Ref. 3, p. 280, by courtesy of Academic Press.)

1. RZ refers only to purity as a protein, not as an enzyme.

2. Absorption at 275 nm is influenced by the presence of inorganic iron (Fe^{3+}), while absorption at 403 nm is influenced by other hemoprotein contaminants.

3. The ratio of absorbance at 403 to 275 nm is different for peroxidases from different sources and for the multiple molecular forms of horseradish peroxidase because of their variable aromatic amino acid and carbohydrate contents. The multiple molecular forms of horseradish peroxidase have RZ values ranging from 2.50 to 4.19 (Table 1).

Protohemin can be removed from the protein at −15°C with acidic acetone without denaturing the protein. The protein is devoid of all peroxidase activity and protohemin has very little activity (<0.001% that of peroxidase). When the two are recombined, most of the original activity can be recovered.

C. Reactions Catalyzed by Peroxidase

Peroxidase catalysis is associated with four types of activity: peroxidatic, oxidatic, catalatic, and hydroxylation.

1. *Peroxidatic*

$$H_2O_2 + 2AH_2 \rightarrow 2H_2O + \text{polymerized products (HAAH)} \quad (3)$$

2. *Oxidatic*

$$2\begin{array}{c}\text{HO—C—COOH}\\ \Vert \\ \text{HOOC—C—OH}\end{array} + O_2 \rightarrow 2\begin{array}{c}\text{O=C—COOH}\\ | \\ \text{O=C—COOH}\end{array} + 2H_2O \quad (4)$$

3. *Catalatic*

$$2H_2O_2 \rightarrow 2H_2O + O_2 \quad (5)$$

4. *Hydroxylation*

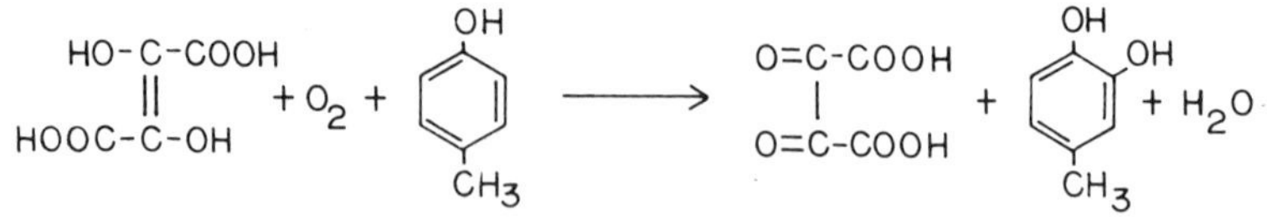

(6)

p-Cresol 4-Methylcatechol

Under the usual assay conditions in vitro where a phenolic substrate is used, only the peroxidatic reaction is of importance. Peroxidatic reactions occur when *p*-cresol, guaiacol, resorcinol, aniline, and so on, are used as substrates. Oxidatic reactions occur when the substrate is dihydroxyfumaric acid, ascorbic acid, hydroquinone, and so on, and they require O_2.

In the absence of a hydrogen donor, peroxidase converts hydrogen peroxide to

Table 1 RZ Values for Isozymes of Horseradish Peroxidase

Isozyme	RZ	Isozyme	RZ
A-1	4.19	C	3.42
A-2	4.12	D	2.57
A-3	3.71	E	2.50
B	3.37		

Source: Ref. 4.

H_2O and O_2 in a catalatic reaction [Eq. (5)]. This reaction is at least 1000 times slower than the peroxidatic and oxidatic reactions.

In the presence of certain hydrogen donors, particularly dihydroxyfumaric acid and molecular oxygen, peroxidase hydroxylates a variety of aromatic compounds, including tyrosine, phenylalanine, *p*-cresol, and benzoic and salicylic acids [Eq. (6)].

At the risk of oversimplification we shall describe a mechanism of action for peroxidase that incorporates all four types of reactions. The basis for such a mechanism includes the spectral changes which are observed on addition of hydrogen peroxide to peroxidase in the absence of a hydrogen donor, which we shall describe first.

D. Spectral Changes of Peroxidase on Combining with Hydrogen Peroxide

Purified horseradish peroxidase is a brown protein with several absorption maxima but the one at 403 nm is the most intense ($\epsilon_m = 1.08 \times 10^5\ M^{-1}\ \text{cm}^{-1}$). When a slight excess of hydrogen peroxide is added to peroxidase in the absence of hydrogen donor, there is rapid formation of a green compound. The green compound in turn is converted to a pale red compound which decays much more slowly [Eq. (7)]. k_1 for formation of Compound I is $9 \times 10^6\ M^{-1}\ \text{sec}^{-1}$ at pH 4.6

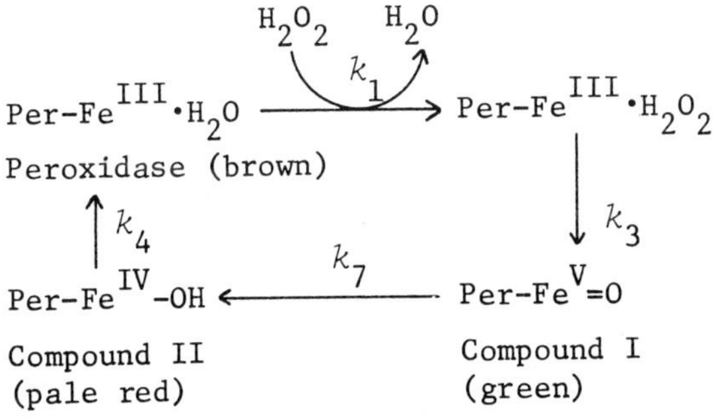

(7)

and 25°C, while k_7 and k_4 are 4.0 and 0.02 sec^{-1}, respectively. k_1 appears to be smaller than k_3 in these reactions. There are marked changes in the spectra of the intermediates formed in the peroxidase reaction (Fig. 2). The absorbance in the Soret region (400 nm) is much higher than that at the higher wavelengths. When Compound I is formed there is a marked decrease in absorbance at 403 and 497 nm and an increase in absorbance above 550 nm (broad peak at 575 nm and a peak at 650 nm). This accounts for the change in color from brown to green on formation of Compound I. There are corresponding distinctive spectral changes in the formation of Compounds II and III. Compound III (bright red) and Compound IV (brilliant green, not shown) are formed at high hydrogen peroxide concentrations. Compounds III and IV are thought not to be involved directly in the mechanism of action of peroxidase, so we shall not mention them further.

E. Mechanism of Action of Peroxidase

On the basis of the experimental data available, the following general mechanism for the action of peroxidase is proposed (Fig. 3). In the first step hydrogen peroxide replaces water at one of the available coordination bonds of the iron of the protohemin to form the enzyme·substrate complex. k_1 is quite fast ($9 \times 10^6\ M^{-1}$ sec^{-1}) and is a second-order rate constant, indicating that it represents the formation of an enzyme·substrate complex (Per-$Fe^{III} \cdot H_2O_2$). k_3, the rate constant for formation of Per-$Fe^{V}=O$ (Compound I), is faster than k_1. k_2, which involves either conversion of Compound I to Per-$Fe^{III} \cdot H_2O_2$ or formation of Per-$Fe^{III} \cdot H_2O$ from Per-$Fe^{III} \cdot H_2O_2$, is very slow and is on the order of 0.1 to 3 sec^{-1}. In the third step of the peroxidatic reaction, Compound I reacts with an exogenous donor, if available, to give Compound II (Per-Fe^{IV}-OH) and a free radical from the hydrogen donor. The rate of this reaction is very rapid and k_7 is usually 40 to 100 times larger than k_4. The rate constant depends on the nature of the hydrogen donor. In the fourth step of the peroxidatic reaction, Compound II reacts with a second molecule of the hydrogen donor to regenerate the original enzyme (Per-$Fe^{III} \cdot H_2O$) and a free radical from the hydrogen donor. The rate constant for this reaction is dependent on the nature of the hydrogen donor and varies from 8×10^7 M^{-1} sec^{-1} for *p*-hydroxydiphenyl to 17 for nitrite. At high concentrations of most donors except *p*-hydroxydiphenyl and *o*-phenylenediamine, the rate-determining step of the reaction is that controlled by k_4 (i.e., the conversion of Compound II to peroxidase).

Although a number of compounds can act as the hydrogen donor, only hydrogen peroxide, methyl peroxide, and ethyl peroxide react with peroxidase to form Compound I. k_1 is 9.0×10^6, 1.5×10^6, and $3.6 \times 10^6\ M^{-1}\ sec^{-1}$ for HOOH, CH_3OOH, and CH_3CH_2OOH, respectively.

In the absence of added hydrogen donor, the rate of conversion by the catalytic pathway of Compound I to Compound II is 4 sec^{-1}, while that for conversion of Compound II to peroxidase is 0.02 sec^{-1}. The catalatic activity of peroxidase

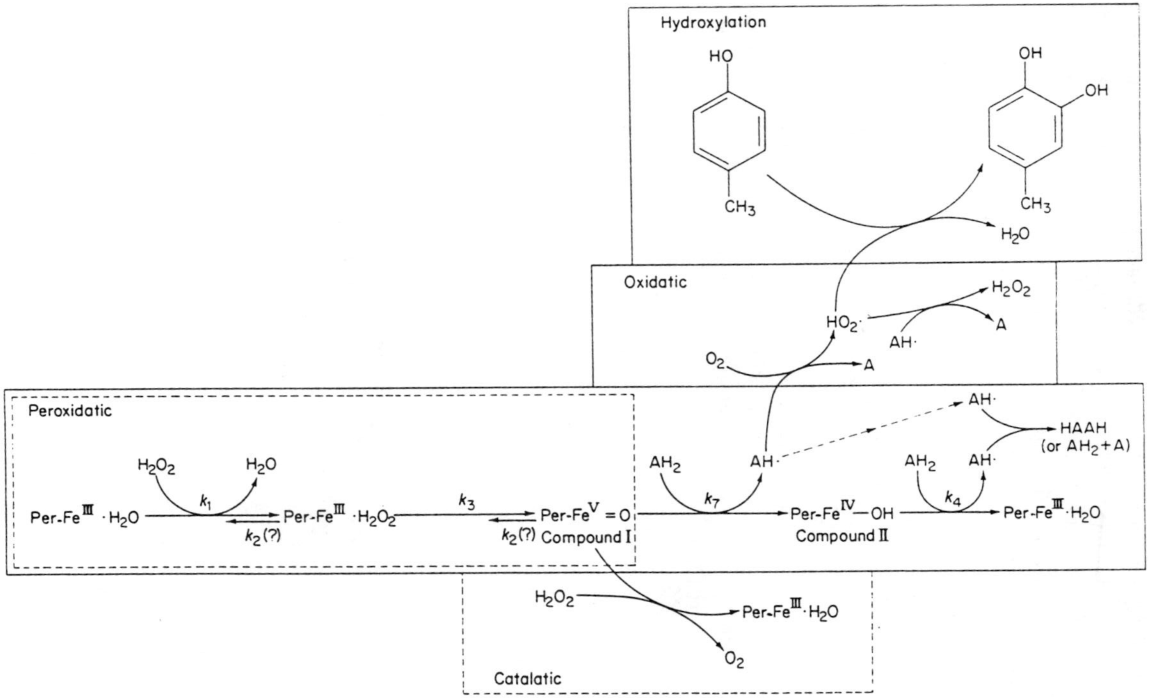

Figure 3 Proposed scheme for mechanism of peroxidase action in the four types of reactions in which the enzyme is involved.

[Eq. (5)] is shown as a second molecule of hydrogen peroxide reacting directly with Compound I to form peroxidase, H_2O, and O_2, as is the case with catalase.

The fate of the free radicals, $AH^{\cdot}$, produced in the peroxidative reaction depends on their nature and the presence of O_2 and phenolic compounds.

1. If AH_2 is guaiacol, for example, the free radicals interact with each other to give a polymerized products shown as HAAH. (HAAH can serve as a hydrogen donor, leading to higher degrees of polymerization, as with production of tetraguaiaiol from guaiaiol.)
2. If AH_2 is ascorbic acid or dihydroxyfumaric acid, for example, free radicals react with each other to give one molecule each of reduced and oxidized compound (AH_2 and A).
3. O_2 may react with the free radical of certain hydrogen donors (e.g., dihydroxyfumaric acid) to give a molecule of oxidized donor and an $HO_2^{\cdot}$ free radical which, in the absence of an aromatic compound, reacts with a second $AH^{\cdot}$ to give $A + H_2O_2$.
4. In the presence of a variety of aromatic compounds, the $HO_2^{\cdot}$ radical can hydroxylate the compound as shown for *p*-cresol. According to this scheme, the oxidatic and hydroxylation activities of peroxidase are not a result of the activity of the enzyme directly but come about through nonenzymatic secondary reactions caused by formation of $HO_2^{\cdot}$ free radicals. Proteins can be cross-linked via tyrosyl residues in vivo and in vitro by peroxidase, according to this reaction.

F. Assay of Peroxidase Activity

Assays for peroxidase activity involve the use of various hydrogen donors, such as guaiacol, pyrogallol, mesidine, cytochrome *c*, uric acid, and dihydroxyphenylalanine. Willstätter developed the pyrogallol reaction, which gives purpurogallin, to a high degree of perfection. However, the involved procedure and lack of information about some of the rate-limiting steps have resulted in its lack of general acceptance. The assay most frequently used in the guaiacol assay.

The guaiacol assay is very simple and readily followed in a continuous fashion in a recording spectrophotometer. Guaiacol and hydrogen peroxide are added to a buffer, usually at pH 6.0 to 7.0, and the reaction is initiated by addition of peroxidase. The increase in absorbance at 470 nm is measured as a function of time. The major product of the reaction is tetraguaiacol [Eq. (8)].

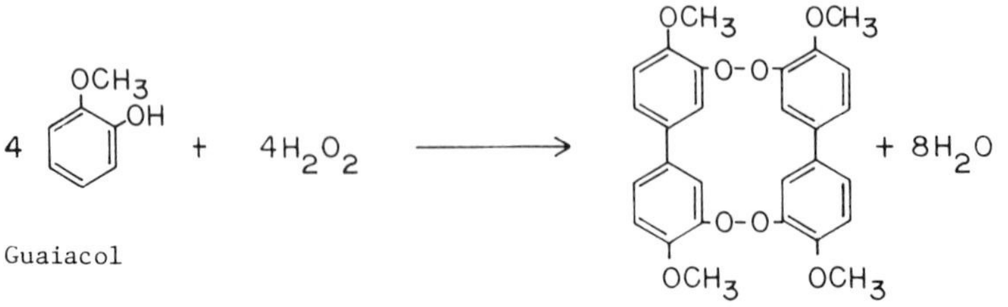

(8)

Guaiacol

Tetraguaiacol

Exercise 1. From the proposed mechanism of action of peroxidase in Fig. 3, write a complete scheme for Eq. (8) showing all the free-radical intermediates that might be formed. What other products might be formed in addition to tetraguaiacol?

From the molar extinction coefficient, ϵ_m, of tetraguaiacol of $2.66 \times 10^4\ M^{-1}\ \mathrm{cm}^{-1}$ at 470 nm and from the change in absorbance, the amount of hydrogen peroxide consumed in the reaction can be calculated. (Note that there are four hydrogen peroxides consumed for each tetraguaiacol produced.)

II. CATALASE

Catalase is found in animals, plants, and microorganisms and has been purified from all three sources. Crystalline beef liver catalase was first prepared by Sumner and Dounce in 1937 [5] and the enzyme has been crystallized from several other sources since that time.

A. Protein and Prosthetic Group

All catalases appear to have molecular weights of approximately 240,000 and to contain four protohemin groups per molecule. The molecule is composed of four subunits each of which contains a protohemin group. The four active sites of catalase appear to function independently of each other. Since the prosthetic group of catalase is ferriprotoporphyrin III (Fig. 1), the enzyme has an absorption maximum at 406 nm. The ratio A_{280}/A_{406} is approximately 1.0 for all the catalases studied.

Catalase, perhaps as a result of its subunit structure, is not as heat stable as peroxidase. It loses activity quite rapidly at 35°C while peroxidase is stable to 65 to 70°C at pH values near 7. Catalase is also less stable at alkaline conditions above pH 9, where the subunits dissociate. Protohemin is not covalently linked to the protein moiety and can be removed by treatment with acidic acetone. The mutual importance of the prosthetic group and protein can be seen from the relative rates of enzyme versus nonenzymatic decomposition of hydrogen peroxide. The turnover numbers (sec^{-1}) are 9×10^4 (at $10^{-2}\ M\ H_2O_2$), 0.05, 1.0, and 22.7 for catalase, hemin phosphate, ferric salts, and the iron chelate of triethylenetetramine, respectively.

B. Reactions Catalyzed

Catalase can catalyze two types of reactions, a catalatic and a peroxidatic reaction [Eqs. (3) and (5)]. Under the usual assay conditions, the catalatic reaction is observed. To see the peroxidatic reaction, very low concentrations of hydrogen peroxide are used in the presence of hydrogen donors such as methanol, ethanol, and phenols.

C. Spectral Changes of Catalase on Combining with Peroxides

When low concentrations of hydrogen peroxide are added to catalase, there is formation of a spectrally different compound that resembles very closely that of Compound I of peroxidase (Fig. 2). Low concentrations of Compound II are also formed in the case of catalase and hydrogen peroxide; with ethyl peroxide, rather high concentrations of Compound II are produced. Compound III is not produced with ethyl and methyl peroxidases alone but is formed when a high concentration of hydrogen peroxide is added along with one of these peroxides. Compound III probably does not play a direct role in the activity of catalase [6].

D. Hydrogen Donors for Catalase; Mechanism of Action

Only HOOH, CH_3OOH, and CH_3CH_2OOH serve as primary substrates for catalase, while a number of compounds serve as hydrogen donors (Fig. 4). We shall discuss further only the D_2 (phenols), D_3 (alcohols), and D_6 (hydrogen peroxide) groups of hydrogen donors.

With HOOH, CH_3OOH, and CH_3CH_2OOH as substrates, k_1 is 6×10^6, 8.5×10^5, and $2 \times 10^4\ M^{-1}\ sec^{-1}$, respectively, for formation of Compound I from catalase at pH 7 and 25°C. k_1 is a second-order rate constant indicating the formation of an enzyme·substrate complex, which then reacts to form Compound I with a first-order rate constant (k_2) larger than k_1. k_1 for catalase and HOOH

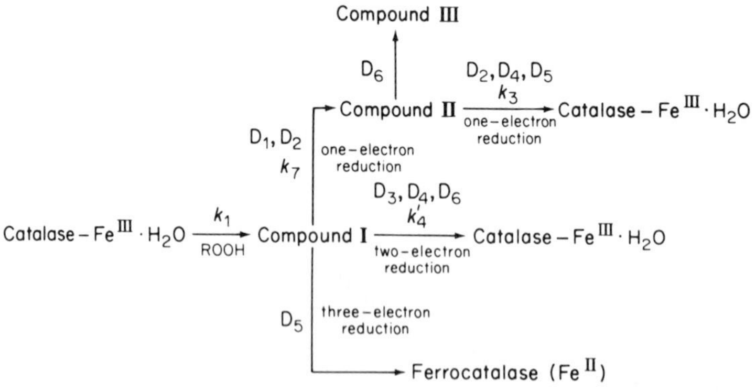

Figure 4 Reactions of catalase with hydrogen donors. The D's refer to the donors: D_1 is ascorbate or ferrocyanide, D_2 is phenols, D_3 is alcohols or formic acid, D_4 is nitrite, D_5 is azide or hydroxylamine, and D_6 is hydrogen peroxide. (From Ref. 7, p. 305, by courtesy of Elsevier Publishing Co.)

($6 \times 10^6\ M^{-1}\ sec^{-1}$) is about the same as that for peroxidase and HOOH ($9 \times 10^6\ M^{-1}\ sec^{-1}$). However, k_1 with catalase is much more dependent on the nature of the peroxide than with peroxidase (2×10^4 versus $3.6 \times 10^6\ M^{-1}\ sec^{-1}$ for catalase and peroxidase with CH_3CH_2OOH, respectively).

When hydrogen peroxide is also the hydrogen donor (the catalatic reaction), the reaction proceeds directly from Compound I to produce water and O_2 and to regenerate the enzyme catalase-$Fe^{III} \cdot H_2O$. The rate constant for this reaction (k_4') is usually about three times k_1, so that in the case of hydrogen peroxide, k_4' is $1.8 \times 10^7\ M^{-1}\ sec^{-1}$. Therefore, both k_1 and k_4' are rate-determining steps in the catalase-catalyzed decomposition of hydrogen peroxide and $k_0 = k_1k_4'/(k_1 + k_4') = 4.5 \times 10^6\ M^{-1}\ sec^{-1}$.

When the reaction is performed at very low concentrations of hydrogen peroxide ($\sim 10^{-9}\ M$) in the presence of alcohols such as methanol or ethanol, there is oxidation of the alcohol to aldehyde (CO_2 and H_2O in the case of methanol). The reaction has been shown to proceed by the same pathway as when hydrogen peroxide is the hydrogen donor. In the case of alcohols as the hydrogen donor, k_4' is $1 \times 10^3\ M^{-1}\ sec^{-1}$ for both methanol and ethanol and is $9 \times 10^5\ M^{-1}\ sec^{-1}$ for formic acid.

Exercise 2. The rate dx/dt for the conversion of Compound I to catalase and products at low H_2O_2 concentrations is given by the expression $dx/dt = k_0D$, where D is the hydrogen donor concentration and k_0 is the observed rate constant. Calculate the concentration of hydrogen peroxide at which the rate of conversion of hydrogen peroxide to H_2O and O_2 would equal the rate of oxidation of $1 \times 10^{-2}\ M$ ethanol to acetaldehyde.

When hydrogen peroxide and catalase are added to a solution containing certain types of aromatic compounds, such as pyrogallol or *p*-cresol, these compounds are oxidized to colored products. These reactions occur in an analogous fashion to those catalyzed by peroxidase [Eq. (9)]. k_3 is 80 and 8 $M^{-1}\ sec^{-1}$

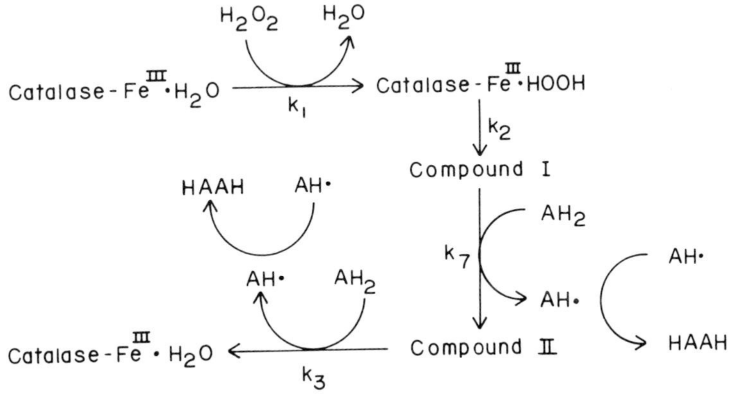

(9)

for pyrogallol and *p*-cresol, respectively, while k_7 is approximately 240 and 24 M^{-1} sec^{-1}, respectively. These rates are about 4×10^4 times lower than when the reaction is catalyzed by peroxidase. Therefore, catalase, while able to catalyze peroxidatic reactions, is not a very efficient catalyst for these types of reactions, just as peroxidase is not a very efficient catalatic catalyst.

E. pH Dependence

The catalatic activity of catalase is independent of pH over the range 3 to 9. Above pH 9 and below pH 3 there is a decrease in activity of the enzyme. At alkaline pHs the enzyme dissociates into subunits, while at acid pH values the protohemin group dissociates from the protein. The steps controlled by k_7 and k_3 (Fig. 4) are also pH independent when neutral donors are used. When donors capable of ionizing within the pH range measured are used, the observed effect of pH is on the donor and not the enzyme. Therefore, it does not appear there are any prototropic groups on the enzyme that influence activity within the pH range where catalase activity is measured.

F. Effect of Temperature

The Q_{10} value of the catalatic reaction above 25°C is 1.1 (Fig. 5). This is the lowest Q_{10} found for an enzyme-catalyzed reaction and is well outside the expected normal range of 2 to 3. k_4' changes with temperature in the expected manner and

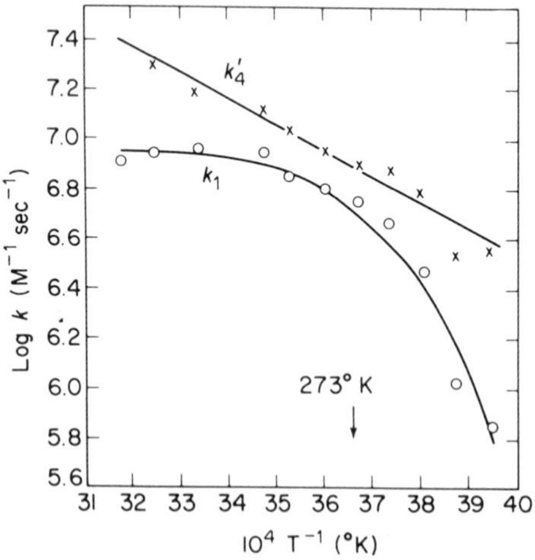

Figure 5 Effect of temperature on k_1 and k_4' in the catalatic reaction for bacterial catalase, pH 7.5, 50% glycerol. (From Ref. 8, p. 322, by courtesy of Elsevier Publishing Co.)

E_a is 5.1 kcal/mol. k_1 is essentially independent of temperature above 25°C ($Q_{10} \sim 1.1$) but shows a marked temperature dependence at lower temperatures (Fig. 5). This must represent a change in the rate-determining step of the reaction. Perhaps the rate-determining step above 25°C is the diffusion-controlled formation of the enzyme·substrate complex. The Q_{10} for such a process is 1.03 since it is dependent only on the change in kinetic energy of the system. Below 25°C perhaps conversion of the proposed enzyme·substrate complex to Compound I becomes rate controlling.

G. Assay of Catalase Activity

Catalase activity can be measured spectrophotometrically, manometrically, polarographically, and titrimetrically. The overall reactions catalyzed in vitro is

$$2H_2O_2 \rightarrow 2H_2O + O_2 \qquad (10)$$

Hydrogen peroxide has a molar extinction coefficient of 52 M^{-1} cm^{-1} at 235 nm and its disappearance can be followed by a continuous spectrophotometric method when purified catalase preparations are used. Since O_2 is a product of the reaction, catalase activity can be determined readily by polarographic (preferred) and manometric techniques. The least desirable, in terms of time involved, is the titrimetric method, which involves removal of aliquots of the reaction and titration of the remaining hydrogen peroxide with standard potassium permanganate solution in an acid solution.

REVIEW QUESTIONS

See the end of Chapter 27.

REFERENCES

General References

Robinson, D. S., Peroxidases and catalases in foods, in *Oxidative Enzymes in Foods* (D. S. Robinson and N. A. M. Eskin, eds.), Elsevier Applied Science, New York, 1991, pp. 1–47.

Schonbaum, G. R., and B. Chance, Catalase, in *The Enzymes*, 3rd Ed. (P. D. Boyer, ed.), Vol. XIII, Academic Press, New York, 1976, pp. 363–408.

Walsh, C., *Enzymatic Reaction Mechanisms*, W.H. Freeman, San Francisco, 1979, pp. 464–500.

Whitaker, J. R., Mechanisms of oxidoreductases important in food component modification, in *Chemical Changes in Foods During Processing* (T. Richardson and J. W. Finley, eds.), AVI, Westport, Conn., 1985, pp. 121–176.

Cited References

1. C. F. Schoenbein, *Kerhandl. Naturforsch. Ges. Basel*, **1**, 339 (1855).
2. H. Theorell, *Arkiv Kemi Mineral. Geol.*, **16A**(2), 1 (1943).
3. P. Nicholls, in *Oxygenases* (O. Hayaishi, ed.), Academic Press, New York, 1962, p. 273.
4. L. M. Shannon, E. Kay, and J. Y. Lew, *J. Biol. Chem.*, **241**, 2166 (1966).
5. J. B. Sumner and A. L. Dounce, *J. Biol. Chem.*, **121**, 417 (1937).
6. P. Nicholls and G. R. Schonbaum, in *The Enzymes* (P. D. Boyer, H. Lardy, and K. Myrbäck, eds.), Vol. 8, Academic Press, New York, 1963, p. 147.
7. D. Keilin and P. Nicholls, *Biochim. Biophys. Acta*, **29**, 302 (1958).
8. G. K. Strother and E. Ackerman, *Biochim. Biophys. Acta*, **47**, 317 (1961).

27

Lipoxygenase (Lipoxidase)

Lipoxygenase (linoleate:oxygen oxidoreductase; EC 1.13.11.12) was discovered in 1928 when Bohn and Haas [1] reported the presence of a carotene-destroying enzyme, "carotene oxidase," in soybeans. In 1932 [2] it was reported that soybeans contain an enzyme, subsequently named "lipoxidase," which oxidizes unsaturated fats. In 1940 [3] it was shown that carotene oxidase and lipoxygenase were the same enzyme.

Lipoxygenase is found in a wide variety of plants, particularly the legumes [4]. It is present in alfalfa; various legumes, including peas, beans, and peanuts; radishes; and potato. While the highest activity is found in soybeans, urd beans, mung beans, and green peas also contain relatively high concentrations of the enzyme (Table 1).

There are a number of lipoxygenases in animal tissues [4] which are being actively investigated. Oxidation of unsaturated fatty acids in animal tissues can also be the result of nonenzymatic hemin catalysis.

There are at least three detrimental effects of the action of lipoxygenase in foods: (a) There is destruction of the essential fatty acids, linoleic, linolenic, and arachidonic acids; (b) the free radicals produced damage other compounds, including vitamins and proteins; and (c) there is a development of off-flavor and odor. In beans and peas this is characterized as a haylike flavor. This is a particularly serious problem in unblanched frozen peas, beans, and corn since lipoxygenase can continue its action even at low temperatures.

Exercise 1. The activation energies for lipoxygenase and nonenzymatically catalyzed oxidation of unsaturated fatty acids have been reported to be 4.3 and

Table 1 Relative Amounts of Lipoxygenase in Several Plants

Source	Relative activity (%)	Source	Relative activity (%)
Soybeans	100	Peas	35
Urd beans	60	Wheat	2
Mung beans	47	Peanuts	1

Source: Ref. 5.

15.2 kcal/mol, respectively [6]. Assuming that an equal amount of oxidation is due to each cause at 25°C, calculate the relative rates of oxidation by the two at −10°C. Based on an assumed value of 0.1% by weight of lipoxygenase in fresh green peas and on a specific activity of pure lipoxygenase (MW = 102,000) of 360 mol of substrate oxidized per mole of lipoxygenase per second at 20°C [7], calculate the theoretical amount of linoleic acid that could be oxidized in 15 days of storage at −10°C. How does this compare with the total amount of linoleic, linolenic, and arachidonic acids in green peas?

Lipoxygenase-1 was crystallized from soybeans in 1947 [7,8] and has a molecular weight of 102,000 and an isoelectric point of 5.4. Its specific activity at 20°C was reported to be 360 moles of substrate oxidized per mole of enzyme per second [7]. The pH optimum is around 9, although it varies somewhat depending on the presence or absence of detergents added to solubilize the substrate. Problems caused by insolubility of substrate and effect of solubilizing agents, such as detergents and organic solvents, on observed enzyme activity are difficult to resolve. Interpretations of data must always take these effects into account. Lipoxygenase-2 and lipoxygenase-3 are also found in soybeans. They are similar in molecular weights to lipoxygenase-1 but are optimally active at pH 7.

I. pH-ACTIVITY PROFILE

It is difficult to interpret the published pH–activity profiles of lipoxygenases, partially due to insolubility of linoleic acid, the usual substrate, below pH 7. Most workers have found bell-shaped curves with maxima near 7.0 to 8.0 [6,9], but there is a possibility that the decrease below 7 is due to insolubility of substrate [7]. Addition of organic solvents or detergents to solubilize the substrate at low pH values often results in peculiar pH–activity profiles because of their effects on the triglyceride micelles, the substrate for lipoxygenases.

Activities on linoleic acid cannot be measured below pH 7 because of insolubility of the substrate. To overcome difficulties due to insolubility of substrate and presence of detergents, it was suggested that octadeca-9,12-dienyl sulfate (linoleyl

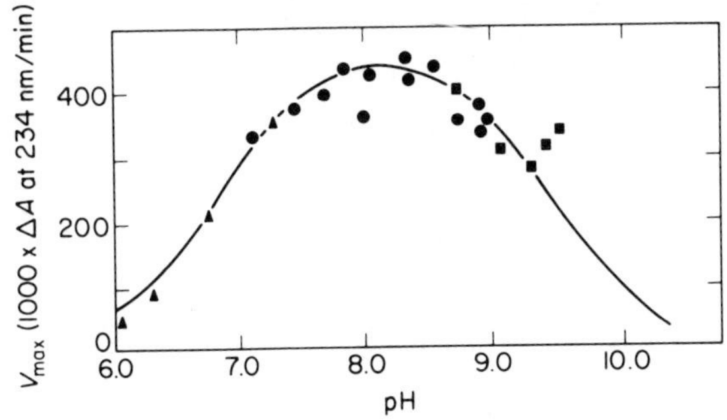

Figure 1 pH–activity profile for soybean lipoxygenase-1 action on octadeca-9,12-dienyl sulfate, a water-soluble substrate. (From Ref. 10, p. 907, by courtesy of the Chemical Society.)

sulfate) be used as a substrate since it is water soluble [10]. Soybean lipoxygenase-1 appears to have the same specific activity on it as on linoleate. With this compound the pH–activity profile for lipoxygenase-1 is a symmetrical bell-shaped curve with a pH maximum of 8.15 and pK_a values of 6.8 and 9.4 (Fig. 1). The availability of this substrate should make kinetic studies on lipoxygenase more meaningful, but it has rarely been used.

II. SPECIFICITY OF LIPOXYGENASE

Soybean lipoxygenase-1 is highly specific for the oxidation of fatty acids (and in triglycerides) which contain a *cis,cis* penta-1,4-diene unit (—CH═CH—CH_2—CH═CH—). The naturally occurring polyunsaturated fatty acids, linoleic, linolenic, and arachidonic acids, contain one or more *cis,cis* penta-1,4-diene units. There are four possible isomers of linoleic acid. Only one of these isomers, the *cis,cis* linoleic acid, serves as a substrate for lipoxygenase, which indicates the stereospecific requirements needed to fit the active site of the enzyme.

Oleic acid, with a single double bond at C(9), is not a substrate for lipoxygenases. The isomer of linoleic acid where the double bonds are conjugated (9,11-octadecadienoic acid) also is not a substrate.

The enzyme requires the susceptible methylene group of the penta-1,4-diene unit to be in the ω-8 position [11]. The enzyme shows specificity from the terminal methyl end of the fatty acid; therefore, it is best to use the ω numbering system, which starts from that end of the chain (Fig. 2). Acids with the methylene group of their penta-1,4-diene units in the ω-10 or ω-11 positions do not serve as substrates for soybean lipoxygenase-1.

Addition of a *cis* double bond at the ω-3 position does not affect specificity of

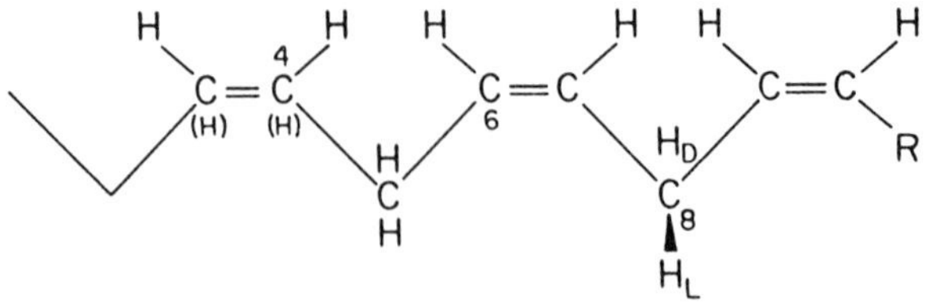

Figure 2 The essential part of a polyunsaturated fatty acid which permits it to be a substrate for lipoxygenase. Required are *cis* double bonds at ω-6 and ω-9 positions, a hydrogen in the H_L position on ω-8 carbon, and hydrogen at the ω-6 carbon. The ω-3–ω-4 position may be saturated or unsaturated.

soybean lipoxygenase-1 for the substrate, as linolenic acid is a good substrate. Addition of other double bonds between ω carbon 10 and the carboxyl end of the fatty acid also does not prevent the compound from being a substrate, as arachidonic acid (5,8,11,14-eicosatetraenoic acid) and 8,11,14-eicosatrienoic acids are substrates. There is no oxidation at the other penta-1,4-diene positions of these fatty acids.

The action of soybean lipoxygenase-1 on several fatty acids is shown in Fig. 3. All fatty acids that have an ω-8 methylene group located in a penta-1,4-diene unit are substrates except for the one acid that is substituted with a methyl group on the ω-6 position. In all cases the hydroperoxy group is introduced at the ω-6 position exclusively, except for a very small amount of ω-10 hydroperoxy acid formed with linoleic acid (which may be due to autoxidation). In the case of autoxidation of

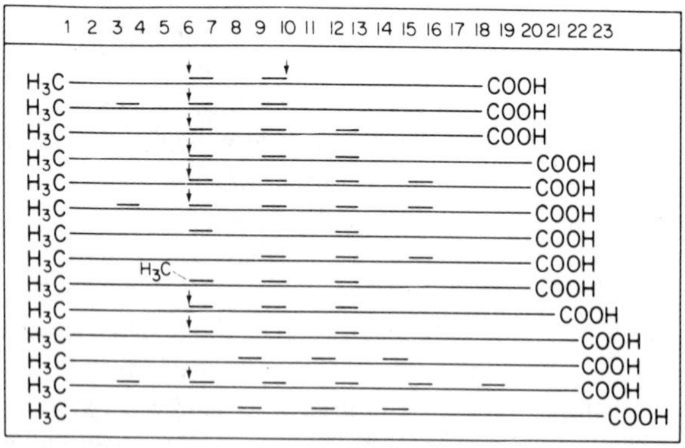

Figure 3 Positional specificity of soybean lipoxygenase–catalyzed reactions. Positions of the oxygen introduced are indicated by the arrows. (From Ref. 11, p. 5334, by courtesy of the American Society of Biological Chemists, Inc.)

linoleic acid (in the absence of lipoxygenase) a 50:50 mixture of the 13-hydroperoxy (ω-6) and 9-hydroperoxy acids (ω-10) is formed.

Exercise 2. Design an experiment that would permit you to determine what percentage of the oxidation of linoleic acid (or other similar acid) in a food such as a green pea is due to autoxidation and how much is due to lipoxygenase action.

Soybean lipoxygenase-1 is stereospecific for the hydrogen atom of L-configuration of the ω-8 methylene group (Fig. 2). This can be shown by labeling the substrate with tritium at either the indicated H_D or H_L position [11]. If labeled at the H_D position, the label is found almost exclusively in the conjugated fatty acid product, while if the label is at position H_L, the tritium is found exclusively in the solvent.

The substrate specificity of lipoxygenases from a number of plant and animal sources have been determined, using arachidonic acid (5,8,11,14-all *cis*-eicosatetraenoic acid) as the substrate, since the animal lipoxygenases have rather low activity on linoleic and linolenic acids [12]. The specificities are shown in Table 2. The nomenclature includes not only the positional specificities but also the stereo and chiral specificities.

Free fatty acid substrates are preferred by soybean lipoxygenase-1, which has a pH optimum of 9. Other isoenzymes of soybean, with pH optimal between 6.5

Table 2 Nomenclature of Various Lipoxygenases

Source	Nomenclature	Short name
Soybean 1	$[C_{n-8}]$-$L_{pro\text{-}S}$:[+2]-L_S	15-LOX
Corn germ	$[C_{11}]$-$D_{pro\text{-}R}$:[−2]-D_S[a]	9-LOX[b]
Blood platelets	$[C_{n-11}]$-$L_{pro\text{-}S}$:[+2]-L_S	12-LOX
Reticulocytes	$[C_{n-8}]$-$L_{pro\text{-}S}$:[+2]-L_S[c]	15-LOX
	$[C_{n-11}]$-L-$_{pro\text{-}S}$:[+2]-L_S[c]	12-LOX
Leukocytes	$[C_7]$-$D_{pro\text{-}S}$:[−2]-D_S[a,c]	5-LOX
	$[C_{n-11}]$-$L_{pro\text{-}S}$:[+2]-L_S[c]	12-LOX
	$[C_{n-8}]$-$L_{pro\text{-}S}$:[+2]-L_S[c]	15-LOX
Potato tuber	$[C_7]$-$D_{pro\text{-}S}$:[−2]-D_S[a,c]	5-LOX
Cyclooxygenase (first step)	$[C_{13}]$-$L_{pro\text{-}S}$:[−2]-L_R[a]	11-LOX
Leukotriene A_4 synthase (5,6-LTA_4)	$[C_{10}]$-$D_{pro\text{-}R}$:[−2][a] hydroperoxy fatty acid/ epoxide synthase	

Source: Ref. 12, p. 289, by courtesy of John Wiley & Sons.

[a]For lipoxygenases with a [−] rearrangement of the fatty acid radical the positional specificity of the initial hydrogen removal has not been studied in detail. Therefore, the position of the hydrogen removal with arachidonic acid is indicated.

[b]With linoleic acid as substrate.

[c]Conclusion from the stereochemistry of the dioxygenation product to the stereoselective hydrogen removal according to the antarafacial character of the lipoxygenase reaction.

Table 3 Substrate Specificity of Lipoxygenase Isoenzyme-1 from English Pea and Soybean Lipoxygenase-1

		Pea isoenzyme-1		Soybean lipoxygenase-1	
Substrate[a]	Detergent or solvent	Activity (mM min^{-1})	%[b]	Activity (mM min^{-1})	%[b]
Linoleic acid	Tween-20	0.197	188	0.467	45.8
Linoleic acid	Ethanol	0.105	100	1.02	100
Linoleic acid	Ethanol + 0.05% BSA	0.048	46	0.277	27.2
Linoleyl alcohol	Ethanol	0.073	70	0.0095	0.93
Linoleyl acetate	Ethanol	0.030	29	0.0047	0.46
Methyl linoleate	Tween-20	0.054	51	0.0026	0.25
Methyl linoleate	Ethanol	0.036	34	0.0066	0.65
Ethyl linoleate	Ethanol	0.019	18	0.0036	0.35
Propyl linoleate	Ethanol	0.023	22	0.0098	0.96
Trilinolein	Ethanol	0.081	77	0.013	1.3

Source: Ref. 13, p. 210, by courtesy of the American Chemical Society.
[a]Substrate concentration was 0.2 mM in all cases, in 0.2 M phosphate buffer, pH 7.0, at 24°C.
[b]Relative to linoleic acid in ethanol as 100%.

and 7, and other higher plant lipoxygenases, also with pH optima near 7, have about equal activity on the free fatty acid, the methyl ester, and the triglyceride forms of the substrate. The substrate specificities of soybean lipoxygenase-1 and English pea lipoxygenase-1 are shown in Table 3 [13]. The ionic nature of the detergent used can affect the results.

III. KINETICS OF LIPOXYGENASE-CATALYZED REACTIONS

With linoleic acid and O_2 as the substrates, detailed kinetic studies indicate that both soybean lipoxygenase-1 (Fig. 4) and English pea lipoxygenase-1 (Fig. 5) follow a Bi Bi Ordered Sequential mechanism, not a Ping-Pong mechanism. The data indicate that both linoleic acid and O_2 bind to the enzyme before any product is formed; the hydrogen must be removed from the ω-8 carbon and rearrangement to the ω-6 free radical must occur before oxygenation step. With soybean lipoxygenase-1 at pH 7, the K_m for linoleic acid was 1.63×10^{-3} M and K_m for O_2 was 1.8×10^{-4} M in the presence of Tween-20. Under the same conditions, but with English pea lipoxygenase-1, K_m values for linoleic acid and O_2 were 2.67×10^{-3} M and 3.6×10^{-4} M, respectively [13].

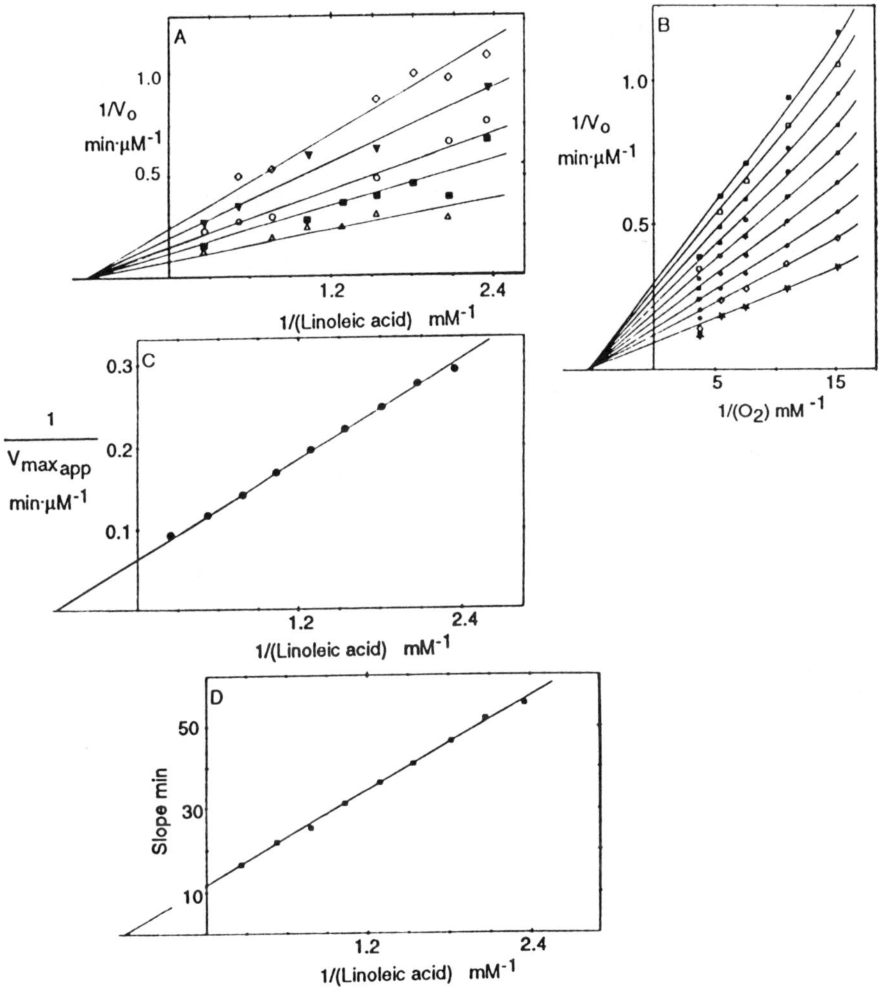

Figure 4 Effect of concentration of linoleic acid and oxygen on initial velocities with soybean lipoxygenase-1. (A), O_2 concentration in each reaction was determined with an O_2 electrode: ◇, 0.066 m*M* O_2; ▼, 0.0925 m*M* O_2; ○, 0.132 m*M* O_2; ■, 0.185 m*M* O_2; △, 0.264 m*M* O_2. (B) Effect of concentrations of linoleic acid and oxygen on initial velocities with soybean lipoxygenase-1 catalyzed reaction. Data were replotted from Fig. 4A. The concentrations of linoleic acid and symbols are ■, 0.43 m*M*; □, 0.48 m*M*; ■, 0.55 m*M*; □, 0.65 m*M*; ★, 0.77 m*M*; ○, 0.96 m*M*; ●, 1.29 m*M*; △, 1.93 m*M*; ▼, 2.42 m*M* linoleic acid. (C) Replot of the *y*-intercepts of the data of Fig. 4B versus 1/[linoleic acid]. (D) Replot of the slopes of the data of Fig. 4B versus 1/[linoleic acid]. (From Ref. 13, p. 210, by courtesy of the American Chemical Society.)

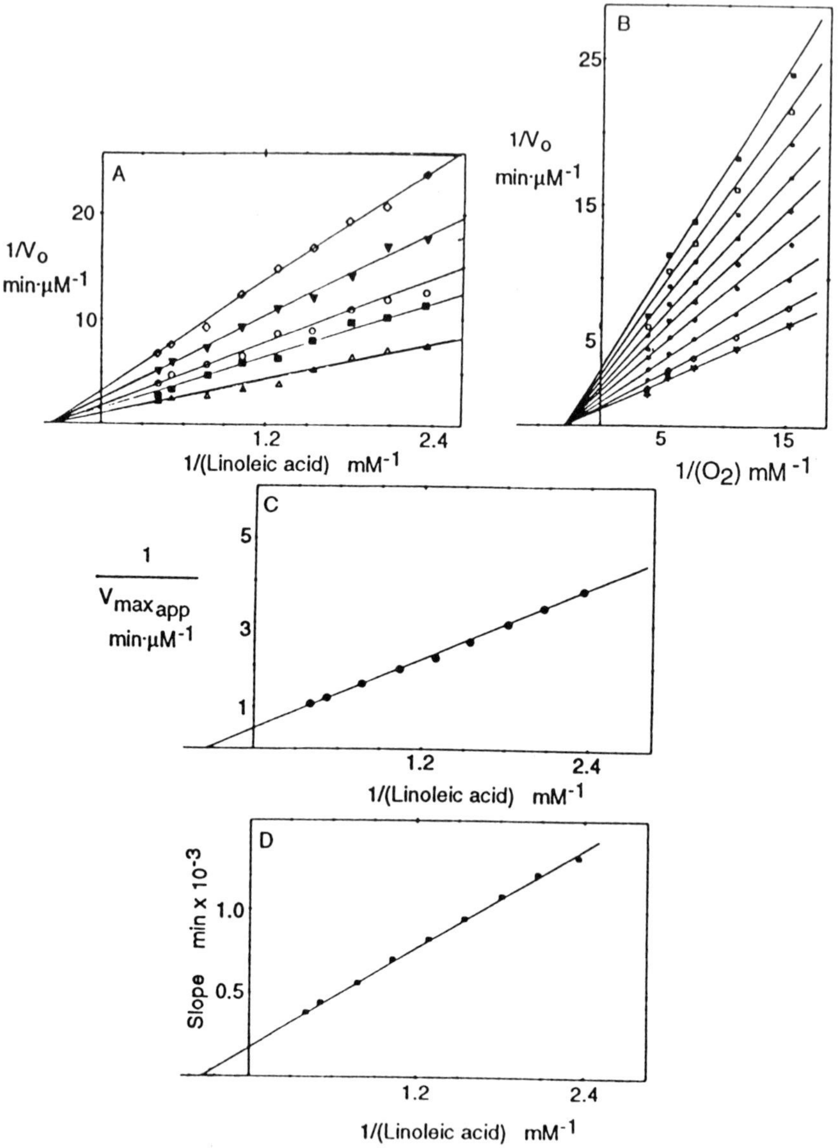

Figure 5 Effect of concentration of linoleic acid and/or oxygen on initial velocities with isolated English pea lipoxygenase isoenzyme-1. The O_2 concentration in each reaction was determined with an O_2 electrode. (A) The O_2 concentrations were the same as in Fig. 4. (B) Effect of concentration of linoleic acid and oxygen on initial velocities with isolated pea lipoxygenase isoenzyme-1. Data were replotted from Fig. 5A. The concentrations of linoleic acid and symbols were as in Fig. 4. (C) Replot of the *y*-intercepts of the data of Fig. 5B versus 1/[linoleic acid]. (D) Replot of the slopes of the data of Fig. 5B versus 1/[linoleic acid]. (From Ref. 13, p. 209, by courtesy of the American Chemical Society.)

IV. PATHWAY AND MECHANISM OF ACTION

The general reaction catalyzed by soybean lipoxygenase-1 is shown in Fig. 6 for 8,11,14-eicosatrienoic acid (linolenic acid).

The total sequence of events may be described in the following manner.

1. The enzyme forms a stereospecific complex with the polyunsaturated fatty acid.
2. The enzyme abstracts either an electron or a hydrogen atom stereospecifically from the ω-8 position, producing a free radical at ω-8 of the fatty acid.
3. While still attached to the enzyme, the fatty acid free radical isomerizes to place the unshared electron at ω-6, causing conjugation and isomerization of the double bond.
4. The enzyme·product combines with O_2 to give a ternary complex (not shown).
5. O_2 reacts with the free radical at ω-6 to give a hydroperoxy free radical.
6. An electron abstracted from the enzyme (see step 2) and a proton from the medium (or an $H^\cdot$ alone) form the 15-hydroperoxy-8,11,13-eicosatrienoic acid, which then dissociates from the enzyme. Alternatively, an electron (or $H^\cdot$) could be extracted from a second molecule of 8,11,14-eicosatrienoic acid, thereby perpetuating the reaction, which is more likely.

Even though lipoxygenase was crystallized from soybeans in 1947 [7,8], no cofactor was found until 1973, when Chan [14] reported that nonheme iron ($Fe^{2+} \rightleftharpoons Fe^{3+}$) is an essential component of the active site. This was soon confirmed by other researchers [15,16]. There are five conserved histidine residues in rabbit reticulocyte (RBC) 15-lipoxygenase [17] and human leukocyte 5-lipoxygenase [18,19] that may be responsible for chelating the iron cofactor. The mechanism proposed by de Groot et al. [20] is generally accepted (Fig. 7). The overall reaction is composed of four parts: (a) the activation of the enzyme, (b) the aerobic pathway, (c) the anaerobic pathway, and (d) the nonenzymatic pathway. According to the proposal, the iron prosthetic group of the "resting" enzyme is in the active E-(Fe^{II})·O_2 form. The resting enzyme requires a small amount of the hydroperoxy fatty acid (shown as 13-L-ROOH) to oxidize the iron to E-(Fe^{III}). E-(Fe^{III}) stereospecifically abstracts a hydrogen from the ω-8 carbon (step 2, Fig. 4) of linoleic acid (for example) to form E-Fe^{II}. . .$R^\cdot$, which, in the aerobic pathway, adds O_2 to give E-Fe^{II}. . .$ROO^\cdot$ and then rearranges to E-Fe^{III}. . .ROO^-. The product E-Fe^{III}. . .$RCOO^-$ is protonated to E-Fe^{III}. . .ROOH; and the E-Fe^{III} is again ready to oxidize another fatty acid molecule. Under anaerobic conditions, the E-Fe^{II}. . .$R^\cdot$, in a lipohydroperoxide-type reaction, catalyzes a number of oxidation, rearrangement, and polymerization reactions [21]. The free radical $ROO^\cdot$ may dissociate from the active site of the enzyme and react nonenzymatically with the polyunsaturated fatty acid to give racemic (S/R) mixtures of products.

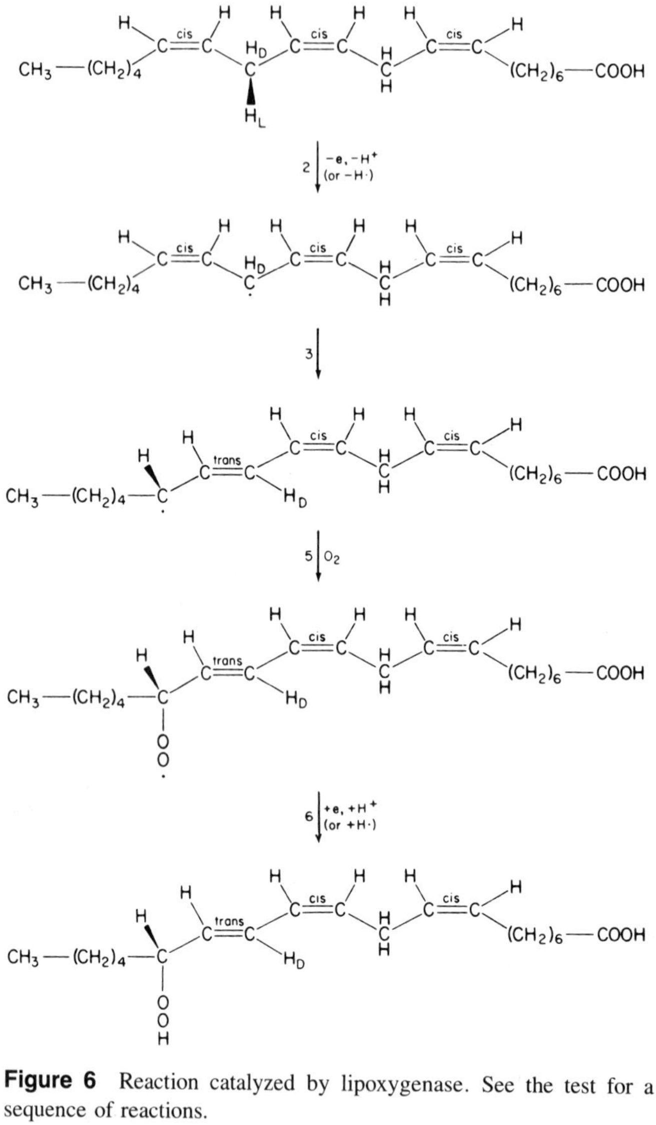

Figure 6 Reaction catalyzed by lipoxygenase. See the test for a description of the sequence of reactions.

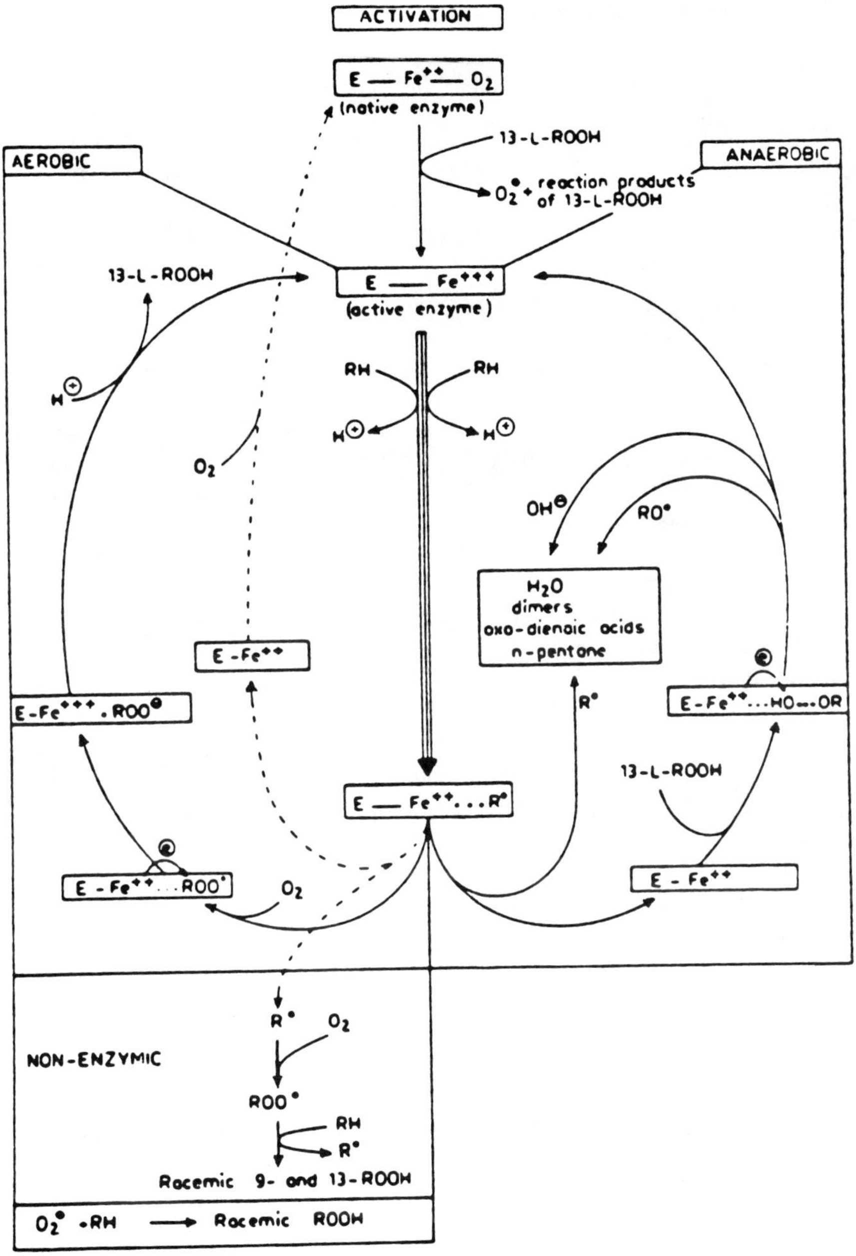

Figure 7 Proposed pathway and mechanism of soybean lipoxygenase-1 action. (From Ref. 20, p. 77, by courtesy of Elsevier Scientific Publishing Co.)

V. ENZYME INHIBITION

Peroxidation by lipoxygenase is inhibited by compounds such as nordihydroguaiacetic acid, propyl gallate, and α-tocopherol, which are lipid antioxidants. Presumably, these act by breaking the chain of events at the free-radical level rather than by direct inhibition of the enzyme. Lipoxygenase is also inhibited by saturated monohydric alcohols, probably through hydrophobic bonding with the enzyme at the site of interaction of the $CH_3(CH_2)_4$ group of the substrate with enzyme.

VI. ASSAY OF LIPOXYGENASE ACTIVITY

As a consequence of the action of lipoxygenase on linoleic acid there are three readily measurable changes that occur: (a) uptake of O_2, (b) a conjugation of double bonds, and (c) formation of a hydroperoxide. Each of these is the basis of an assay for lipoxygenase activity. Oxygen uptake can be measured manometrically (Warburg apparatus) but even better in a continuous fashion with an O_2-sensitive electrode.

Because of conjugation of double bonds there is a new absorbance peak formed with a maximum at 232.5 nm and $\epsilon_m = 2.8 \times 10^4 \, M^{-1}$ cm^{-1} (observed ϵ_m is dependent on side reactions). Therefore, the activity can be measured by observing the increase in absorbance at 232.5 nm.

The rate of the reaction can also be followed by measuring the rate of hydroperoxide formation by any one of a number of methods; for example, the oxidation of I^- to I_2 in an acid solution by hydroperoxide is a convenient method. This method has the disadvantage that aliquots must be removed from the reaction periodically, placed in an acid containing KI, and assayed for hydroperoxide content.

A coupled reaction involving the bleaching of the yellow color of carotene may also be used. The reaction contains linoleate, carotene, and enzyme at an appropriate pH. There is oxidation of carotene to a colorless derivative by the free radicals formed. Since the reaction is not stoichiometric, it must be performed under well-controlled conditions.

REVIEW QUESTIONS

1. A protein preparation having catalytic activity was found to contain 0.1% iron. Inorganic iron salts are catalysts for the decomposition of H_2O_2. Suggest several experiments that might prove whether the catalytic activity was due to an iron-containing protein (lipoxygenase?) or to iron salts adsorbed to a protein.
2. Discuss the use of an O_2-sensitive electrode for measuring the activity of an enzyme. Include a discussion of the conditions under which it can be used, the nature of the apparatus, calibration, and so on.

3. Discuss the functions of the apoenzymes of peroxidase and catalase.

4. How would you determine whether an enzyme is a dehydrogenase or an oxygenase?

5. Under suitable conditions both peroxidase and polyphenol oxidase can hydroxylate monophenols. How can you tell experimentally which of these enzymes you have in a preparation? How does the formation of *o*-diphenols by these two enzymes differ?

6. Glucose oxidase is relatively specific for β-D-glucose. However, if either α- or β-D-glucose is dissolved in water, an equilibrium mixture of approximately 35% α- and 64% β-glucose is formed. The half-life for this first-order process is approximately 70 min at 25°C in distilled water. Discuss the effect that this mutarotation has on the quantitative determination of glucose by the coupled glucose oxidase-peroxidase method where the incubation time is 10 min.

7. Outline in detail a method for determining the activity of each of the following enzymes: polyphenol oxidase, glucose oxidase, lipoxygenase, peroxidase, catalase, xanthine oxidase, and lactate dehydrogenase. For each indicate the theoretical basis of the assay method, what is to be measured, and how you measure it. Draw a rough graph of the expected data with properly labeled axes.

8. For each of the enzymes discussed in Chapters 22 to 27, summarize, by the use of chemical structures, the essential features of a compound which permits it to be a substrate for the enzyme.

9. For each of the enzymes discussed in Chapter 22 to 27, list several ways in which it is or may be important in food production and preservation.

10. How would you distinguish between

(a) glucose oxidase from *Penicillium notatum* and glucose dehydrogenase of bovine liver?

(b) lipoxygenase- and hematin-catalyzed oxidation of unsaturated fatty acids?

11. For each of the enzymes discussed in Chapters 22 to 27, suggest at least three inhibitors. For each inhibitor indicate whether it is an irreversible or reversible inhibitor (kinetically) and if the latter, indicate whether you would expect it to be a competitive or noncompetitive inhibitor.

12. How would you determine whether the observed decreasing rate of product formation for an enzyme such as polyphenol oxidase is due to limitations of substrate, general enzyme instability, or specific inhibition of the enzyme by the products?

13. In an experiment involving the action of glucose oxidase on glucose the rate of the reaction was followed by two methods:

(a) by coupling the reaction to peroxidase and a hydrogen donor (*o*-dianisidine);

(b) by a pH Stat method.

The second method gave an appreciably lower rate than did the first. Why?

14. The peroxidase activity of a blanched food was zero 5 min after heating for 10 min at 70°C. However, after storage at 2°C for 3 days, the material was found to have 15% of the original peroxidase activity. Explain.

15. Describe the theoretical basis of electron paramagnetic resonance spectroscopy. Under what conditions can it be used effectively in the study of enzyme reactions and mechanisms?

REFERENCES

General References

Chan, H. W.-S., The mechanism of oxidation, in *Autoxidation of Unsaturated Lipids* (H. W.-S. Chan, ed.), Academic Press, New York, 1987, pp. 1–16.

Eriksson, C. E., Oxidation of lipids in food systems, in *Autoxidation of Unsaturated Lipids* (H. W.-S. Chan, ed.), Academic Press, New York, 1987, pp. 207–231.

Gardner, H. W., Effect of lipid hydroperoxides on food components, in *Xenobiotics in Foods and Feeds* (J. W. Finley and D. E. Schwass, eds.), ACS Symposium Series 234, American Chemical Society, Washington, D.C., 1983, pp. 63–84.

Gardner, H. W., Lipoxygenase pathways in cereals, in *Advances in Cereal Science and Technology* (Y. Pomeranz, ed.), Vol. IX, American Association of Cereal Chemists, St. Paul, Minn., 1988, pp. 161–215.

Whitaker, J. R., Lipoxygenases, in *Oxidative Enzymes in Foods* (D. S. Robinson and N. A. M. Eskin, eds.), Elsevier Applied Science, New York, 1991, pp. 175–215.

Cited References

1. R. M. Bohn and L. W. Haas, cited in J. B. Sumner and R. J. Sumner, *J. Biol. Chem.*, **134**, 531 (1940).
2. E. André and K. Hou, *Compt. Rend.*, **194**, 645 (1932).
3. J. B. Sumner and R. J. Sumner, *J. Biol. Chem.*, **134**, 531 (1940).
4. J. R. Whitaker, Lipoxygenases, in *Oxidative Enzymes in Foods* (D. S. Robinson and N. A. M. Eskin, eds.), Elsevier Applied Science, New York, 1991, pp. 175–215.
5. A. M. Siddiqi and A. L. Tappel, *J. Am. Oil Chemists' Soc.*, **34**, 529 (1957).
6. A. M. Siddiqi and A. L. Tappel, *Arch. Biochem. Biophys.*, **60**, 91 (1956).
7. R. T. Holman, *Arch. Biochem.*, **15**, 403 (1947).
8. H. Theorell, R. T. Holman, and A. Åkeson, *Acta Chem. Scand.*, **1**, 571 (1947).
9. M. G. Dillard, A. S. Henick, and R. B. Koch, *J. Food Sci.*, **25**, 544 (1960).
10. J. C. Allen, *Chem. Commun.*, **1969**, 906 (1969).
11. M. Hamberg and B. Samuelsson, *J. Biol. Chem.*, **242**, 5329 (1967).
12. H. Kühn, T. Schewe, and S. M. Rapoport, *Adv. Enzymol.*, **58**, 273 (1986).
13. A. O. Chen and J. R. Whitaker, *J. Agr. Food Chem.*, **34**, 203 (1986).
14. H. W.-S. Chan, *Biochim. Biophys. Acta*, **327**, 32 (1973).
15. M. Roza and A. Franke, *Biochim. Biophys. Acta*, **327**, 24 (1973).
16. E. K. Pistorius and B. Axelrod, *Federation Proc.*, **32**, 544 (1973).

17. J. Fleming, B. J. Thiele, J. Chester, J. O'Prey, S. Janetzki, A. Aitken, I. A. Anton, S. M. Rapoport and P. R. Harrison, *Gene*, **79**, 181 (1989).
18. R. A. F. Dixon, R. E. Jones, R. E. Diehl, C. D. Bennett, S. Kargman, and C. A. Rouzer, *Proc. Natl. Acad. Sci. USA*, **85**, 416 (1988).
19. T. Matsumoto, C. D. Funk, O. Radmark, J.-O. Hoog, H. Jornvall, and B. Samuelsson, *Proc. Natl. Acad. Sci. USA*, **85**, 26 (1988).
20. J. J. M. C. de Groot, G. A. Veldink, J. F. G. Vliegenthart, J. Boldingh, R. Wever, and B. F. van Gelder, *Biochim. Biophys. Acta*, **337**, 71 (1975).
21. H. W. Gardner, Lipoxygenase pathways in cereals, in *Advances in Cereal Science and Technology* (Y. Pomeranz, ed.), Vol. IX, American Association of Cereal Chemists, St. Paul, Minn., pp. 161–215.

Index